# Biological Ice Nucleation and Its Applications

Edited by

**Richard E. Lee, Jr.**
Department of Zoology, Miami University
Oxford, Ohio, U.S.A.

**Gareth J. Warren**
Department of Biochemistry
Imperial College of Science,
Technology and Medicine
London, England

**Lawrence V. Gusta**
Crop Development Centre
University of Saskatchewan
Saskatoon, Saskatchewan, Canada

APS PRESS
The American Phytopathological Society
St. Paul, Minnesota

Cover reprinted by permission from P. J. Ward and P. J. DeMott, 1989, Journal of Weather Modification 21:9-13.

Library of Congress Catalog Card Number: 95-75812
International Standard Book Number: 0-89054-172-8



Printed in the United States of America on acid-free paper

The American Phytopathological Society
3340 Pilot Knob Road
St. Paul, MN 55121-2097, USA

# Contributors

**Soichi Arai,** Department of Agricultural Chemistry, University of Tokyo, Bunkyo-ku, Tokyo 113, Japan

**Edward N. Ashworth,** Department of Horticulture, Purdue University, West Lafayette, Indiana 47907, U.S.A.

**Michael J. Burke,** Department of Horticulture, Oregon State University, Corvallis, Oregon 97331, U.S.A.

**T. H. H. Chen,** Department of Horticulture, Oregon State University, Corvallis, Oregon 97331, U.S.A.

**Jon P. Costanzo,** Department of Zoology, Miami University, Oxford, Ohio 45056, U.S.A.

**John G. Duman,** Department of Biological Sciences, University of Notre Dame, Notre Dame, Indiana 46556, U.S.A.

**Gregory M. Fahy,** Transfusion Medicine Research Program, Naval Medical Research Institute, 8901 Wisconsin Ave., Bldg. 29, Bethesda, Maryland 20889, U.S.A.

**Ray Fall,** Department of Chemistry and Biochemistry, University of Colorado, Boulder, Colorado 80309, U.S.A.

**Rachel A. Fallon,** Amersham International, White Lion Road, Amersham, Bucks, HP7 9LL, U.K.

**Robert L. Green,** Idetek, Inc., 1245 Reamwood Ave., Sunnyvale, California 94809, U.S.A.

**Lawrence V. Gusta,** Crop Development Centre, University of Saskatchewan, Saskatoon, Saskatchewan S7N 0W0, Canada

**Susan S. Hirano,** USDA Agricultural Research Service and Department of Plant Pathology, University of Wisconsin, Madison, Wisconsin 53706, U.S.A.

**Fred Jerva,** Department of Pharmacology, Yale University, New Haven, Connecticut 06511, U.S.A.

**Andrey V. Kajava,** ISREC, Chemin des Boveresses 155, CH-1066 Epalinges, Lausanne, Switzerland

**Thomas L. Kieft,** Department of Biology, New Mexico Institute of Mining and Technology, Socorro, New Mexico 87801, U.S.A.

**Richard J. LaDuca,** Genencor International, Inc., 180 Kimball Way, South San Francisco, California 94080, U.S.A.

**Marcia R. Lee,** Department of Microbiology, Miami University, Oxford, Ohio 45056, U.S.A.

**Richard E. Lee, Jr.,** Department of Zoology, Miami University, Oxford, Ohio 45056, U.S.A.

**Charlotte Lindhardt,** Amersham International, Forest Farm Estate, Whitchurch, Cardiff CF4 7YT, U.K.

**Steven Lindow,** Department of Plant Pathology, University of California, Berkeley, California 94720, U.S.A.

**T. Mark Olsen,** Department of Biological Sciences, University of Notre Dame, Notre Dame, Indiana 46556, U.S.A.

**Nikolas J. Panopoulos,** Institute of Molecular Biology and Biotechnology, Foundation for Research and Technology–Hellas, P.O. Box 1527, 71-110 Heraklion, Crete, Greece

**Harvey A. Quamme,** Agriculture Canada Research Station, Summerland, British Columbia V0H 1ZO, Canada

**A. Franklin Rice,** Genencor International, Inc., 1879 South Winton Road, Rochester, New York 14618, U.S.A.

**Anthony J. Smith,** Amersham International, White Lion Road, Amersham, Bucks, HP7 9LL, U.K.

**Janet M. Strong-Gunderson,** Environmental Science Division, Oak Ridge National Laboratory, Oak Ridge, Tennessee 37831, U.S.A.

**William T. Tucker,** DNA Plant Technology, 6701 San Pablo Ave., Oakland, California 94608, U.S.A.

**Christen D. Upper,** USDA Agricultural Research Service and Department of Plant Pathology, University of Wisconsin, Madison, Wisconsin 53706, U.S.A.

**Gabor Vali,** Department of Atmospheric Science, University of Wyoming, Laramie, Wyoming 82071, U.S.A.

**Caroline A. Vance,** DNA Plant Technology, 6701 San Pablo Ave., Oakland, California 94608, U.S.A.

**Patrick J. Ward,** Genencor International, Inc., 1700 Lexington Ave., Rochester, New York 14606, U.S.A.

**Gareth J. Warren,** Department of Biochemistry, Imperial College of Science, Technology and Medicine, Exhibition Road, London SW7 2AZ, England

**Nancy M. Watanabe,** Scios-Nova, Inc., 2450 Bayshore Parkway, Mountain View, California 94043, U.S.A.

**Michiko Watanabe,** Food Science Laboratory, Tokyo Gakugei University, Koganei-shi, Tokyo 184, Japan

**Michael Wisniewski,** USDA Agricultural Research Service, Appalachian Fruit Research Station, 45 Wiltshire Road, Kearnysville, West Virginia 24530, U.S.A.

**Paul Wolber,** Idetek, Inc., 1245 Reamwood Ave., Sunnyvale, California 94809, U.S.A.

**King Lun Yeung,** Department of Chemical Engineering, University of Notre Dame, Notre Dame, Indiana 46556, U.S.A.

# Preface

The phenomenon of ice nucleation has both intrigued and demanded the attention of biologists from a wide range of disciplines. Plant physiologists and crop scientists found it relevant because of the agricultural importance of frost-sensitivity. When ice nucleators in this context were found to be a small group of epiphytic bacteria—the most potent of heterogeneous ice-nucleating agents—they excited the interest of microbiologists and biochemists. Applications suggested themselves, ranging from snowmaking, to food technology, to the use of *ina* genes as reporters of transcription and transduction. Moreover, bacterial epiphytes were discovered to be an important source of ice nuclei in the atmosphere, and thus of interest to meteorologists.

Ice nucleation also governs the winter survival of some ectothermic animals. Many freeze-intolerant insects avoid lethal freezing by depressing the temperature at which internal ice nucleation can occur; conversely, some freeze-tolerant forms synthesize ice-nucleating proteins to ensure that freezing happens at high subzero temperatures. Similarly, the temperature of nucleation is a critical factor in the cryopreservation of cells and tissues.

Although the phenomenon of heterogeneous ice nucleation had long been recognized, it was not known until the 1970s that the most active ice nuclei in nature were of biological origin. This discovery led to a series of conferences on this subject (1982, San Francisco, California, organized by L. M. Kozloff, S. E. Lindow, and R. C. Schnell; 1984, Flagstaff, Arizona, by G. Caple and G. Layton; 1987, Newport, Oregon, by M. J. Burke and S. E. Lindow; 1989, Saskatoon, Canada, by L. V. Gusta; 1991, Madison, Wisconsin by S. S. Hirano and C. D. Upper; 1993, Laramie, Wyoming, by G. Vali). The idea and impetus for this book originated, at least in part, from discussions at these conferences.

The scientific literature related to ice nucleation in biological systems is unusually scattered owing to the diversity of disciplines with interest in this subject. Reports have come from the fields of meteorology, bacteriology, plant physiology, crop science, physiology of cold tolerance in ectothermic animals (particularly insects), and the application of ice nucleation to medicine, cryobiology, food science, and snowmaking. The purpose of this book is to integrate for the first time information from each of these areas to serve not only as a reference for researchers in the field but also one that will appeal to those with a more peripheral interest in the topic. We hope that this book will aid in the synthesis of principles of ice nucleation spanning a range from theoretical to applied aspects in bacterial, plant, and animal systems. In addition, students new to this subject may use it as an accessible starting point.

Since each of these disciplines has developed its own, largely independent, body of literature, complete with different sets of terms for phenomena that are common to all, we have included a glossary as an aid to interpretating the literature. During the preparation of this book, we soon recognized, however, that it was not possible to reconcile all points of view with common definitions.

We thank our colleagues for their contributions. Christina Vertucci made valuable suggestions for improving the glossary. We especially appreciate the efforts of our editor, Joyce Loper, for her thorough review and cogent suggestions for improving this book. We also gratefully acknowledge Crop Development Centre, DNA Plant Technology Corp., Genencor International, Inc., National Institutes of Health, National Science Foundation, and the United States Department of Agriculture for their support of research in our laboratories.

*Richard E. Lee, Jr., Gareth J. Warren, and Lawrence V. Gusta*
July, 1994

# Contents

# Biological Ice Nucleation and Its Applications

CHAPTER 1

# Principles of Ice Nucleation

**Gabor Vali**

## Homogeneous and Heterogeneous Ice Nucleation

Thermodynamic equilibrium between different phases of water exists along well-known pressure-temperature lines. Changes of phase (liquid ↔ vapor, vapor ↔ solid, and liquid ↔ solid) take place at conditions slightly, or significantly, away from these lines. The deviations are usually minor for changes from the more ordered to the less ordered states (liquid → vapor, solid → liquid, and solid → vapor) but can be very large for the reverse processes, even though these changes are toward states of lower free energy. Phase changes toward lower energy states occur via metastable states; the most significant ones, and the only ones addressed in this discussion, are supersaturated vapor and supercooled liquid water.

In general, metastable states are reached from the liquid either by lowering temperatures or by changes in pressure. However, the melting point of ice varies so little with pressure (for the range of pressures encountered at the surface of the earth or in the atmosphere) that the metastable liquid state is reached, in practical terms, only by cooling. Hence it is a very good approximation to refer to water at temperatures below 0°C as supercooled. Water vapor becomes supersaturated when cooled or raised in pressure past the equilibrium line, with the same term used to describe this state whether the stable phase is the liquid (above 0°C) or the solid (below 0°C). Figure 1 depicts these phase changes and the names most commonly used to describe them.

The process of conversion from a metastable state to the stable phase is initiated by *nucleation*, the first appearance of a very small volume of the new phase. Following that, the phase change is completed by the growth of the stable phase until either all of the mass is converted or the temperature stabilizes at the phase equilibrium point. The growth process is controlled by the rate of removal of the latent heat being released. The nucleation step is the topic of this chapter.

The key to envisaging how nucleation takes place at the molecular level is the fact that even in dispersed phases there are small, transient aggregates of the condensed phase. For example, in a vapor, small clusters of molecules exist in addition to the single molecules making up most of the gaseous state. The number distribution of the aggregates of different sizes is defined as a function of the free energy of the clusters by a Boltzmann type distribution. As a first approximation, the clusters can be considered as minute entities of liquid. Similarly, in a liquid there are clusters of molecules in a solidlike configuration. Since the free energy of a cluster is higher than the sum of the energies of the molecules making up the cluster, there is a natural tendency for them to disintegrate, leading to very short lifetimes.

The energetics of cluster formation changes for vapor or liquid in a metastable state. The bulk free energy per molecule is then lower for molecules inside the cluster than outside it. The lower bulk energy is opposed by the energy involved in creating the interface between the cluster and the dispersed phase, but the balance changes in favor of the bulk energy as the embryo becomes larger. Beyond a certain size, further increases actually lead to lowering total free energies, so that growth becomes more likely than diminution. Because of this possibility, clusters in a metastable phase are called *embryos*. The size at which growth becomes more likely than decay is called the *critical size*; the embryo at that point becomes a *nucleus* for the further growth of the condensed phase; and the event is called *nucleation*.

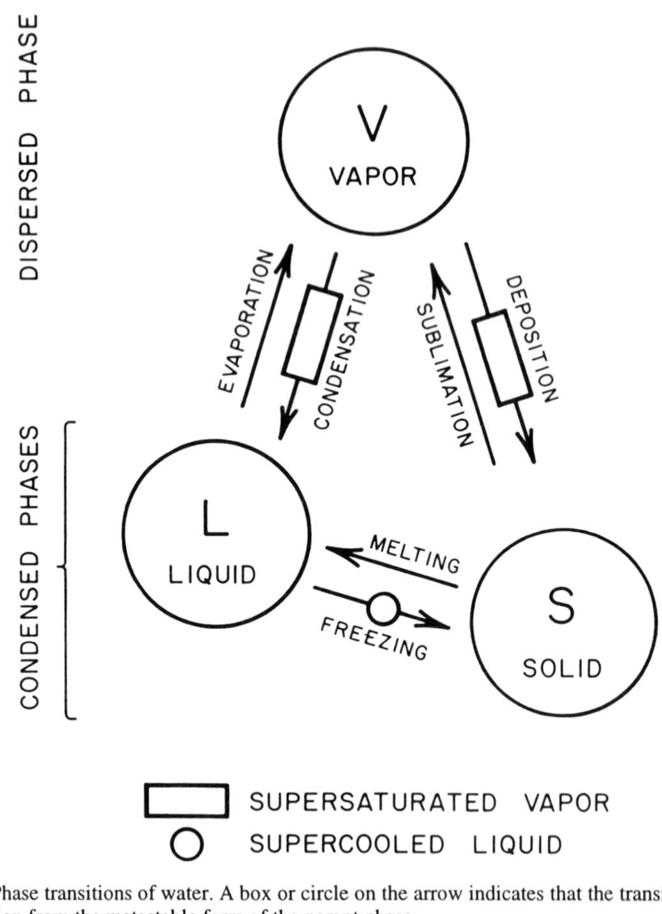


**Figure 1.** Phase transitions of water. A box or circle on the arrow indicates that the transition proceeds via nucleation from the metastable form of the parent phase.

The critical question regarding nucleation, then, is What controls the growth of embryos in the metastable phase? Two cases must be distinguished. The first is the simple case of having only molecules of the substance (water) to consider. In this case, the formation and growth of embryos is characterized by an increase in their average size as conditions move further from the phase equilibrium curve and by additional random fluctuations in embryo size due to the rapid attachment and detachment of molecules associated with thermal motion. Formation of a critical size embryo in this case is called *homogeneous nucleation*. The other situation to consider is that in which embryos form on the surface of some foreign material with which the metastable substance (supercooled water or supersaturated vapor) is in contact. Attachment of an embryo to a foreign surface, a *substrate*, can increase its stability by increasing the volume-to-surface ratio and by replacing part of the embryo–parent phase interface with an embryo–substrate interface. This is the case of *heterogeneous nucleation*. The role of the substrate is to make the formation of a critical embryo possible at a smaller supercooling, or smaller supersaturation, than that needed for homogeneous nucleation.

In both the homogeneous and the heterogeneous case, the growth of ice embryos is governed by the dynamics of additions of water molecules to the embryo. The process is not monotonic, but one of fluctuations: molecules attach and detach from the embryo at roughly equal and very rapid rates. Most molecules get added or removed singly, although some groups of few molecules might attach or detach as a unit. This random process gives a finite probability for large fluctuations from the average size to occur and for the critical size for nucleation to be reached and exceeded, even though a decrease in size is more likely, on the average, than an increase. Once the critical size has been reached, the energy decrease associated with an increasing volume makes the probability of growth more likely but still not absolutely certain.

The quantitative description of the nucleation process outlined above consists of two parts: calculation of the free energy change associated with the formation of an embryo, and the calculation of the rate of critical embryo formation (the probability of nucleation). The free energy of an embryo has a negative term proportional to the volume of the embryo and a positive term proportional to the surface area of the embryo. A difference in dependence of these two terms on the number of molecules in the embryo (to the third and second powers, respectively, for homogeneous nucleation) leads to a maximum in the free energy vs. size curve—the energy barrier to nucleation—at the critical size. For heterogeneous nucleation the formulation of the energy term is more complex.

With the energy vs. embryo size function established, the *nucleation rate* (the number of embryos per unit volume reaching critical size in a unit of time) is calculated as the product of a preexponential factor that depends on the kinetics of molecular attachments to the embryo, and a factor that contains the energy of a critical embryo in the exponent. In schematic form,

$$\Delta E = \text{volume energy} + \text{surface energy} \tag{1}$$

and

$$\text{nucleation rate, } J = \text{kinetic factor} \cdot e^{-\Delta E^*/kT} \tag{2}$$

with $\Delta E^*$ denoting the maximum of equation 1, $k$ the Boltzmann constant, and $T$ temperature in degrees Kelvin. Explicit forms of these equations can be found in many texts (e.g., Abraham, 1974; Fletcher, 1962, 1970; Hobbs, 1974; Pruppacher and Klett, 1978) using different assumptions and refinements. In general, one can refer to this theoretical formulation as the Gibbs-Volmer theory or, in a generic sense, as the thermodynamic/kinetic approach. The main ideas of the theory were developed around the turn of the century. It provides a useful framework for the examination of nucleation phenomena, though in the case of heterogeneous nucleation so many variables must be introduced that results of general validity are rarely obtained. Fundamental limitations of the thermodynamic/kinetic theory arise from having to prescribe specific embryo geometries (spheres for the homogeneous case, spherical caps or disks for the heterogeneous case) and from the use of volume and surface energy values, which are, strictly speaking, valid only for the bulk phases. Other approaches to the theory of nucleation are being developed, principally by extending molecular interaction models to multimolecular clusters, but the application of such models to water and ice is still very limited.

## Empirical Results on Homogeneous Ice Nucleation

The supersaturation required for condensation via homogeneous nucleation is around 450% (i.e., a saturation ratio 4.5 times higher than the equilibrium value at the same temperature). Homogeneous freezing nucleation takes place at approximately –40°C. Homogeneous deposition (from vapor, at temperatures colder than 0°C), in fact, takes place with the sequence of homogeneous condensation followed by homogeneous freezing nucleation. Of these processes, only homogeneous freezing nucleation is of practical importance, since the –40°C temperature required is commonly found at the earth surface and in the atmosphere and can be readily produced in the laboratory.

The requirements quoted above for homogeneous ice nucleation need to be qualified in two ways. First, their meaning must be made more precise by specifying them in terms of a nucleation rate. Second, disagreements among experimental results and uncertainties in theoretical values need to be considered.

Nucleation rate can be viewed either as the number of critical embryos that form in a volume over a given length of time or as the probability per unit time and unit volume that a critical embryo will form. The latter view is more useful, since the formation of one embryo is usually sufficient for phase transformation to proceed in the entire volume. Initial ice growth in supercooled water is so rapid that the formation of critical embryos beyond the first is practically excluded, except for samples cooled extremely fast. Thus, the probability that a volume $V$ of supercooled water will freeze in time $\Delta t$ is $J(T) \cdot V \cdot \Delta t$, where $J(T)$ is the nucleation rate at temperature $T$. The consensus of empirical results is characterized (Götz et al., 1991, 142) by the equation

$$J(T) = 6.8 \times 10^{-50}\, e^{-3.9T} \quad \textbf{(3)}$$

where the $J$ is in units of cubic meters per second and $T$ in degrees Celsius. As the equation demonstrates, the rate of increase of $J$ with decreasing temperature is very rapid, increasing at a factor of nearly 50 for each 1°C lowering of temperature. This is why a single temperature, –40°C, is quite useful for defining the point of homogeneous ice nucleation, and why there will be only small, though quantifiable, changes in that temperature due to variations in sample volume or in the rate of cooling. The *threshold temperature* is customarily defined as the temperature at which $J = 1\ cm^{-3}\ s^{-1} = 10^6\ m^{-3}\ s^{-1}$.

Experimental techniques used for the study of homogeneous ice nucleation will be discussed later in this chapter. At this point, only the caveat will be made that the threshold temperature of homogeneous nucleation is known within, perhaps, ±3°C. The uncertainty arises from difficulties in relating observations to a nucleation rate, from corrections for the volume distributions of emulsions, from cooling rate effects, and from a variety of artifacts.

If homogeneous freezing is to be considered in a practical situation, application of equation 3, and the definition of $J$ preceding it in the text, offer a basis for predicting the rate of nucleation for a given temperature. Strictly speaking, that temperature cannot be reached instantaneously, and that makes the application of equation 3 more difficult. However, the strong variation of $J$ with $T$ makes refinements aimed at allowing for the transients less important. As a simpler result, it is usually sufficient to view the temperature of –40°C as the limit for the supercooling of pure water.

It is instructive to consider the physical scale on which the nucleation event takes place (i.e., the critical embryo size). For homogeneous freezing at –40°C, –20°C, and –5°C, the calculated values of the critical radius are 0.8, 1.8, and 7.0 nm, respectively, with roughly 70, 650, and 45,000 molecules of water in the embryo. The large number of molecules in the critical embryo at warmer temperatures makes it evident why the probability of homogeneous nucleation becomes negligibly small at just a few degrees above –40°C.

As already mentioned, homogeneous ice nucleation is conditioned on having no foreign material in the sample of liquid or vapor—or, to relax that condition a little, none that could influence the formation of ice embryos. That is a stringent requirement, one that can hardly ever be satisfied without question. It is most likely to be satisfied for small volumes of water in some inert gas or liquid. Cloud droplets in the upper troposphere and carefully prepared laboratory samples may come close. In biological systems, the internal water or fluid volumes are inevitably in contact with other surfaces, so that nucleation is justifiably assumed to be heterogeneous.

## Modes of Heterogeneous Ice Nucleation

When ice forms at temperatures above –40°C, or at supersaturations less than 450%, it is due to the presence of some material other than water (i.e., by heterogeneous nucleation). Foreign bodies serve as sites on which ice embryos grow more readily than purely by the random aggregation of water molecules with one another. For embryo formation on a foreign surface, the temperature or supersaturation required for ice nucleation depends mostly on the interaction of water molecules with the foreign surface. The location at which an ice embryo forms on a surface, with the potential for that embryo to grow to critical size, is called a *nucleating site*. Similarly to the homogeneous case, heterogeneous nucleation is governed by two major factors: the free energy change involved in forming the embryo and the dynamics of fluctuating embryo growth. Importantly, in heterogeneous nucleation, the configuration and energy of interaction at the nucleating site become the dominating influences.

A simple description of heterogeneous nucleation is that the embryo forms a spherical cap on a flat and uniform surface, much like a drop of water would sit on a hydrophobic surface. With that model, the energy of interaction between the embryo and the surface can be characterized by the contact angle. Note that this model does not involve specific sites. The spherical cap model is readily fitted to equation 1: the volume and volume energy of the embryo are adjusted for the new geometry, and the surface energy term is divided into two parts, namely, embryo-to–mother phase and embryo-to–foreign surface, with the size of each surface calculated from the geometry of the spherical cap. The result is pleasingly simple: the critical free energy change (the maximum in the energy vs. size curve) is a fraction of that required for homogeneous nucleation, and the fraction is a unique function of the contact angle. The smaller the contact angle, the lower the energy barrier to nucleation. The homogeneous case appears as a limit for a contact angle of 180°. The impact of a small contact angle can be illustrated by comparing the critical embryo size of 600 molecules (assuming a 30° contact angle) required for nucleation at –5°C with the previously cited figure of 45,000 molecules required for homogeneous freezing.

More complex forms of the theory assign different interaction energies to the nucleating site than to the rest of the surface, allow the surface to be curved, use different assumptions for the embryo shape, and include surface irregularities (steps, dislocations, cavities, etc.) as possible nucleation sites. As mentioned before, such elaborations, while conceptually appealing, usually involve so many degrees of freedom in the assumptions that no generalizations are reached. Hence the remainder of this discussion focuses on empirical results.

Although freezing (the liquid-to-solid transition) is the only homogeneous ice nucleation process of practical interest in the biosphere and in the atmosphere, a variety of heterogeneous processes may be considered. Both the liquid-solid and vapor-solid transitions are of practical interest, and observations have shown that freezing nucleation depends on how the supercooled liquid and the nucleating particle come together. Hence the pathways, or *ice nucleation modes,* have to be considered. Definitions of these modes follow, quoted from a "Nucleation Terminology" (see J. Aerosol Sci. 16[1985]:575-576; Götz et al., 1991, 265-267) first adopted by the atmospheric science community but certainly not restricted to atmospheric processes. (Two word changes, in parentheses, have been made so that the definitions may apply beyond the atmospheric situation in which the nuclei are aerosol particles.)

1. *Deposition nucleation*—The formation of ice in a (supersaturated) vapor environment
2. *Freezing nucleation*—The formation of ice in a (supercooled) liquid environment
   2.1 *Condensation freezing*—The sequence of events whereby a condensation nucleus initiates freezing of the condensate
   2.2 *Contact freezing*—Nucleation of a supercooled droplet subsequent to a nucleating particle coming into contact with it.
   2.3 *Immersion freezing*—Nucleation of supercooled water by a nucleus suspended in the body of water.

When the mode of nucleation is known, it is advantageous to refer to that specific mode by employing the corresponding term. If the mode is unknown, or a collective description is desired, one can refer to "ice nuclei," or "ice nucleation."

The main characteristic of deposition nucleation is that the number of nucleating sites (per unit surface area of the material, or per aerosol particle) increases rapidly with supersaturation. The relevant parameter is supersaturation over ice. Temperature per se does not have any influence, except through the dependence of the saturation vapor pressure of ice. Some activity is present even below water saturation, and there is no discontinuous increase in the number of nuclei at water saturation. The dependence on supersaturation over ice is very strong: a power law relationship between number and supersaturation has exponents ranging from 4 to 12. These relationships are valid below a temperature that appears to be characteristic for each substance. Above that temperature, ice nucleation requires that the vapor pressure exceed saturation with respect to water, suggesting that the mode of nucleation even from the vapor environment is freezing, with a transient condensation of water. This limiting temperature was determined for $CdI_2$ (–22°C) by Bryant et al. (1959), for leucine (–20°C) by Maybank and Barthakur (1966), and for silver iodide (–8°C), kaolinite (–13°C), and other substances by Schaller and Fukuta (1979).

Immersion freezing is commonly observed very near 0°C for large volumes of water in common containers or in puddles, lakes, rivers, etc. Small volumes of water under laboratory conditions can be observed to freeze as a result of nuclei suspended in the water at temperatures ranging from –1°C to near –30°C, depending on the type and quantity of suspended matter. Condensation freezing and contact freezing have been found to be as or more effective than immersion freezing for some nucleating materials in aerosol form, but these experiments have been restricted by experimental difficulties to temperatures at less than –10°C. It is not known how activities of materials differ with respect to the the three freezing modes at smaller supercoolings. This is an important point, since differences among the freezing modes can yield useful information on the factors that govern the formation of ice embryos on the particular surfaces. Possible explanations for enhanced activity in the contact mode are that nucleating sites may partially dissolve when immersed in water, that embryos in equilibrium with the vapor environment are large enough to serve as freezing nuclei, or that the special characteristics of the surface layer of water enhance embryo formation. Information about the condensation freezing sequence is so scant that there is little basis for theorizing; it is worth mentioning, though, that perhaps some connection exists between condensation freezing and evaporation freezing, the latter having been observed when droplets are in their final stages of evaporation.

In biological systems, immersion freezing is the most likely path of ice initiation. So far, the distinction among the freezing modes has been shown to be important only in laboratory tests mimicking atmospheric conditions (i.e., with aerosol particles and droplets of water). It is conceivable that contact freezing also arises in biological systems—for example, by the growth of a water volume until it just comes into contact with the nucleating site, or by actual movement of either the water volume or the nucleating surface. Condensation freezing could be taking place just as in the atmosphere. It is also possible that the set of definitions given above will have to be broadened to include specific circumstances applicable to biological systems.

In view of the known role of immersion freezing in biological systems, attention in subsequent sections of this overview will focus on that mechanism.

## Historical Survey of Freezing Nucleation Experiments

The supercooling of water was observed in 1721 by Fahrenheit (to a lowest temperature of –9°C) in a sealed glass vessel—an experiment frequently repeated afterwards, with the usual result that the water froze at the instant the vessel was opened or moved. These observations linked freezing to mechanical disturbances and showed that the crystallization rate of water, once started, is so rapid that it is impossible to count the centers of crystallization in the same way as in experiments with supersaturated solutions. These were the best known observations relating to the freezing of water for over 200 years; however, as Altberg (1938) reviews, experiments as early as 1788 already gave indications that the degree of supercooling achieved for water is limited by impurities suspended in it. That fact became more firmly established with other substances through the nineteenth century, until experimenters finally returned to water in the 1930s and 1940s. Research with water during that period was part of a broad increase in phase change experiments and of significant developments in nucleation theory.

Modern nucleation experiments with water have two traditional roots. Experiments with bulk supercooled water followed work on the solidification of supersaturated solutions and supercooled liquids, as mentioned above. Samples were held in test tubes or similar vessels. Solidification was observed as a function of the composition of the sample, the time or the temperature at which solidification took place, the effects of shaking, the length of storage of the sample, etc. Another line of investigation has its origins in the use of the cloud chamber invented by C. T. R. Wilson around 1895 for making the tracks of elementary particles visible. This apparatus led to the series of "cloud chamber experiments," which have the essential feature of tiny droplets of water suspended in air during the observations.

Cloud chamber experiments are of primary relevance to atmospheric problems, and they may be useful in the testing of biological ice nuclei if the nuclei are dispersed along with the water to form the cloud. This is not very desirable in general, and the results of the experiments can be difficult to interpret. It will suffice here to mention that the principal advantage of the cloud chamber technique is that the individual droplets are only a few micrometers in diameter, and hence the probability for the droplets to contain foreign matter (potential nuclei) is quite small. As a result, homogeneous nucleation can be investigated by this approach, as seen both in the early experiments (Cwilong, 1947; Schaefer, 1948; Fournier d'Albe, 1949; Mossop, 1955) and recently (Hagen et al., 1981). The low temperatures needed for these experiments were reached by rapid expansion of humidified air; the degree of expansion determines the lowest temperature reached. A concomitant disadvantage of the rapid expansion is that the sample remains at the lowest temperature only for a brief interval of time. The derivation of a nucleation rate from these experiments therefore depends on the application of some theoretical model to the data.

Greater control over the variables can be achieved in experiments with volumes of water supported in vessels, suspended in liquids, supported on surfaces, etc., at the cost of increasing the possibility of interference by the supporting surfaces. There is an intermediate approach, that of letting drops fall into cold air (e.g., Kuhns and Mason, 1968). Overall, solid and liquid supports have been most widely employed. Development of this approach took off with the work of Meyer and Pfaff (1935) and of Tammann and Büchner (1935) and got a major impulse from the work of Dorsey (1938, 1948). Dorsey went considerably beyond his predecessors in the variety and number of water samples tested, all in sealed glass "bulbs." Dorsey's results clearly confirmed that the freezing (nucleation) temperature depends on the source of the water (distilled, conductivity, Washington City, aquarium, stream, or snow and ice); that mechanical disturbances rarely result in freezing; and that repeated tests with a sample result in frequent repetitions of the previously observed freezing temperatures, although not without large unpredictable changes.

When Schaefer (1946) and Vonnegut (1947) reported that ice crystals can be produced in a laboratory cloud and that silver iodide (AgI) particles added to the cloud cause ice to form abundantly, research on the nucleation of ice accelerated enormously. By the end of the 1950s, essentially all experimental methods now known had been introduced, and theoretical developments from physical chemistry and metallurgy had been rapidly applied to ice. Many of the practical and theoretical problems that had not been solved by then remain unsolved.

One of the factors guiding the development of new experimental methods was the realization that the freezing of any one sample reveals information only about the most active nucleus in it, the one producing nucleation at the warmest temperature. To gain a more complete idea of the nuclei contained in a sample, and to improve the statistical validity of any result, it was necessary to deal with large numbers of subunits of any sample. This fact forced experimenters to go beyond test tubes, sealed glass tubes, etc. Rau (1944) and Dorsch and Hacker (1950) produced extensive, if somewhat confusing, data sets with drops placed on metal plates that were cooled at rates of the order of one degree per minute. Bigg (1953) and Langham and Mason (1958) protected their sample drops from possible contamination and from the influence of a metal support by suspending them at the interface of two immiscible liquids. Much attention was paid in these experiments to the effect of drop volume on the temperature of nucleation and to the statistical distribution of freezing temperatures. Also, of course, experiments were carried out with potential nucleating substances; with silver iodide above all, but with many naturally occurring and synthetic substances as well. Interpretations of the results in terms of the thermodynamic/kinetic theory applied to ice nucleation (Fletcher, 1958) met with some success but usually remained short of general validity.

Another direction of nucleation studies focused on how foreign surfaces induce ice nucleation. Instead of dealing with particles suspended in water, a direct examination of the surfaces was desired. This was achieved by placing carefully prepared crystal surfaces on a microscope cold stage and enclosing the sample in a small chamber in which humidity could be controlled. Such studies on well-characterized surfaces (Montmory, 1956; Bryant et al., 1959; Fukuta and Mason, 1963, Caslavsky and Vedam, 1971) led to the recognition that surface features such as steps and screw dislocations are favored sites for nucleation. These findings led to much speculation about the microscopic factors influencing ice nucleation, but the relevance of these studies is limited by the fact that nucleation takes place on a scale much smaller than the surface features that can be identified by direct examination.

The developments described above were motivated by questions related to the atmosphere: the initiation of ice in clouds, the formation of precipitation, and artificial cloud seeding. The overlap of interests with the biological sciences became established—perhaps not uniquely, but importantly—through the work of Salt (1958, 1966, and others), who realized the applicability of new findings on heterogeneous ice nucleation to the winter survival of insects. The experiments of Salt established that the fundamental aspects of ice nucleation clearly manifest themselves in biological systems as well, although a number of additional complicating factors also have to be dealt with. Nonetheless, it is now clear from cumulative evidence that the principles discussed in this chapter have general applicability to atmospheric, biological, or other systems of comparable temperature and pressure regimes.[1]

## Practical Methods of Measurement of Immersion Freezing Nuclei

Many versions of the so-called drop-freezing assay[2] have been developed. The essence of these tests is that the nucleating material is dispersed in water; the water sample is divided into many subunits (drops); the set of drops is cooled; and their freezing is observed. The basic reason for testing many subunits of a sample has been mentioned before. The tests provide information about the abundances of nucleating sites with different characteristic temperatures (a precise definition of the term will follow). These tests, in themselves, do not reveal any other property of the nucleating material.

The systems to be examined here have certain basic characteristics, and their use implies certain assumptions. Perhaps of broadest significance is the assumption that the test systems involve essentially "bulk" water, with widely dispersed nucleating material. The properties of the water are not altered by a prevalence of interfacial forces, as would be the case, for example, if one dealt with adsorbed water layers. This requirement does not exclude the possibility of deliberately adding dissolved materials, nor does it signify that the embryo is assumed to have the properties of bulk water (the contrary is, in fact, the likely case). The nucleating material has to be dispersed among the drops. Given the small sizes of nucleating sites (<1 nm), pieces of the nucleating material much larger than the nucleating sites can be suspended in water drops of tens of microns to millimeters in size without violating the assumption of bulk water properties. Clearly, the possibility for cooperative action among pieces of the nucleating substance becomes limited; there are some hints, but no clear evidence, that cooperative action of some kind may be important with some samples. In order to allow quantitative interpretation of the results, the dispersion of the nuclei should be macroscopically uniform (i.e., each drop should contain the same amount of nucleating material). As was shown in a number of tests with different materials, the degree of dispersion does not influence the results derived from the tests (Vali, 1971; Rogers et al., 1987; Dubrovský and Skoloud, 1989; see also comments below on the volume effect). From a practical point of view, the general concern with the drop-freezing assays is that the immediate environment of the drops not influence the results (i.e., that the supporting or carrying

[1]The emphasis on ice formation within the atmospheric sciences is derived from the theories of Bergeron (1935) and Findeisen (1938) connecting precipitation to ice formation in clouds. World War II and the attendant increase in aviation, including flights in inclement weather, brought these questions to greater importance and also brought to light problems related to the icing of aircraft in supercooled clouds.

[2]The term *drop-freezing assay* will be employed here, even though samples may, in fact, be held in test tubes or other containers or may be other entities. The essential feature—that there be many identical units of the sample—is well conveyed by the reference to drops.

medium be inert, neither increasing nor decreasing the inherent activity of the nuclei suspended in the sample). Beyond this requirement, the system must be stable over the period of testing (tens of minutes to hours).

Another general question is the role of time versus temperature in leading to a nucleation event. As discussed in the first section of this chapter, a nucleating site can be characterized in terms of its rate of nucleation as a function of temperature. Since this definition is for a rate, time is an intrinsic variable in the process. However, it has been shown that nucleating sites can be defined by a *characteristic temperature,* the temperature at which the nucleation rate has some fixed value[3]. This is similar to the definition of threshold temperature for homogeneous nucleation. The increase of the nucleation rate is so rapid with decreasing temperatures that the characteristic temperature is a reasonably good predictor of the actual temperature of nucleation for wide variations in time-temperature history. Thus, observed freezing temperatures can be taken as the first approximation to the characteristic temperatures for experiments performed with continuous cooling at some moderate rate. A better approximation can be achieved by replacing the observed freezing temperature, $T$, by the adjusted value $T' = T - 0.65 \log_{10} r$, where $r$ is the absolute value of the rate of cooling in degrees Celsius per minute. For time-temperature histories more complex than a linear one, the interpretation of the results is less straightforward. The basis for the cooling-rate correction and the general issue of time dependence are treated in more detail by Vali (1994).

Drop-freezing assays can be carried out in a number of different forms. There are differences in the volumes of the samples used, in the way the drops are supported or suspended, in the way cooling is effected, and in the way freezing is detected. Perhaps the simplest method is to place drops of, approximately, millimeter size (tens of microliters in volume) on solid surfaces, the temperature of which can be varied in a controlled way. Alternatively, the water may be contained in small centrifuge tubes or larger test tubes. The latter is especially useful if large pieces of material, such as whole leaves, are to be tested. Cooling of the tubes is accomplished by immersing them in a temperature-controlled bath. At the other end of the scale, the water sample may be dispersed as an emulsion, yielding the advantage of a compact sample and a reduced danger of interference from the supporting surface or container.

Detection of freezing can be based on the change in opacity, release of latent heat, volume expansion, or the change in electric conductivity that accompanies the phase change. The signals derived from any of these physical changes are roughly proportional to the amount of ice forming during the brief initial period when the latent heat release is much more rapid than the rate of transfer of heat from the drop to its surroundings. After that initial freezing, the rate of solidification is governed by the rate of heat removal from the drop. Since the initial ice fraction is proportional to the temperature (in degrees Celsius), detection of freezing at small supercoolings is more difficult than detection at large supercoolings.

A few examples of specific experimental arrangements will serve to illustrate how drop-freezing assays are done in practice.

---

[3]For heterogeneous nucleation, the term *onset temperature* is used to describe the highest temperature at which some material exhibits appreciable nucleating activity. The term has not been given a rigorous definition. In contrast, *characteristic temperature* relates to a specific nucleating site. A given substance may have a large range of different sites, with corresponding characteristic temperatures.

**Drops on Plates**

Rau (1944), Brewer and Palmer (1951), and Vali and Stansbury (1966) describe some examples of the application of this method. In comparison with earlier methods of enclosing the samples in glass vessels, tubes, or capillaries, placing drops on a plate is simpler. The plate can be cooled in any number of ways, the drops are readily observed, and many experiments can be carried out in a relatively short time due to the ease of preparation. The supporting plate, of course, has to be inert (not causing nucleation); this appears to be possible, to temperatures of about –20 or –25°C, by careful polishing and cleaning of chrome or gold surfaces, or by coating of the surfaces with thin layers of oils or varnishes. The drops can be produced by condensation from a humid airstream or by some mechanical dispenser. The latter is preferable for controlling the volumes of the drops and their spacing. Visual detection of freezing is the simplest, but is difficult at temperatures close to 0°C. Figure 2 shows a set of drops on a plate at a point when some have frozen and some are still supercooled. Depending on the sophistication of the detection method and the data processing—video image interfaced to a computer is the most practical at this time—it is possible to test several samples at the same time, for comparison or control. If freezing events are recorded with reference to the location of each drop, the data can be separately analyzed for each sample.

In addition to ensuring that the supporting surface does not cause nucleation at temperatures comparable to that of the nuclei contained in the samples, care is re-

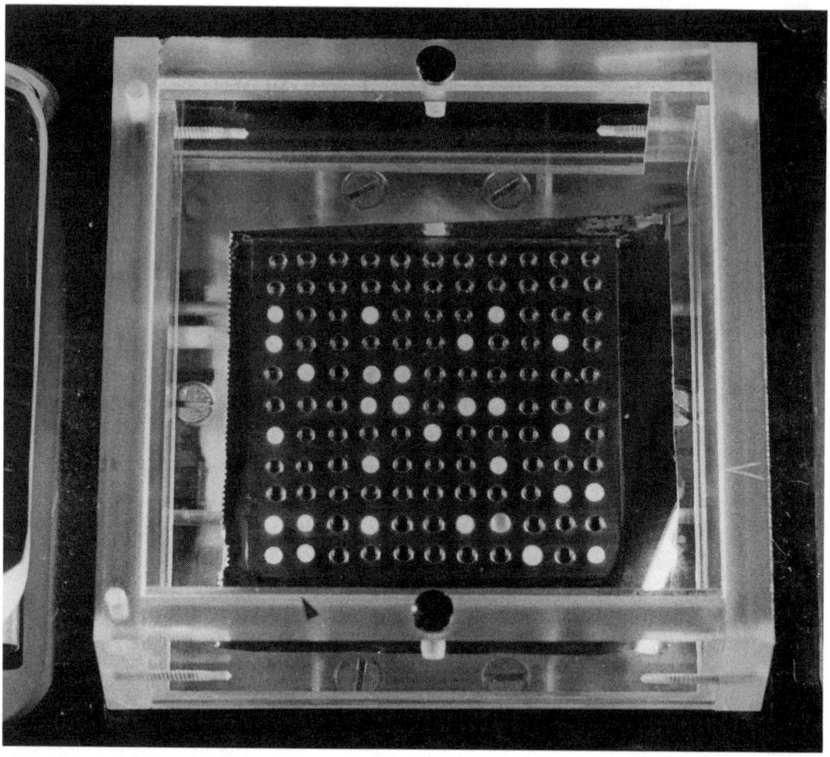

**Figure 2.** A set of drops, some still supercooled, others (white) frozen. All drops are from the same sample. Drops are 0.10 μl in volume.

quired to ensure that ice does not spread from one frozen drop to another. The danger of such an interaction among the sample drops is one of the major disadvantages of the method. Other risks, though minor, are evaporation of the drops during the tests, airborne particles falling onto the drops, and cooling rates so fast that significant temperature gradients arise within the drops.

The use of a solid plate as the supporting surface is a common arrangement. A plate of about 5 cm square can hold up to 100 drops without apparent danger of interference. The drops are either placed on the polished metal surface, or on a thin plastic film or aluminum foil stretched over the plate. In all cases, a hydrophobic coating is needed to ensure that the drops do not spread and to minimize the possibility of the solid supporting surface nucleating the drop. The film can be replaced for each experiment. The plate is cooled by a circulating liquid, thermoelectric coolers, or conduction to a heat sink (dry ice or liquid nitrogen). The temperature of the plate is taken as the drop temperature. For drops no greater than a few millimeters, the temperature gradients in the drops are only a few hundredths of a degree, although a thin layer at the drop surface may have a greater temperature difference.

A version of the drop-freezing assay, requiring little specialized equipment, was described by Lindow (1982). Drops are placed on a sheet of paraffin-coated aluminum foil whose edges are bent up to form a tray. The tray is then placed on an open cold bath. This method is limited to a single temperature or to a set of fixed temperatures.

A particularly elegant version of the drop-freezing assay is that of Parody-Morreale et al. (1986). Drops are placed on a thermopile, and the whole assembly is cooled by liquid nitrogen via a long metal column. Freezing of each drop is accompanied by a voltage pulse from the thermopile. This system has great sensitivity, so that the freezing of drops as small as 1 μl can be detected. A limitation of the method is that it can be used only with drops of a single sample and of the same volume in any given run. The data produced contains the time and temperature at which the drops froze, but the time at which each particular drop froze cannot be recovered from the data.

**Test Tubes**

A combination of the early experiments with sealed glass vessels and of the drops-on-a-plate technique is the placement of the samples in test tubes that are cooled in a liquid bath. This method offers the advantage of the multiple subunits of the samples, as with the drops, and at the same time provides greater isolation of the individual units. Hirano and Upper (1986) describe the use of test tubes to hold small leaves, bean pods, cereal heads, etc., immersed in 0.01 M buffer. The tubes were tested at a series of temperatures, with unfrozen samples transferred to baths at successively lower temperatures. Makino (1982) sealed samples into 10-μl micropipettes, submerged the tubes in a cooling bath, and detected the freezing events by the change in volume within the tubes as the temperature of the bath was gradually lowered.

An apparatus using sample tubes is shown in Figure 3. The sample containers are microcentrifuge tubes. Within the holder plate, each tube has a thermocouple touching its side; these sense the latent heat of fusion and provide signals for recording the freezing events. The array of tubes is immersed in a bath, the temperature of which is controlled by the programmed mixing of two streams of liquids, one cooled to about −35°C and the other kept at about +5°C. In general, this method offers flexibility for testing different kinds of samples and provides a simple and sound procedure, but the cost of the equipment is a drawback.

**Emulsions**

Dispersing the water in the form of an emulsion fulfills, in a very effective way, the principal requirement stated earlier that the sample be subdivided in many small units. Especially for homogeneous nucleation tests, or for others involving temperatures of −25°C or lower, the small droplet sizes of emulsions is a major advantage. Critical factors in the method are the choice of the carrier fluid and the method of preparation. Silicone oils, other oils, and heptane have been used as carrier fluids, usually with a small amount of added dispersing agent (sorbitan tristearate). The emulsion is prepared with strong mechanical agitation. The resulting droplets are from a few µm to about 10 µm in diameter. Clearly, only small particles of a nucleating material can be contained in the drops, although Rasmussen et al. (1975), Franks et al. (1983), and Clausse et al. (1991) succeeded in dispersing cell suspensions. A well-prepared emulsion can remain stable for many hours. The small volumes of the emulsions allow cooling in any number of ways. Detection of freezing can also take different forms. Wood and Walton (1970) placed the emulsion in a small cavity in a metal plate attached to a heat sink. Freezing of the droplets was recorded through a microscope. More recently, calorimeters (Taborek, 1975) or commercially available differential thermal analyzers (e.g., Rasmussen et al., 1975) and differential scanning calorimeters (e.g., Michelmore and Franks, 1982) have been used.

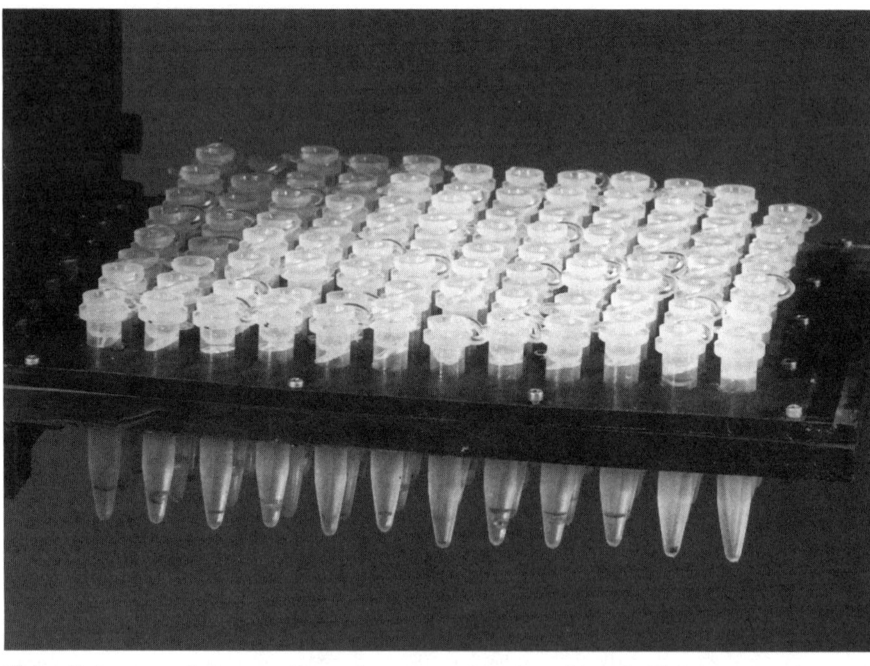

**Figure 3.** An array of microcentrifuge tubes supported in plate with embedded thermocouples next to each tube. Tubes contain 250 µl of water. For a test, this assembly is immersed in a cooling liquid.

**Other Forms of Drop-Freezing Assay**

From the point of view of biological ice nucleation, it is of special importance that the essence of the drop-freezing assay can be captured with the testing of a large number of externally identical specimens of any shape or form. Disks cut from leaves, larvae, buds, or other entities can be treated exactly as "drops" in the analyses of the data. Of course, the manner of cooling of the units and the detection of freezing have to be matched to the type of samples tested. Strictly from the point of view of the measurement, neglecting any possible physiological influences, the same units can be immersed in clean water drops or tubes filled with water. The additional water volume acts as a signal amplifier, making the detection of freezing easier while leaving the initiation of freezing unaffected.

**Evaluation of Drop-Freezing Assays**

In essence, the techniques described can be viewed as consisting of $N_0$ drops of volume $V$, all drawn from the same sample, and subjected to temperature $T$ as a function of time $t$. The observations yield $N(T)$ or $N(t)$, the number of drops remaining unfrozen as a function of temperature or of time.

There are numerous alternative measures for characterizing the results of a freezing experiment. One can note the range of temperatures over which the nucleation events took place. One can obtain the average or modal freezing temperature, or the temperatures at which given percentiles of the drops were observed to be frozen. Frequently, an onset temperature (at which freezing is first observed) is reported. One can present a histogram of the freezing events or plot the fraction of the sample remaining unfrozen as a function of temperature or time. Tests done at one or a few fixed temperatures yield the fraction of drops frozen at those temperatures.

Each of the descriptors mentioned can serve specific purposes of evaluation, but all are dependent on the drop volumes used and on the proportions of nucleating material to water. Tests with different volumes or different dilutions cannot be quantitatively compared using these measures.

With the caveat that time-dependent effects are treated as second-order, a more general interpretation of the drop-freezing assays can be obtained by using the differential and cumulative nucleus concentrations introduced by Vali (1971). The differential concentration spectrum $k(T)$ describes the concentration of nuclei active within unit temperature interval of $T$, and the cumulative spectrum $K(T)$ describes the concentration of nuclei active at temperatures warmer than $T$:

$$k(T) = \frac{1}{V} \frac{1}{N(T)} \frac{\Delta N}{\Delta T} \quad \textbf{(4)}$$

and

$$K(T) = -\frac{1}{V} \log_e \frac{N(T)}{N_0} \quad \textbf{(5)}$$

where $N(T)$ is the number of drops still unfrozen at temperature $T$, $\Delta N$ is the change in the number of unfrozen drops within the interval $\Delta T$ (note that both $\Delta N$ and $\Delta T$ are negative), and $N_0$ is the total number of drops. The units of $k(T)$ are expressed as $cm^{-3}\,°C^{-1}$, or equivalent, and the units of $K(T)$ are in $cm^{-3}$. As their names imply, the differential concentration represents the volume density of nuclei active per unit temperature interval about $T$, and the cumulative spectrum describes the volume density of nuclei active at temperatures warmer than $T$. Naturally, the cumulative spectrum is the integral of the differential spectrum from $T$ to 0°C.

Strictly speaking, equation 4 is valid only for infinitesimally small values of $\Delta T$ and $\Delta N$. For finite values, there is an error of approximately 5% for $\Delta N/N = 0.1$, and ~10% for $\Delta N/N = 0.2$. If the intervals of observations in an experiment are large enough to approach these values, or if the total number of drops in the experiment is small, a more accurate value of $k(T)$ can be obtained (Vali, 1971) using

$$k(T) = -\frac{1}{V\Delta T}\log_e\left(1-\frac{\Delta N}{N(T)}\right) \tag{6}$$

with $N(T)$ taken at the beginning of the interval.

The range of values of $k(T)$ and of $K(T)$ obtained from an experiment depends on the volume of the drops and on the number of drops used in the experiment; for $k(T)$, the range also depends on the widths of the temperature intervals used. The minimum and maximum values obtained from an experiment are

$$\begin{aligned} [k(T)]_{min} &= \frac{1}{N_0 V\Delta T} \quad \text{and} \quad [k(T)]_{max} = \frac{1}{V\Delta T} \\ [K(T)]_{min} &= \frac{1}{VN_0} \quad \text{and} \quad [K(T)]_{max} = \frac{\log_e N_0}{V} \end{aligned} \tag{7}$$

It may be noted the range of differential concentrations covered is equal to the number of drops in the test. The cumulative concentrations cover a somewhat larger range. For example, with $N_0 = 100$, $V = 0.03$ cm$^3$, and $\Delta T = 0.2$°C, $k(T)$ will range from 1.7 to 170 cm$^{-3}$ °C$^{-1}$, and $K(T)$ from 0.33 to 153 cm$^{-3}$.

General expressions for the accuracies of $k(T)$ and $K(T)$ are rather complex. It is clear that the lower and upper ends of the spectra have greater error bands, due to the small numbers of drops involved in the calculations there. The simple assumption can be made that the number of drops freezing per interval would have a Poisson distribution in many repeated tests. On that basis, the standard deviation of $k(T)$ values is

$$\sigma_k = k(T)\left(\frac{1}{\Delta N}+\frac{1}{N(T)}\right)^{0.5} \tag{8}$$

demonstrating that small values of d$N$ and of $N(T)$ lead to larger uncertainties in $k(T)$. Since, for given $N_0$, the expected values of d$N$ are tied to the magnitudes of the temperature intervals used to group the data, aiming for high resolution in temperature can lead to large uncertainties in the values of $k(T)$. Similarly, the errors in $K(T)$ can be estimated as

$$\sigma_K = \begin{cases} \dfrac{1}{\sqrt{N(T)}} \\ \\ \dfrac{1}{\sqrt{\Sigma \Delta N}} \end{cases} \tag{9}$$

using whichever of the two expressions is larger. The summation in the denominator of the lower expression is taken from the beginning of the experiment until the temperature in question has been reached. At the beginning of the experiment, when few of the drops have yet frozen, the larger source of uncertainty is associated with that number. Toward the end of the experiment, when few drops remain unfrozen, the uncertainty arises from the variability of that number (upper expression in equation 9).

In equation 5, the fraction of drops remaining unfrozen can be replaced by the somewhat more convenient fraction frozen, and the equation can be represented by a simple nomogram, as shown in Figure 4. For any given temperature, the corresponding fraction of frozen drops can be converted with Figure 4 into the appropriate value of $K(T)$. For example, 35% of drops of 0.01 $cm^3$ volume frozen at some temperature indicate ~100 $cm^{-3}$ nuclei active above that temperature. The diagram can also be used to predict the fraction of drops that will be frozen as a result of some nucleus concentration.

The concentrations (per unit volume of water) given by equations 4 and 5 can readily be converted to numbers of nuclei per unit mass of nucleating material if

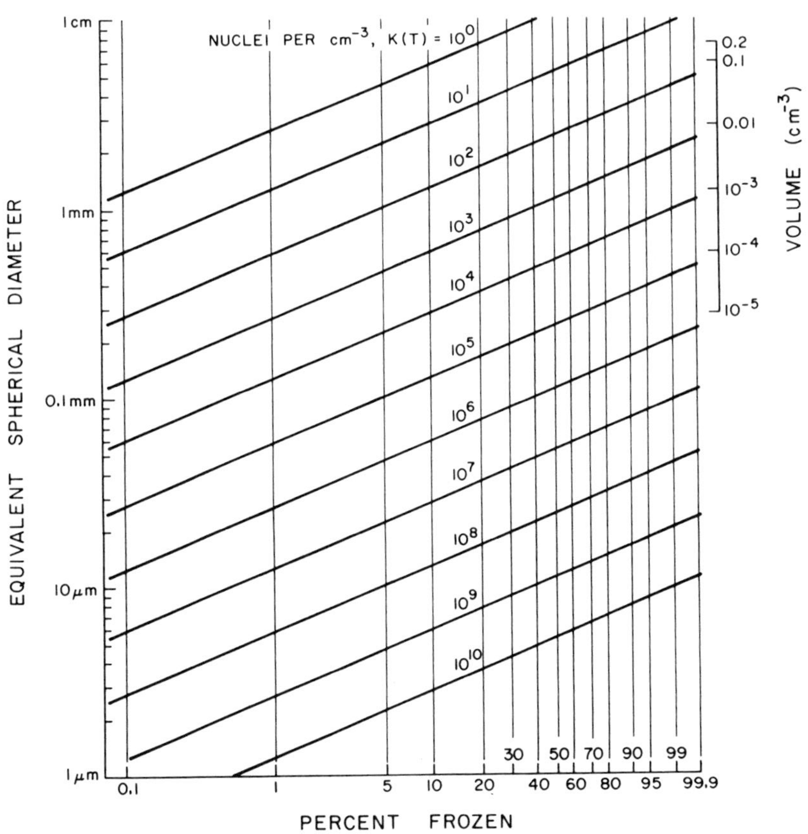


**Figure 4.** Nomogram relating the cumulative nucleus concentration $K(T)$ to drop diameter and volume and to the fraction of drops frozen at $T$.

the ratio of water volume to mass of nucleating material in the drops is known, by the substitution

$$V' = V\,m \tag{10}$$

where $m$ is the mass of nucleating material per unit volume of water. The units of the corresponding spectra $k'$ and $K'$ change to $g^{-1}$ °$C^{-1}$ and $g^{-1}$. Similarly, the spectra can be expressed with reference to the number of cells, or other units, that are suspended in the water at known concentrations.

Equations 4 and 5 are written for the case in which all drops in the test are of the same volume. If that is not the case, the derivation of the spectra becomes considerably more cumbersome; the appropriate formulas are given by Vali (1971).

As already mentioned, applying equations 4 and 5 to an experiment with, say, 100 drops will yield nucleus concentrations over two orders of magnitude. To cover a larger range, a series of dilutions of the original sample can be tested. In that case, the drop volumes in equations 4 and 5 have to be multiplied by the dilution factors.

It is evident from the definition that the differential concentration $k(T)$ can be determined only from observations made as the temperature of a sample is gradually lowered and if $\Delta N$ and $N(T)$ are noted at intervals. Cumulative concentrations can be obtained from measurements at discrete temperatures, with the number of temperatures depending on the application. Observation even at a single temperature can be used to determine the concentrations of nuclei for temperatures warmer than the test temperature. Experiments with emulsions in DTA or DSC instruments yield signals that are approximately proportional to d$N$ times the mean droplet volume. Neglecting the variation in droplet volumes, these data can be used to generate $N(T)$:

$$N(T) = N_0 - \int_0^T \mathrm{d}N \tag{11}$$

and so enable equations 4 to 6 to be applied.

Observed nucleus spectra often can be approximated by exponential and power law functions. Exponential functions appear to provide reasonably good fit, for both k($T$) and $K(T)$, when the temperature range covered is far below 0°C, say, less than −10°C. At warmer temperatures, most samples exhibit a rapid rise in nucleus concentrations, so that a power law function gives a better description. Some samples, in fact, exhibit distinct peaks in $k(T)$, so that simple functions do not provide good fits. The abrupt rise at a warm temperature, or the existence of a peak there, lends some basis for the use of "onset temperatures" to describe the activities of samples. Similarly, some authors adopted the designation Type 1 or Type 2 for samples in which the initial rise of spectra are found near −5°C or near −8°C. It must be recognized, though, that there are some pitfalls associated with such designations, since the portion of the spectrum actually examined in a given experiment (depending on sample volume) influences the temperature at which the first freezing events are observed, and statistical significance for peaks in the spectra can usually be established only with quite large sample sizes.

## Measurements of Other Modes of Activity of Nuclei

Although not expected to be of appreciable significance for ice nucleation in biological systems, some mention should be made of the methods of measurement for deposition, condensation freezing, and contact freezing nuclei.

Deposition nucleation takes place from a vapor environment. From experience, the range of vapor densities of interest is limited to those between saturation with respect to ice and saturation with respect to water. This amounts to supersaturations of 5.1 to 21.5% with respect to ice, for temperatures of –5 to –20°C. Above water saturation, it is more likely that the vapor would condense as liquid water and then freeze. For hygroscopic materials, the formation of liquid water can start even before water saturation is reached. For a meaningful measurement, the saturation ratio of the vapor and the temperature must be controlled. This can be achieved in so-called thermal diffusion chambers or thermal saturation chambers. In both kinds of chambers, the sample at temperature $T$ is near an ice surface at temperature ($T + \delta T$), with the value of $\delta T$ determining the resulting supersaturation with respect to ice. Once ice is nucleated, it grows, and its presence is visually detected.

Condensation freezing is almost certainly applicable to aerosols only, so it is of concern here only in connection with the activity of biological nuclei in the atmosphere, a characteristic specifically considered by Levin et al. (1987) for the application of bacteria as cloud seeding agents. Unambiguous measurement of this mode of activity is quite problematic. It can be attempted in chambers used for deposition, but if the test is made directly at subzero temperatures the two modes are very difficult to distinguish; the presence of microscopic amounts of bulk water would have to be detected. With condensation allowed to take place above 0°C, the process is really freezing nucleation, although the combined ability of a particle to initiate both processes is significant in applications such as that of Levin et al. (1987).

Contact freezing nucleation arises at the surface of a supercooled volume of water when coming into contact with the nucleating particle. This kind of sequence is considerably more complex to observe than the ones discussed so far. In general outlines, the measurements proceed by either letting supercooled drops fall onto the potential nucleating particles, or the particles are brought into contact with the drops via diffusion, electric forces, etc. These measurement techniques and results for atmospheric aerosol are given in Götz et al. (1991, pages 165-171).

## Macroscopic Factors in Freezing Nucleation

The description of freezing nucleation and the determination of nucleus concentrations has been presented so far with some idealizations and simplifications. In the following, some additional factors known to influence freezing nucleation will be summarized.

### Volume Dependence

The variation of nucleation temperature with the volume of water being tested was of interest early on because of the indications it provides concerning the mechanism of nucleation, especially the distinction between homogeneous and heterogeneous nucleation. Volume dependence is now well clarified, as expressed by the nucleation rate per unit volume for homogeneous nucleation, and by equations 4 and 5 for heterogeneous nucleation. Both predict proportionality of nucleation probability with volume. For homogeneous nucleation with fixed cooling rates, the mean freezing temperatures of different volumes will decrease linearly with the logarithm of the volume in response to the exponential temperature dependence of equation 3. Simple proportionality to volume becomes obscured by the chance allocation of nucleating particles in the heterogeneous case. The mean freezing temperatures of drops of different volumes drawn from the same sample will vary, depending on the shape of the nucleus spectrum $k(T)$. If that function can be approximated by an exponential, as is often the case when the nuclei do not have a single origin, then the volume dependence approaches the same form as for homogeneous nucleation, a linear decrease in temperature for a logarithmic decrease in volume.

The fact that volume variations are properly accounted for by the nucleus spectra $k(T)$ and $K(T)$ is demonstrated by the example shown in Figure 5. Drops taken from the same sample but differing in volume by a factor of 4 yielded distributions of freezing temperatures shown in the smoothed histograms on the left of Figure 5. However, when the differential spectrum is calculated for each group, the three sets of points define a single curve, well within experimental variability. Clearly, in this case, the random distribution of nuclei of all activities within the sample was a valid assumption. In fact, such a test is a good way to establish whether the sample support or containers influence the nucleation temperatures; since those effects are proportional to the contact surface area, $k(T)$ would not be independent of $V$.

The same argument can be used to check on the randomness of distribution of nuclei within various volumes. For example, the dependence of freezing temperature on the length of conifer needles (Kaku and Salt, 1968) could be quantitatively analyzed if the frequency distributions of freezing temperatures for different needle lengths were converted to differential spectra. Average freezing temperatures by themselves do not allow quantitative deductions to be made.

**Time Dependence**

This topic has many facets and therefore has given rise to a number of debates. Here it will have to suffice to state two fundamental aspects of the problem. First, as is evident from the discussions in the introductory section, the process of embryo growth to critical size is governed by random addition and removal of molecules from the embryo whether reference is made to the homogeneous case or to embryo growth on a nucleating site. Thus, nucleation is an inherently time dependent, stochastic event; it is described by a probability per unit time. Second, because nucleation probability for a site is so strongly dependent upon temperature, time dependence manifests itself only if the temperature changes very slowly with time. These points are developed in greater detail by Vali (1994). The practical consequences of time dependence for the determination of nucleus concentrations for cooling rates of from degrees per hour to several degrees per minute are described earlier in this chapter.

For samples held at a constant temperature, the number of freezing events is the sum of $K(T)$ times the sample volume (which equals the number freezing during cooling to $T$) plus an additional number that grows at a diminishing rate with time. The total number approaches asymptotically to $K(T-\zeta)$, where $\zeta$ is about 1°C. This is an approximation only; a rigorous theoretical treatment of the problem has not been developed yet.

For samples brought to some subzero temperature very suddenly, the number of

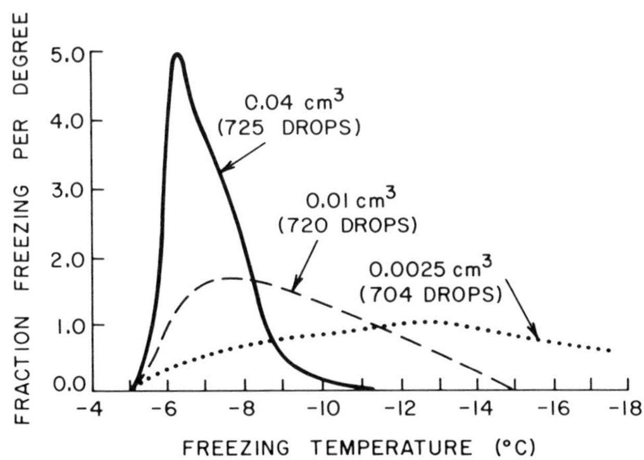


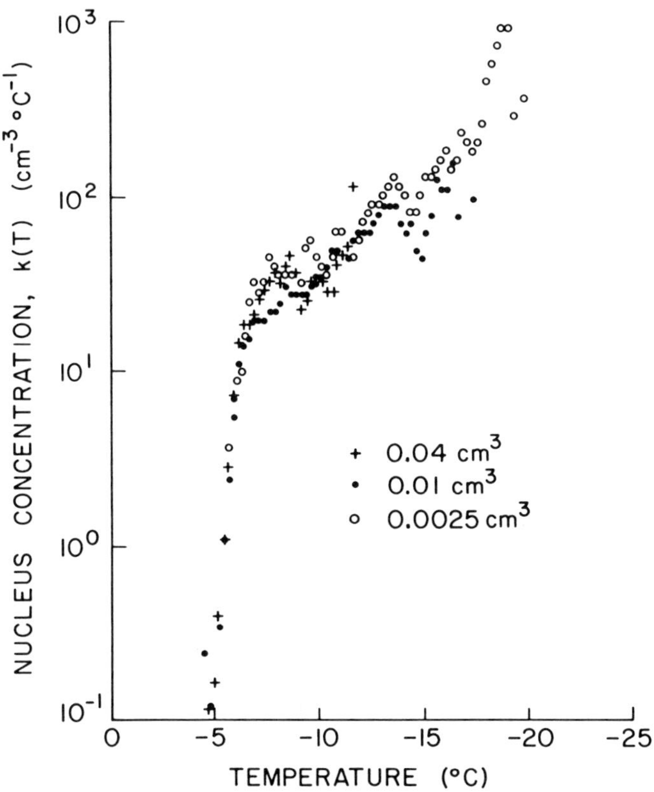


**Figure 5**. Distribution of freezing temperatures and differential nucleus spectra for three sets of drops of different volumes taken from the same sample.

freezing events was reported by several investigators to be higher than it would have been for moderate cooling. The magnitude of the effect has not been established, nor is the reason for this effect known. Nonetheless, some caution seems to be warranted when making determinations of $K(T)$ by sequential transfer of a sample from one temperature to another, since in such cases cooling to each new temperature is very rapid.

The time dependences discussed so far relate to nucleating sites that can be assumed to be fully stable. There are a variety of other time-dependent phenomena that affect the sites themselves. Sites can be formed by etching of a surface; they can be destroyed by dissolution; they can become contaminated by adsorption of solute molecules; and so on. These effects are often difficult to identify, but when present add to the time dependence for stable sites. With cloud droplets and aerosol nuclei, one must also consider the rates at which the aerosol nuclei are captured by the droplets.

A fairly large number of publications deal with the "time lag" in nucleation. Unfortunately, this term has been used in many different contexts. As discussed above, there are two main sources of time lags—the inherent dynamics of cluster growth and what may be viewed as varying boundary conditions. It is often difficult to separate the many time dependent factors that may be present, but without that distinction misleading conclusions might result.

**Chemical Factors**

The basic situation envisaged when discussing heterogeneous ice nucleation is a system composed of water and a nucleating material. In many situations there is a third component to be dealt with—namely, some material dissolved in the water. The solute may interact with the nuclei or may influence the properties of the water. In the latter case, the minimum effect is the depression of the melting point. The magnitude of the depression is fairly well predicted for dilute (<1 M) solutions, and in the majority of cases the nucleation temperatures, both homogeneous and heterogeneous, are depressed by an equal amount. For higher-solute concentrations the homogeneous nucleation temperature is depressed approximately twice as much as the melting point (Rasmussen and MacKenzie, 1972); the actual values depend on the specific solute. For heterogeneous nucleation, some nonlinearities have been found even for dilute solutions, giving evidence of interactions of the solute molecules with the nucleating sites (e.g., Reischel and Vali, 1975; Reischel, 1984, 1987).

Ice nuclei exposed to vapors show varied reactions. Both increases and decreases in activity have been observed. Most work has been done with silver iodide nuclei and with natural atmospheric nuclei. Amines cause minor losses of activity; ethanol and other alcohols produce variable results; and ammonia can lead to increased activity. Whereas some researchers thought the effects to be the result of irreversible chemisorption, others found the effects reversible.

**Memory Effect, Preconditioning**

The activity of nuclei were found to be influenced, for some materials, by previous exposure of the nuclei to freezing, or just to subzero temperatures and high humidities. These effects have been given considerable attention because they are revealing about the mechanisms of embryo formation and about the nature of nucleating sites. However, the variability of the phenomenon is so great that no simple generalizations can be made about the ways in which it occurs. Broadly speaking, explanations of these effects center on the notion that some ice patches may be retained on the surfaces of materials at irregularities such as crystal lattice steps or cavities, or perhaps precisely at potential nucleation sites because of the stronger interactions there with water molecules. The effect may disappear completely after warming or drying beyond some threshold values. Below those thresholds the effects gradually diminish with time. Not all materials exhibit memory effects or are subject to preconditioning. In fact, most are not, but there are also some surprisingly large effects. For example, nucleation temperatures were noted to rise by as much as 8°C (from –20°C to –12°C) on minerals after the first formation of ice (Roberts and Hallett, 1968); mercuric iodide will nucleate ice at –1 to –2°C after original nucleation at –13°C and if not warmed above +4°C (Edwards et al., 1970).

**Electric Effects**

The orienting effects of electric fields was thought from the very beginning of ice nucleation studies to be a possible key factor. Experimental evidence, however, was difficult to provide. Imposed external fields proved to be of little effect. Fields existing at the substrate surface because of its crystal structure were found to have an appreciable influence, and electric resistance was shown to be negatively correlated with nucleating activity for doped silicon. Recently, Gavish et al. (1992) showed evidence of strong local fields within cracks of a nucleating surface; they speculate that electric fields at polar surfaces of crystals may account for ice-nucleating activity in silver iodide and in bacteria as well.

## Characteristics of Effective Freezing Nuclei

General ideas about both the conditions governing nucleation and the behavior of nucleating materials can be made more concrete by examining some specific examples. As mentioned before, many questions about the interactions of materials with ice embryos still remain unanswered, but there are definite factors that make embryo growth favorable. Here the topic is discussed from a practical point of view; a good overview of the theoretical background is given by Knight (1979).

**Silver Iodide (AgI)**

This ice-nucleating material has been studied more extensively than any other. Since its first identification as an effective ice nucleus (Vonnegut, 1947), it has been the subject of studies aimed at unraveling the basics of ice nucleation and at developing the use of AgI as a cloud seeding agent. Much progress was made on both fronts, experimentally as well as theoretically.

AgI is highly insoluble in water and has a crystal structure very close to that of ice. On this basis was its ice-nucleating potential first suggested, and since then the importance of that similarity has been well demonstrated. The match of crystal structures, or good epitaxial fit, was proven to be dominant by the use of solid solutions which introduce slight variations in the fit (Vonnegut and Chessin, 1971; Palanasamy et al., 1986). The role of epitaxy can be vividly demonstrated for deposition nucleation: ice crystals growing on the surface of an AgI crystal have the same orientation as the substrate. Strong as the evidence is for the importance of epitaxy, the need for specific sites on the surface of AgI is also well demonstrated. Also, evidence exists that pure silver iodide is not effective, and that trace contaminants are needed for it to be a good ice nucleator.

As a freezing nucleus, AgI is effective at temperatures cooler than –5°C; the fraction of surface area, or the fraction of particles that are active, increases rapidly beyond that. For cloud seeding, a major advantage of AgI is that it can be produced as aerosol particles smaller than 0.01 μm, although there is an inverse relationship between the fraction of particles active at some temperature and their average size. As a colloid, smaller sizes are more difficult to obtain, resulting in a certain gap in relating cloud tests to the more controlled immersion freezing tests in the laboratory.

It is noteworthy that some other iodine compounds are also effective ice nuclei. Lead iodide ($PbI_2$), cupric iodide (CuI), and mercuric iodide ($HgI_2$) are among the dozen or so compounds that nucleate ice at temperatures above or near –6°C.

### Minerals

The abundance of mineral dusts in the atmosphere and the ubiquity of minerals at the surface of the earth raise an important question as to their ice-nucleating abilities. The role of minerals as atmospheric ice nuclei has been extensively investigated and given some support by identifying minerals in snow crystals and by correlating increased ice nucleus counts in the air to the transport of dusts from desert areas. Yet these ties are elusive, so that mineral particles are now considered only as partial contributors, not as dominant sources of atmospheric ice nuclei.

Covellite (CuS) and vaterite ($CaCO_3$) are the most active minerals; their reported onset temperatures are around –4 to –5°C. Both CuS and $CaCO_3$ have a hexagonal structure, but the fit to the ice structure is not good enough to make a strong case for epitaxy to be the main reason for their effectiveness. Of the more abundant minerals, kaolinite is most effective, though only at around –10°C. Quartz, montmorillonite, and many other minerals show even lower effectiveness. A unique example of a very effective mineral was located by Passarelli et al. (1974). Miersite, from one location in Australia, is a solid solution of CuI in AgI; it has excellent epitaxial fit to ice and produces ice nucleation at temperatures as warm as –1.4°C. A corollary to ice nucleation by minerals is that a large fraction of them exhibit memory effects or respond to preconditioning.

It became apparent in the early experiments with minerals that the particular source of a particular sample had a strong influence on its observed ice-nucleating ability compared to other samples of the same composition. Later work showed that the organic components of soils in fact had a stronger influence on ice-nucleating ability than the mineral component (see Chapter 2). This shifted the focus of inquiry to the organics and eventually led to the identification of bacteria as a possible major source of these nuclei. However, the details of the connection between ice-nucleating bacteria and ice nuclei found in soils has not yet been fully clarified, and the possibility of still unrecognized sources of ice nuclei in soils has to be left open. The relationship of the nuclei to the mineral base of the soils has not been explored either, even though there are, from the atmospheric point of view, important questions about how these materials get lofted into the air, what the lifetimes of the nuclei are, and so on. One must also guard against the simplification that because minerals are not as active as the organics, they are not important; their abundance argues against that assumption.

**Organic compounds.** A great variety of reports of ice nucleation by organics is available. The search for organic ice nucleators began very soon after the discovery of the nucleating activity of silver iodide. Films of egg albumin were observed by

Schaefer (*personal communication*) and by Zettlemoyer et al. (1961) to nucleate ice at small supercoolings. Sano et al. (1960) reported appreciable nucleating activity for urea. Head (1961, 1962a) found activity at temperatures between –2 and –6°C for about twelve different steroids and for some substituted fluorenes. Komabayasi and Ikebe (1961) examined benzenoid compounds. Head (1962b) found that diazines provide good ice nuclei, with α-phenazine being most effective. Power and Power (1962) identified four amino acids that nucleate ice at temperatures warmer than –10°C. Fukuta (1963) added metaldehyde to the list, with activity reported at –0.4°C. Using the work of Bashkirova and Krasikov (1957) as a basis, Langer et al. (1963) tested the activity of phloroglucinol and found the activity promising enough for field tests in cloud seeding (Braham, 1963).

Further studies, mostly expanding on the findings just described, were reported by Fukuta and Mason (1963), Barthakur and Maybank (1963), Parungo and Lodge (1965), Maybank and Barthakur (1966), Fukuta (1965, 1966), Fletcher (1972), and Langer et al. (1978). The activities of the amino acids and some steroids were confirmed, and some other compounds (1,5-dihydroxynaphtalene, acetoacetanilide with high activity) were added to the list. No general argument emerged that allowed the relative activities of the substances to be related to their structure. Factors that showed at least some success in providing an explanation are the spacing of OH groups (Fukuta and Mason, 1963), rotational symmetry (Fukuta, 1966), and group free energy (Parungo and Lodge, 1965).

A new group of ice-nucleating substances, with very interesting properties, was described by Gavish et al. (1990). Aliphatic alcohol ($C_nH_{2n+1}OH$) monolayers placed on droplets of water were found to induce ice nucleation at temperatures that increase monotonically with chain length and approach 0°C for the longest ($n = 31$) chain. A systematic difference was found between the even and odd carbon number alcohols, with the even *n* series being much less active. The structure of the monolayer, as revealed by X-ray diffraction observations, has OH groups in a hexagonal pattern similar to that of ice, suggesting a possible basis for the nucleating activity.

Ice nucleation by bacteria, lichens, and fungi provided the latest opportunities for the study of organic ice nuclei. The great deal of material on these topics is detailed in subsequent chapters.

It is apparent that a great variety of organic materials are capable of serving as ice nuclei at temperatures just a few degrees below 0°C. This relationship between ice and other hydrogen-bonded substances is perhaps not surprising; learning about the factors governing the interaction of the organic molecules with the ice embryos will constitute important progress.

## Looking Ahead

Even with considerable progress over the past century, ice nucleation studies are still largely phenomenological. The thermodynamic/kinetic theory proved to be a useful reference framework, but it falls short of providing useful descriptions of the great variety of phenomena associated with heterogeneous ice nucleation. Experimentally, examination of surfaces on the scales of ice embryos is just barely possible, and no method is yet capable of direct examination of the ice embryos themselves. Their presence and action must be deduced from the phase transitions they eventually trigger. Thus, in total, we lack the tools of analysis for the behavior and interactions of the precursors of nucleation.

In spite of these limitations, however, the situation is constantly improving. New approaches are developing on a number of fronts. Theoretically, better understanding of liquids (density fluctuations, cell models), computer simulations of interfaces and clusters, and the application of nonequilibrium statistical thermodynamics promise advances. In experiments, the control of conditions is extending to an ever-widening set. In addition to prevailing macroscopic factors like temperature, humidity, and time, new possibilities exist in the characterization of colloids and aerosols. Also, there are new techniques for the examination of surfaces and the preparation of new molecules. The study of ice nucleation in biological systems and the application of molecular biology to ice nuclei have broadened the horizon considerably.

While progress in developing more complete theories of ice nucleation is not likely to be very fast in the foreseeable future, practical knowledge about ice nucleation and ice nuclei can be expected to find a rapidly broadening circle of applications. The last two decades has brought to light the role of ice nucleation in living systems; added to the atmospheric role, this new dimension will undoubtedly spur both research and application. Ice nucleation by organic molecules will, in all likelihood, be recognized as having even wider-ranging importance than now known.

## Literature Cited

Abraham, F. F. 1974. Homogeneous Nucleation Theory. Academic Press, New York.

Altberg, I. V. 1938. O tsentrakh ili iadrakh kristaalizatsii vody. Meteor. i Gidrol. 3-12. Translated as "On centers or nuclei of water crystallization," Cold Reg. Res. Eng. Lab. TL294, 1972.

Barthakur, N., and Maybank, J. 1963. Anomalous behaviour of some amino-acids as ice nucleators. Nature 200:866-868.

Bashkirova, G. M., and Krasikov, P. N. 1957. Experiments with certain substances as crytallization agents for supercooled fogs. Tr. Gl. Geofiz. Obs. 72:118-126.

Bergeron, T. 1935. On the physics of clouds and precipitation. Proc. Assem. U.G.G.I., Lisbon. 2:156-173.

Bigg, E. K. 1953. The supercooling of water. Proc. Phys. Soc. B 66:688-694.

Braham, R. R., Jr. 1963. Phloroglucinol seeding of undercooled clouds. J. Atmos. Sci. 20:563-568.

Brewer, A. W., and Palmer, H. P. 1951. Freezing of supercooled water. Proc. Phys. Soc. B 64:765-773.

Bryant, G. W., Hallett, J., and Mason, B. J. 1959. The epitaxial growth of ice on single-crystalline substrates. J. Phys. Chem. Solids 12:189-195.

Caslavsky, J. L., and Vedam, K. 1971. Epitaxial growth of ice crystals on the muscovite cleavage plane and their relation to partial dislocations. J. Appl. Phys. 42:516-520.

Clausse, D., Bouoabdillah, D., Cochet, N., Luquet, M. P., and Pulvin, S. 1991. Ice crystallization by silver iodide and bacteria in microsize droplets dispersed within emulsions. Pure Appl. Chem. 63:1491-1494.

Cwilong, B. M. 1947. Sublimation in a Wilson chamber. Proc. R. Soc. London 190:137-143.

Dorsch, R. G., and Hacker, P. T. 1950. Photomicrographic investigation of the spontaneous freezing temperatures of supercooled water droplets. Natl. Advis. Comm. Aeronaut. Tech. Notes 2142.

Dorsey, N. E. 1938. Supercooling and freezing of water. J. Res. Natl Bur. Stand. 20:799-808.

Dorsey, N. E. 1948. The freezing of supercooled water. Am. Phil. Soc. 38:247-328.

Dubrovský, M., and Skoloud, O. 1989. Measurement of a freezing nucleus content in a water sample: Evaluation of the method. Stud. Geophys. Geod. 33:287-304.

Edwards, G. R., Evans, L. F., and Zipper, A. F. 1970. Two-dimensionsal phase changes in water adsorbed on ice-nucleating substrates. Trans. Faraday Soc. 66:220-234.

Findeisen, W. 1938. Der Aufbau der Regenwolken. 2. f. Angew. Meteorol. 55:208-225.

Fletcher, A. 1972. High-temperature contact nucleation of supercooled water by organic chemicals. J. Appl. Meteorol. 11:988-993.

Fletcher, N. H. 1958. Size effect in heterogeneous nucleation. J. Chem. Phys. 29:572-576.

Fletcher, N. H. 1962. The Physics of Rainclouds. Cambridge University Press, Cambridge. 386 pp.

Fletcher, N. H. 1970. The Chemical Physics of Ice. Cambridge University Press, Cambridge. 271 pp.

Fournier d'Albe, E. M. 1949. Some experiments on the condensation of water vapour at temperatures below 0°C. Q. J. R. Meteorol. Soc. 75:1-14.
Franks, F., Mathias, S. F., Galfre, P., Webster, S. D., and Brown, D. 1983. Ice nucleation and freezing in undercooled cells. Cryobiology 20:298-309.
Fukuta, N. 1963. Ice nucleation by metaldehyde. Nature 199:475-476.
Fukuta, N. 1965. Activated ice nucleation by sprayed organic solutions. J. Atmos. Sci. 22:207-211.
Fukuta, N. 1966. Experimental studies of organic ice nuclei. J. Atmos. Sci. 23:191-196.
Fukuta, N., and Mason, B. J. 1963. Epitaxial growth of ice on organic crystals. J. Phys. Chem. Solids 24:715-718.
Gavish, M., Popovitz-Biro, R., Lahav, M., and Leiserowitz, L. 1990. Ice nucleation by alcohols arranged in monolayers at the surface of water drops. Science 250:973-975.
Gavish, M., Wang, J. L., Eisenstein, M., Lahav, M., and Leiserowitz, L. 1992. The role of crystal polarity in α-amino acid crystals for induced nucleation of ice. Science 256:815-818.
Götz, G., Mészáros, E., and Vali, G. 1991. Atmospheric Particles and Nuclei. Akadémiai Kiadó, Budapest. 274 pp.
Hagen, D. E., Anderson, R. J., and Kassner, J. J. L. 1981. Homogeneous condensation- freezing nucleation rate measurements for small water droplets in an expansion cloud chamber. J. Atmos. Sci. 38:1236-1243.
Head, R. B. 1961. Steroids as ice nucleators. Nature 191:1058-1059.
Head, R. B. 1962a. Ice nucleation by some cyclic compounds. J. Phys. Chem. Solids 23:1371-1378.
Head, R. B. 1962b. Ice nucleation by α-phenazine. Nature 198:736-737.
Hirano, S. S., and Upper, C. D. 1986. Bacterial nucleation of ice in plant leaves. Meth. Enzymol. 127:730-738.
Hobbs, P. V. 1974. Ice Physics. Oxford University Press, Oxford. 837 pp.
Kaku, S., and Salt, R. W. 1968. Relation between freezing temperature and length of conifer needles. Can. J. Bot. 46:1211-1213.
Knight, C. A. 1979. Ice nucleation in the atmosphere. Adv. Colloid Interface Sci. 10:369-395.
Komabayasi, M., and Ikebe, Y. 1961. Organic ice nuclei: Ice-forming properties of some aromatic compounds. J. Meteorol. Soc. Jpn 39:82-95.
Kuhns, I. E., and Mason, B. J. 1968. The supercooling and freezing of small water droplets falling in air and other gases. Proc. R. Soc. A. 302:437-452.
Langer, G., Rosinski, J., and Bernsen, S. 1963. Organic crystals as icing nuclei. J. Atmos. Sci. 20:557-562.
Langer, G., Cooper, G., Nagamoto, C. T., and Rosinski, J. 1978. Ice nucleation mechanism of submicron monodispersed silver iodide, 1,5-dihydroxynaphtalene and phloroglucinol aerosol particles. J. Appl. Meteorol. 17:1039-1048.
Langham, E. J., and Mason, B. J. 1958. The heterogeneous and homogeneous nucleation of supercooled water. Proc. R. Soc. A. 247:493-505.
Levin, Z., Yankofsky, S. A., Pardes, D., and Magal, N. 1987. Possible application of bacterial condensation freezing to artificial rainfall enhancement. J. Clim. Appl. Meteorol. 26:1188-1197.
Lindow, S. E. 1982. Population dynamics of epiphytic ice nucleation active bacteria on frost sensitive plants and frost control by means of antagonistic bacteria. Pages 395-416 in: Plant Cold Hardiness and Freezing Stress, Vol. 2. A. Sakai and P. H. Li, eds. Academic Press, New York.
Makino, T. 1982. Micropipette method, a new technique for detecting ice- nucleation activity of bacteria and its application. Ann. Phytopathol. Soc. Jpn 48:452-457.
Maybank, J., and Barthakur, N. 1966. The ice nucleation behavior of amino-acid particles. Can. J. Phys. 44:2431-2445.
Meyer, v. J., and Pfaff, W. 1935. Zur Kenntnis der Kristallisation von Schmelzen. III. Z. Anorg. Allg. Chem. 224:305-314.
Michelmore, R. W., and Franks, F. 1982. Nucleation rates of ice in undercooled water and aqueous solutions of polyethylene glycol. Cryobiology 19:163-171.
Montmory, R. 1956. La congelation de l'eau surfundue sur une surface crystalline. Bull. Obs. Puy de Dome 126.
Mossop, S. C. 1955. The freezing of supercooled water. Proc. Phys. Soc. B. 68:193-208.
Palanasamy, M., Thangaraj, K., Gobinathan, R., and Ramasamy, P. 1986. X-ray diffraction and ice nucleation studies of AgI-AgCl solid solutions. J. Cryst. Growth 79:1005-1009.
Parody-Morreale, A., Bishop, G., Fall, R., and Gill, S. J. 1986. A differential scanning calorimeter for ice nucleation distribution studies: Application to bacterial nucleators. Anal. Biochem. 154:682-690.
Parungo, F. P., and Lodge, J. P., Jr. 1965. Molecular structure and ice nucleation of some organics. J.

Atmos. Sci. 22:309-313.

Passarelli, J. R. E., Chessin, H., and Vonnegut, B. 1974. Ice nucelation by miersite. J. Appl. Meteorol. 13:610-612.

Power, B. A., and Power, R. F. 1962. Some amino-acids as ice nucleators. Nature 194:1170-1171.

Pruppacher, H. R., and Klett, J. D. 1978. Microphysics of Clouds and Precipitation. D. Reidel Publishers, Dordrecht. 714 pp.

Rasmussen, D. H, and MacKenzie, A. P. 1972. Effect of solute on ice-solution intrefacial free energy: Calculation from measured homogeneous nulceation temperatures. Pages 126-145 in: Water Structure at the Water-Polymer Interface. H. H. G. Jellinek, ed. Plenum Press, New York.

Rasmussen, D. H., Macaulay, M. N., and Mackenzie, A. P. 1975. Supercooling and nucleation of ice in single cells. Cryobiology 12:328-339.

Rau, W. 1944. Gefriervorgänge des Wassers bei tiefen Temperaturen. Schr. Dtsch. Akad. Luftfahrforsch. 8:65-84.

Reischel, M. T. 1984. Freezing-nucleation in aqueous solutions of clathrate forming gases. Tellus 36(B):73-84.

Reischel, M. T. 1987. Variation of the activity of ice nuclei upon exposure to ammonium ion and iodine. Tellus 39(B):363-373.

Reischel, M. T., and Vali, G. 1975. Freezing nucleation in aqueous electrolytes. Tellus 27:414-427.

Roberts, P., and Hallett, J. 1968. A laboratory study of the ice nucleating properties of some mineral particulates. Q. J. R. Meteorol. Soc. 94:25-34.

Rogers, J. S., Stall, R. E., and Burke, M. J. 1987. Low-temperature conditioning of the ice nucleation active bacterium, *Erwinia herbicola*. Cryobiology 24:270-279.

Salt, R. W. 1958. Application of nucleation theory ot the freezing of supercooled insects. J. Insect Physiol. 2:178-188.

Salt, R. W. 1966. Effect of cooling rate on the freezing temperatures of supercooled insects. Can. J. Zool. 44:655-659.

Sano, I., Fujitani, Y., and Maena, Y. 1960. The ice-nucleating property of some substances and its dependence on particle size. Mem. Kobe Mar. Obs. 14:107-118.

Schaefer, V. J. 1946. The production of ice crystals in a cloud of supercooled water droplets. Science 104:457-459.

Schaefer, V. J. 1948. The production of clouds containing supercooled water droplets or ice crystals under laboratory conditions. Bull. Am. Meteorol. Soc. 29:175-182.

Schaller, R. C., and Fukuta, N. 1979. Ice nucleation by aerosol particles: Experimental studies using a wedge-shaped thermal diffusion chamber. J. Atmos. Sci. 36:1788-1802.

Taborek, P. 1975. Nucleation in emulsified supercooled water. Phys. Rev. B 32:5902-5906.

Tammann, G., and Büchner, A. 1935. Die Unterkühlungsfähigkeit des Wassers und die lineare Kristallisationsgeschwindigkeit des Eises in wässerigen Lösungen. Z. Anorg. Chem. 222:371-381.

Vali, G. 1971. Quantitative evaluation of experimental results on the heterogeneous freezing nucleation of supercooled liquids. J. Atmos. Sci. 28:402- 409.

Vali, G. Freezing rate due to heterogeneous nucleation. J. Atmos. Sci. 51:1843-1856.

Vali, G., and Stansbury, E. J. 1966. Time-dependent characteristics of the heterogeneous nucleation of ice. Can. J. Phys. 44:477-502.

Vonnegut, B. 1947. The nucleation of ice formation by silver iodide. J. Appl. Phys. 18:593-595.

Vonnegut, B., and Chessin, H. 1971. Ice nucleation by coprecipitated silver iodide and silver bromide. Science 174:945-946.

Wood, G. R., and Walton, A. G. 1970. Homogeneous nucleation kinetics of ice from water. J. Appl. Phys. 41:3027-3036.

Zettlemoyer, A. C., Tcheurekdjian, N., and Chessick, J. J. 1961. Surface properties of silver iodide. Nature 192:653.

CHAPTER 2

# The Discovery of Bacterial Ice Nucleation and its Role in the Injury of Plants by Frost

**Christen D. Upper and Gabor Vali**

Twenty-five years ago, the existence of biological ice nucleation, and its effects on living organisms at subfreezing temperatures, was unsuspected. At that time, two groups with different backgrounds, interests, and goals were pursuing separate lines of research that would lead to discovery of the phenomenon. One of these groups searched for sources of ice nuclei that act in clouds to make rain, snow, and hail. The other investigated the way in which a fungus could apparently affect the susceptibility of corn to frost injury. The two groups independently discovered biological ice nucleation within a short time of each other. Neither group had anticipated the existence of the phenomenon. This chapter relates both histories of its discovery.

## The Discovery of Bacterial Ice Nuclei
*(as told by Gabor Vali)*

The fact that supercooling may be limited to –3 to –5°C or even less in some natural water samples, but distilled water can supercool to –12 to –20°C in the same experimental arrangement, was first thoroughly demonstrated by Ernest Dorsey at the National Bureau of Standards during the 1930s and 1940s (Dorsey, 1948). Similarly, the relatively early formation of ice in clouds was also suspected by the 1950s but not clearly established. On the other hand, direct measurements of atmospheric ice nuclei (in chambers designed to simulate natural cloud conditions) indicated a surprising paucity of ice nuclei. Attempts to identify the particles that might have initiated snow crystals pointed to the general prevalence of minerals, although microorganisms were also detected in a small number of crystals (Soulage, 1957). Unfortunately, such results were ambiguous as to whether the particles indeed nucleated the crystals or were merely trapped by them. Laboratory experiments were conducted with a wide variety of materials suspected to make up the majority of nonsoluble atmospheric aerosol; these experiments were generally discouraging, since supercoolings of −10°C or greater were needed for nucleation. A few substances rarely found in the atmosphere showed activity at temperatures up to about −5°C, most notably silver iodide, cupric sulfate, calcium carbonate, *l*-leucine, *dl*-alanine, and phloroglucinol. Therefore, the general belief was that ice nuclei in the atmosphere are mineral particles. Ideas from the 1960s about ice nuclei in the atmosphere are well summarized in, for example, the review of Mossop (1963).

A line of investigation started at McGill University by J. S. Marshall in 1958 or 1959 focused on ice nuclei (specifically, immersion-freezing nuclei) that could be detected in precipitation samples—rain, snow, and especially, hailstones, because of the interest in hail prevention via cloud seeding in the Canadian province of Alberta. This work was part of my Ph.D. research. The main task was to determine the nucleus content of hailstones by a drop-freezing technique, which I inherited in part and improved (Vali and Stansbury, 1966). For comparison, I also looked at rain and snow collected in clean plastic bags. Although minerals were then known to be poor ice nucleators (and, therefore, soil particles were low on my list of probable atmospheric ice nuclei, in comparison to meteoritic dust and other candidates favored at the time), on the spur of the moment I collected, probably in the spring of 1963, a sample of very dirty snow. It came from under my daughters' swing set in my backyard as the snowpack of the Canadian winter was disappearing. This went against the grain, since it was usual to take extreme precautions to collect samples in a clean manner. The dirty snow sample produced the warmest nucleation temperatures I had seen up to that point and caused a revision in my thinking about the possible role of soils (as opposed to isolated mineral components) in ice nucleation. Prompted by this observation, it seemed appropriate to examine the nucleus contents of soils, in addition to the precipitation samples, in search of possible connections. As a first step, samples from the collection of the McGill soil science department were tested. Samples from different regions were selected, and as it turned out, for some of the locations samples were available of both the surface soil and the underlying clay substrate. Tests with such pairs of samples showed that the surface soils invariably had higher ice nucleus contents than the purer clays. Thus the focus fell on the organic components of soils, and the search was broadened to encompass a range of soil types in this regard. With peat leading and volcanic ash at the other extreme, a correlation emerged between the organic mass fraction of soils and their ice nucleation temperatures (Vali, 1968). No attempt was made to identify the chemical composition of the nucleating material, because of the likely difficulty of purification of the minute quantities of ice nuclei from soil without loss of activity.

In 1969 I joined the faculty of the University of Wyoming, and Russell Schnell came there to study atmospheric science. He had had some exposure to cloud and precipitation research as a summer student with the hail project in Alberta. His background in biology was excellent preparation for investigating how ice nuclei get into surface soils. We were, at the time, unaware of the connection made by Lucas (1954) between ice nucleation and the freezing of plants at temperatures as warm as −1.4°C. In retrospect, that and other work should have alerted the atmospheric scientists to the presence of very active nuclei on plants and should have caused them to consider the converse of what Lucas thought was happening (namely, the introduction of the ice nuclei to plants from the atmosphere).

Russell quickly confirmed the propensity of soils with higher organic content to have higher ice-nucleating activity by collecting a diverse set of soil samples from the Wyoming–Alberta mountain and prairie area. He found further evidence of the connection to organics, in that heating of the clays to 200°C reduced their nucleus contents by factors of about 10. The major discovery made by Russell was that decaying grass and leaves are the likely source of ice nuclei in humus. He showed this, first, by soaking fresh (green) and old (brown) leaves from poplar in distilled water, with the results that within 7 days for the brown leaves and within 22 days for the green leaves, the nucleus content of the water increased dramatically, with nucleation initiating as warm as −1.5°C. The maximum effect lasted only for a short time. Aerobic conditions were found to be more conducive to the appearance of nuclei than anaerobic conditions. Russell next showed that decayed litter from the vicinity of poplars had a stable nucleus content and nucleation threshold temperatures only slightly lower than those of decaying leaves. The freezing nuclei from decaying leaves were named *leaf-derived nuclei* (LDN). The possible atmospheric role of LDN was investigated by Russell, via laboratory-scale dispersion experiments and calculations. He adds the following anecdote:

> In the summer of 1970… I was testing fresh leaf and grass samples for ice nucleus content by soaking the materials and testing the washes. In a rush to go to a party, I left a grass sample in the wash bag untested. Ten days later, as the drop freezer was being prepared for return to Laramie, I tested the now turbid water from the well-decayed grass cuttings. The water froze at −1.3°C. At that point I suspected that the most active ice nuclei might be related to living microorganisms.

The possibility that fungi or bacteria might produce the LDN led us to ask (in the fall of 1970) for the help of Professor Martha Christensen of the botany department at the University of Wyoming. She, in turn, set Richard Fresh to work on the problem as a master's project. The results on LDN, as well as earlier work that showed some ice nucleation activity by phenolic and benzoic acids (Komabayashi and Ikebe, 1961) and by amino acids (Power and Power, 1962), lent motivation for the work. Dick Fresh collected leaf samples from trees at monthly intervals. Suspensions of macerated leaf samples were tested for ice nucleus content when collected and at regular intervals afterwards. The most interesting results came from *Alnus tenuifolia* (alder). Within two to five days of collection, these samples yielded ice nuclei active at temperatures above −5°C. Samples from later in the growing season became the most active, and those later samples retained the activity almost without change. For comparison, sterilized samples had freezing temperatures of around −16°C.

With this mounting evidence for biological production of ice nuclei, the work turned to isolating the specific agents. Dick Fresh and Martha Christensen isolated a large number of fungi (65) and bacteria (100) from one of the decaying alder leaf samples that showed high activity. The major fungal isolates and the overall fungal composite were then used to inoculate sterilized leaf samples, and the ice nucleus content was determined after some time. None of these samples produced appreciable increases in nucleation temperatures over 17 days. The bacterial composite, however, showed activity similar to the original leaf sample. Pairs of bacterial isolates were tested next, after 7 and 14 days. Of the 25 pairs tested, only one pair showed any significant increase in nucleation temperatures. Further tests showed that one of that pair of isolates (labeled C-9) was responsible for all the effect. The activity produced by C-9 proved, in many different tests, to be nearly identical to that of entire decaying leaves, so one could conclude that this bacterium was producing the activity observed in the *A. tenuifolia* decay process. This was in the fall of 1972.

At this point, the focus of activity shifted to an expert on bacteria, Leroy Maki, Professor of Microbiology at the University of Wyoming. He, and students working with him, identified bacterium C-9 as a *Pseudomonas syringae* strain, provided detailed taxonomy for the strain, and determined under what conditions this bacterium exhibited ice-nucleating ability. They showed that the ice-nucleating ability of *P. syringae* was critically dependent on temperature and nutritional conditions, that the activity was associated with the cell membrane, and that preserving activity in parts of the cells was difficult. Leroy and his students also widened the search for other ice nucleation–active bacteria, but found none to equal *P. syringae* C-9, and very few others that had any appreciable activity at all. Still not knowing whether these were the same nuclei as those in the various LDN samples, these nuclei were named bacteria-derived nuclei (BDN).

Thus, by the end of 1972, it was clear that *P. syringae* has a special relation to the nucleation of ice. That a bacterium could serve as a very effective ice nucleus, exhibiting activity at even warmer temperatures than the most effective inorganic compound, silver iodide, was startling, but the effect was so clearly reproducible that there was no room for doubt. When the connection appeared between the frost damage studies at the University of Wisconsin (detailed below) and the work at Wyoming, the horizon of implications of the discovery broadened even further. From there on, the story of discoveries moved to the realm of the biological sciences, as detailed in subsequent chapters of this volume. Finding that most surprising ice nucleator, *P. syringae,* prompted many more questions: What is the precise nature of the ice-nucleating site? Why does this microorganism produce the nuclei? How, and under what conditions, is the nucleating ability developed? Are there other bacteria with similar properties? and so on. At the present time, many of these questions have been answered, but yet deeper questions remain to be pursued. The scope of research on biological ice nuclei has broadened to include lichens, fungi, and specific proteins. It appears likely that the study of biological ice nuclei will contribute to the clarification of such fundamental aspects of ice nucleation as the chemical and structural requirements for heterogeneous nucleating sites and will help to unravel the role of ice nuclei as a biosphere—atmosphere link.

The majority of the results described here are from work performed in relation to the M.S. theses of Russell C. Schnell (June, 1971), Richard W. Fresh (December, 1972), Elizabeth Galyan (December, 1973), Mei-Mon G. Chang-Chien (December, 1974), and Karen Willoughby (August, 1976), and the Ph.D. dissertation of Russell C. Schnell (May, 1974). Publications that describe the early work at Wyoming are Schnell and Vali (1972, 1973, 1975, 1976), Maki et al. (1974), Schnell (1975), Vali et al (1976), and Maki and Willoughby (1978).

## The Discovery of Bacterially Incited Frost Injury
## (*as told by Chris Upper*)

Corn is generally regarded as a "healthy" crop. Although a number of pathogens affect the U.S. corn crop, overall effects on yield are considered relatively minor in most seasons. However, unexpected epidemics sometimes occur because of a new pathogen, a new genotype of a common pathogen, or unfavorable growing conditions. Following such epidemics, there is heightened interest in control of the disease in question. One common response by plant pathologists is to search through collections of germ plasm for sources of resistance to the causal pathogen. If resistance can be identified, it is included in breeding programs to produce crop plants that are resistant to the pathogen. In response to an epidemic of northern corn leaf blight, Paul Hoppe had a sizable planting of corn lines in plots at the University of Wisconsin's Arlington Experimental Farms in the spring of 1964. In early June he inoculated plants in the appropriate plots with *Helminthosporium turcicum* (now *Setosphaeria turcica* [*Exserohilum turcicum*] Leonard and Suggs) to screen for resistance to the disease. His inoculation technique was to dry infected corn leaves and sprinkle the powdered preparation on plants in the early morning while they were still wet with dew. When Hoppe returned to the plots several days later, he was at first disappointed to find that his experiment had been complicated by an unseasonable frost. But careful inspection changed disappointment to excitement. He invited Deane Arny and J. W. Martens to verify his observations and to assist with quantitation of the amount of frost injury. Control plots that had not been inoculated were largely undamaged by the frost, whereas those that had been inoculated were extensively damaged (Hoppe et al., 1964). Hoppe showed that the experiment could be repeated in the growth chamber. Differences in the amount of frost injury were greatest if the plants were held in a mist chamber in the dark for 24 hours between inoculation and exposure to about –4°C. Autoclaved powder did not increase the sensitivity of corn to frost. Although the results of these additional experiments were not published, a demonstration of the effect of *E. turcicum* on frost injury was added to the Plant Pathology course to illustrate how changes that occurred early in the infection process resulted in a decrease in tolerance of plants to frost.

Sometime in late 1970 or early 1971, Deane Arny convinced me that this might be a suitable system in which to search for early physiological changes in the host during the infection process. After all, didn't the change in frost tolerance indicate that something was happening before symptoms developed? Our first experiment was not successful, because the growth chamber in which we attempted to expose the corn plants to freezing temperatures would not go below about 1°C. In search of better physical facilities, we took advantage of an introductory offer from the University of Wisconsin Biotron—reduced rates to encourage scientists to use the facility. If I recall correctly, the first two weeks were free. We devised an experiment that could be completed in that period. During that time we learned two things: that 1) inoculation of corn seedlings with washed spores of *E. turcicum* did not affect the likelihood of frost damage to corn seedlings, and 2) temperatures within controlled environmental facilities are not always as advertised. In one experiment, the temperature was slowly decreased to between –7 and –8°C, at which point symptoms of freezing were present in one third to one half of the leaves. The room was then slowly warmed to above freezing. Although frost injury was equally frequent on inoculated and control leaves, there was a distinct band of injured plants diagonally across the room. A third point—a strong clue to the nature of the phenomenon we were studying—was clearly illustrated by these results. We failed to recognize the significance of the relatively low temperature required to damage laboratory-grown corn until several years later.

Following the experiments in the Biotron, I was completely convinced that there was no effect of *E. turcicum* on frost injury to corn. I expressed this opinion in a conversation with Frank Vojtic, the person in the Department of Plant Pathology with responsibility for maintaining the departmental plant growth facilities. Frank took strong exception to my statement. Among his duties was the preparation of materials for the undergraduate courses in the department. He had successfully demonstrated the effect of *E. turcicum* on frost injury to the beginning plant pathology course every year for nearly a decade!

Frank did two things differently from what we had done. First, he did the experiment in a walk-in growth chamber in the department. The chamber had been designed to operate between 4 and 40°C. Asking it to operate at –4°C was risky, but at least excessive cooling rates were never a problem. Experiments could only be done during relatively dry weather—if it was too humid, the coils would become covered with ice, and the temperature in the chamber would not drop below freezing. The problem of temperature gradients was solved by placing three or four additional fans in the chamber to provide better mixing of the air and maintain the total range of temperatures in the chamber within 0.5°C.

The other change was the inoculation technique. Vojtic inoculated the corn seedlings by sprinkling dried, powdered, infected corn leaves on them, as Hoppe had done. I had insisted that this unnecessary source of variability be eliminated, and we had used a washed spore suspension in the Biotron. Much to my chagrin, the results were exactly as Frank had predicted: inoculated plants were significantly more damaged than controls.

Frank's procedure became the basis of the assay that we used for further study of the phenomenon. We found that the dried, powdered, infected corn leaves could increase the susceptibility of lettuce (a nonhost of *E. turcicum*) to frost. But dried, powdered, healthy corn leaves were nearly as effective as infected ones. Thus, we concluded that we were dealing with some compound in corn powder, not with a fungal infection at all (Upper et al., 1972).

When a new graduate student joined the group in the fall of 1973, the suggested thesis project was to identify the compound in dried, powdered corn leaves that could increase the likelihood that corn seedlings and other plants would be damaged by frost. Steve Lindow began his thesis research by showing that leaves of a number of other plants also contained the frost injury–increasing activity. Attempts to extract the active material met with mixed success. Aqueous extracts were often highly active; solvent extracts, however, were not. Procedures designed to purify the active material often resulted in loss of activity. Despite Steve's incredible energy level and work schedule, progress was slow. Intensely self-motivated, Steve rarely took time off work for more than classes, food, and a little sleep. On one rare occasion during the winter of 1974–1975, faced with the choice of continuing a purification or a weekend of cross-country skiing, he put the active fractions in the coldroom and went skiing. When Steve returned on Monday, the extract had become turbid. Steve's comments reflected rather unkindly upon the obvious microbial contamination of an extract in which he had invested much time and effort. He decided to assay it anyway, on the chance that it was still active.

The assay that Steve used was laborious. The amount of plant material required, and the size of the growth chamber, severely limited the number of treatments that could be included in each assay. The test material was dusted or sprayed on corn seedlings, and the seedlings were placed in a mist chamber, where they were kept wet but not dripping for 24 hours. Then they were allowed to dry in a hallway for about half an hour, and then placed in the growth chamber, which was just above 0°C. The temperature was decreased at 0.1°C per minute to about –3°C, and then half that rate until about half of the plants appeared to be frozen. When this procedure was finally completed, and Steve counted the number of damaged leaves in each treatment, he was surprised to learn that the turbid fraction was far more active than it had been the preceding week! The extract was plated onto bacteriological medium, and suspensions of the individual colonies were tested for an effect on the frost sensitivity of corn. One, isolate 31, was active. With some guidance from Arthur Kelman, Steve quickly identified the bacterium as *Pseudomonas syringae*. At this point we were elated! We (Steve) had identified the active entity that caused the increase in frost injury to corn.

In retrospect, what we had accomplished at that point was to replace a fungus with a bacterium. We had eliminated the possibilities that we were dealing with a compound in the corn leaf itself, and, of course, that we were studying an early response to the infection process. But were we any closer to understanding the mechanism? In May of 1975 we submitted an abstract for a paper to be presented at the annual meeting of the American Phytopathological Society to announce that the change in frost injury was due to *P. syringae*. Although we reported that *P. syringae* (all of eleven strains tested), but no other bacteria, was active, no mention of a possible mechanism was made in the abstract (Lindow et al., 1975).

Steve stored all of the random cultures he had made from the extract (about 200 in total) on agar slants in the refrigerator. The collection contained several replicate slants of isolate 31 to make sure that he wouldn't lose it. The refrigerator apparently malfunctioned, its temperature dropping below 0°C. The tubes containing isolate 31 (and only 31) had frozen. We noted the event with interest, but without explanation. However, it did prepare us for the next development.

A connection to the work at Wyoming was made by Richard Barchett, an assistant professor of meteorology. He had visited me in 1975 to explore the possibility of research projects of interest to scientists in agriculture. He showed interest in the issue of how a bacterium might be associated with frost injury to corn. Some cloud physicists in Wyoming were actively searching for sources of naturally occurring, highly efficient ice nuclei. He thought that they had recently found that a bacterium was active as an ice nucleus. Ice nucleation! The idea had not occurred to us.

We added salt to an ice bath to depress the temperature to about –5°C, placed some test tubes of sterile buffer in it, and waited until they had time to cool to below 0°C. Steve prepared a dense suspension of his bacterium and carefully pipetted a few microliters of the suspension into the supercooled buffer in one of the tubes. Ice formed instantly where the bacterial suspension touched the supercooled buffer, and it spread smoothly throughout the tube within a few seconds.

Steve found the paper by Maki et al. (1974) describing *P. syringae* as an ice nucleus. Indeed, one of the strains that Maki and co-workers had used for comparison during identification of their ice-nucleating bacterium had been supplied to them by Don Hagedorn from our department! Not only had Steve used the same strain as a reference for identification of his isolate 31, but he had also shown that it was highly active in increasing the amount of frost injury of corn. The pieces began to fall into place rapidly. We finally realized that what was unusual about our frost assay was not that the treated plants did freeze, but that the controls did not. In the field, where ample bacteria were available to provide ice nuclei, corn and other sensitive plants are normally damaged at temperatures above –4°C. Only in our growth chamber assay (or in the Biotron experiment), where control plants were free of ice nucleation–active (INA) bacteria, did sensitive plants escape damage at these temperatures. The system worked because of what was left out when plants were grown in the growth chamber. Later experiments established that numbers of INA bacteria are relatively small on corn early in the growing season, which explained Hoppe's original observation of less damage to uninoculated plants.

We first met Gabor Vali, Russ Schnell, and Leroy Maki at a meeting of the International Geophysical Union in San Francisco in December, 1975. A session had been organized dealing with aerobiology, and the group in Wyoming had encouraged us to submit an abstract (Lindow et al., 1976). We found it reassuring that the concept that bacterial ice nucleation could affect frost injury to corn was fairly well received by this group, composed mostly of physical scientists.

Convincing biologists provided a different kind of a challenge. Our first manuscript was sent to *Science*, from where it returned rejected. In retrospect, the manuscript that was submitted was not well written, and the revised manuscript that *Nature* rejected was much improved. The technical issues raised by reviewers were addressable. The comment that we didn't understand was that the manuscript was not of sufficient general interest to warrant publication in *Nature*. Deane called the editor and (to the credit of both) was able to convince him that there should be quite broad interest in the concept, and the manuscript was accepted for publication. When the paper appeared, a misunderstanding by the typesetter had changed the order of the authors from Lindow, Arny and Upper to Arny, Lindow and Upper (Arny et al., 1976). Subsequent publications invariably encountered the comment in review that additional ice nuclei were not necessary for frost injury because, under conditions of radiative frost, plants are invariably covered with hoarfrost. The idea that ice nucleation was necessary to initiate formation of hoarfrost was difficult for us to communicate to editors.

## Atmospheric Sciences
## (*as told by Gabor Vali*)

It is worth returning briefly to the other thread of the story, the one that initiated my involvement and motivated our work at the University of Wyoming. This was the question of where atmospheric ice nuclei originate. The search for nuclei, first in precipitation and then in soils, led to the discovery of ice-nucleating bacteria, but the reverse path—determining what fraction of the ice nuclei found in soils are bacterially mediated, how these nuclei might enter the atmosphere, and the broader question of what overall role biogenic nuclei play in the atmosphere—presented very difficult questions that are still without conclusive answers.

Some important elements of the atmospheric connection were established in Russell Schnell's Ph.D. work. He continued to examine the development of ice nuclei in natural forests, principally alder and spruce, but also several other species. These studies confirmed the earlier results, showed their broader applicability, and set the stage for the question of how extensive LDN might be worldwide. Roving the Northern Hemisphere, Russ collected samples from different climate zones and showed a strong correlation with ice nucleus content: leaf litters from tropical climates had the lowest, and litters from temperate climates the highest, concentrations of ice nuclei. Mountains, irrigation, and other local factors interfered to some extent with the general trend. Importantly, when examining the ice nucleus content of precipitation from different climate zones, it was found to follow the pattern found in leaf litters. Also, analysis of the results of an independent, worldwide set of ice nucleus measurements on filter samples revealed the samc trend.

A yet further connection to biogenic ice nuclei emerged when the question was addressed of whether a major difference exists between atmospheric concentrations of ice nuclei over land surfaces and over oceans. The affirmative answer would have had to hold if indeed LDN were the major source of atmospheric ice nuclei. To examine this question, Russ collected a small number of samples of ocean water. One sample of high phytoplankton content contained ice nuclei active at −8°C, and material obtained with a plankton net in the same area had nuclei active at −3°C. This was a meager but significant demonstration of the possibility that ocean waters are also sources of ice nuclei, and that these nuclei too are associated with biological activity. (Brier and Kline [1959] also saw evidence for oceanic sources for ice nuclei, but at the cold temperatures [<−15°C] and with the type of chamber they used, the effect could well have been due to humidity and aerosol changes rather to ice nuclei.) For lack of more specific information, these nuclei were termed ocean-derived nuclei (ODN). Support for the idea of biogenic sources of atmospheric ice nuclei over oceans came from the data of Bigg (1973) collected during a scientific cruise in the Southern Ocean. Examination of these data revealed a correlation between the ice nucleus concentrations in the air and the productivity in the water.

This recollection would be incomplete if no mention were made of a couple of other intellectual threads of the story. In view of the role of ice nuclei in precipitation initiation, a cause-and-effect connection can be envisaged between the removal of vegetation (and hence the elimination of LDN) and drought in the Sahel and other like regions via the reduction of precipitation efficiency. To demonstrate this connection, the important point to establish would be that a decrease in precipitation efficiency accompanied droughts rather than, or in addition to, a decrease in cloud frequency. However, to show a change in precipitation efficiency a long-term cloud physics study would be needed, involving radars, instrumented aircraft, and so on. There are, of course, other atmospheric changes related to droughts—principally in global-scale circulation—so the link to ice nuclei is less likely to be the dominant factor. As a result, the question has not been systematically addressed so far.

Another important aspect of ice nucleation in the atmosphere is the use of ice nuclei as cloud-seeding agents for the increase of precipitation, reduction of hail damage, etc. Therefore, some motivation and funding for the work with biogenic nuclei came from the vision that biogenic nuclei might provide a useful alternative to silver iodide as a cloud-seeding agent. (Most of the work at Wyoming was funded by the National Science Foundation's Atmospheric Sciences Division, through the then-current Research Applied to National Needs [RANN] program.) Silver iodide is not effective at temperatures warmer than −5°C; also, during the late 1960s and early 1970s, there were concerns about the possible environmental impact of the substance. The prospect for cloud seeding was pursued to the point of exploratory experiments in the field (dispersing ground leaf litter into convective clouds) and abandoned in view of the difficulties encountered in separating either LDN or BDN from the bulk of the soil or from the bacterial cells. Having to disperse large quantities of materials into clouds is a major disadvantage from a practical perspective. Also, the utility of cloud seeding as an effective tool for altering cloud processes was increasingly brought into question by the mid-1970s.

As the foregoing paragraphs indicate, the last quarter century produced little progress toward characterizing and understanding the origin of atmospheric ice nuclei. The suggestion of a connection to biogenic nuclei appears strong, but providing rigorous proof and quantifying the links are yet to be addressed with full force. Similarly, the impact of atmospheric ice nuclei on the biosphere is largely unknown, even though it is clear that nuclei are transported to the surface by precipitation and are deposited from the air. All these questions—the atmospheric connection—can be expected to lead to many interesting findings.

## Conclusions

Thus it was that two independent groups with very different interests found that they were both searching for ice-nucleating bacteria and discovered that *P. syringae* was active as an ice nucleus. One of these groups was part of a plant pathology department, where expertise in microbiology is abundant and expected. The other was part of an atmospheric sciences department, where microbiology is of somewhat less interest. Yet it was the cloud physicists who arrived at the answer first. In this day, when so many biologists emphasize the importance of methods, why was it that the group with all the appropriate methods and facilities had to wait for the group of physical scientists to discover that the activity was due to a bacterium? Consider the importance of the question that was being asked. The Wisconsin group was searching for the explanation of a biological phenomenon whose mechanism was unclear. The Wyoming group was looking for ice nuclei.

## Acknowledgments

In addition to the names appearing in this account, we wish to acknowledge contributions of others at the University of Wyoming, particularly the wide-ranging support and encouragement given to the work by D. L. Veal, head of the Department of Atmospheric Science in the 1970s; the contributions of J. Howatson, S. Forenza, and R. D. Kelly of the Department of Chemistry in checking out innumerable leads about the nature of the nuclei; and the advice of. D. Caldwell on matters related to cell physiology. We also acknowledge with appreciation Deane Arny and Steve Lindow for their comments on, and additions to, drafts of this chapter.

## Literature Cited

Arny, D. C., Lindow, S. E., and Upper, C. D. 1976. Frost sensitivity of corn increased by application of *Pseudomonas syringae*. Nature 262:282–284.

Bigg, E. K. 1973. Ice nucleus measurements in remote areas. J. Atmos. Sci. 30:1153–1157.

Brier, G. W., and Kline, D. B. 1959. Ocean water as a source of nuclei. Science 130:717–718.

Chang-Chien, M. G. 1974. Taxonomy and ice nucleation properties of a fluorescent pseudomonad, *Pseudomonas syringae* strain C-9. M.S. thesis. University of Wyoming, Laramie.

Dorsey, N. E. 1948. Freezing of supercooled water. Trans. Am. Phil. Soc. 38:247-328.

Fresh, R. W. 1972. Microbial production of freezing nuclei from decomposing tree leaves. M.S. thesis. University of Wyoming, Laramie. (Also published as Report AR106, February 1973.)

Galyan, E. L. 1973. Ice nucleation produces by a *Pseudomonas* isolate, Strain C-9. M.S. thesis. University of Wyoming, Laramie.

Hoppe, P. E. , Arny, D. C., and Martens, J. W. 1964. Frost susceptibility in corn increased by leaf blight infections. Plant Dis. Rep. 48:815-816.

Komabayashi, M., and Ikebe, Y. 1961. Organic ice nuclei: Ice-forming properties of some aromatic compounds. J. Meteorol. Soc. Jpn. 39:82-95.

Lindow, S. E., Arny, D. C., and Upper, C. D. 1975. Increased frost sensitivity of maize in the presence of *Pseudomonas syringae*. Proc. Am. Phytopathol. Soc. 2:158.

Lindow, S. E., Arny, D. C., Barchet, W. R., and Upper, C. D. 1976. The relationship between populations of bacteria active in ice nucleation and frost sensitivity in herbaceous plants. (Abstr.) E.O.S.

Trans. Am. Geophys. Union 56:994.
Lucas, J. W. 1954. Subcooling and ice nucleation in lemons. Plant Physiol 29:245–251.
Maki, L. R., Galyan, E. L., Chang-Chien, M., and Caldwell, D. R. 1974. Ice nucleation induced by *Pseudomonas syringae*. Appl. Microbiol 28:456–459.
Maki, L. R., and Willoughby, K. J. 1978. Bacteria as biogenic sources of freezing nuclei. J. Appl. Meteorol. 17:1049–1053.
Mossop, S. C. 1963. Atmospheric ice nuclei. J. Angew. Math. Phys. 14:456–486.
Power, B. A., and Power, R. F. 1962. Some amino acids as ice nucleators. Nature 194:1170–1171.
Schnell, R. C. 1971. Freezing nuclei from decomposing tree leaves. M.S. . University of Wyoming, Laramie. (Also published as Report AR102, April 1972).
Schnell, R. C., and Vali, G. 1972. Atmospheric ice nuclei from decomposing vegetation. Nature 236:163–165.
Schnell, R. C., and Vali, G. 1973. World-wide source of leaf--derived freezing nuclei. Nature 246:212–213.
Schnell, R. C. 1974. Biogenic sources of atmospheric ice nuclei. Ph.D. dissertation. University of Wyoming, Laramie. (Also published as Report ARlll, May 1974.)
Schnell, R. C. 1975. Ice nuclei produced by laboratory cultured marine phytoplankton. Geophys. Res. Lett. 2:500–512.
Schnell, R. C., and Vali, G. 1975. Freezing nuclei in marine waters. Tellus 27:321–323.
Schnell, R. C., and Vali, G. 1976. Biogenic ice nuclei. Part I. Terrestrial and marine sources. J. Atmos. Sci. 33:1554-1564.
Soulage, G., 1957. Les noyaux de congelation de l'atmosphere. Ann. Geophys. 13:1–32.
Upper, C. D., Arny, D. C., and Vojtik, F. 1972. A reexamination of the relationship between *Helminthosporium turcicum* infection and frost sensitivity of maize. (Abstr.) Phytopathology 62:794.
Vali, G., and Stansbury, E. J. 1966. Time-dependent characteristics of the heterogeneous nucleation of ice. Can. J. Phys. 44:477–502.
Vali, G. 1968. Ice nucleation relevant to hail formation. Ph.D. dissertation. McGill University, Montreal. 129 pp. (Also Report MW-58, Stromy Weather Group, McGill University, Montreal.)
Vali, G., Christensen, M, Fresh, R. W., Galyan, E. L., Maki, L. R., and Schnell, R. C. 1976. Biogenic ice nuclei: Part II. Bacterial sources. J. Atmos. Sci. 33:1565-1570.
Willoughby, K. J. 1976. Ice nucleation of supercooled water by microorganisms. M.S. thesis. University of Wyoming, Laramie.

CHAPTER 3

# Ecology of Ice Nucleation–Active Bacteria

**Susan S. Hirano and Christen D. Upper**

The discovery of ice-nucleating bacteria raised several new questions. How do they do it? Where, when, and why do they do it? Who are "they"? Biochemical and genetic approaches have made substantial progress in understanding how ice nucleation–active (INA) bacteria initiate the liquid-to-crystalline phase transition (see Chapters 4, 5, and 6). In this chapter, we discuss what is known about the ecology of INA bacteria—the where, when, and who. As with any retrospective question of evolution, the *why* can only be answered speculatively. However, by understanding the life strategies of these bacteria, we may gain insights relevant to possible selective advantages that the Ice⁺ phenotype could confer on INA bacteria.

## Species and Sources of INA Bacteria

At least 10 species of bacteria are known to include INA strains. The species and sources from which they have been isolated are summarized in Table 1. It is important to recognize that not all strains belonging to these species are INA. Indeed, a significant proportion of strains of most of these species that have been tested are not INA (cf. Hirano et al., 1978; Paulin and Luisetti, 1978). Additionally, although all cells of an INA strain carry the necessary gene(s) for production of ice nuclei, the phenotype is actually expressed relatively infrequently (cf. Maki et al., 1974; Lindow et al., 1978a). Bacterial ice nucleation activity is often expressed as the ratio of numbers of ice nuclei to numbers of viable cells (i.e., nucleation frequency) (Lindow et al., 1982a,b). The nucleation frequency of a particular strain grown under a specific set of conditions increases with decreasing temperature of measurement but rarely approaches unity (cf. Lindow et al., 1978a, 1982b; Hirano et al., 1978). Hence, the detection and enumeration of INA bacteria from natural sources may be influenced by the probability that each individual cell will express an active ice nucleus at any particular time and temperature.

To date, most of the INA bacteria that have been isolated have come from leaves and other aerial parts of plants. Indeed, the sources from which INA bacteria were first isolated were leaves (see Chapter 2), albeit dead ones: decaying alder (*Alnus tenuifolia*) leaves (Maki et al., 1974) and dried, powdered corn (*Zea mays*) leaves

**Table 1.** Species and sources of ice nucleation–active bacteria

| Source | Species[a] | Reference |
|---|---|---|
| Plant | | |
| Leaves and other aerial parts of plants | *Pseudomonas syringae* | Maki et al., 1974; Arny et al., 1976; Lindow et al., 1978b; others[2] |
| | *P. fluorescens* | Kaneda, 1986; Baca et al., 1987; Lindemann and Suslow, 1987a; Luisetti and Gaignard, 1987; Mazarei and Kerr, 1987; Goto et al., 1989; Montesinos and Vilardell, 1991; Morris et al., 1992 |
| | *P. viridiflava* | Paulin and Luisetti, 1978; Anderson and Ashworth, 1986; Minsavage and Stall, 1987; Young, 1987; Goto et al., 1989; Obata et al., 1989; Montesinos and Vilardell, 1991; Morris et al., 1992 |
| | *P. chlororaphis* | Schnell et al., 1991 |
| | *Erwinia herbicola* | Lindow et al., 1978a–c; Minsavage and Stall, 1987; Phelps et al., 1986; Gusta and O'Connor, 1987 |
| | *E. ananas* | Paulin and Luisetti, 1978; Newton and Hayward, 1986; Goto et al., 1988 |
| | *E. uredovora* | Newton and Hayward, 1986; Obata et al., 1990 |
| | *Xanthomonas campestris pv. translucens* | H. K. Kim et al., 1987; Azad and Schaad, 1988 |
| | *X. campestris* | Minsavage and Stall, 1987; Goto et al., 1988 |
| | Unidentified spp. | Yankofsky et al., 1981; Makino, 1983 |
| Xylem of roots of alfafa (and adjacent soil) | *P. syringae, P. fluorescens, E. herbicola, unidentified pseudomonads* | Richard et al., 1987 |
| Phytoplankton | | |
| Marine dinoflagellate | *P. fluorescens* | Fall and Schnell, 1985 |

*(continued on next page)*

[a] Not all strains within a species are active as ice nuclei (cf. Hirano et al., 1978; Paulin and Luisetti, 1978). Additionally, the nucleation frequencies (i.e., ratio of ice nuclei to viable cells) of INA strains within a species may vary signficantly (Maki et al., 1974; Lindow et al., 1978a; 1982b). Species designations according to *Bergey's Manual of Systematic Bacteriology* by Krieg and Holt, 1984: *Pseudomonas syringae* van Hall 1902; *P. fluorescens* (Trevisan 1889) Migula 1895; *P. viridiflava* (Burkholder 1930) Dowson 1939; *P. chlororaphis* (Guignard and Sauvageau 1894) Bergey, Harrison, Breed, Hammer and Huntoon, 1930; *P. putida* (Trevisan 1889) Migula 1895; *Erwinia herbicola* (Löhnis 1911) Dye 1964; *E. ananas* Serrano 1928; *E. uredovora* (Pon, Townsend, Wessman, Schmitt and Kingsolver 1954) Dye 1963; *Xanthomonas campestris* (Pammel 1895) Dowson 1939; *Xanthomonas campestris* pv. *translucens* (Jones, Johnson and Reddy 1917) Dye 1978.

[b] Hirano et al., 1978; Paulin and Luisetti 1978; Latorre and Jones 1979; Lindow 1982a; Gross et al., 1983; Jones et al., 1983; Butera et al., 1984; Lindow and Connell 1984; Newton and Hayward 1986; Baca and Moore 1987; Baca et al., 1987; Gusta and O'Connor 1987; Y. H. Kim et al., 1987; Luisetti and Gaignard 1987; Marshall 1988; Mazarei and Kerr 1987; Minsavage and Stall 1987; Olive and McCarter 1988; Goto et al., 1988; Goto et al., 1989; H. K. Kim et al., 1987; Leonardi et al., 1989; Lindow 1989; Colangeli et al., 1990; Cirvilleri et al., 1991; Constantinidou et al., 1991; Klement and Rozsnyay 1991; Montesinos and Vilardell 1991; Whitesides and Spotts 1991; Georgakopoulos and Sands 1992.

[c] *Enterobacter agglomerans* is taxonomically synonymous with *Erwinia herbicola.*

[d] Tentatively identified as *E. herbicola.*

**Table 1.** *Continued*

| Source | Species | Reference |
|---|---|---|
| Animals | | |
| Insect gut | *Enterobacter agglomerans*[3], *Ent. taylorae* | R. E. Lee et al., 1991 |
| Insect pupae | *E. herbicola* | Kaneko et al., 1991a,b |
| Amphibian gut | *P. fluorescens, P. putida, Ent. agglomerans* | M. R. Lee et al., 1991 |
| Environment | | |
| Soil | *E. herbicola*-like | Lindemann et al., 1982 |
| Rain | *P. syringae* | Sands et al., 1982; Constantinidou et al., 1990 |
| Snow | *P. fluorescens* | Maki and Willoughby, 1978 |
| Streams/lakes | *P. fluorescens* | Maki and Willoughby, 1978 |
| Ocean | *Erwinia* spp.[4] | Fall and Schnell, 1985 |
| Air | *P. syringae, E. herbicola* | Lindemann et al., 1982; Jayaweera and Flanagan, 1982; Sands et al., 1982 |

(Arny et al., 1976). On the basis of what we know today, the most abundant and widely distributed INA bacteria appear to be strains of *Pseudomonas syringae*—the first INA species to be identified. Ice$^+$ strains of *P. syringae* have been isolated from a wide range of plant species including monocots and dicots, annuals, and perennials growing in many different parts of the world, including North and South America, Australia, Europe, and Asia (references cited in Table 1). Some plant species, such as almond (*Prunus dulcis*), pear (*Pyrus communis*), oats (*Avena sativa*), and snap bean (*Phaseolus vulgaris*) tend to harbor larger numbers of *P. syringae* than others (e.g., *Citrus* spp., avocado [*Persea americana*], potato [*Solanum tuberosum*], and gymnosperms) (cf. Lindow et al., 1978b; Lindow, 1982a). *P. syringae* is not only the most widely distributed INA bacterium in nature but also among the most active (cf. Lindow, 1982a,b, 1983b; Paulin and Luisetti, 1978; Hirano et al., 1978, 1985). Nucleation is detectable near −2°C for many strains, and nucleation frequencies for occasional strains approach unity at −10°C (cf. Lindow, 1982a, 1983b; Hirano et al., 1978).

Ice$^+$ strains of *P. fluorescens* and *P. viridiflava* have been reported from aerial parts of a number of plant species (references cited in Table 1). These bacteria may be the most abundant INA bacteria on some plant species—for example, *P. viridiflava* on Chinese cabbage (*Brassica campestris* L. ssp. *pekinensis*) (Morris et al., 1992) and kiwifruit (*Actinidia deliciosa*) (Young, 1987). However, they are more commonly found in relatively small numbers on leaf habitats. *P. fluorescens* is more commonly considered an inhabitant of soil or aquatic habitats than of leaves (Starr et al., 1981; Krieg and Holt, 1984).

Three species of *Erwinia* include strains that are Ice$^+$: *E. herbicola*, *E. ananas*, and *E. uredovora*. Ice$^+$ strains of *E. ananas* and *E. uredovora* have been isolated from buds of tea (*Thea sinensis*) plants grown in Japan (Goto et al., 1988; Obata et al., 1990). INA bacteria with nutritional and biochemical profiles similar to *E. ananas* and *E. uredovora* have been identified from wheat (*Triticum aestivum*), barley (*Hordeum vulgare*), and certain grasses in Australia (Newton and Hayward, 1986). Ice$^+$ strains of *E. herbicola* have been reported as present on a much broader range of plants (Lindow et al., 1978b; Minsavage and Stall, 1987; Phelps et al., 1986; Gusta and O'Connor, 1987). Caution should be exercised in interpreting these results, however, because many of the reports of the broad distribution of *E. herbicola* predate description of INA strains of the other two species. Thus, although some strains have been identified as *E. herbicola*, others have undoubtedly been reported to be this species but identified only to the level of genus under the assumption that INA Erwineae were *E. herbicola* (Lindow et al., 1978b). Ice-nucleating strains of *Erwinia* are often present at relatively low levels on many plants and hence are probably less abundant than INA *P. syringae*. In general, they are also less active ice nucleators than *P. syringae* (cf. Lindow et al., 1978b; Paulin and Luisetti, 1978; Hirano et al., 1985).

Reports of the presence of INA strains of *Xanthomonas campestris* pv. *translucens* are restricted to barley and wheat, plants that are susceptible to disease caused by this bacterium (H. K. Kim et al., 1987; Azad and Schaad, 1988). Thus, although numbers of INA *X. campestris* pv. *translucens* may be very large on individual leaves, the total numbers in nature may be modest if they are restricted to relatively few plant species. INA strains of *X. campestris* that were found not to be pathogenic to the plants from which they were isolated have been reported from citrus in Florida (Minsavage and Stall, 1987) and tea in Japan (Goto et al., 1988).

There are a few reports on the isolation of INA bacteria from some rather interesting sources, including the xylem of roots of alfalfa and adjacent soil (Richard et al., 1987), a marine dinoflagellate (Fall and Schnell, 1985), and the gut of insects (Kaneko et al., 1991a, 1991b; R. E. Lee et al., 1991) and frogs (M. R. Lee et al., 1991) (Table 1). INA strains of *Enterobacter taylorae* and *Pseudomonas putida* have been uniquely isolated from nonplant sources. The detection of INA bacteria in aerosols over plant canopies and in rain and snow is of particular interest because of the possible, but yet unproven, role that bacterial ice nuclei may play in precipitation processes (Maki and Willoughby, 1978; Jayaweera and Flanaga, 1982; Lindemann et al., 1982; Sands et al., 1982; Constantinidou et al., 1990).

Although the majority of sources of INA bacteria to date have been aerial parts of plants, it is quite possible that this is more closely related to where scientists have looked than to where the preponderance of these bacteria live. Since the isolations of INA bacteria from the gut of invertebrates and vertebrates are relatively recent reports, it will be interesting to note within the next decade or two whether these sources are shown to be common habitats for INA bacteria. For phyllosphere bacteria (i.e., those inhabiting leaves and other aerial parts of plants), the number of plant species from which INA strains other than *P. syringae* and *E. herbicola* have been isolated is relatively small. It is not apparent whether this may be due to their more limited distributions on plant species, smaller population sizes, and/or lower nucleation frequencies, which would render the Ice$^{+}$ phenotype more difficult to detect. Perhaps no one has searched intensively and extensively for their presence on a large number of plant species in a study comparable to that of Lindow et al. (1978b).

## The Phyllosphere as a Habitat for INA Bacteria: General Considerations

Although the phyllosphere was recognized as a habitat suitable for diverse communities of bacteria in the early years of this century, little effort was devoted to studying these communities and the bacterial species they comprised. Indeed, the phyllosphere has been described as a "neglected milieu" by Ruinen (1961). The discovery of bacterial ice nucleation and its relationship to frost injury to plants (see Chapter 13) was an important factor in renewing interest in the microorganisms, particularly bacteria, that inhabit the phyllosphere.

It is intriguing that eight of the ten species of bacteria known to include Ice$^+$ strains share a common habitat—the phyllosphere. The species of INA bacteria that inhabit the phyllosphere are not unique in this regard. Many species of bacteria that are not active in ice nucleation are successful (defined in terms of abundance) inhabitants of the phyllosphere. What are the evolutionary and ecological implications of the Ice$^+$ phenotype appearing so frequently in residents of this particular habitat?

Unlike other habitats such as aquatic systems or the rhizosphere, the phyllosphere is an open habitat that exists in a continuously changing environment. Daily cycles in temperature, radiation, relative humidity, wind velocity, and leaf wetness are common. Slightly longer-term changes occur as weather systems pass. Seasonal climatic changes impose a still longer cycle. Rain may provide sudden, possibly catastrophic, perturbations of the system. The time scale of these changes relative to growth rates of phyllosphere bacteria dictates that different generations of the same species must survive and compete under very different conditions. Microorganisms have undoubtedly evolved a number of distinct strategies for existence in such a habitat.

At any given time, the structure of bacterial communities on leaf habitats is a function of four population processes: growth, death, immigration, and emigration. Populations increase as immigrants arrive and reproduce; they decrease through death and emigration. The rates with which these processes occur may be regulated by a number of interacting factors, including parameters related to the physical (e.g., weather conditions), biological (e.g., plant genotype and phenology, other microbes), and chemical (e.g., nutrient status) environments. The cumulative effect of changes in the rates of the population processes is seen in temporal changes in the relative abundance of INA bacteria. Clearly, the system is complex. Nonetheless, some understanding of the factors that regulate the dynamics of INA bacterial populations can be gleaned from an examination of how these populations change with time.

## Population Dynamics of INA Bacteria on Leaf Habitats

### Relative Abundance of INA Bacteria on Individual Leaf Habitats

Before discussing temporal changes in INA bacterial population sizes, we will consider the relative abundance of these bacteria on individual leaf habitats at a single time. A striking property of the population sizes of leaf-inhabiting bacteria is the amount of variability among apparently comparable leaves (Fig. 1). Population sizes of INA bacteria (and other components) frequently varied more than 1,000-fold among individual leaves (or leaflets) that were sampled on the basis of apparent similar size (and therefore, presumably, similar age) and position in a plant canopy (Hirano et al., 1982, 1985). These bacterial population sizes can be adequately modeled by the lognormal frequency distribution (Hirano et al., 1982). Thus, despite the substantial variability among such populations, and the skewed nature of their distribution, it is relatively simple to describe the relative abundance of INA bacteria among populations of leaf habitats in terms of the population mean and variance of the lognormal.

An unfortunate consequence of this type of bacterial population size distribution among leaves is that population means, determined from bulk samples (wherein several leaves are pooled to form a single sample prior to estimation of bacterial numbers), are difficult to interpret. Further, estimations of the probabilities that such means differ, by analysis of variance, are not reliable (Kinkel, 1992). Because many studies of INA bacteria were completed before this problem was adequately appreciated, some results in the literature should be interpreted with caution.

## Seasonal Trends

Large seasonal variations in relative abundance of INA bacteria have been measured on a number of woody perennials and herbaceous annuals growing in different locations (Lindow et al., 1978a–c; Lindow, 1982a, 1983a, 1987; Gross et al., 1983; Lindow and Connell, 1984; Kaneda, 1986; Luisetti and Gaignard, 1987; Mazarei and Kerr, 1987; Minsavage and Stall, 1987; Lindow and Panopoulos, 1988; Olive and McCarter, 1988; Colangeli et al., 1990; Constantinidou et al., 1991; Montesinos and Vilardell, 1991; Georgakopoulos and Sands, 1992; Morris et al., 1992). Although many of these studies relied on bacterial population estimates from bulk samples taken at weekly to monthly intervals, some overall generalizations can be made regarding seasonal trends in INA bacterial population sizes. On deciduous fruit trees in California (Lindow, 1982a; Lindow and Connell, 1984) and the Pacific Northwest (Gross et al., 1983), populations of INA bacteria (primarily

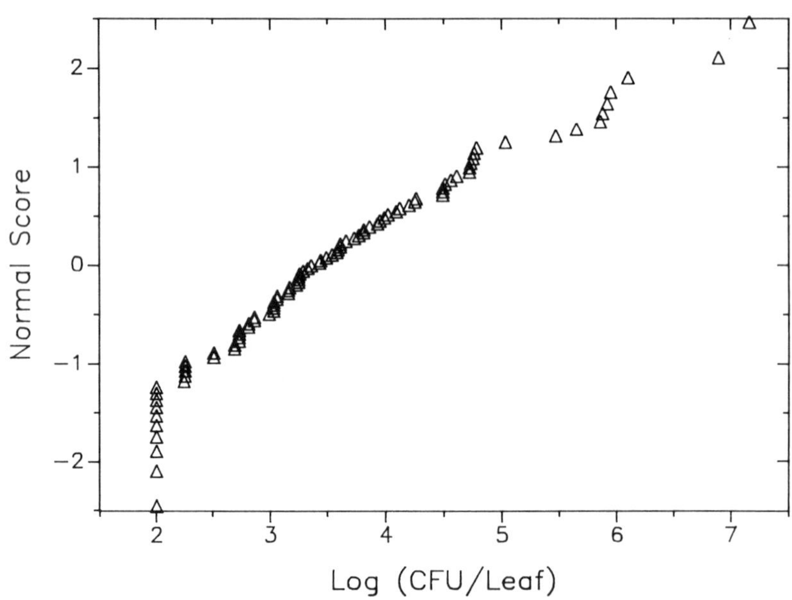


**Figure 1.** Frequency distribution of ice nucleation–active (INA) bacterial population sizes on each of 90 individual oat leaves. Leaves of similar size and age were collected at random from similar positions in the plant canopy. INA bacterial population sizes were determined by replica-freezing of dilution plates seeded with appropriate dilutions of individual leaf washings. INA bacteria were predominantly *Pseudomonas syringae*. Bacterial population sizes, expressed as the logarithm (base 10) of the number of colony-forming units (CFU) per leaflet were ranked and plotted against their corresponding normal scores (i.e., average deviates from each rank order in a sample of $n$ observations drawn at random from a normal population with $\mu = 0$ and $\sigma = 1$. (Reprinted, with permission, from Hirano et al., 1985)

*P. syringae*) tended to increase rapidly, and abundance was greatest during bud development and fruit set. This period coincided with the cooler and wetter months of spring (i.e., March and April). INA *P. syringae* attained mean population sizes on the order of $10^5$ to $10^6$ colony-forming units (CFU) per gram fresh weight of tissue and constituted a significant proportion of the total bacteria present on aerial plant parts. Population sizes of INA *P. syringae* invariably decreased with the onset of the warmer and drier months of summer. Gross et al. (1983) observed a gradual increase in population sizes of INA *P. syringae* on pear leaves in a single orchard as rainfall increased and temperatures decreased with the onset of autumn. To complete the cycle, Gross et al. (1983) also noted that INA bacteria could be found in dormant buds, but their relative abundance decreased until the subsequent spring.

In the study of Gross et al. (1983), INA bacteria were monitored on fruit trees grown in two regions of the Pacific Northwest. Whereas large populations of INA *P. syringae* were found in orchards at one location (Hood River valley, Oregon), these bacteria were much less frequently detected on fruit trees at another location (Yakima valley, Washington). The authors noted that the annual rainfall in the Hood River valley generally exceeds 75 cm but is less than 20 cm for most of the Yakima valley. The authors suggested that differential amounts of rain and possibly prevailing temperatures (slightly cooler in the Hood River valley during the growing season) may account for the differences in relative abundance of INA bacteria on fruit trees in the Pacific Northwest. Although plant phenology influences the relative abundance of INA bacteria, parameters associated with the physical environment appear to play a greater role in regulating the dynamics of these bacterial populations.

Lindow et al. (1978a,b) examined the relative abundance of INA bacteria on several annual plant species, including corn, snap bean, soybean (*Glycine max*), pumpkin (*Cucurbita maxima*), and tomato (*Lycopersicon esculentum*) grown in Wisconsin. In general, population sizes of INA bacteria were relatively small on young plants and increased to substantial levels as the season progressed. INA bacteria frequently constituted 1% or less of the total bacterial populations present on the leaf samples (Lindow et al., 1978b). In contrast, population sizes of INA bacteria (*P. syringae* and *P. fluorescens*) on pea (*Pisum sativum*) in South Australia decreased as the season progressed (July to October) (Mazarei and Kerr, 1987). The maximum mean population sizes measured in July were $10^2$ to $10^3$ per gram fresh weight, significantly less than those measured in Wisconsin on annual crops such as bean or tomato (Lindow et al., 1978b; Lindow, 1982a). Additionally, Mazarei and Kerr (1987) noted that there was no obvious influence of rainfall on the relative abundance of INA bacteria on pea in South Australia.

### Short-Term Changes in Population Sizes of *P. syringae*

Doubling times of approximately 1.3 h have been reported for strains of *P. syringae* growing in broth cultures at 28°C (Young et al., 1977). If similar generation times occur on leaves under field conditions, then sampling frequencies on a time frame of hours are necessary to estimate the rates of change in bacterial population sizes. A 2-h sampling interval was selected to examine short-term changes in population sizes of *P. syringae* on snap bean leaflets under field conditions (Hirano and Upper, 1989). *P. syringae* is the dominant INA species on bean leaves in Wisconsin (Hirano et al., 1987a). The intense sampling was conducted during each of three 26-h periods selected in advance and thus unbiased by weather conditions. At

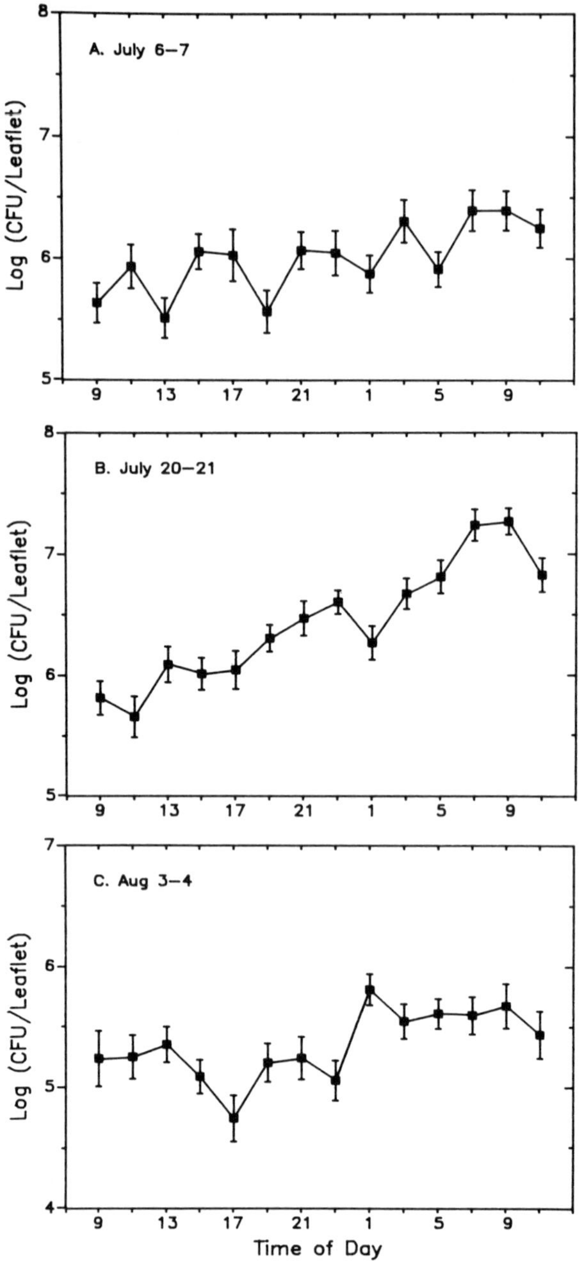


**Figure 2.** Diel variations in population sizes of *Pseudomonas syringae*. Population sizes of *P. syringae* were estimated for each of 30 individual snap bean leaflets per sampling time. Each datum point represents the mean $\log_{10}$ *P. syringae* population for each set of 30 leaflets harvested on 6 and 7 July, 30–31 days after planting (DAP) (**A**); 20 and 21 July, 44–45 DAP (**B**); and 3 and 4 August, 58–59 DAP (**C**). Error bars represent standard error of the mean. (Reprinted, with permission, from Hirano et al., 1989)

each sampling time, a tube nucleation test was used to assay 100 individual leaflets for ice nuclei, and dilution plating was used to determine population sizes of *P. syringae* on a subset of 30 individual leaflets.

The overall dynamics in *P. syringae* populations differed during each of the 26-h periods (Fig. 2). During the first 26-h period, substantial short-term fluctuations in *P. syringae* population sizes occurred, precluding meaningful estimates of rates of change (Fig. 2A). Overall, there was a 5.8-fold increase in the population mean from 9 a.m. on the first day to 9 a.m. on the second. During the second 26-h period, an almost continuous exponential increase occurred in population sizes of *P. syringae* (Fig. 2B). Twenty-four hours after the start of the experiment, the population size of *P. syringae* had increased 28.5-fold. This increase was orders of magnitude too large to be attributed to immigration (Lindemann and Upper, 1985; Hirano and Upper, 1991). If we assume that the increase was due solely to growth, then populations of *P. syringae* doubled approximately every 4.9 h. If death and emigration occurred, then the doubling time would be even shorter. During the third 26-h period, population sizes of *P. syringae* decreased during the day, then increased at night for no significant net change (Fig. 2C).

The diel patterns in ice nucleation activity did not parallel the short-term changes in population sizes of *P. syringae*. Bacterial ice nucleation activity, expressed as the cumulative percentage of bean leaflets that froze at each of three test temperatures, was lower during the day than during the night for all three sampling periods (Fig. 3). The frequencies with which leaflets froze at the test temperatures were inversely correlated with the average ambient temperature 30 min prior to leaf harvest and not with population sizes of INA *P. syringae* (Hirano and Upper, 1989). As ambient temperature decreased, nucleation frequencies of INA *P. syringae* on leaves increased. Similar effects of temperature on ice nucleation activity have been reported for pure cultures of INA bacteria in suspensions (Anderson et al., 1982; Rogers et al., 1987). When leaf samples are collected at a given time or at similar times on different days (e.g., early morning), bacterial ice nucleation activity is highly dependent on population sizes of INA bacteria (Hirano et al., 1987a; Lindow, 1988).

The three 26-h periods described above provided merely a glimpse of the short-term dynamics of population sizes of *P. syringae* on snap bean leaflets. There are approximately 55–60 such periods during the life span of the bean crop from plant emergence to pod harvest. A second sampling strategy, which involved almost daily (5 or 7 days per week) collections of leaflets from plant emergence to pod harvest, was used to determine short-term changes in *P. syringae* population sizes over a large number of 24-h periods. A 24-h sampling interval is too long to allow estimates of rates of change. However, estimates of net changes in bacterial population sizes over many such intervals have provided information on the frequencies with which large increases (e.g., as measured during one of the 26-h periods, Fig. 2B) and decreases occur during the life span of the host plant. Additionally, such a sampling strategy has been useful in identifying those weather conditions that have a significant effect on the population dynamics of *P. syringae* on bean leaflets.

An example of daily changes in population sizes of *P. syringae* on bean leaflets is presented in Figure 4. From this and several other data sets, the following picture has emerged. Most day-to-day changes in population sizes of *P. syringae* are relatively small (less than 5-fold). However, increases and decreases from 100- to 1,000-fold occur occasionally within periods of 24 h. Rainfall, either its occurrence

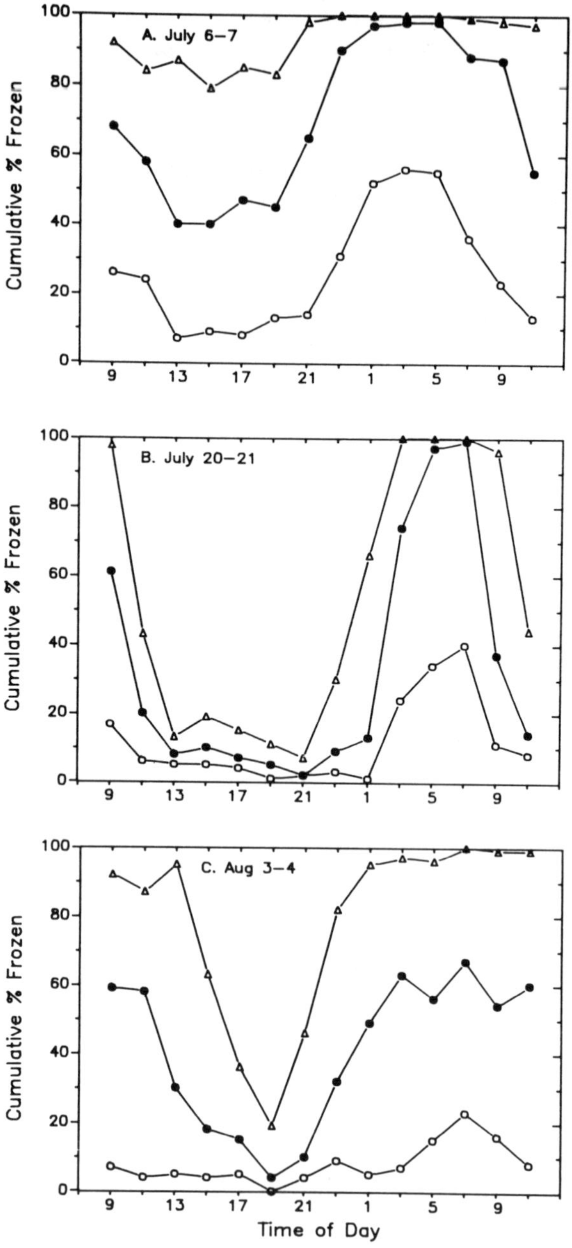


**Figure 3.** Diel variations in bacterial ice nucleation activity. A total of 100 individual snap bean leaflets was harvested at each sampling time and assayed with a tube ice nucleation test (Hirano et al., 1985). Samples were collected on 6 and 7 July (A), 20 and 21 July (B), and 3 and 4 August (C). Bacterial ice nucleation activity is expressed as the cumulative percentages of bean leaflets frozen at –2.0 (O), –2.2 (●), and –2.5 (Δ). (Reprinted, with permission, from Hirano et al., 1989)

or its absence, is the environmental factor that has the greatest effect on the population dynamics of *P. syringae* on bean leaflets. The immediate effect of rain is a decrease in population sizes of the bacterium, presumably due to wash-off (emigration). The decrease is followed by a period of increase in population sizes of *P. syringae* (Fig. 4, approximately 34 to 41 days after planting). Indeed, the nearly exponential increase in population sizes of *P. syringae* shown in Figure 2B was associated with a rainfall (22 mm) that occurred in the 12-h period prior to the start of the experiment. Populations of *P. syringae* tend to decrease (presumably due to death) during periods without rain (Fig. 4, approximately 43 to 49 days after planting), even when leaves are wet with dew for several consecutive nights.

A number of microclimate modification studies have been conducted in the field to explore the association of growth of *P. syringae* with rain. Alberga (1987) demonstrated that growth of *P. syringae* in response to rain is not associated with the duration of leaf wetness, temperature during rain, acidity, or volume of rain. The momentum of raindrops does appear to be important for initiating the onset of rapid growth of *P. syringae* (Hirano et al., 1987b). Under field conditions, the momentum of raindrops falling on a bean canopy was absorbed by allowing the rain to fall onto a fine-meshed inert fiberglass screen before dripping onto the leaves. Increases in population sizes of *P. syringae* were 1,000-fold less on leaflets that were under the screens compared to those that were naturally rained upon. Although an equivalent amount of water of comparable quality fell on both groups of plants, the velocity and, therefore, the momentum of raindrops impacting the leaves was less for plants under the screens than for those that were totally exposed. The mechanism by which raindrop momentum triggers the growth of *P. syringae* is not known.

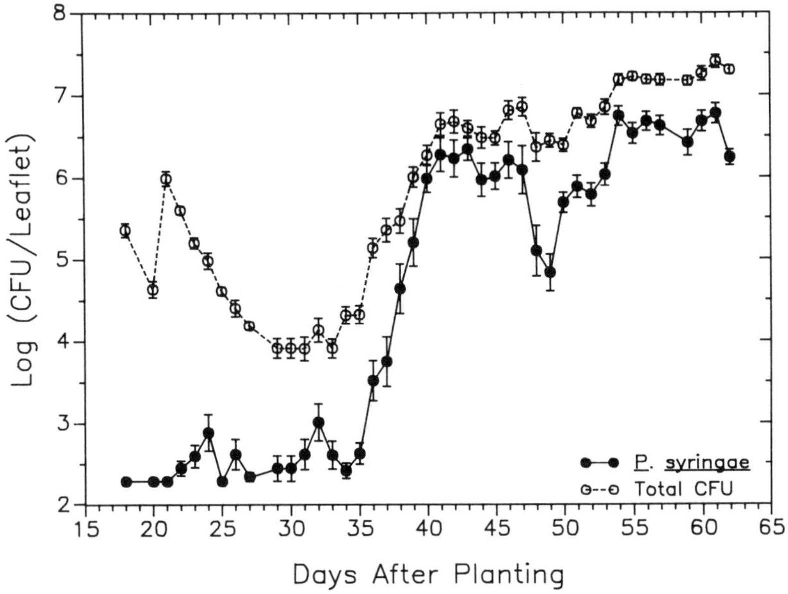


**Figure 4.** Temporal changes in population sizes of *Pseudomonas syringae* and total bacteria on snap bean leaflets. Thirty leaflets were collected at 8:00 a.m. at each sampling time. Each leaflet was processed individually by dilution plating of leaflet homogenates. Total bacterial populations represent total colony-forming units (CFU) on King's medium B (King et al., 1954). Each datum point represents the mean $\log_{10}$ CFU per leaflet and standard error of the mean.

In summary, sampling strategies that were selected to optimize measurements of short-term changes in population sizes of *P. syringae* on leaves under field conditions have demonstrated the inherently dynamic nature of leaf habitats. Population sizes of INA bacteria and bacterial ice nucleation activity change in response to changes in the physical environment. The development of large populations of INA bacteria is driven by growth. Growth, in turn, is dependent on the initial arrival of immigrants to leaf habitats.

## Dispersal of INA Bacteria

Leaf habitats are open, exposed systems. There is ample opportunity for facile exchange of bacteria, including INA bacteria, between the leaf surface and surrounding air. That this is indeed the case is supported by the findings of Lindemann et al. (1982), Lindemann and Upper (1985), and Constantinidou et al. (1990).

### Dispersal During Dry Conditions

Viable microbial impaction samplers were used to measure concentrations of airborne bacteria and vertical flux of bacteria over soil and over canopies of corn, winter wheat, pea, bean, and alfalfa (Lindemann et al., 1982). Concentrations of total airborne bacteria ranged from 46 CFU $m^{-3}$ over young corn to 6,500 CFU $m^{-3}$ over winter wheat. Concentrations of airborne INA bacteria ranged from undetectable over young corn to 178 CFU $m^{-3}$ over winter wheat. Population sizes of total and of INA bacteria were higher in soil from a fallow field than on leaves of alfalfa plants growing in an adjacent field. However, concentrations of both groups of bacteria were higher in the air over the alfalfa canopy than over the bare soil. The plant canopy was a stronger source of airborne bacteria than soil. Of 28 INA strains isolated from the air samples, 75% were identified as *P. syringae*; the remaining isolates resembled *E. herbicola*.

In a more extensive examination of the relationship between parameters of the physical environment and aerial dispersal of phyllosphere bacteria over bean canopies, Lindemann and Upper (1985) found that a net upward flux of total bacteria was nearly always present on dry sunny days when the humidity was below saturation and wind speeds exceeded about 1 m $s^{-1}$. Rates of emigration (measured as net upward flux) were highest at midday and were frequently on the order of about 100 CFU $m^{-2}$ $s^{-1}$, which corresponds to the loss of about $1.8 \times 10^3$ bacteria per hour per bean leaflet (Lindemann and Upper, 1985; Hirano and Upper, 1991). Upward flux of bacteria was never observed during periods of darkness, when the humidity was saturated or leaves were wet, during rain, or when wind speeds were lower than about 1 m $s^{-1}$ (Lindemann and Upper, 1985). Immigration rates (measured as deposition) were highly correlated with aerosol concentrations and were typically on the order of 1–10 CFU $m^{-2}$ $s^{-1}$. This corresponds to approximately 18–180 bacterial immigrants arriving per hour on a bean leaflet with an area of 50 $cm^2$. These estimates are for total bacteria and would be much less for component species such as INA bacteria. Hence, the cumulative number of immigrants arriving on a bean leaflet during a 24-h period is orders of magnitude too low to explain the large increases in *P. syringae* populations discussed earlier. During periods of dispersal, deposition was slightly less than 10% of upward flux, indicating that about 10% of the bacteria that were emigrating from bean leaves (estimated as upward flux) were immigrating (estimated as deposition) to other leaves in the immediate vicinity. The remainder may be swept aloft to die or be transported some unknown distance.

### Dispersal During Rain

During rain, substantial numbers of bacteria are removed from leaf habitats (cf. Lindemann and Upper, 1985). Some of the bacteria may be redistributed from one leaf to another by rain splash (Lindemann and Upper, 1985; Butterworth and McCartney, 1991). However, because there is a strong net downward flux of bacteria during rain, most of the bacteria that are removed from the phyllosphere are deposited on the soil. Lindemann and Upper (1985) estimated that approximately $10^5$ bacteria were lost per bean leaflet in 15 min during a rain of modest intensity. Hence, large numbers of bacteria emigrate from leaves during rain.

In addition to examining the direction of net fluxes of bacteria during rain, Constantinidou et al. (1990) also assessed the potential importance of rain in dispersal of bacteria for distances greater than those effected by rain splash. During each of two rain events, Andersen six-stage microbial impaction samplers and deposition plates (petri plates filled to the rim with appropriate medium) were deployed at two (1.27 and 2.18 m) or three (1.27, 3.76, and 10.21 m) heights above ground in a soybean field. The height of the soybean canopy was 0.73 m. Interestingly, the ice nucleation activity and phytopathogenic host range of strains of *P. syringae* that were isolated from rainwater and aerosols collected during the rainfalls differed from those isolated as epiphytes from the soybean leaflets. With respect to ice nucleation activity, 80% ($n$ = 50) of the isolates from air and rain were active at −5°C, whereas only 16.7% ($n$ = 54) of the isolates from soybean leaves were active at −5°C. Of the isolates from air/rain and leaves, 12% and 83.3%, respectively, were active only at −10°C. With respect to pathogenicity, none of the strains isolated from soybean leaves were pathogenic to snap bean, but 81.5% were pathogenic to soybean. However, among the isolates from air and rain, 10% and 6% were pathogenic to snap bean and soybean, respectively. The observations that there was a net downward flux of bacteria during the rain events and that bacteria isolated from rain and air over a soybean canopy were different from those on the soybean leaves were taken as evidence that the bacteria being delivered to the soybean canopy originated from a source some distance away from the soybean field.

## Role of the Ice$^+$ Phenotype in the Ecology of INA Bacteria

Organisms live by effecting chemical reactions for their own benefit. Altering the likelihood of the phase transition of water to ice is one of many chemical transformations that living organisms have evolved to effect. What advantage(s) might the Ice$^+$ phenotype confer upon bacteria that could have led to its selection? An interesting model has been proposed for the evolution of ice nucleation genes (Wolber and Warren, 1991; see also Chapter 5). The model postulates that the Ice$^+$ phenotype probably evolved only once by natural selection and was horizontally dispersed via conjugation between members of the different species of INA bacteria that inhabit the phyllosphere. The basis for the model is the observed similarities in nucleotide and corresponding predicted protein sequences among ice nucleation genes isolated from *P. syringae, P. fluorescens, E. herbicola,* and *E. ananas.* If the model is correct, it implies that the Ice$^+$ phenotype can confer a selective advantage across the broad range of life strategies exhibited by the various species that exhibit this phenotype. Its possible roles in bacterial life strategies are discussed in the following sections.

## Modification of the Habitat

### Frost Injury

By causing frost injury to leaves at near-zero subfreezing temperatures, bacteria expressing the Ice$^+$ phenotype are capable of what might be viewed as extensive, irreversible habitat modification. The injured leaf may have a larger carrying capacity for INA bacteria, possibly due to the release of a pool of nutrients from the injured cells. Multiplication of INA bacteria would ensue, resulting in increased relative abundance.

Under laboratory conditions, increases in INA *P. syringae* were noted on leaves of oat plants (Buttner and Amy, 1989) and bark of apricot stems (Klement et al., 1984) subjected to subzero temperatures. In the experiments of Buttner and Amy (1989), oat plants were inoculated with either an Ice$^+$ or Ice$^-$ (deletion mutant) strain of *P. syringae.* A 15-fold increase in population sizes of the Ice$^+$ strain was measured 18 h after the oat plants incurred frost injury at –5°C. The oat plants inoculated with the Ice$^-$ strain were not injured at this temperature, and there was no increase in population sizes of the Ice$^-$ strain. Population sizes of both strains increased on plants that were frost-damaged at –12°C. Thus, the authors suggested that Ice$^+$ strains may have a competitive advantage over Ice$^-$ strains under conditions of relatively mild frosts. These experiments demonstrate that *P. syringae* can grow better on frost-injured tissues than on uninjured ones. If further experiments with mixed bacterial populations demonstrate an increase in the proportion that is Ice$^+$, selective enrichment of the phenotype associated with frost can be inferred. It is important that similar experiments be conducted under field conditions in the presence of the natural microflora that colonize leaf surfaces.

In the field, Kaneda (1986) found that total numbers of bacteria on bean leaves increased approximately 10-fold and 200-fold during the first and second weeks, respectively, following a natural frost. Changes in numbers of INA bacteria within these populations were not reported. A 10-fold increase in population sizes of INA *P. syringae* was measured on apricot blossoms following a natural frost event (minimum temperature of –4.7°C) (Gross et al., 1983). Population sizes of total bacteria were not presented. Thus, enrichments for the Ice$^+$ phenotype were not reported by either of these authors. Constantinidou et al. (1991) reported that the ratio of INA to non-INA *P. syringae* on citrus leaves in Greece decreased from 5.2 to 0.6 following frost events. Hence, the Ice$^+$ phenotype was not enriched in this case.

### Disease

Phytopathogenicity is a frequent phenotype among strains of most of the species of INA bacteria that inhabit the phyllosphere (cf. Dye et al., 1980; Krieg and Holt, 1984). Is it possible that the Ice$^+$ phenotype provides a competitive advantage to bacteria by increasing the likelihood of infection and hence increasing relative fitness? As is the case with ice nucleation activity, pathogenicity is not a universal characteristic of all strains of these species. Even before bacteria were shown to be active ice nucleators and incitants of frost damage, frost injury was frequently reported to be associated with increased amounts of disease. For example, Panagopoulos and Crosse (1964) examined the effect of subzero temperatures (–1 to –2°C) on the severity of bacterial blossom blight of pear caused by *P. syringae* pv. *syringae*. Pear trees that were supercooled until an exotherm was noted and then sprayed with a suspension of the pathogen had a higher percentage of diseased blossoms than trees that were inoculated but not subjected to the frost conditions. The authors reported that frost injury was a predisposing factor in blossom blight of pear.

An association between ice formation and increased disease has been reported for a number of other systems including bacterial leaf spot of sour cherry (Süle and Seemüller, 1987), bacterial canker of peach (Weaver, 1978), and bacterial canker of apricot (Klement et al., 1984), diseases caused by *P. syringae* pv. *syringae*; and black chaff of wheat, caused by *X. campestris* pv. *translucens* (Azad and Schaad, 1988). Vigouroux (1989) suggested that freezing and thawing facilitates the ingress and spread of *P. syringae* into stems of fruit trees which, in turn, promotes disease development. In this study, young peach and apricot trees were subjected to a temperature of –4°C for 24 h, then warmed to about 20°C in the laboratory. After thawing, the stems were superficially punctured with a needle (1 mm), and a drop (5 µl) of a suspension of *P. syringae* pv. *persicae* or black ink was micropipetted over each puncture. Within 1–2 min following inoculation, the mean extent of spread of the pathogen (and ink) ranged from 11 to 24 mm (the longer of two directions measured) for trees that were subjected to a freeze-thaw cycle, but only 1–3 mm for control trees. Passive ingress facilitated by freezing and thawing has also been demonstrated for *P. syringae* pv. *syringae* in sour cherry leaves (Süle and Seemüller, 1987). Many of the pathovars used in the above studies are regarded as relatively "weak" pathogens, particularly those belonging to the pathovar *syringae*. Enhanced access to susceptible tissue may provide a greater advantage to relatively weak pathogens than to relatively aggressive ones.

Can modification of habitats through ice nucleation activity provide the necessary selection for the Ice$^+$ phenotype? Studies such as those described above suggest that the Ice$^+$ phenotype may enhance the growth and pathogenic potential of these bacteria, but a clear selection for the Ice$^+$ phenotype remains to be demonstrated. Further, many strains of INA bacteria are probably not pathogenic at all. Others that may be pathogenic cause disease to frost-sensitive tissues. The killing of these tissues by frost would preclude an opportunity for enhanced pathogenicity. Thus, although increased pathogenic fitness through frost-assisted habitat modification may provide a selective advantage for the Ice$^+$ phenotype in some strains, such a selection is not available to others. It seems likely that other mechanisms exist for selection of this phenotype.

### Enhanced Survival Through Protection of Habitat from Freezing Injury

A selection pressure for the Ice$^+$ phenotype may be achieved by protection of its habitat from freezing injury. Such a mechanism may function in a range of habitats, from frogs and insects to leaves. There have been suggestions that some hardy plants are able to survive to lower temperatures if supercooling is prevented by ice formation at temperatures close to zero (Rajashekar et al., 1983). Krog et al. (1979) reported that the African alpine plant *Lobelia telekii* accumulates large quantities of water in the central portion of its inflorescence. The water freezes at relatively warm temperatures, and the heat of fusion released by ice formation protects the vital parts of the flower from injury. Ice-nucleating *P. chlororaphis* has been isolated from *L. telekii* and is presumed to be the ice nucleus that initiates freezing of this water (Schnell et al., 1991).

## Enhanced Survival in Environments Subject to Frequent Freeze-Thaw Cycles

Selection for the Ice$^+$ phenotype may occur through improved survival rather than habitat modification. Lindow (1989) sampled plants from approximately 9,000 to 14,000 ft on two volcanic peaks in Central America. Larger population sizes of INA *P. syringae* were measured on vegetation at the higher than the lower elevations. In addition, the fraction of the total *P. syringae* strains recovered that were Ice$^+$ was greater in the zones that were frequently subjected to freezing temperatures than at the lower elevations.

The effect of freeze-thaw cycles on bacterial survival has been examined using nearly isogenic strains of Ice$^+$ and Ice$^-$ (deletion mutants) of *P. syringae*. Lindow (1985) found no difference in relative survival of Ice$^+$ wild-type and Ice$^-$ mutants when the strains were subjected to freeze-thaw cycles in phosphate buffer. However, Lindemann and Suslow (1987b) reported that Ice$^+$ parent strains survived freezing on strawberry blossoms better than the Ice$^-$ deletion mutants.

## Modification of the Environment

### Enhanced Accumulation of Water from Cold Air

Kieft (1988) has suggested that the most likely advantage of ice nucleation to lichens is to increase the accumulation of water from the atmosphere at the lichen surface. The lower vapor pressure over ice would increase the amount of atmospheric water accumulating on frozen as opposed to unfrozen surfaces during periods when the air temperature was below freezing and near saturation, such as during radiative frost events. A similar advantage could accrue to all INA bacteria that overwinter on surfaces such as buds of woody plants or within wettable matrices such as dead leaves in areas where winters include frequent periods of subfreezing temperatures with humidities near saturation. Bacterial survival might be substantially enhanced on those dead leaves which froze quickly when the temperature dropped slightly below freezing and hence accumulated more moisture from the air. The advantage to the relatively quiescent bacteria could accrue because the accumulated ice prevented desiccation while frozen or at a later time when the temperatures rose above freezing. Such a mechanism might function on almost any surface.

### The Role of Rain in the Life Cycle of INA Bacteria

As discussed above, *P. syringae* modifies the physical environment through ice nucleation, and the physical environment, most extensively through the momentum of rain, modifies the population dynamics of this bacterium. A plausible role of the Ice$^+$ phenotype of this INA bacterium in adaptation to life in the phyllosphere may involve both the physical modification of the biosphere and the biological modification of the physical environment.

A scenario for the meso- and long-range dispersal of phyllosphere bacteria may be described as follows. Bacteria, including INA bacteria, are readily dispersed from plant canopies during dry sunny weather (Lindemann et al., 1982; Lindemann and Upper, 1985). One process by which colloidally stable clouds are converted into precipitation is dependent on the presence of ice nuclei (cf. Pruppacher and Klett, 1980). Even if there are insufficient INA bacteria in the atmosphere to provide nuclei to initiate a rainstorm, when rain does begin, those bacteria that have the capability to initiate formation of a raindrop may have an increased likelihood of returning to a plant surface. Although it is not known if INA bacteria participate in precipitation processes, they are present in the air at high altitudes (Jayaweera and Flanagan, 1982; Sands et al., 1982). They are also present in rain (see Table 1); Ice$^+$ *P. syringae* were abundant in rain and aerosols during rain (Constantinidou et al., 1990). The atmosphere, and not the plant canopy, appeared to be the source of *P. syringae* in falling rain. Thus, the INA bacteria arriving on the phyllosphere had been removed from the atmosphere by the rain. The selective advantage provided by the Ice$^+$ phenotype may be to rescue bacteria from the troposphere and return them to the phyllosphere under conditions conducive to their growth. If INA bacteria actually initiate precipitation, then both enhanced transport and multiplication due to rain may provide selective advantages of the Ice$^+$ phenotype.

## Conclusions

There are several plausible suggestions for the selective advantage bestowed by the Ice$^+$ phenotype, but none can be identified as the most likely candidate for selection of this phenotype in all species. The relative ease of studying phyllosphere bacteria in nature, and their rapid growth rates, should make it possible to test at least some of the proposed hypotheses experimentally. We look forward to following the evolution of the solution to this puzzle.

## Literature Cited


Alberga, A. H., 1987. Microclimate influences on *Pseudomonas syringae* bacteria on snap beans. Ph.D. thesis. University of Wisconsin, Madison.

Anderson, J. A., Buchanan, D. W., Stall, R. E., and Hall, C. B., 1982. Frost injury of tender plants increased by *Pseudomonas syringae* van Hall. J. Am. Soc. Hortic. Sci. 107:123-125.

Anderson, J. A., and Ashworth, E. N., 1986. The effects of streptomycin, desiccation, and UV radiation on ice nucleation by *Pseudomonas viridiflava*. Plant Physiol. 80:956-960.

Arny, D. C., Lindow, S. E., and Upper, C. D., 1976. Frost sensitivity of *Zea mays* increased by application of *Pseudomonas syringae*. Nature 262:282-284.

Azad, H., and Schaad, N. W., 1988. The relationship of *Xanthomonas campestris* pv. *translucens* to frost and the effect of frost on black chaff development in wheat. Phytopathology 78:95-100.

Baca, S., Canfield, M. L., and Moore, L. W., 1987. Variability in ice nucleation strains of *Pseudomonas syringae* isolated from diseased woody plants in Pacific Northwest nurseries. Plant Dis. 71:412-415.

Baca, S., and Moore, L. W., 1987. Variations in *Pseudomonas syringae* isolated from grass species occurring in woody plant nurseries in the Pacific Northwest. Plant Dis. 71:724-726.

Butera, N., Alliata, V., Billitteri, L., and Catara, A., 1984. Individuazione di ceppi criogeni di *Pseudomonas syringae* in Sicily. Inf. Fitopatol. 34:51-54.

Butterworth, J., and McCartney, H. A., 1991. The dispersal of bacteria from leaf surfaces by water splash. J. Appl. Bacteriol. 71:484-496.

Buttner, M. P., and Amy, P. S., 1989. Survival of ice nucleation-active and genetically engineered non-ice nucleating *Pseudomonas syringae* strains after freezing. Appl. Env. Microbiol. 55:1690-1694.

Cirvilleri, G., Catara, V., and Marino, R., 1991. Variability in ice-nucleation active isolates of *Pseudomonas syringae* from citrus orchards. Pages 281-292 in: *Pseudomonas syringae* Pathovars. Proc. Int. Work. Group on *Pseudomonas syringae* Pathovars, 4th. Stamperia Granducale, Florence, Italy.

Colangeli, A. M., McAuley, L., and Owens, J. N., 1990. Seasonal occurrence of potential ice-nucleating bacteria on Douglas fir foliage and seed cones. New For. 4:55-61.

Constantinidou, H. A., Hirano, S. S., Baker, L. S., and Upper, C. D., 1990. Atmospheric dispersal of ice nucleation-active bacteria: The role of rain. Phytopathology 80:934-937.

Constantinidou, H. A., Menkissoglu, O., and Stergiadou, H. C., 1991. The role of ice nucleation active bacteria in supercooling of citrus tissues. Physiol. Plant. 81:548-554.

Dye, D. W., Bradbury, J. F., Goto, M., Hayward, A. C., Lelliott, R. A., and Schroth, M. N., 1980. International standards for naming pathovars of phytopathogenic bacteria and a list of pathovar names and pathotype strains. Rev. Plant Pathol. 59:153-168.

Fall, R., and Schnell, R. C., 1985. Association of an ice-nucleating pseudomonad with cultures of the marine dinoflagellate, *Heterocapsa niei*. J. Mar. Res. 43:257-265.

Georgakopoulos, D. G., and Sands, D. C., 1992. Epiphytic populations of *Pseudomonas syringae* on barley. Can. J. Microbiol. 38:111-114.

Goto, M., Goto, T., and Inaba, T., 1989. Identification of ice nucleation-active bacteria isolated from frost-damaged vegetable leaves. Ann. Phytopathol. Soc. Jpn. 55:330-335.

Goto, M., Huang, B. L., Makino, T., Goto, T., and Inaba, T., 1988. A taxonomic study on ice nucleation-active bacteria isolated from gemmisphere of tea (*Thea sinensis* L.), phyllosphere of vegetables, and flowers of *Magnolia denudare* Desr. Ann. Phytopathol. Soc. Jpn. 54:189-197.

Gross, D. C., Cody, Y. S., Proebsting, E. L., Jr., Radamaker, G. K., and Spotts, R. A., 1983. Distribution, population dynamics, and characteristics of ice nucleation-active bacteria in deciduous fruit tree orchards. Appl. Environ. Microbiol. 46:1370-1379.

Gusta, L. V., and O'Connor, B. J., 1987. Frost tolerance of wheat, oats, barley, canola and mustard and the role of ice-nucleating bacteria. Can. J. Plant Sci. 67:1155-1165.

Hirano, S. S., Baker, L. S., and Upper, C. D., 1985. Ice nucleation temperature of individual leaves in relation to population sizes of ice nucleation active bacteria and frost injury. Plant Physiol. 77:259-265.

Hirano, S. S., Maher, E. A., Kelman, A., and Upper, C. D., 1978. Ice nucleation activity of fluorescent plant pathogenic pseudomonads. Pages 717-724 in: Proc. Int. Conf. Plant Pathogenic Bacteria, 4th. Institut National de la Recherche Agronomique, Beaucozé, France.

Hirano, S. S., Nordheim, E. V., Arny, D. C., and Upper, C. D., 1982. Lognormal distribution of epiphytic bacterial populations on leaf surfaces. Appl. Environ. Microbiol. 44:695-700.

Hirano, S. S., Rouse, D. I., and Upper, C. D., 1987a. Bacterial ice nucleation as a predictor of bacterial brown spot disease on snap beans. Phytopathology 77:1078-1084.

Hirano, S. S., Tanner, C. B., and Upper, C. D., 1987b. Rain-triggered multiplication of *Pseudomonas syringae* on snap bean leaflets. (Abstr.) Phytopathology 77:1694.

Hirano, S. S., and Upper, C. D., 1989. Diel variation in population size and ice nucleation activity of *Pseudomonas syringae* on snap bean leaflets. Appl. Environ. Microbiol. 55:623-630.

Hirano, S. S., and Upper, C. D., 1991. Bacterial community dynamics. Pages 271-294 in: Microbial Ecology of Leaves. J. H. Andrews and S. S. Hirano, eds. Springer-Verlag, New York.

Jayaweera, K., and Flanagan, P., 1982. Investigations on biogenic ice nuclei in the Arctic atmosphere. Geophys. Res. Lett. 9:94-97.

Jones, J. B., Gitaitis, R. D., and McCarter, S. M., 1983. Evaluation of indirect immunofluorescence and ice nucleation activity as rapid tests for identifying foliar diseases of tomato transplants incited by fluorescent pseudomonads. Plant Dis. 67:684-687.

Kaneda, T., 1986. Seasonal population changes and characterization of ice-nucleating bacteria in farm fields of central Alberta. Appl. Environ. Microbiol. 52:173-178.

Kaneko, J., Kita, K., and Tanno, K., 1991a. INA bacteria isolated from diamondback moth *Plutella xylostella* L. pupae. Jpn. J. Appl. Entomol. Zool. 35:7-12.

Kaneko, J., Yoshida, T., Owada, T., Kita, K., and Tanno, K., 1991b. *Erwinia herbicola* ice nucleation active bacteria isolated from diamondback moth *Plutella xylostella* L. pupae. Jpn. J. Appl. Entomol. Zool. 35:247-251.

Kieft, T. L., 1988. Ice nucleation activity in lichens. Appl. Environ. Microbiol. 54:1678-1681.

Kim, H. K., Orser, C., Lindow, S. E., and Sands, D. C., 1987. *Xanthomonas campestris* pv. *translucens* strains active in ice nucleation. Plant Dis. 71:994-997.

Kim, Y. H., Kim Y. C., Cho, B. H., and Kim, K. C., 1987. Ice-nucleation activity of *Pseudomonas syringae* isolated in Korea. Korean J. Plant Pathol. 3:180-186.

King, E. O., Ward, M. K., and Raney, E. D., 1954. Two simple media for the demonstration of pyocyanin and fluorescin. J. Lab. Clin. Med. 44:301-307.

Kinkel, L. L., 1992. Statistical consequences of combining population samples. (Abstr.) Phytopathology 82:1168.

Klement, Z., and Rozsnyay, Zs. D., 1991. The role of the ice-minus nucleation active bacteria in the development of canker and dieback diseases of apricots. Pages 53-60 in: *Pseudomonas syringae* Pathovars. Proc. Int. Work. Group on *Pseudomonas syringae* Pathovars, 4th. Stamperia Granducale, Florence, Italy.

Klement, Z., Rozsnyay, Z. D., Balo, E., Panczel, M., and Prileszky, Gy., 1984. The effect of cold on development of bacterial canker in apricot trees infected with *Pseudomonas syringae* pv. *syringae*. Physiol. Plant Pathol. 24:237-246.

Krieg, N. R., and Holt, J. G., ed., 1984. Bergey's Manual of Systematic Bacteriology, Vol. 1. Williams & Wilkins, Baltimore.

Krog, J. O., Zachariassen, K. E., Larsen, B., and Smidsrod, O., 1979. Thermal buffering in Afro-alpine plants due to nucleating agent-induced water freezing. Nature 282:300-301.

Latorre, B. A., and Jones, A. L., 1979. *Pseudomonas morsprunorum*, the cause of bacterial canker of sour cherry in Michigan, and its epiphytic association with *P. syringae*. Phytopathology 69:335-339.

Lee, M. R., Lee, R. E., and Strong-Gunderson, J. M., 1991. Isolation of ice nucleating active bacteria from a freeze-tolerant frog: Identification of *Pseudomonas putida* strains active in ice nucleation. Int. Conf. on Biological Ice Nucleation, 5th. Madison, WI.

Lee, R. E., Jr., Strong-Gunderson, J. M., Lee, M. R., Grove, K. S., and Riga, T. J., 1991. Isolation of ice nucleating active bacteria from insects. J. Exp. Zool. 257:124-127.

Leonardi, O., Leonardi, M., and Catara, A., 1989. Antagonistic bacteria as a biological control against ice nucleation active *Pseudomonas* on strawberry. Acta Hortic. 265:513-520.

Lindemann, J., Constantinidou, H. A., Barchet, W. R., and Upper, C. D., 1982. Plants as sources of airborne bacteria, including ice nucleation-active bacteria. Appl. Environ. Microbiol. 44:1059-1063.

Lindemann, J., and Suslow, T. V., 1987a. Competition between ice nucleation-active wild type and ice nucleation-deficient deletion mutant strains of *Pseudomonas syringae* and *P. fluorescens* Biovar I and biological control of frost injury on strawberry blossoms. Phytopathology 77:882-886.

Lindemann, J., and Suslow, T V., 1987b. Characteristics relevant to the question of environmental fate of genetically engineered INA$^-$ deletion mutant strains of *Pseudomonas*. Pages 1005-1012 in: Plant Pathogenic Bacteria. E. L. Civerolo, A. Collmer, R. E. Davis, and A. G. Gillaspie, eds. Martinus Nijhoff, Dordrecht, Netherlands.

Lindemann, J., and Upper, C. D., 1985. Aerial dispersal of epiphytic bacteria over bean plants. Appl. Environ. Microbiol. 50:1229-1232.

Lindow, S. E., 1982a. Population dynamics of epiphytic ice nucleation active bacteria on frost sensitive plants and frost control by means of antagonistic bacteria. Pages 395-416 in: Plant Cold Hardiness and Freezing Stress. Mechanisms and Crop Implications. P. H. Li and A. Sakai, eds. Academic Press, New York.

Lindow, S. E., 1982b. Epiphytic ice nucleation-active bacteria. Pages 335-362 in: Phytopathogenic Prokaryotes. M. S. Mount and G. H. Lacy, eds. Academic Press, New York.

Lindow, S. E., 1983a. Methods of preventing frost injury caused by epiphytic ice-nucleation-active bacteria. Plant Dis. 67:327-333.

Lindow, S. E., 1983b. The role of bacterial ice nucleation in frost injury to plants. Annu. Rev. Phytopathol. 21:363-384.

Lindow, S. E., 1985. Ecology of *Pseudomonas syringae* relevant to the field use of Ice$^-$ deletion mutants constructed in vitro for plant frost control. Pages 23-35 in: Engineered Organisms in the Environment: Scientific Issues. H. O. Halvorson, D. Pramer, and M. Rogul, eds. American Society for Microbiology, Washington, D.C.

Lindow, S. E., 1987. Competitive exclusion of epiphytic bacteria by Ice$^-$ *Pseudomonas syringae* mutants. Appl. Environ. Microbiol. 53:2520-2527.

Lindow, S. E., 1988. Construction of isogenic Ice$^-$ strains of *Pseudomonas syringae* for evaluation of specificity of competition on leaf surfaces. Pages 509-515 in: Microbial Ecology. F. Megušar and M. Gantar, eds. Slovene Society for Microbiology, Ljubliana, Yugoslavia.

Lindow, S. E., 1989. Ice nucleation active strains of *Pseudomonas syringae* associated with high elevation Central American volcanic peaks. Int. Conf. on Biological Ice Nucleation, 4th. Saskatoon, SK, Canada.

Lindow, S. E., Arny, D. C., Barchet, W. R., and Upper, C. D., 1978a. The role of bacterial ice nuclei in frost injury to sensitive plants. Pages 249-263 in: Plant Cold Hardiness and Freezing Stress. P. Li, ed. Academic Press, New York.

Lindow, S. E., Arny, D. C., and Upper, C. D., 1978b. Distribution of ice nucleation-active bacteria on plants in nature. Appl. Environ. Microbiol. 36:831-838.

Lindow, S. E., Arny, D. C., Upper, C. D., 1978c. *Erwinia herbicola*: a bacterial ice nucleus active in increasing frost injury to corn. Phytopathology 68:523-527.

Lindow, S. E., Arny, D. C., and Upper, C. D., 1982a. Bacterial ice nucleation: A factor in frost injury to plants. Plant Physiol. 70:1084-1089.

Lindow, S. E., and Connell, J. H., 1984. Reduction of frost injury to almond by control of ice nucleation active bacteria. J. Am. Soc. Hortic. Sci. 109:48-53.

Lindow, S. E., Hirano, S. S., Barchet, W. R., Arny, D. C., and Upper, C. D., 1982b. Relationship between ice nucleation frequency of bacteria and frost injury. Plant Physiol. 70:1090-1093.

Lindow, S. E., and Panopoulos, N. J., 1988. Field tests of recombinant Ice$^-$ *Pseudomonas syringae* for biological frost control in potato. Pages 121-138 in: Release of Genetically-Engineered Micro-Organisms. M. Sussman, C. H. Collins, F. A. Skinner, and D. E. Stewart-Tull, eds. Academic Press, New York.

Luisetti, J., and Gaignard, J. L., 1987. Bacterial populations in buds and probability of frost injury on grapevine. Pages 983-993 in: Plant Pathogenic Bacteria. E. L. Civerolo, A. Collmer, R. E. Davis, and A. G. Gillaspie, eds. Martinus Nijhoff, Dordrecht, Netherlands.

Maki, L. R., Galyan, E. L., Chang-Chien, M. M., and Caldwell, D. R., 1974. Ice nucleation induced by *Pseudomonas syringae*. Appl. Microbiol. 28:456-459.

Maki, L. R., and Willoughby, K. J., 1978. Bacteria as biogenic sources of freezing nuclei. J. Appl. Meteorol. 17:1049-1053.

Makino, T., 1983. Ice-nucleation activity of bacteria isolated from gemmisphere of tea trees. Ann. Phytopathol. Soc. Jpn. 49:32-37.

Marshall, D. A., 1988. A relationship between ice-nucleation-active bacteria, freeze damage and genotype in oats. Phytopathology 78:952-957.

Mazarei, M., and Kerr, A., 1987. Identification and seasonal population changes of ice nucleation active bacteria in a pea field in South Australia. Australas. Plant Pathol. 16:13-15.

Minsavage, G. V., and Stall, R. E., 1987. Ice-nucleation-active bacteria on citrus in Florida. Pages 994-1004 in: Plant Pathogenic Bacteria. E. L. Civerolo, A. Collmer, R. E. Davis, and A. G. Gillaspie, eds. Martinus Nijhoff, Dordrecht, Netherlands.

Montesinos, E., and Vilardell, P., 1991. Relationships among population levels of *Pseudomonas syringae*, amount of ice nuclei and incidence of blast of dormant flower buds in commercial pear orchards in Catalunya, Spain. Phytopathology 81:113-119.

Morris, C. E., Wen, A.-M., Xu, X.-H., and Di, Y. B., 1992. Ice nucleation-active bacteria on Chinese cabbage in northern China: Population dynamics and characteristics and their possible role in storage decay. Phytopathology 82:739-746.

Newton, D., and Hayward, A. C., 1986. Ice nucleation activity of some reference cultures of *Pseudomonas syringae* and field isolates of bacteria from wheat and barley in Queensland. Australas. Plant Pathol. 15:71-73.

Obata, H., Nakai, T., Tanishita, J., and Tokuyama, T., 1989. Identification of an ice-nucleating bacterium and its ice nucleation properties. J. Ferment. Bioeng. 67:143-147.

Obata, H., Takinami, K., Tanishita, J. I., Hasegawa, Y., Kawate, S., Tokujama, T., and Ueno, T., 1990. Identification of a new ice nucleating bacterium and its ice nucleation properties. Agric. Biol. Chem. 54:725-730.

Olive, J. W., and McCarter, S. M., 1988. Occurrence and nature of ice nucleation-active strains of *Pseudomonas syringae* on apple and peach trees in Georgia. Plant Dis. 72: 837-843.

Panagopoulos, C. G., and Crosse, J. E., 1964. Frost injury as a predisposing factor in blossom blight of pear caused by *Pseudomonas syringae* van Hall. Nature 202:1352.

Paulin, J.-P., and Luisetti, J., 1978. Ice nucleation activity among phytopathogenic bacteria. Pages 725-731 in: Proc. Int. Conf. Plant Pathogenic Bacteria, 4th. Institut National de la Recherche Agronomique, Beaucozé, France..

Phelps, P., Giddings, T. H., Prochoda, M., and Fall, R., 1986. Release of cell-free ice nuclei by *Erwinia herbicola*. J. Bacteriol. 167:496-502.

Pruppacher, H. R., and Klett, J. D., 1980. Microphysics of Clouds and Precipitation. D. Reidel Publishing Co., Dordrecht, Netherlands.

Rajashekar, C., Li, P. H., and Carter, J. V., 1983. Frost injury and heterogeneous ice nucleation in leaves of tuber-bearing *Solanum* species. Plant Physiol. 71:749-755.

Richard, C., Gagne, S., and Antoun, H. A., 1987. A note on the presence of ice nucleation-active bacteria in roots of alfalfa grown in Quebec. Phytoprotection 68:127-129.

Rogers, J. S., Stall, R. E., and Burke, M. J., 1987. Low-temperature conditioning of the ice nucleation active bacterium, *Erwinia herbicola*. Cryobiology 24:270-279.

Ruinen, J., 1961. The phyllosphere. I. An ecologically neglected milieu. Plant Soil 15:81-109.

Sands, D. C., Langhans, V. E., Scharen, A. L., DeSmet, G., 1982. The association between bacteria and rain and possible resultant meteorological implications. Idöjárás 86:148-152.

Starr, M. P., Stolp, H., Truper, H. G., Balows, A., and Schlegel, H. G., eds., 1981. The Prokaryotes, Vol. I. Springer-Verlag, New York.

Schnell, R. C., Fall, R., Nemecek-Marshall, M., Sweeting, K., and LaDuca, R., 1991. A new INA bacterium from high altitude equatorial vegetation. Int. Conf. on Biological Ice Nucleation, 5th. Madison, WI.

Süle, S., and Seemüller, E., 1987. The role of ice formation in the infection of sour cherry leaves by *Pseudomonas syringae* pv. *syringae*. Phytopathology 77:173-177.

Vigouroux, A., 1989. Ingress and spread of *Pseudomonas* in stems of peach and apricot promoted by frost-related water-soaking of tissues. Plant Dis. 73:854-855.

Weaver, D. J., 1978. Interaction of *Pseudomonas syringae* and freezing in bacterial canker on excised peach twigs. Phytopathology 68:1460-1463.

Whitesides, S. K., and Spotts, R. A., 1991. Frequency, distribution, and characteristics of endophytic *Pseudomonas syringae* in pear trees. Phytopathology 81:453-457.

Wolber, P. K., and Warren, G. J., 1991. Evolutionary perspective on the ice nucleation gene-encoded membrane protein. Pages 315-330 in: Microbial Ecology of Leaves. J. H. Andrews and S. S. Hirano, eds. Springer-Verlag, New York.

Yankofsky, S. A., Levin, Z., and Moshe, A., 1981. Association with citrus of ice-nucleating bacteria and their possible role as causative agents of frost damage. Curr. Microbiol. 5:213-217.

Young, J. M., 1987. Ice nucleation on kiwifruit. Ann. Appl. Biol. 111:697-704.

Young, J. M., Luketina, R. C., and Marshall, A. M., 1977. The effects of temperature on growth in vitro of *Pseudomonas syringae* and *Xanthomonas pruni*. J. Appl. Bacteriol. 42:345-354.

CHAPTER 4

# Biochemistry of Bacterial Ice Nuclei

**Ray Fall and Paul K. Wolber**

This chapter focuses on the biochemical investigation of bacterial ice nuclei. A number of reviews of the biology and molecular biology of ice nuclei have appeared (Lindow, 1982, 1983a,b; Warren, 1987; Warren et al., 1987; Warren and Wolber, 1991; Wolber and Warren, 1989, 1991; Margaritis and Bassi, 1991). More recently, Wolber (1993) reviewed the molecular aspects of microbial ice nucleation.[1] Each of these reviews covers some of the biochemical aspects of ice nuclei but generally considers broader topics. This review will attempt to draw together the biochemical evidence that has accumulated on these macromolecular structures, focusing on additional biochemical aspects not covered in detail earlier as well as a few more recent developments.

## The Difficulty of Biochemical Characterization of Bacterial Ice Nuclei

A detailed biochemical characterization of bacterial ice nuclei has been hampered by a variety of factors, including the following. First, usually not all the cells in Ice$^+$ cultures express ice nuclei. Second, the ice nuclei active at the "warmest" temperatures (i.e., −1.5 to −5°C) are destroyed by cell disruption, complicating their isolation. Third, the ice nucleation protein is a minor cell protein. Each of these points merits a brief discussion.

Numerous workers have commented on the relatively low frequency of ice nuclei active at "warm" temperatures in cultures of Ice$^+$ bacteria (Maki et al., 1974; Yankofsky et al., 1981; Lindow et al., 1982). That is, the ice nucleation frequency, defined as the ratio of the number of ice nuclei to the number of bacterial cells in a culture (Lindow et al., 1982), is less than unity, and in the case of the "warm" ice nuclei is often only a small fraction of the cell population. For example, Figure 1 (left-hand curve) presents a typical cumulative ice-nucleation spectrum for an Ice$^+$ *Pseudomonas syringae* strain. As can be seen from the figure, the ice nucleation frequencies for intact cells in the −1.5 to −2°C range are very low (e.g., 1 in 1 million cells) and are orders of magnitude lower than at colder nucleation temperatures such as −8°C. In addition, the ice nucleation frequency is dramatically affected by growth conditions (Maki et al., 1974; Yankofsky et al., 1981; Lindow et al., 1982;

[1]Much of the text of this chapter is reprinted from Wolber (1993) by permission of Academic Press.

Pooley and Brown, 1991). More recently, different growth conditions have raised the nucleation frequency closer to unity (Deininger et al., 1988; Lindow et al., 1989), and in some *P. syringae* strains careful control of these conditions allows formation of "warm" ice nuclei in all cells in the culture (Nemecek-Marshall et al., 1993; see also Chapter 19).

Attempts to extract and purify the most active ice nuclei from cells have been especially frustrating. Pooley and Brown (1990) emphasized the fact that earlier efforts at cell fractionation by Maki et al. (1974), Yankofsky et al. (1981), and Lindow et al. (1989) resulted in the loss of at least 99.9% of the more active ice nuclei (i.e., –4°C nuclei). This can be seen in Figure 1 by comparing the cumulative ice nucleation spectrum of outer membrane fragments isolated from *P. syringae* (right-hand curve) with that for the parent cells (left-hand curve). Ice nuclei with "colder" thresholds (i.e., <–7°C) seem to be much more stable (Fig. 1). It may be necessary to have intact cells (or intact outer membrane vesicles, as described below) in order to maintain the structural integrity of the more active ice nuclei.

Quantitation of the recovery of more stable, –9°C ice nuclei in cellular fractions after breakage of *P. syringae* spheroplasts was carried out by Lindow et al. (1989). Essentially all of the starting –9°C ice nuclei were recovered in the total membrane fraction; essentially none were recovered in the soluble cell fraction. It is noteworthy that cell disruption did not result in a large increase in the total number of –9°C ice nuclei, suggesting that there may only be one or a few ice nucleation sites per

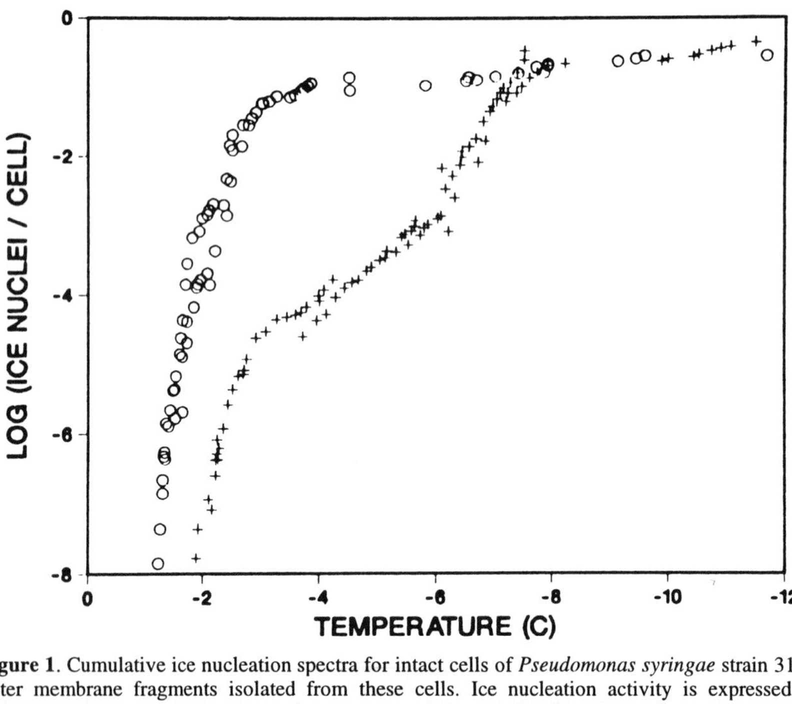


**Figure 1**. Cumulative ice nucleation spectra for intact cells of *Pseudomonas syringae* strain 31R1 and outer membrane fragments isolated from these cells. Ice nucleation activity is expressed as the logarithm of the cumulative ice nucleus content of the serially diluted samples at the temperatures shown divided by the cell concentration for the intact cell samples or the starting cell concentration for the outer membrane fragments. O = Intact cells; + = outer membrane fragments. (Reprinted, by permission, from Lindow et al., 1989)

cell rather than multiple sites per cell. As mentioned below, it is not possible to measure more than one ice nucleus per cell by drop-freezing assays. Some of the loss of ice nuclei after cell disruption might be due to proteolysis of the ice nucleation protein, a point also addressed below.

Another complication for biochemical analysis of bacterial ice nuclei is that the ice nucleation protein is relatively rare. Data for ice nucleation protein copy numbers using immunoblot techniques imply maximum levels of about $10^4$ per cell in natural strains (Deininger et al., 1988). This compares to copy numbers of major outer membrane proteins in gram-negative bacteria that approach $10^5$ per cell (Nikaido and Vaara, 1987). The combination of low level and low frequency of ice nucleation protein expression has hindered isolation and biochemical analysis of ice nucleation sites. No one has yet reported the isolation of a constitutive mutant that overexpresses the ice nucleation protein. Overexpression of the protein in *Escherichia coli* or *P. syringae* with the use of plasmid expression vectors has led to much higher copy numbers but not to the presence of "warm" nuclei in every cell (Southworth et al., 1988).

## The Assay for Ice Nuclei

The method usually employed for quantifying ice nuclei in biochemical research is the drop-freezing assay described in Chapter 1. It is important to note that the activity measured by this assay is cumulative: at each temperature, the assay estimates those nuclei active at *or above* the test temperature. Simple subtraction or more complicated formulae can be used to approximate the differential frequency of ice nuclei (with the concentration of nuclei becoming active in a particular range of temperature). However, the accuracy of such calculations may be very limited, because a particular droplet can only freeze once. Therefore, although a droplet may contain additional nuclei that would have become active at lower temperatures, they are hidden from experimental measurement (Vali, 1971). If the differential frequency of nucleation increases strongly with decreasing temperature, and all nuclei sort into droplets in an independent manner, this effect is negligible. However, it is difficult to detect small numbers of colder-threshold nuclei in the presence of a higher concentration of warmer-threshold nuclei. To do so requires testing very many droplets at a dilution where the probability of a droplet containing more than one nucleus is very low. If the nuclei do not sort independently (e.g., if multiple nuclei are bound to a bacterial cell), then only the nucleus active at the warmest temperature is experimentally observable. Measurements of ice nucleation activity should be interpreted with this in mind.

## Temperature Classes of Bacterial Ice Nuclei

It is obvious from Figure 1 that bacterial ice nuclei are active at widely varying frequencies and temperatures. Traditionally, this variation has been used to define three classes of bacterial ice nuclei (Yankofsky et al., 1981): Type 1 nuclei, active at temperatures between –2°C and –5°C; Type 2 nuclei, active between –5°C and –7°C; and Type 3 nuclei, active below –7°C. More recently, Turner et al. (1990) presented additional evidence of three classes of bacterial ice nuclei. The three classes were differentiated by several criteria, including freezing-difference spectra in $D_2O$ versus $H_2O$. The nucleation of supercooled $D_2O$ and $H_2O$ by Ice$^+$ *P. syringae* strain C-9 are shown in Figure 2. The difference in freezing temperatures between these two curves revealed classes of ice nuclei: class A, active above –4.4°C;

class B, active between –4.8 and –5.7°C; and class C, active at –7.6°C or colder. Presumably Turner and co-workers deviated from standard nomenclature (Types 1, 2, and 3) to emphasize the new basis for distinguishing classes A, B, and C; however, the two nomenclatures seem to describe roughly the same classes of nuclei. Some investigators have not noticed distinct classes of nucleators, and it is possible that under many conditions a continuum of ice nucleator thresholds may be present.

One of the central quests of bacterial ice nucleation research has been to discover the mechanisms responsible for the perplexing variation observed in nucleation temperature and frequency. The source of the variation does not appear to be genetic; it is observed in clonal populations and at all stages of bacterial growth (Warren and Wolber, 1991). The theory of heterogeneous nucleation (see Chapter 1) predicts that such variation might be caused by variations in $m$, a parameter that defines the intrinsic quality of the nucleating surface, or variations in the size of the nucleation site or in the "coherence length" of the nucleation site if nucleation occurs by coherent template matching.

Some insight has been gained into the source of nucleation site variations by means of two simple physical experiments. First, frozen or freeze-dried Ice⁺ *P.*

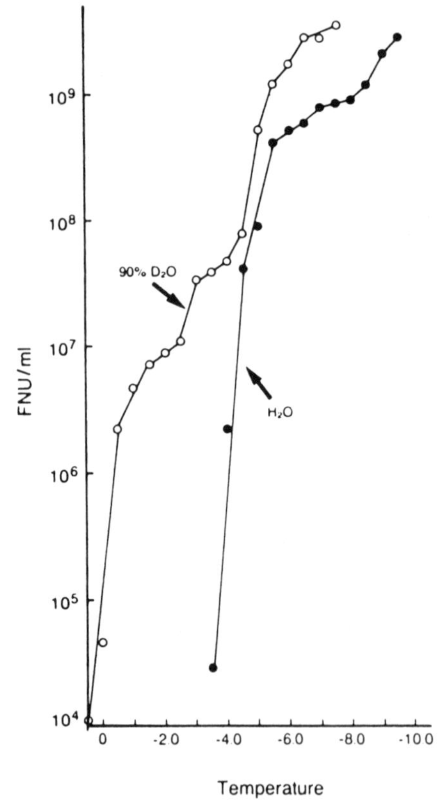


**Figure 2.** Freezing spectra for suspensions of *Pseudomonas syringae* C-9 in supercooled $H_2O$ and 90% $D_2O$. FNU/ml is the number of freezing nucleus units (i.e., ice nuclei) per milliliter of the suspension. (Reprinted, by permission, from Turner et al., 1990)

*syringae* cells have been irradiated with varying doses of $^{60}$Co γ-radiation, the effects on ice nucleation activity at various temperatures measured, and the results interpreted according to target theory (Govindarajan and Lindow, 1988a). Such experiments have shown that the target size varies log-linearly with nucleation temperature (Fig. 3). When coupled with biochemical data that indicate that the nucleation site is membrane-bound and protease-sensitive (Phelps et al., 1986; Govindarajan and Lindow, 1988b), the data have been interpreted to indicate that the different classes of nucleation sites are of different sizes, varying from more than $2 \times 10^7$ Da (active at −3°C) to approximately $10^5$ Da (active at −13°C). It should be noted that variations in coherence length (or effective sizes of the nucleation sites) are also compatible with the data; however, variations in the intrinsic quality of the nucleating surface are incompatible with these results. It is interesting to note that the measured target sizes agree with theoretical predictions for the size of a disk-shaped heterogeneous nucleation site with "perfect" surface properties (i.e., with $m = 1$) (Burke and Lindow, 1990). More recently, Kieft and Ruscetti (1992) used similar γ-radiation techniques and concluded that the masses of ice nuclei from *P. syringae* Cit7 were approximately threefold smaller than those measured by Govindarajan and Lindow (1988a).

The second set of physical experiments (Lindow, 1983c) is based on the observation that bacterial ice nuclei are temperature-labile: nuclei with warmer threshold temperatures are preferentially destroyed by briefly heat-shocking the host bacteria but reappear upon subsequent incubation at the normal growth temperature

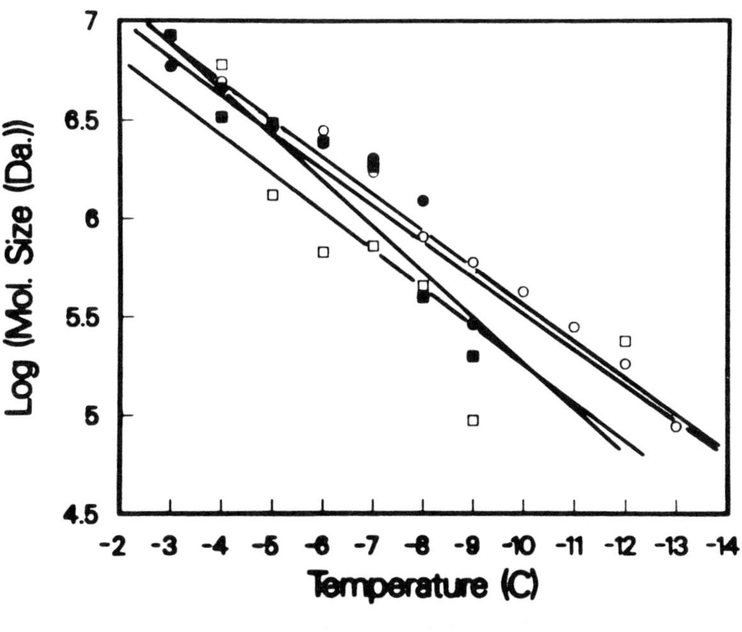


**Figure 3**. Molecular mass (Da) of ice-nucleating structures of various INA bacterial strains active within different 1°C intervals estimated by the sensitivity to gamma radiation at 22°C. The logarithm of the size of ice nucleation sites for *Pseudomonas syringae* strain 31R1 (■), *P. syringae* strain Cit7 (○), *Erwinia herbicola* strain 26SR6-2 (□), and *Escherichia coli* HB101 (pICE1.1) (●) active between temperatures $t + 1$ and $t$ is depicted as a function of temperature ($t$). (Reprinted, by permission, from Govindarajan and Lindow, 1988a)

(~20°C). This reappearance is independent of new protein synthesis. The destruction kinetics are first-order, while the reappearance kinetics are of mixed order. These data have been used to argue that the ice nucleation sites are membrane-bound aggregates that disaggregate at higher temperatures by sequentially losing individual molecules and reaggregate at normal growth temperatures by a cooperative assembly process. Again, the data are also compatible with changes in the effective size of the nucleation site if individual molecules in an aggregate undergo independent "melting" to yield an inactive conformation and cooperative "crystallization" to yield an active conformation. In either case, the data are difficult to explain via variations in the intrinsic quality of the nucleation site, and they point toward a structure that is an aggregate of parts that can, under certain circumstances, behave independently of one another.

## Low-Temperature Conditioning

An additional complication for assay of cellular ice nuclei is the phenomenon of low-temperature conditioning described by Rogers et al. (1987). Low-temperature conditioning occurs when Ice$^+$ bacteria grown at a relatively high temperature (e.g., 30°C) are shifted to a lower temperature (e.g., 5°C), which triggers the appearance of new ice nuclei active at temperatures warmer than –5°C. This conditioning process is blocked by protein synthesis inhibitors, such as streptomycin, indicating

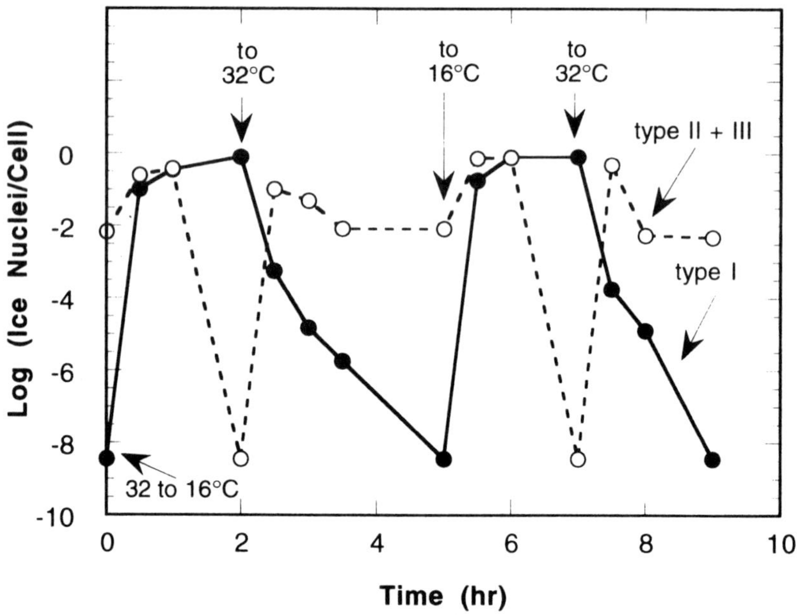


**Figure 4.** Type 1 ice nuclei in stationary *Pseudomonas syringae* T1 cells can be repeatedly induced by low-temperature/high-temperature cycles. Strain T1 was grown to stationary phase in a low-nitrogen, low-phosphate, sorbitol medium at 32°C, the culture was cycled between 16 and 32°C, and samples were removed for determination of ice nucleation activity at times indicated on the figure. Type 1, Type 2, and Type 3 ice nuclei are those nuclei that catalyze nucleation at –2 to –5°C, –5 to –7°C, and –7 to –10°C, respectively (Yankofsky et al., 1981). (Unpublished data from M. Nemecek-Marshall and R. Fall, based on results shown in Nemecek-Marshall et al., 1993)

new protein synthesis during the appearance of ice nuclei. Conditioning has been modeled as the aggregation of the ice nucleator protein into larger structures that are more effective nucleator templates (Burke and Lindow, 1990).

An additional aspect of cold conditioning is illustrated in Figure 4, which shows that the appearance and disappearance of ice nuclei in *P. syringae* T1 is under thermal control (Nemecek-Marshall et al., 1993). When cells were grown to stationary phase under nutrient-limited conditions at 32°C, they contained only Type 3 ice nuclei and not Type 2 or 1 (time zero on the figure). After a shift to a lower temperature (e.g., 18°C), Type 1 ice nuclei rapidly appeared in the culture and persisted for hours (or days, in some cases) until the culture was shifted back to 32°C, in which case Type 1 ice nuclei completely disappeared with a half-life of about 1 hr. As shown in Figure 4, cycles of low-temperature induction and high-temperature disappearance of Type 1 ice nuclei could be repeated several times in the same culture of nongrowing cells. Under these same conditions, only small changes in the number of Types 2 and 3 ice nuclei were detected. Low-temperature induction was completely prevented by protein-synthesis inhibitors.

These large changes in ice nucleation numbers and thresholds in nongrowing cells argue for careful handling of intact cell samples if uniform assays are to result. The addition of protein-synthesis inhibitors to cell dilution media would be recommended to prevent the appearance of new ice nuclei during handling prior to drop-freezing determinations.

## Biochemistry and Immunology of Ice Nucleation Proteins

The biochemistry and immunology of bacterial ice nuclei have been extensively studied, in both naturally occurring Ice$^+$ bacteria and recombinant bacteria. The goals of such studies have included 1) localization of ice nucleating sites in or on the bacterial cell, 2) examination of the role of membrane phospholipids, 3) studies of the sensitivity of ice nuclei to small molecule inhibitors and enzymes, 4) inhibition of ice nucleation by antibodies or antifreeze glycopeptides, 5) analysis of the assembly and turnover of ice nuclei, and 6) purification and characterization of ice nucleation proteins.

### Localization of Ice Nuclei to Cell Membranes

Lindow et al. (1989) reviewed the early evidence implicating cellular membranes in ice nucleus function. Essentially, it was found that ice nucleation is associated with the cell envelope and not the soluble cell contents, and that ice nucleation in intact cells is sensitive to agents such as lipases, phospholipases, detergents, and solvents that disrupt membranes. More direct evidence has come from analysis of ice nuclei shed by certain species of *Erwinia herbicola* (Phelps et al., 1986) and fractionation of natural Ice$^+$ bacteria (Lindow et al., 1989), which demonstrated that "natural" bacterial ice nuclei are localized on the gram-negative outer membrane. Studies of overexpression of the *inaZ* gene in *E. coli* have also demonstrated that active ice nuclei can be assembled on the inner membrane (Wolber et al., 1986).

Phelps et al. (1986) detected the spontaneous release of cell-free ice nuclei into the growth medium by *E. herbicola* strains. These ice nuclei could be concentrated by centrifugation of the cell-free growth medium at 200,000 × *g* or by co-precipitation with calcium phosphate (see Ruggles et al., 1991). Analysis of concentrated cell-free ice nuclei on Percoll gradients demonstrated that the nuclei copurified with outer membrane vesicles, which were visualized by electron microscopy. Figure 5 shows electron micrographs of outer membrane vesicles detected in Percoll-isolated ice nuclei and on the surface of intact cells. Many gram-negative strains spontaneously release similar vesicles from the cell surface, perhaps because the outer membrane grows more rapidly than the underlying peptidoglycan layer (Mayrand and Grenier, 1989; Fukuoka et al., 1992)

For unknown reasons, *P. syringae* strains generally do not shed ice nucleation vesicles into the growth medium. Pooley and Brown (1990) described the use of lysozyme and EDTA to release a small percentage (3.6%) of –5°C ice nuclei from cells, presumably as membrane fragments, accompanied by an outer membrane marker KDO. These cell-free nuclei have not been further characterized (T. Brown, *personal communication*). Obata et al. (1990) were able to release some ice nucleation activity from *P. viridiflava*, active in the range of –5 to –9°C, with a combination of EDTA and the detergent Triton X-100; whether these nuclei were associated with membrane fragments, or represent solubilized nuclei, was not reported. A process for the commercial production of cell-free ice nuclei was developed recently (Wight et al., 1992).

Lindow et al. (1989) fractionated *P. syringae* cells to determine the subcellular location of ice nucleation activity. Differential centrifugation showed that –9°C ice nuclei copurify with total cell membranes. These same workers used isopycnic sucrose density gradient to resolve the total membrane fraction into inner and outer membranes using Ice$^+$ *P. syringae* and *E. coli* containing a cloned *iceC* gene. Figure 6 shows the results of the experiment with *E. coli* membranes; –9°C ice nuclei were detected in membrane fractions with buoyant densities similar to outer membranes or intermediate between inner and outer membranes. Most of the ice nucleation activity cofractionated with 2-keto-3-deoxyoctonate (a marker for the lipopolysaccharide component of the outer membrane) but not with NADH oxidase (a marker for the inner membrane). The results were less ambiguous in *E. coli* membranes than in *P. syringae* membranes, perhaps because the membrane frac-

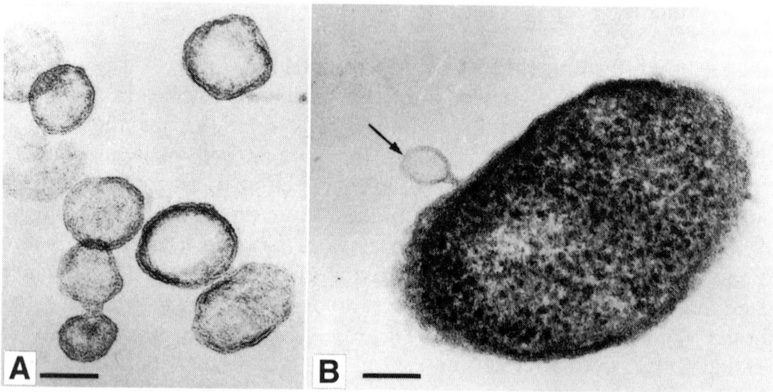


**Figure 5.** Ice$^+$ *Erwinia herbicola* strains release ice nucleation–active membrane vesicles from the outer membrane surface. **A,** Electron micrograph of a preparation of ice nucleation vesicles isolated from the cell medium of *E. herbicola* after fractionation on a Percoll gradient. **B,** Thin-section electron micrograph of *E. herbicola* M1, forming an outer membrane bleb (arrow). Scale bars = 100 nm. (Reprinted, by permission, from Phelps et al., 1986)

tionation technique has been largely developed and optimized for fractionation of *E. coli*, but not *Pseudomonas*, inner and outer membranes. Immunoblots showed that most of the IceC protein was associated with the outer membrane fractions, and was not present in the soluble cell fraction or inner membrane, confirming that ice nuclei are located in the outer membrane of source strains as well as heterologous hosts (Lindow et al., 1989)

## Requirement of Membrane Phospholipids for Ice Nucleation Activity

Lipid extraction and reconstitution experiments have shown that membrane lipids are required for ice nucleation activity (Govindarajan and Lindow, 1988b). Treatment of *P. syringae* outer membrane preparations with agents that destroy or remove phospholipids, such as detergents or phospholipase A2, led to a loss of most of the –9°C ice nuclei. Ice nucleation activity in such delipidated membranes could be restored by reconstitution with various phospholipids, as shown in Table 1. Restoration of activity was obtained by addition of purified phospholipids, with phosphatidylinositol and phosphatidylethanolamine yielding the largest restoration and phosphatidylserine and cardiolipin yielding little or no restoration. Phosphatidylethanolamine is slightly enriched in the outer membrane of gram-negative bacteria (Nikaido and Vaara, 1987). Lipopolysaccharide, a major component of the outer leaflet of outer membranes of gram-negative bacteria (Nikaido and Vaara, 1987), is apparently not required for ice nucleation activity. Extraction of 70% of the cellular lipopolysaccharide from an Ice⁺ *E. coli* strain by EDTA treatment did not significantly reduce ice nucleation activity (Govindarajan and Lindow, 1988b). Phelps (1987) also noted that the isolation of cell-free ice nuclei from *E. herbicola*

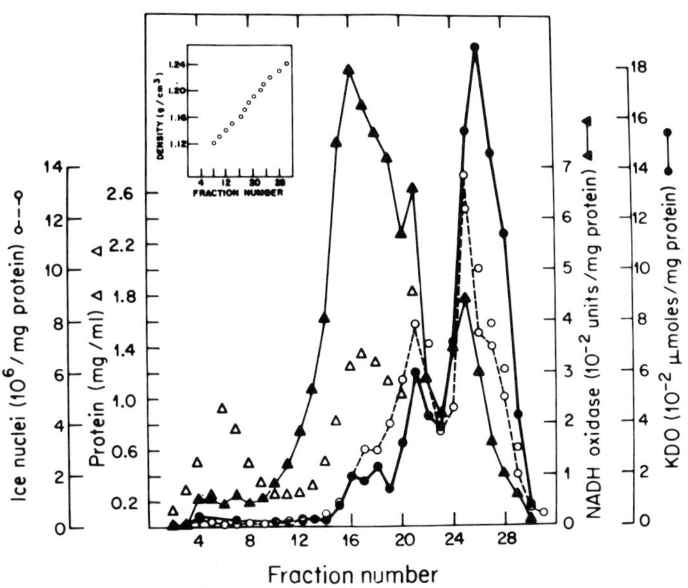


**Figure 6.** Separation of membrane fragments of *Escherichia coli* HB101 (pICE1.2) by isopycnic sucrose density centrifugation. Cell membranes were separated according to density (insert) and collected in 0.6-ml fractions (abscissa). Fractions were then assayed for NADH oxidase-specific activity (▲), protein content (Δ), 2-keto-3-deoxyoctonate content (●), and ice nucleation activity (O). (Reprinted, by permission, from Lindow et al., 1989)

was enhanced by treatment with high concentrations of EDTA, which would be expected to remove lipopolysaccharide from these outer membrane vesicles, and that treatment of vesicles with the detergent Triton X-100 removed lipopolysaccharide and increased ice nucleation activity.

Kozloff and his co-workers have made the case that a particular phospholipid, phosphatidylinositol, is associated with the ice nucleation site (Kozloff et al., 1984, 1991b). This is unusual because heretofore phosphatidylinositol has not been detected among bacterial phospholipids. Four types of evidence suggest its involvement. First, low but detectable phosphatidylinositol synthase activity is present in $Ice^+$ strains; the activity is even lower in $Ice^-$ strains. Second, phosphatidylinositol can be detected as a minor phospholipid in $Ice^+$ strains, ranging from about $1 \times 10^4$ to $2 \times 10^5$ molecules per cell. For comparison, total cellular phospholipid is about $1.3–5.6 \times 10^7$ molecules per cell, and $Ice^-$ strains contain $2 \times 10^2$ to $3 \times 10^3$ phosphatidylinositol molecules per cell. Third, addition of phosphatidylinositol vesicles to an $Ice^+$ *E. coli* strain leads to a 70-fold increase in class A ice nucleation activity, as shown in Figure 7. Finally, treatment of cells with phospholipase CII, which is specific for phosphatidylinositol, leads to destruction of class A but not class B or C ice nuclei. Phospholipase CII treatment inhibited 38–90% of class A ice nuclei in four different $Ice^+$ strains under these conditions with little loss of cell viability. Since this phospholipase is reputed to be active on phosphatidylinositol anchors of eukaryotic proteins, Kozloff and co-workers argue for a similar role in ice nucleation–active bacteria. This is further discussed below. Detection of phosphatidylinositol covalently bound to the Ina protein was unsuccessful, perhaps because of the small amount of phosphatidylinositol present and the possibility that only a few Ina proteins are modified.

## Disruption of Ice Nucleation Activity by Chemical Agents and Enzymes

Several chemical agents known to disrupt membranes have also been shown to decrease or abolish bacterial ice nucleation activity. These include dyes that stain bacterial cell walls (Maki et al., 1974), cationic detergents (Maki et al., 1974; Watanabe et al., 1988a, 1988b), and membrane fluidizing agents such as 2-phenylethanol (Lindow, 1983c). Other chemical agents, such as the protein denaturant urea (Lindow, 1983d), sulfhydryl-modifying reagents, and carbohydrate-reactive

**Table 1.** Phospholipid requirement for reconstitution of in vitro ice nucleation activity[a]

| | | Ice nucleation activity of membrane reconstituted with phospholipid[b] | |
|---|---|---|---|
| **Phospholipid added** | **Source** | **Ice nuclei/ml** | **% of original membrane** |
| None | | $8.2 \times 10^7$ | 47 |
| Phosphatidylserine | Bovine heart | $9.2 \times 10^7$ | 53 |
| Cardiolipin | Bovine heart | $1.3 \times 10^8$ | 75 |
| Phosphatidylglycerol | Egg yolk | $1.5 \times 10^8$ | 85 |
| Phosphatidylcholine | Egg yolk | $1.5 \times 18^8$ | 87 |
| Mixed phospholipids | *Pseudomonas syringae* | $1.6 \times 10^8$ | 93 |
| Phosphatidylinositol | Soybean | $1.9 \times 10^8$ | 107 |
| Phosphatidylethanolamine | *Escherichia coli* | $2.0 \times 10^8$ | 115 |

[a]Adapted from Govindarajan and Lindow (1988b).

[b]Figures show the recovery of ice nucleation activity of cholate-solubilized outer membranes of *Pseudomonas syringae* after reconstitution with various phospholipids.

compounds, such as lectins and borates (Kozloff et al., 1983), have also been shown to decrease bacterial ice nucleation activity. However, in some cases chemical inactivation studies have been conducted with whole bacteria (as opposed to cell-free ice nuclei). These earlier studies did not adequately control for cell death after treatment with a given agent. Subsequent experiments with cell-free nuclei have shown that in at least one case (with the sulfhydryl-modifying reagent *N*-ethylmaleimide) the observed effects were entirely due to cell death and were not observed when cell-free ice nuclei were substituted for whole cells (Phelps et al., 1986).

In light of these limitations, the inhibitor data obtained using whole cells is of uncertain value. Table 2 summarizes the types of reagents that inhibit cellular ice

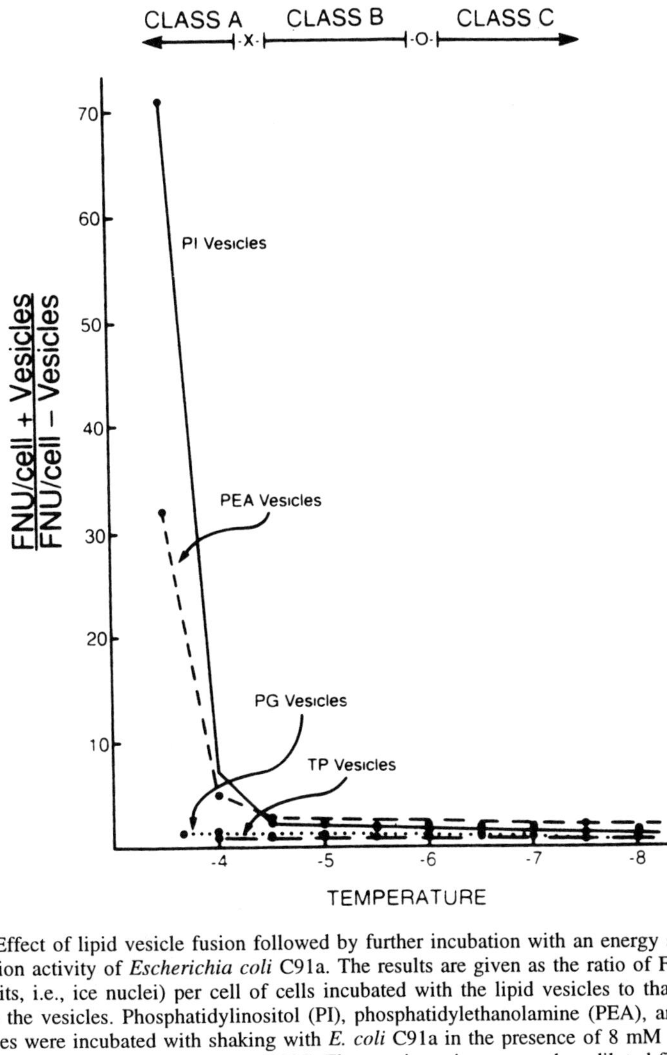


**Figure 7**. Effect of lipid vesicle fusion followed by further incubation with an energy source on the ice nucleation activity of *Escherichia coli* C91a. The results are given as the ratio of FNU (freezing nucleus units, i.e., ice nuclei) per cell of cells incubated with the lipid vesicles to that of cells not exposed to the vesicles. Phosphatidylinositol (PI), phosphatidylethanolamine (PEA), and tripalmitin (TP) vesicles were incubated with shaking with *E. coli* C91a in the presence of 8 mM spermine and 0.5 mM vanadate at pH 7.5 for 30 min at 30°C. The reaction mixture was then diluted fivefold with a glycerolphosphate solution and incubated for 20 min prior to ice nucleation assays. (Reprinted, by permission, from Kozloff et al., 1991b)

nuclei or cell-free ice nuclei and the presumed site or mode of inhibition. As can be seen from the table, ice nucleation activity in intact cells and cell-free ice nuclei is susceptible to a wide variety of reagents that react with proteins, membrane lipids, and carbohydrates, presumably either by direct reaction with the Ina protein in the outer membrane or by alteration of cellular membranes.

It is noteworthy that the reagents listed in Table 2 may not react equally well in all Ice⁺ strains. For example, Ruggles (1991) has shown that a variety of proteases (including pronase, proteinase K, and subtilisin) that degrade ice nucleation activity in cell-free ice nuclei from *E. herbicola* have little or no effect on the ice nucleation activity of intact *E. herbicola* or *P. syringae* cells. This could be explained if the

**Table 2.** Small molecules and enzymes that inactivate cellular or cell-free ice nuclei

| Reagent | Material[a] | Presumed reaction or cellular target |
|---|---|---|
| Reaction with proteins | | |
| Heavy metals (e.g., Cu, Zn, Hg, Cd, Ni salts) | IC | Reaction with cysteine-SH residues (Lindow et al., 1978; Maki et al., 1974) |
| *N*-Ethylmaleimide, *p*-hydroxymercuribenzoate, iodoacetamide | IC, CFN | Reaction with cysteine-SH residues (Kozloff et al., 1983; Phelps et al., 1986) |
| Proteases (e.g., trypsin, pronase proteinase K, subtilisin) | IC, CFN | Protein degradation (Phelps et al., 1986; Lindow et al., 1989; Ruggles, 1991) |
| Lactoperoxidase | IC, CFN | Iodination of protein tyrosine residues (Ruggles, 1991) |
| *N*-Bromosuccinimide, 2-hydroxy-5-nitrobenzyl bromide | CFN | Reaction with tryptophan (Phelps, 1987; Obata et al., 1990) |
| Tetranitromethane | CFN | Reaction with tyrosine (Phelps, 1987) |
| Reaction with lipids | | |
| Cationic detergents (e.g., cetylpyridinium chloride, benzethonium chloride, alkylbenzyldimethylammonium salts) | IC, CFN | (Maki et al., 1974; Lindow et al., 1978; Phelps, 1987; Watanabe et al., 1988a,b; Okada et al., 1990) |
| Organic solvents (e.g., dioxane, dimethylsulfoxide, chloroform, phenylethanol) | IC | Disrupts membrane fluidity (Lindow, 1983c; Turner et al., 1990) |
| Phospholipase A2 | IC | Hydrolysis of most phospholipids (Govindarajan and Lindow, 1988b) |
| Phospholipase CII | IC | Specific for phosphatidylinositol; inhibits class A but has little effect on class B and C ice nuclei (Turner et al., 1991) |
| Reaction with carbohydrates | | |
| Borate, phenylboronate, *m*-nitrophenylboronate | IC, CFN | Bind to *cis*-diols (Kozloff et al., 1983; Phelps, 1987; Turner et al., 1991) |
| Nitrous acid | IC | Reaction with glucosamine of phosphatidylinositol-anchored proteins (see Turner et al., 1991); may also react with proteins (see text) |
| α-Mannosidase, β-mannosidase, β-galactosidase, *N*- and *O*-glycanases | IC | Degrades mannan-containing membrane anchors or glycan side chains (Turner et al., 1991; Kozloff et al., 1991a) |

[a] Most of these experiments were carried out with intact cells (IC) of Ice⁺ *Pseudomonas* or *Erwinia* strains; it should be noted that not all reagents react identically in these different bacteria. All experiments with isolated cell-free ice nuclei (CFN) used *E. herbicola* as the source strain, except for the work of Obata et al. (1990) where CFN were isolated by detergent/EDTA treatment of *P. viridiflava*.

cell surface lipopolysaccharide layer masked ice nuclei from the added proteases. Treatment of *E. herbicola* cells with EDTA to remove lipopolysaccharide did not enhance access of proteases. In contrast, ice nuclei in *P. syringae* treated with EDTA were completely degraded by these three proteases.

Turner et al. (1991) found that nitrous acid treatment destroyed class A and B and some class C ice nuclei in many Ice$^+$ cell types, with little loss of cell viability, and argued that nitrous acid could degrade an essential membrane anchor component in the ice nucleation site. Nitrous acid is known to react with glucosamine moieties of eucaryotic membrane-anchoring structures (Ferguson, 1988). However, nitrous acid also has well-known reactivity with proteins, catalyzing reactions with α-amino and lysine ε-amino groups as well as reaction with the ring nitrogen of tryptophan (Kurosky and Hofmann, 1972). Another reagent that reacts with tryptophan, *N*-bromosuccinimide, is extremely reactive with ice nuclei, destroying most ice nuclei in all types of Ina$^+$ strains at micromolar levels (Phelps, 1987; M. Nemecek-Marshall and R. Fall, *unpublished*). This suggests that one or more tryptophan residues are essential for proper ice nucleation activity. In this regard it is interesting that several tryptophan residues, three in the *N*-terminal domain and one in the C-terminal domain, are conserved in the InaA, InaE, InaV, InaW, InaX, and InaZ proteins (Wolber, 1993; see also Chapter 5).

Future studies with isolated, purified Ina proteins should reveal the sites of reaction with the reagents listed in Table 2.

## Inhibition of Ice Nuclei with Specific Antibodies and Antifreeze Glycopeptides

Several different antisera raised against ice nucleation proteins or their component peptides have been used to probe the location of these proteins on immunoblots and in cells (see Chapter 5). These antibodies have been shown to inhibit ice nucleation by cell-free ice nuclei isolated from *E. herbicola* (Ruggles et al., 1991). The concentration of antibody needed to inhibit nucleation varied with the nucleation threshold temperature. Type 1 nuclei were most sensitive to inhibition, and Type 3 nuclei were least sensitive (Fig. 8). These results have been interpreted according to a model in which antibody binding breaks a large coherent template into smaller templates by physically blocking part of the water-binding surface (Ruggles et al., 1991). Such blockage would be predicted to lower the nucleation threshold temperature by decreasing the real or effective template size and would be expected to affect Type 1 ice nuclei most severely (since they afford a larger target for binding), in agreement with the results observed. In a related experiment, antifreeze glycopeptides, which are believed to prevent ice crystallization by binding to the surface of ice crystal embryos (DeVries, 1986), have been shown to inhibit ice nucleation by cell-free ice nuclei from *E. herbicola* (Parody-Morreale et al., 1988). The pattern of inhibition closely paralleled that seen with antibodies. Warmer threshold nucleation sites were more sensitive than colder threshold sites. This result provides strong evidence that bacterial ice nuclei function by a coherent template mechanism, since the only property relating bacterial ice nuclei and antifreeze glycopeptides is their shared postulated affinity for ice.

## Assembly and Turnover of Ice Nuclei

As stated previously, studies of γ-ray inactivation of ice nuclei have demonstrated that the target size for radiation damage increases log-linearly with temperature of nucleation and is always greater than or equal to the average predited size of ice nucleation proteins (about 120 kDa; Govindarajan and Lindow, 1988a). Studies of the dependence of ice nucleation frequency on the concentration of ice nucleation protein have shown that nucleation frequency at all temperatures increases nonlinearly with this concentration, indicating that two or three monomers must assemble cooperatively to initiate formation of an active nucleus (Southworth et al., 1988; see also Chapter 5).

Measurements of the rates of assembly of ice nuclei have confirmed cooperative initiation of nucleus assembly and have shown that nuclei active at warmer temperatures are assembled more slowly than nuclei active at colder temperatures (Watanabe et al., 1990). This difference in assembly rates may indicate that Type 1 nuclei are formed by addition of monomers of ice nucleation protein to Type 2 or 3 nuclei. Since heterologously expressed ice nuclei appear on both the inner and outer gram-negative bacterial membranes (Wolber et al., 1986), this may simply be evidence that different temperature classes of nucleation sites are assembled at different cellular locations in "unnatural" hosts.

As mentioned above, Turner et al. (1990) differentiated three types or classes of ice nuclei on the basis of their $D_2O$-induced shifts in nucleation thresholds. These results, coupled with the differential effects of phospholipase CII (Kozloff et al., 1984; Turner et al., 1990) and other enzymatic and chemical probes of glycosylation (Kozloff et al., 1991a; Turner et al., 1991) on various classes of ice nuclei, have been used to argue that the ice nucleation protein in Type 1 nuclei has been modified by stepwise glycosylation and addition of a phosphatidylinositol membrane anchor. This proposed sequence is shown in Figure 9. If verified, this would be a unique finding, since such membrane anchors have only been detected in eu-

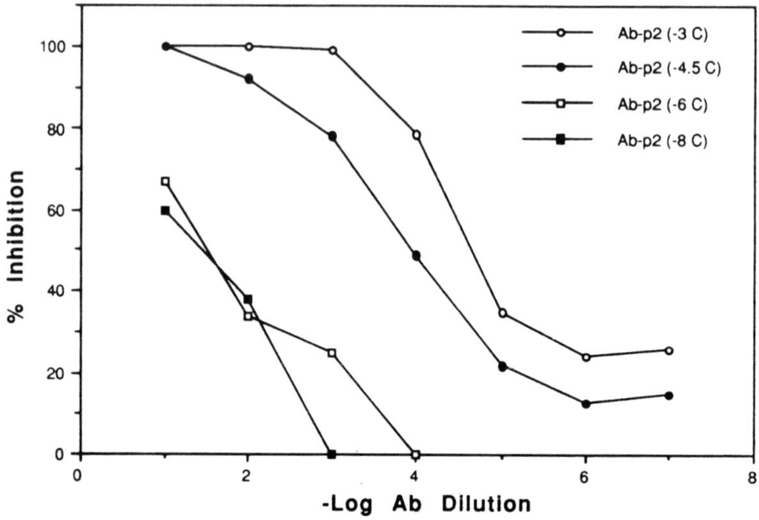


**Figure 8.** Antibody inhibition of ice nuclei present in cell-free ice nuclei from *Erwinia herbicola*. Antibody Ab-p2, prepared against the synthetic peptide LTAGYGSTQTC, which contains a sequence repeated multiple times in the InaZ protein sequence, was incubated for 30 min on ice with a suitable dilution of cell-free ice nuclei, and freezing spectra were obtained with a differential scanning calorimeter (Parody-Morreale et al., 1986). Threshold nucleation temperatures ranging from −3 to −8°C are indicated on the figure. (Reprinted, by permission, from Ruggles et al., 1991)

karyotic systems (Ferguson, 1988), not in any prokaryotic cells. Unfortunately, no one has yet produced direct biochemical confirmation of this intriguing hypothesis. Such confirmation may be difficult to obtain, since ice nucleation proteins are minor proteins (as discussed above); in addition, since ice nuclei appear to be homoaggregates of ice nucleation protein, posttranslational modification of a minority of protein monomers may be sufficient to change the nucleation threshold by such means as increasing the coherence length of a template array.

Measurements of the rates of destruction of ice nuclei assembled in natural hosts show that Type 1 ice nuclei disappear more rapidly than Type 2 or 3 nuclei after shifts to high temperature (Fig. 10; Ruggles et al., 1993). In contrast, the rates of destruction of ice nuclei at 28°C in *E. coli*, a heterologous host, are the opposite; Type 1 nuclei turn over much more slowly than Type 2 or 3 nuclei (Watanabe et al., 1990). This turnover could be due to 1) disaggregation of membrane-bound ice nucleation protein complexes, and/or 2) proteolytic degradation of ice nucleation protein. Evidence of extensive proteolysis of the ice nucleation protein has been presented by Deininger et al. (1988) and Watanabe et al. (1990). As shown in Figure 11, immunoblots of sodium dodecyl sulfate polyacrylamide gels of *P. syringae* total cell membranes revealed the presence of the intact InaZ protein (about 120

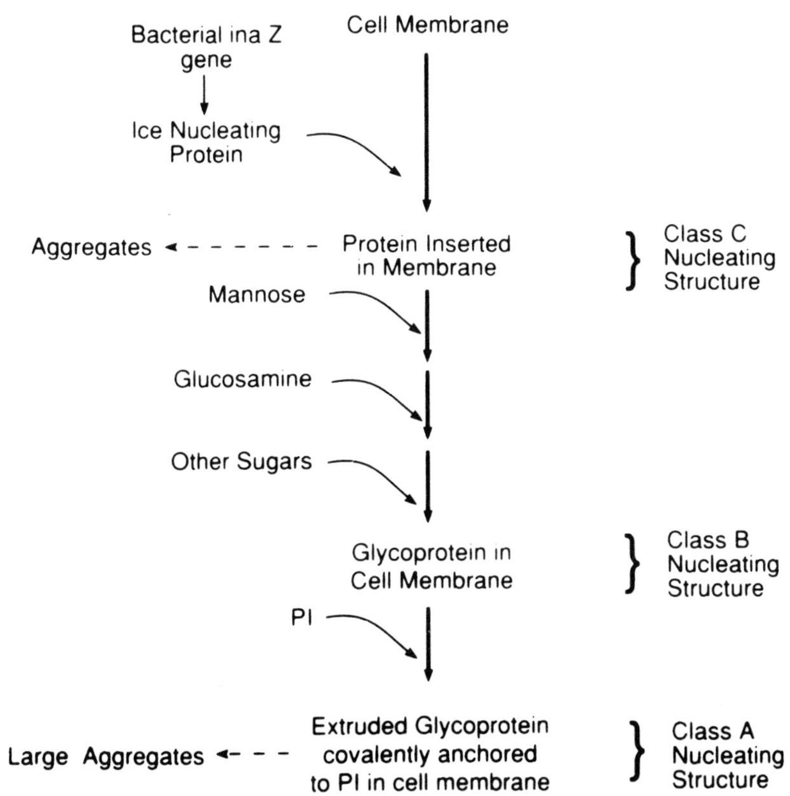


**Figure 9**. Suggested sequential formation of the most effective ice-nucleating structure from class C (Type 3) to class B (Type 2) to class A (Type 1). Side-to-side aggregation is thought to involve the substituted sugars, and phosphatidyinositol (PI) provides a mobile anchor to the cell membrane. (Reprinted, by permission, from Kozloff et al., 1991a)

kDa) as well as a series of cross-reacting bands of lower molecular weight. The pattern of InaZ protein fragments shows a marked threefold periodicity in band intensity, with heavier staining of every third band; this may correspond to some multiple of the conserved 48-amino acid repeats of the InaZ protein. Immunoblots of membrane proteins from other Ice$^+$ strains have also revealed fragments of ice nucleation proteins, which presumably are products of proteolysis (Deininger et al., 1988; Watanabe et al., 1990; Lindow et al, 1989; Ruggles, 1991; Turner et al., 1991). The protease(s) responsible for the turnover of Ina proteins is unknown, but it is well known that outer membranes of gram-negative bacteria contain proteases (Maurizi, 1992).

## Purification and Characterization of Ice Nucleation Proteins

There is strong evidence that the Ice$^+$ phenotype is dependent on the presence of an Ina protein, and various Ina proteins have been isolated from *E. coli* strains overexpressing *ina* genes (see Chapter 5). It is significant that the isolated InaZ (Wolber et al., 1986; Kim et al., 1989) and InaA (Watanabe et al., 1993) proteins, solubilized and purified to homogeneity in the presence of detergents, express ice nucleation activity without added membrane components. Although the ice nucleation thresholds of these proteins occur at much lower temperatures (i.e., –6 to –14°C) than in their native or heterologous host membranes, these results demonstrate that isolated ice nucleation proteins have the intrinsic ability to form nucleation centers.

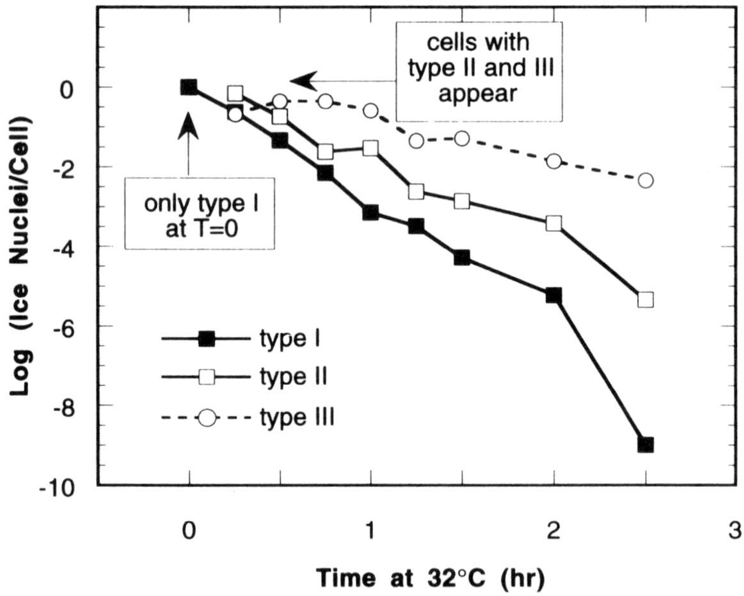


**Figure 10.** The rates of disappearance of Types 1, 2, and 3 ice nuclei in *Pseudomonas syringae* T1 as demonstrated by shifting fully induced cultures (containing essentially one Type 1 ice nucleus per cell) from 16 to 32°C. Cultures were grown and induced as described in Nemecek-Marshall et al. (1993). The appearance of Types 2 and 3 after the temperature shift could be due to "unmasking" of these nuclei by Type 1 nuclei or degradation of Type 1 nuclei, or both. (Redrawn from Ruggles et al., 1993)

## A Biochemical Model for an Ice Nucleation Site

The secondary and tertiary structures of the N- and C-terminal domains of ice nucleation proteins have not been modeled in any detail. Such modeling is very difficult, since there are neither structural data nor constraints imposed by a periodic sequence to limit the possibilities (see Chapter 6). One aspect of ice nucleation proteins that seems particularly mysterious, when all of the domains are considered together, is their strong association with the bacterial membrane and lipid requirement for activity. The amino acid sequence of an ice nucleation protein is rather hydrophilic (Warren, 1987; Wolber, 1993). The only portion of the molecule that contains hydrophobic stretches that are candidates for conventional, transmembrane α-helices is the N-terminal unique domain (Warren, 1987). Thus, the N-terminal domain may contain the membrane anchor for ice nucleation proteins, and the template formed by the repeating domain may be associated only with the membrane surface. Such a location would fit the postulated function of the repeating domain: organizing water. Alternatively, ice nucleation proteins could be anchored to lipid molecules via some form of secondary chemical modification (Turner et al., 1990., 1991; Kozloff et al., 1991a).

It is not known how ice nucleation proteins are translocated to the outer membrane; as pointed out by Warren and Wolber (1991), the N-termini of Ina proteins do not contain normal signal peptides for targeting outer membrane insertion. Ina proteins might contain unusual signal features, unlike those of most outer membrane proteins (Izzard and Kendall, 1994).

There is a general consensus on the quaternary structure of ice nucleation proteins: most investigators now believe that the protein forms homoaggregate arrays that serve as water-binding templates for ice formation (Warren et al., 1986; Warren, 1987; Govindarajan and Lindow, 1988a; Mueller et al., 1990; Kozloff et al.,

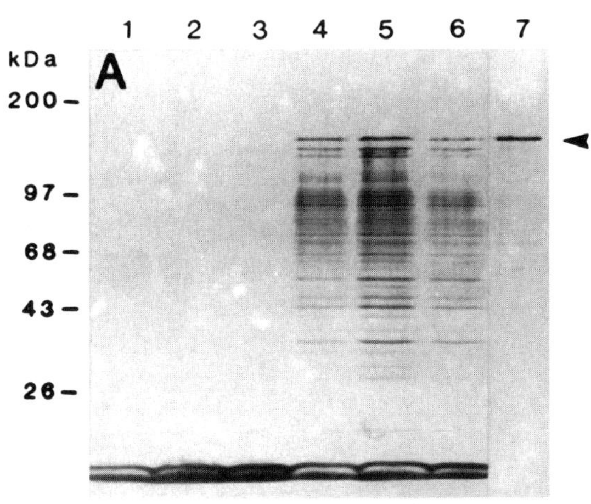


**Figure 11.** Detection of InaZ protein and proteolytic fragments in *Pseudomonas syringae* S203 membranes by immunoblotting. Immunoblot analysis of sodium dodecyl sulfate-polyacrylamide gels of the membrane proteins extracted from cells harvested after 14, 18, 22, 37.5, 44 and 62.8 hr of growth at 24°C (lanes 1 to 6, respectively); lane 7, 720 ng of InaZ protein (arrow). As discussed in this chapter, the InaZ protein and ice nucleation activity were not highly expressed until the culture reached the stationary phase of growth. (Reprinted, by permission, from Deininger et al., 1988)

1991a; Ruggles et al., 1991; Kajava and Lindow, 1993; see also Chapter 6). A schematic representation of this model is shown in Figure 12. The probable points of attack of various agents that inhibit or destroy bacterial ice nuclei are also noted. The model proposes that subunits of the ice nucleation protein aggregate in the plane of the outer membrane to form these arrays, and that the resulting ice template is topologically oriented exofacially (i.e., it is in direct contact with external solution surrounding the cell). This latter point is partially supported by experimental probes of an external surface exposure of the ice nucleation protein in *E. herbicola* and *P. syringae* (Ruggles, 1991); the nonpermeable enzymatic probe, lactoperoxidase, inactivates virtually all Type 1 ice nuclei in intact *E. herbicola* and *P. syringae* cells, and several proteases inactivate Type 1 ice nuclei in lipopolysaccharide-depleted *P. syringae* (but not *E. herbicola*). Evidence has not yet been obtained to indicate that these enzymatic probes react only with the ice nucleation protein and not some other membrane protein that is essential for the aggregation or maintenance of the ice nucleation template.

Significant remaining points of uncertainty in this model are the balance of forces that drive aggregate formation and the molecular sources of those forces. One force is probably the fluidity of the membrane, which may be an important determinant of homoaggregate formation and ice nucleation activity (see discussion by Govindarjan and Lindow, 1988b). Clarification of these forces might be obtained in experiments in which purified ice nucleation protein is reconstituted in

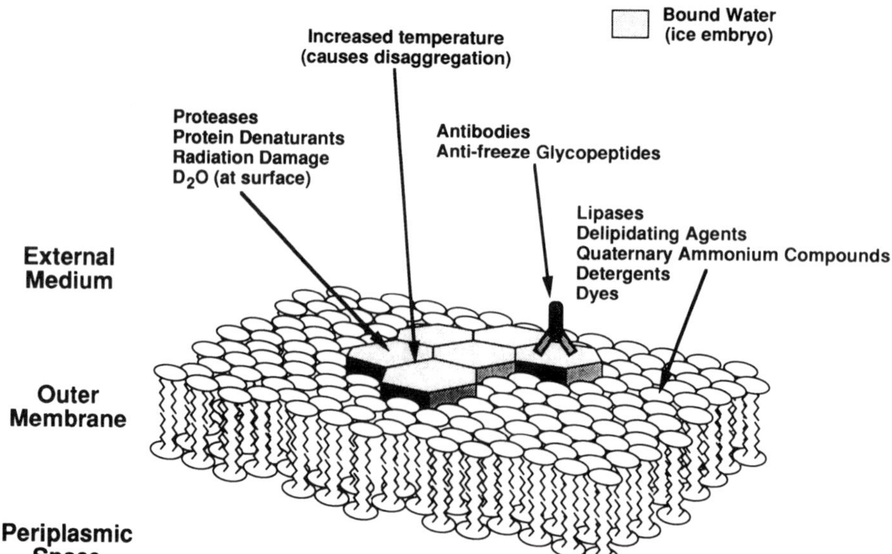


**Figure 12.** A model of the quaternary structure of bacterial ice nuclei in the outer membrane. Copies of the ice nucleation protein are shown as hexagonal prisms in order to emphasize their role as an ice template and their propensity to aggregate into (presumably periodic) membrane-bound arrays. The arrays are shown to be facing the external medium, although as discussed in the text, this topology has not been explicitly proven. For simplicity, the outer membrane bilayer is shown as a simple phospholipid bilayer; in actual outer membranes of gram-negative bacteria, the outer leaflet of the bilayer is enriched with lipopolysaccharide. The postulated sites of action of various treatments that degrade bacterial ice nuclei are noted on the figure. (Redrawn from Wolber, 1993)

phospholipid bilayers; so far, such attempts have not yielded active Type 1 ice nuclei, but only nuclei with colder threshold temperatures (i.e., –9°C; Govindarajan and Lindow, 1988b).

## Concluding Comments

The last 15 years have witnessed an explosive growth in the body of knowledge concerning bacterial ice nuclei. Several former mysteries—the molecular nature of the ice nucleation site, the sizes and membrane locations of the nucleation sites, and the source of heterogeneity of nucleation threshold temperatures exhibited by clonal populations of bacteria—are essentially solved. Recent information is addressing the question of the physiological regulation of *ina* genes, and evidence is accumulating to suggest that expression of *ina* genes is under control of starvation signals, at least in some Ice$^+$ strains (see Chapter 19). However, the old questions are being replaced by new ones, including some of the following. What is the three-dimensional structure of an ice nucleation protein? And how does the folded structure allow insertion and aggregation in the outer membrane? Do ice nucleation proteins undergo posttranslational modification? If so, how do these modifications relate to the assembly and function of ice nucleation sites? How do ice nucleating proteins specifically associate with membrane phospholipids such as phosphatidylinositol? How is the Ina protein translocated to the outer membrane? How many Ina protein monomers are present in different classes of ice nuclei, especially Type 1? Is there a continuum of nucleation site sizes, or are there discrete size classes? What is the mechanism for turnover of ice nuclei, and how is it regulated? Are ice nucleation proteins related to any other bacterial proteins?

These and other questions concerning bacterial ice nuclei should provide new challenges for the future, especially for the next generation of biochemists, biophysicists, and molecular biologists who are attracted to this intriguing area of investigation.

## Literature Cited

Burke, M. J., and Lindow, S. E. 1990. Surface properties and size of the ice nucleation site in ice nucleation active bacteria: theoretical considerations. Cryobiology 27:80-84.

Deininger, C. A., Mueller, G. M., and Wolber, P. K. 1988. Immunological characterization of ice nucleation proteins from *Pseudomonas syringae*, *Pseudomonas fluorescens*, and *Erwinia herbicola*. J. Bacteriol. 170:669-675.

DeVries, A. L. 1986. Antifreeze glycopeptides and peptides: Interactions with ice and water. Methods Enzymol. 127:293-303.

Ferguson, A. J. 1988. Cell-surface anchoring of proteins via glycosyl-phosphatidylinositol structures. Annu. Rev. Biochem. 57:285-320.

Fukuoka, S., Kamishima, H., Tamiya, E., and Karube, I. 1992. Spontaneous release of outer membrane vesicles by *Erwinia carotovora*. Microbios 72:167-173.

Govindarajan, A. G., and Lindow, S. E. 1988a. Size of bacterial ice-nucleation sites measured in situ by radiation inactivation analysis. Proc. Natl. Acad. Sci. USA 85:1334-1338.

Govindarajan, A. G., and Lindow, S. E. 1988b. Phospholipid requirement for expression of ice nuclei in *Pseudomonas syringae* and *in vitro*. J. Biol. Chem. 263:9333-9338.

Izard, J. W., and Kendall, D. A. 1994. Signal peptides: Eququisitely designed transport promoters. Mol. Microbiol. 13:765-773.

Kajava, A. V., and Lindow, S. E. 1993. A model of the three-dimensional structure of ice nucleation proteins. J. Mol. Biol. 232:709-717.

Kieft, T. L., and Ruscetti, T. 1992. Molecular sizes of lichen ice nucleation sites determined by gamma radiation inactivation analysis. Cryobiology 29:407-413.

Kim, J., Park, S.-H., Han, M. H., and Hahm, K.-S. 1989. Purification and characterization of ice nucleating protein from *Pseudomonas syringae*. Korean Biochem. J. 22:73-77.

Kozloff, L. M., Lute, M., and Westaway, D. 1984. Phosphatidylinositol as a component of the ice nucleating site of *Pseudomonas syringae* and *Erwinia herbicola*. Science 226:845-846.

Kozloff, L. M., Schofield, M. A., and Lute, M. 1983. Ice nucleating activity of *Pseudomonas syringae* and *Erwinia herbicola*. J. Bacteriol. 153:222-231.

Kozloff, L. M., Turner, M. A., and Arellano, F. 1991a. Formation of bacterial membrane ice-nucleating lipoglycoprotein complexes. J. Bacteriol. 173:6528-6536.

Kozloff, L. M., Turner, M. A., Arellano, F., and Lute, M. 1991b. Phosphatidylinositol, a phospholipid of ice-nucleating bacteria. J. Bacteriol. 173:2053-2060.

Kurosky, A., and Hofmann, T. 1972. Kinetics of the reaction of nitrous acid with model compounds and proteins, and the conformational state of N-terminal groups in the chymotrypsin family. Can. J. Biochem. 50:1282-1296.

Lindow, S. E. 1982. Epiphytic ice nucleation-active bacteria. Pages 335-362 in: Phytopathogenic Prokaryotes. M. S. Mount and G. H. Lacy, eds. Academic Press, New York.

Lindow, S. E. 1983a. The importance of bacterial ice nuclei in plant frost injury. Curr. Top. Plant Biochem. Physiol. 2:119-128.

Lindow, S. E. 1983b. The role of bacterial ice nucleation in frost injury to plants. Annu. Rev. Phytopathol. 21:363-384.

Lindow, S. E. 1983c. Kinetics of changes in ice nucleation activity of *Pseudomonas syringae* following temperature shifts. (Abstr.) Phytopathology 73:809.

Lindow, S. E. 1983d. Methods of preventing frost injury caused by epiphytic ice nucleation active bacteria. Plant Dis. 67:327-333.

Lindow, S. E., Arny, D. C., Barchet, W. R., and Upper, C. D. 1978. Bacterial ice nucleation inhibitors and reduction of frost damage to plants. (Abstr.) Phytopathol. News 12:138.

Lindow, S. E., Hirano, S. S., Barchett, W. R., Arny, D. C., and Upper, C. D. 1982. Relationship between ice nucleation frequency of bacteria and frost injury. Plant Physiol. 70:1090-1093.

Lindow, S. E., Lahue, E., Govindarajan, A. G., Panopoulos, N. J., and Gies, D. 1989. Localization of ice nucleation activity and the *iceC* gene product in *Pseudomonas syringae* and *Escherichia coli*. Mol. Plant-Microbe Interact. 2:262-272.

Maki, L. R., Galyan, E. L., Chang-Chien, M. M., and Caldwell, D. R. 1974. Ice nucleation induced by *Pseudomonas syringae*. Appl. Microbiol. 28:456-459.

Margaritis, A., and Bassi, A. S. 1991. Principles and biotechnological applications of bacterial ice nucleation. Crit. Rev. Biotechnol. 11:277-295.

Maurizi, M. R. 1992. Proteases and protein degradation in *Escherichia coli*. Experientia 48:178-201.

Mayrand, D., and Grenier, D. 1989. Biological activities of outer membrane vesicles. Can. J. Microbiol. 35:607-613.

Mueller, G. M., Wolber, P. K., and Warren, G. J. 1990. Clustering of ice nucleation protein correlates with ice nucleation activity. Cryobiology 27:416-422.

Nemecek-Marshall, M., LaDuca, R., and Fall, R. 1993. High level expression of ice nuclei in a *Pseudomonas syringae* strain is induced by nutrient limitation and low temperature. J. Bacteriol. 175:4062-4070.

Nikaido, H., and Vaara, M. 1987. Outer membrane. Pages 7-22 in: *Escherichia coli* and *Salmonella typhimurium*. Cellular and Molecular Biology, Vol. 1. F. C. Neidhardt, ed.. American Society for Microbiology, Washington, D.C.

Obata, H., Takeuchi, S., and Tokuyama, T. 1990. Release of cell-free ice nuclei from *Pseudomonas viridiflava* with a Triton X100-EDTA system and their nucleation properties. J. Ferment. Bioeng. 70:308-312.

Okada, K., Takeuchi, Y., Watanabe, M., and Arai, S. 1990. *n*-Alkylphenyldimethylammonium salts as inhibitors of the ice nucleating activity of *Erwinia ananas*. Agric. Biol. Chem. 54:549-550.

Parody-Morreale, A., Bishop, G., Fall, R., and Gill, S. J. 1986. A differential scanning calorimeter for ice nucleation distribution studies: Application to bacterial nucleators. Anal. Biochem. 154:682-690.

Parody-Morreale, A., Murphy, K. P., DiCera, E., Fall, R., DeVries, A. L., and Gill, S. J. 1988. Inhibition of bacterial ice nucleators by fish antifreeze glycoproteins. Nature 333:782-783.

Phelps, P., Giddings, T. H., Prochoda, M., and Fall, R. 1986. Release of cell-free ice nuclei by *Erwinia herbicola*. J. Bacteriol. 167:496-502.

Phelps, P. A. 1987. The expression of ice nuclei in bacteria and the nature of the ice nucleation site. Ph.D. thesis. University of Colorado, Boulder.

Pooley, L., and Brown, T. A. 1990. Preparation of active cell-free ice nuclei from *Pseudomonas syrin-*

*gae*. Proc. R. Soc. Lond. Ser. B Biol. Sci. 241:112-115.

Pooley, L., and Brown, T. A. 1991. Effects of culture conditions on expression of the ice nucleation phenotype of *Pseudomonas syringae*. FEMS Microbiol. Lett. 77:2-3.

Rogers, J. S., Stall, R. E., and Burke, M. J. 1987. Low temperature conditioning of the ice nucleation active bacterium, *Erwinia herbicola*. Cryobiology 24:270-279.

Ruggles, J. A. 1991. Bacterial ice nucleation proteins: Kinetics of assembly and membrane locations of functional ice nuclei. Ph.D. thesis. University of Colorado, Boulder.

Ruggles, J. A., Nemecek-Marshall, M., and Fall, R. 1991. Antibody induced shift of bacterial ice nucleation freezing spectra. Cryo Lett. 12:245-252.

Ruggles, J. A., Nemecek-Marshall, M., and Fall, R. Kinetics of appearance and disappearance of classes of bacterial ice nuclei support an aggregation model for ice nucleus assembly. J. Bacteriol. 175:7216-7221.

Southworth, M. W., Wolber, P. K., and Warren, G. J. 1988. Nonlinear relationship between concentration and activity of a bacterial ice nucleation protein. J. Biol. Chem. 263:15211-15216.

Turner, M. A., Arellano, F., and Kozloff, L. M. 1990. Three separate classes of bacterial ice nucleation structures. J. Bacteriol. 172:2521-2526.

Turner, M. A., Arellano, F., and Kozloff, L. M. 1991. Components of ice nucleation structures in bacteria. J. Bacteriol. 173:6515-6527.

Warren, G. J. 1987. Bacterial ice nucleation: Molecular biology and applications. Biotechnol. Genet. Eng. Rev. 5:107-135.

Warren, G., and Wolber, P. 1991. Molecular aspects of microbial ice nucleation. Mol. Microbiol. 5:239-243.

Warren, G., Coroto, L., and Wolber, P. 1986. Conserved repeats in diverged ice nucletion structual genes from two species of Pseudomonas. Nucleic Acids Res. 14:8047-8060.

Warren, G., Wolber, P., and Green, R. 1987. Functional significance of oligonucleotide repeats in a bacterial ice nucleation gene. Pages 1013-1017 in: Plant Pathogenic Bacteria. E. L. Civerolo, A. Collmer, R. E. Davis, and A. G. Gillaspie, eds. Martinus Nijhoff Publishers, Dordrecht.

Watanabe, M., Makino, T., Okada, K., Hara, M., Watabe, S., and Arai, S. 1988a. Alkylbenzyldimethylammonium salts as inhibitors for the ice nucleating activity of *Erwinia ananas*. Agric. Biol. Chem. 52:201-206.

Watanabe, M., Watabe, S., and Arai, S. 1988b. Interaction of an antinucleating chemical and an ice nucleation active bacterium: A case study with an *n*-octylbenzyldimethylammonium salt and *Erwinia ananas*. Agric. Biol. Chem. 52:1869-1871.

Watanabe, N. M., Southworth, M. W., Warren, G. J., and Wolber, P. K. 1990. Rates of assembly and degradation of bacterial ice nuclei. Mol. Microbiol. 4:1871-1880.

Watanabe, S., Abe, K., Hirata, A., Emori, Y., Watanabe, M., and Arai, S. 1993. Large-scale production and purification of an *Erwinia ananas* ice nucleation protein and evaluation of its ice nucleation activity. Biosci. Biotech. Biochem. 57:603-606.

Wight, D. R., Newman, W. R., Ward, P. J., and Pochodylo, K. 1992. Cell-free and whole cell ice nucleators and process for their production. International patent WO 92/14833.

Wolber, P. K. 1993. Bacterial ice nucleation. Adv. Microb. Physiol. 34:203-237.

Wolber, P. K., Deininger, C. A., Southworth, M. W., Vandekerckhove, J., VanMontagu, M., and Warren, G. J. 1986. Identification and purification of a bacterial ice-nucleation protein. Proc. Natl. Acad. Sci. USA 83:7256-7260.

Wolber, P., and Warren, G. 1989. Bacterial ice-nucleation proteins. Trends. Biochem. Sci. 14:179-182.

Wolber, P. K., and Warren, G. J. 1991. Evolutionary perspective on the ice nucleation gene-encoded membrane protein. Pages 315-330 in: Microbial Ecology of Leaves. J. H. Andrews and S. S. Hirano, eds. Springer-Verlag, New York.

Yankofsky, S. A., Levin, Z., Bertold, T., and Sandlerman, N. 1981. Some basic characteristics of bacterial freezing nuclei. J. Appl. Meteorol. 20:1013-1019.

CHAPTER 5

# Identification and Analysis of *ina* Genes and Proteins

**Gareth J. Warren**

## *ina* Gene and Protein Sequences

### Identification of Genes

Bacterial *ina* genes confer an Ice⁺ phenotype when cloned and expressed in *Escherichia coli* (Orser et al., 1983)—this has provided the ability to detect the appropriate clones in libraries of genomic DNA from Ice⁺ species. In fact, the phenotype can also be expressed in gram-positive bacterial species (Hottinger et al., 1991; see also Chapter 16) and even in higher plants (Baertlein et al., 1992) when the *ina* gene is linked to sequences appropriate for transcription and translation in the foreign host. Genes have now been identified in cloned DNA from *Erwinia ananas*, *E. herbicola*, *Pseudomonas fluorescens*, *P. syringae*, *P. viridiflava*, and *Xanthomonas campestris* (Table 1). In each case, the *ina* gene was approximately 4 kb in length.

The highest threshold temperature for ice nucleation in recombinant *E. coli* usually differs from that in the naturally occurring source of the *ina* gene: it is lower by some fraction of a degree Celsius. The difference is not attributable to differences in expression levels, since manipulation of *E. coli* to produce higher levels of protein than are seen in *Pseudomonas* does not completely close the gap. Do the naturally occurring Ice⁺ bacteria contain other, superior *ina* genes, or cofactors that help to raise threshold temperatures, or do they simply differ in some quantitative trait (e.g., the fatty acid composition of their membranes) that affects the assembly of ice nucleation sites? The first possibility (alternative *ina* genes) can be eliminated. In *P. syringae* and *P. fluorescens*, defective alleles of the native *ina* genes were introduced via homologous recombination (Lindow, 1987; Warren et al., 1987; see also Chapter 13); this caused complete loss of the Ice⁺ phenotype. Both of the other explanations remain plausible. Rigorous attempts to identify a cofactor gene have not been made.

### Sequences and Reading Frames

Sequencing of *ina* genes (see Table 1 for references) has revealed a series of homologous genes whose predicted translation products are without significant homology to other known protein sequences. The largest portion of each sequence is made up of a hierarchy of repeated motifs 24, 48, and 144 nucleotides in length. Although the fidelity of repetition is greatest at the 144-nucleotide periodicity (which is how it can be discerned overlying the shorter periodicities), the repeat fidelity nonetheless varies between genes and between regions within a gene. High repeat fidelity (i.e., almost perfect repetition) introduces ambiguity into the process of aligning overlaps for assembling the data from individual sequencing gels. Therefore, sequencing start points must be mapped with better than 72-nucleotide resolution in order to permit accurate sequence assembly; also, it is vital that the protocol for generating the sequence start points should be incapable of producing rearrangements. Because of the pitfalls in accurate assembly of the *ina* gene sequences, the reader is advised to retain a little skepticism about the accuracy of the published sequences.

Each *ina* gene sequence contains a single long open reading frame (ORF). Nucleotides at the third positions of codons in this frame are most variable between repeats (Warren et al., 1986), which supports the assignment of the ORF. Might a second reading frame running in the opposite direction, embedded within the first and having the same variable third position, be necessary to encode the Ice$^+$ phenotype? While fairly long open reading frames exist in the reverse direction, they are not conserved between *ina* genes, suggesting that the rarity of stop codons in the reverse direction is an accidental consequence of the repetition and has not been subject to functional selection.

Edwards et al. (1994) have observed that species closely related to Ice$^+$ bacteria lack nucleotide homology to the *ina* gene. They suggest that this implies horizontal dissemination of *ina* genes. It has been observed that horizontally transmitted genes commonly confer phenotypes for which the selective advantage is "locally restricted" (Eberhard, 1989). In the case of *ina* genes, this fits with the seasonal occurrence of frosts and the concomitant seasonal fluctuation in the proportion of Ice$^+$ bacteria in plants' epiphytic flora (see Chapters 3 and 13).

## Organization of Repeats

Each Ina protein sequence (as predicted from the known DNA sequence) contains a nonrepetitive aminoterminal portion between 161 and 203 residues in length, a repetitive core between 960 and 1,296 residues in length, and a nonrepe-

**Table 1.** References for the isolation and sequencing of *ina* genes from various bacterial species

| Species | Isolation | Sequencing | Gene |
|---|---|---|---|
| *Erwinia ananas* | Arai et al. (1989) | Abe et al. (1989) | *inaA* |
| *E. herbicola* | Orser et al. (1983) | Warren and Corotto (1989) | *iceE* |
| *Pseudomonas fluorescens* | Corotto et al. (1986) | Warren et al. (1986) | *inaW* |
| *P. syringae* | Orser et al. (1985) | Green and Warren (1985) | *inaZ* |
| | | D. Pridmore, *pers. comm.* | *inaV* |
| *P. viridiflava* | Hasegawa et al. (1990) | ... | ... |
| *Xanthomonas campestris* | Zhao and Orser (1990) | Zhao and Orser (1990) | *inaX* |

**Note added in proof:** An *ina* gene from *Erwinia uredovora* has recently been described (Michigami, Y., Watabe, S., Abe, K., Obata, H., and Arai, S. 1994. Cloning and sequencing of an ice nucleation active gene of *Erwinia uredovora*. Biosci. Biotechnol. Biochem. 58:762-764.)

titive carboxyterminal portion between 41 and 68 residues in length. (InaX protein is longest in all categories!) The repetitive portion contains both 8- and 16-residue motifs (corresponding to 24- and 48-nucleotide repeats) with their periodicity unbroken throughout. A further 48-residue (144-nucleotide) repeating motif is recognizable in most of the repetitive portion, but its periodicity is sometimes broken: one break each in InaW, InaZ, and InaV, and three breaks in InaA. These observations are illustrated in Figure 1, where large triangles represent the 48-residue and and small triangles the 16-residue motifs.

Figure 1 is divided into blocks in reference to differences between repeat sequences. At the level of 16-residue periodicity, repeats within Blocks 2 and 3 differ in their consensus from those of Block 4, and at the level of 48-residue periodicity, differences in consensus can be observed between Blocks 2 and 3. Figure 2 displays the relationships between the 8-, 16- and 48-residue repeat types; these relationships *may* reflect the way in which the various repeat types evolved. In all the Ina proteins, the four blocks are distinguishable and occur in the same order. However, the relative sizes of the domains comprising Block 2 and Block 3 can be seen to vary widely between Ina proteins (Fig. 1).

## Generation and Loss of Repeats

Three lines of evidence point toward an ongoing generation and loss of repeated sequences in *ina* genes. Most obvious, from the foregoing, is that the comparison between proteins indicates that the repeat content of each domain has been flexible during evolution. Second, the overwhelming bias toward the AGPy (as opposed to TCN) type of serine codon, at a repeating position where serine is strongly conserved, argues for the original derivation of all repeats by the amplification of an archetype (Green and Warren, 1985). Third, one particular *ina* gene (*inaW*) contains almost perfect repetition at the third codon position in Block 2 (Warren et al., 1986). Since this cannot be due to selection for maintenance of function, it must be ascribed to a recent saltatory event(s) wherein the majority of 144-nucleotide elements were replicated (or corrected) using one copy as a template.

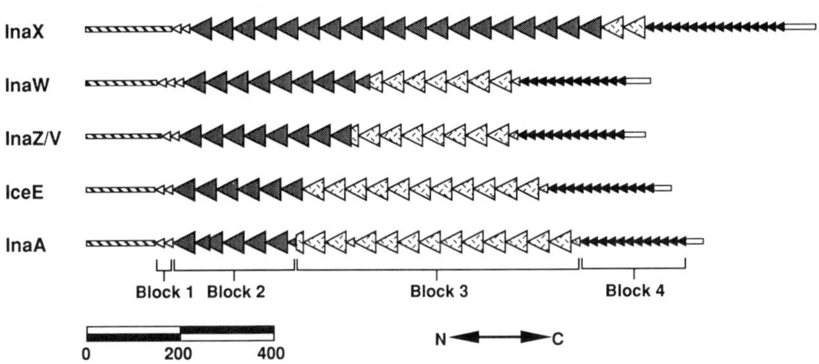


**Figure 1.** Organization of Ina proteins. The scale, in number of aminoacyl residues, and the orientation (N vs. C terminal) are shown at the bottom. Nonrepetitive regions are represented by bars, repetitive regions by strings of triangles. Each large triangle represents a unit of 48-residue repetition; portions of these triangles represent corresponding fractional portions of a 48-residue unit. Where no 48-residue periodicity is present, small triangles represent units of 16-residue repetition. Repetition of the four types (Blocks 1, 2, 3 and 4, as defined in Fig. 2) is coded by different patterns of shading; the key is indicated by reference to the Blocks in InaA protein. InaZ and InaV have been combined for the purpose of this representation because their organization is identical by the criteria used here.

The generation and loss of repetitive sequences is usually thought to occur through unequal crossing-over or gene conversion. But whatever its mechanism(s), it reduces the amount of information that can be derived by sequence comparisons. *Colinear* alignment of sequences becomes arbitrary, since it does not guarantee alignment of the most closely related repeat elements. (I did not fully appreciate this point when presenting a colinear consensus of entire Ina proteins in previous publications.) Potentially, a second consequence of the generation and loss of repeats is that the repetitive sequences might evolve at an anomalous rate. Therefore, *any* comparison of the repeats that are subject to generation and loss (whether or not the comparison is colinear) will give suspect results from the standpoint of inferring evolutionary relatedness of the genes.

Repeat Block 4 has the lowest repeat fidelity and no discernible 144-nucleotide periodicity. If a colinear alignment is made between the Block 4 repeats of all genes except *inaX*, there is less divergence between genes than between repeats within a gene. This indicates little if any rearrangement of Block 4 repeats, since the divergence of *inaA*, *iceE*, *inaV*, *inaW* and *inaZ*. *inaX* differs from the other five genes in the length of Block 4, indicating some generation or loss of repeats since its divergence from the other proteins. However, at least the last 32 codons in repeat Block 4 appear to have escaped rearrangement since before all the genes' divergence.

## Alignments and Comparisons

Comparison of homologous positions between sequences allows inferences to be drawn both about evolutionary relatedness and about requirements for functionality. For these purposes, I have compared those nonrepetitive regions where a strong colinear homology has been demonstrated between all the sequences, plus the last 32 codons in repeat Block 4. Figure 3 utilizes the aligned sequences to infer two types of protein consensus: the majority consensus for single amino acids and the absolute consensus for amino acid types.

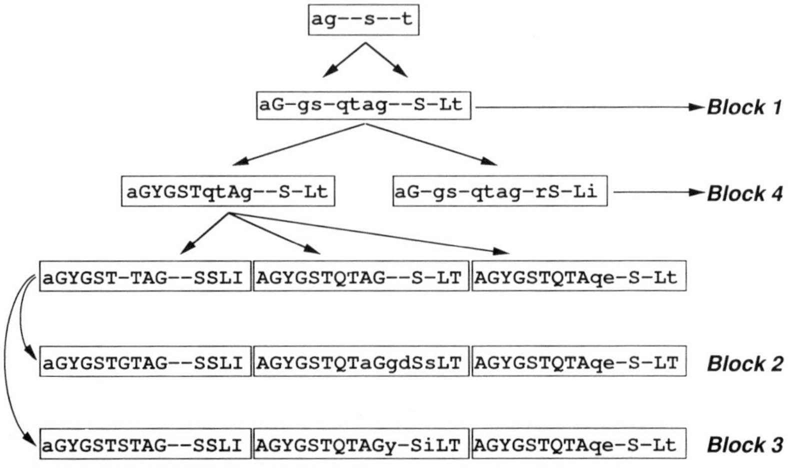


**Figure 2.** Hypothetical derivation of the archetypal repeats of Blocks 1, 2, 3, and 4 during four rounds of amplification. Aminoacyl residues are shown as upper- or lowercase letters to indicate greater or lesser conservation, respectively, in the derivative repeats. Hyphens indicate positions poorly conserved in the derivative repeats.

What can be inferred about the evolutionary relationships of the six genes? In Table 2, the regions shown in Figure 3 are compared for aminoacyl conservation in all pairwise combinations of the six proteins. While this quantifies the close relationship of *inaA* to *iceE*, and that of *inaV* to *inaZ*, the aminoacyl homologies of all other pairings cluster in the range of 60–67%. I believe that this is too small a range from which to construct with any confidence an evolutionary tree—either the four groups' divergences took place at approximately the same time, or they simply happened too long ago. It is worth noting that Wolber (1993) obtained higher measures of aminoacyl homology, because the comparisons included (and were dominated by) homologies between repetitive regions. Zhao and Orser (1990) obtained lower measures of aminoacyl homology in the nonrepetitive portions, because the comparisons included regions where colinear homology between the sequences is not discernible.

**A**

```
Z/V: MNχDKALVLRTCANNMADHCGLIWPASGTVESπYWQSTRRHENGLVGLLWGAGT  54
E/A: MKEDKVLILRTCANNMADHGGIIWPLSGIVECKYWKPVKGFENGLTGLIWGKGS
 W:  MKSEKVLVLRTCANNMTDHCGLVWPILGLVECKFWEPTIKLENGLTGALWGQGS
 X:  MIVEKILALRTCANNMADHCGLIWPMAGPVECKFWKPSGLHENGLTGLLWGKGV
con: M——K-L-LRTCANNMaDHcGliWP—G-VEck-W-p———ENGLtGllWG-G-
con: χ—ηπχχχχπλσχλλχ-ηλλλχχρχχ-λ-χησπρρλ———ηλλχ-λχχρλ-λ-

Z/V: SAFLSVHADARW-VCEVAVADIIλLEEPGMVKFPRAEVVHVGDRISASHFISA  107
E/A: DSPLSLHADAR-VVAEVDADECIAIETHGWIKFPRAEVLHVGTλλSAMQFILH
 W:  SAQLSMNADAKWVVCEVTMGDLIFLENNEGVKFPRAEVVHVGTRSSALGYISD
 X:  GAHLSAHADARWVVCEVAVDEMLQLAEEGMIKFPRATVLFVGNRSQALDYIAA
con: -a-LS-hADArwvVcEV———i-le—g—KFPRAeV-hVG-r-sA——I—
con: λ—χσχλχηχπ—χ-ηχ-χ-η-χ-χ-λ-λ-χπρχπχλχχ-χλλλ-λχ-λρχ—
```

**B**

```
Z/V: AGKNSχLTAGARSKLIGSEGSTLSAGEDS-LI 1151
E/A: AGINSILTAGCRSKLIGSNGSTLTAGENSVLI
 W:  AGKNCVLTAGADSRLIGSLGSTLSGGENSTLI
 X:  AGKNSVLTAGANSRLIGSLGSTLSGGENSTLI
con: AGkNs-LTAGa-S-LIGS-GSTLs-GEnS-LI
con: χλ-λσχχλχλ-λσπχχλσ-λσλχλ-ληλσ-χχ

Z/V: FRLWDGKRYRQLVARTGENGVEADIPYYVNEDDDIVDKPDE 1192
E/A: FRCWDGKRYTNVVAKTGKGGIEADMPYQMDEDNNIVNKPEE
 W:  FRCWDGKRYTNVVVKTGTDEVEADVPYQIDEDSNVLIKAED
 X:  FRCWDGERYTNLVVRTGEQGVESDIPYQVDDEGNLVGKADD
con: FRcWDGkRYtn-V—TG—gvEaD-PYq-ded-n-v-K——
con: ρπ-ρηλλπρλλχχχπλλλλχη-ηχχρ-χληηλλχχ-πχηη
```

**Figure 3.** Alignment of Ina protein sequences. The very similar proteins InaZ and InaV are combined and shown as their consensus (Z/V); likewise, IceE and InaA are combined (E/A). Two types of consensus (con) are shown below the aligned sequences—first, a majority consensus for single amino acids, where lowercase letters denote a three-fourths majority and uppercase letters denote absolute consensus; and second, an absolute consensus for aninoacyl types, with types denoted as follows: σ = serine shape (C,S), η = acidic (D,E), π = basic (K, R), λ = hydrophilic (above plus G, H, N, Q, T), χ = hydrophobic (A, I, L, M, P, V), ρ = aromatic (F, W, Y). Positions at which no consensus is found are denoted by a dash. **A,** Aminoacyl residues 1–107 of each protein are aligned; **B,** 73 contiguous residues from the region homologous to InaZ residues 1,120 –1,192 are aligned.

How significant is the conservation among Ina proteins of each amino acid or aminoacyl type? Consider two homologous genes, so diverged that any unselected codons are completely randomized (except that nonsense codons are avoided). Conservation at each aminoacyl position may be classified as conservation of an identical residue, or of a serine-shaped residue (C or S), an acidic residue (D or E), a basic residue (K or R), a hydrophilic residue (G, H, N, Q, T, and all amino acids of the previous three categories), a hydrophobic residue (A, I, L, M, P, or V), or an aromatic residue(F, W, or Y). Twenty-five codons (out of a total of 61 available sense codons) encode hydrophobic residues, so the likelihood of merely chance conservation of a hydrophobic residue is 25/61; similarly, that for an acidic residue is 4/61. If an acidic residue is conserved in *three* mutually randomized proteins, the likelihood of its conservation being accidental drops to $(4/61)^2$ (for four mutually randomized proteins, it becomes $(4/61)^3$, and so forth).

There are six *ina* sequences to be compared. Certain of these are very closely related, and even for those that are not, we need an objective measure of their divergence. My suggestion is to derive measures of nucleotide randomization between pairwise combinations of the *ina* genes, combine these conservatively to obtain an estimate of the total nucleotide randomization between all genes, and apply this to estimating the significance of each type of aminoacyl conservation.

*Nucleotide* randomization ($R_n$) has been calculated from the frequency of codon third-base substitutions, taking complete conservation of third bases to indicate no randomization ($R_n = 0$) and 25% conservation to indicate complete randomization ($R_n = 1$). Randomization is therefore calculated as $R_n = (1 - \text{conserved frequency}) \times 100/75$. This formula is conservative (underestimating $R_n$), because not all third-base changes are selectively neutral; some third-base changes alter the identity of the encoded amino acid, and even synonymous changes may not be neutral (as

**Table 2.** Comparison of 180 codons each from *inaA*, *iceE*, *inaV*, *inaW*, *inaX*, and *inaZ*[a]

| Comparison | Homology: Amino acid[b] | Homology: Codon 3rd pos.[c] | $R_n$[d] | $R_c$[d] |
|---|---|---|---|---|
| *inaA* and *iceE* | 177 (98.3%) | 177 (98.3%) | 0.02 | <0.01 |
| *inaA* and *inaV* | 113 (62.8%) | 56 (31.1%) | 0.92 | 0.78 |
| *inaA* and *inaW* | 110 (61.1%) | 64 (35.6%) | 0.86 | 0.64 |
| *inaA* and *inaX* | 109 (60.6%) | 66 (36.7%) | 0.84 | 0.59 |
| *inaA* and *inaZ* | 111 (61.7%) | 60 (33.3%) | 0.89 | 0.70 |
| *iceE* and *inaV* | 114 (63.3%) | 57 (31.7%) | 0.91 | 0.75 |
| *iceE* and *inaW* | 111 (61.7%) | 64 (35.6%) | 0.86 | 0.64 |
| *iceE* and *inaX* | 110 (61.1%) | 65 (36.1%) | 0.85 | 0.61 |
| *iceE* and *inaZ* | 112 (62.2%) | 61 (33.9%) | 0.88 | 0.68 |
| *inaV* and *inaW* | 112 (62.2%) | 72 (40.0%) | 0.80 | 0.51 |
| *inaV* and *inaX* | 117 (65.0%) | 72 (40.0%) | 0.80 | 0.51 |
| *inaV* and *inaZ* | 174 (96.7%) | 162 (90.0%) | 0.13 | <0.01 |
| *inaW* and *inaX* | 119 (66.1%) | 76 (42.2%) | 0.77 | 0.46 |
| *inaW* and *inaZ* | 111 (61.7%) | 76 (42.2%) | 0.77 | 0.46 |
| *inaX* and *inaZ* | 116 (64.4%) | 71 (39.4%) | 0.81 | 0.53 |

[a] These are drawn from colinear homologies in two regions of each gene, corresponding to codons 1–107 and 1,120–1,192 of *inaZ*.

[b] The number of codons for which the encoded amino acid is identical.

[c] The number of codons for which the third nucleotide is identical (without regard to identity of the other positions).

[d] Randomization of nucleotides ($R_n$) and of codons ($R_c$) are calculated as described in the text.

**Table 3.** Calculation of significance of conservation of amino acids (Roman letters) and aminoacyl types (Greek letters)[a]

| | A | C | D | E | F | G | H | I | K | L | M | N | P |
|---|---|---|---|---|---|---|---|---|---|---|---|---|---|
| Codon[b] | 4 | 2 | 2 | 2 | 2 | 4 | 2 | 3 | 2 | 6 | 1 | 2 | 4 |
| Significance[c] | 0.985 | 0.995 | 0.995 | 0.995 | 0.995 | 0.985 | 0.995 | 0.990 | 0.995 | 0.97 | 0.998 | 0.995 | 0.985 |

| | Q | R | S | T | V | W | Y | σ | η | π | λ | χ | ρ |
|---|---|---|---|---|---|---|---|---|---|---|---|---|---|
| Codon | 2 | 6 | 6 | 4 | 4 | 1 | 2 | 8 | 4 | 8 | 34 | 22 | 5 |
| Significance | 0.995 | 0.97 | 0.97 | 0.985 | 0.985 | 0.998 | 0.995 | 0.95 | 0.985 | 0.95 | 0.41 | 0.79 | 0.98 |

[a] See Figure 3 for type definitions.
[b] The number of codons ($c$) capable of encoding the amino acid or group.
[c] The significance of conservation as given by the formula $s = (1 - y^{1.53})$, where $y$ is the fraction of the total number of sense codons that can encode the amino acid or group (i.e., $y = c/61$), and 1.53 is the sum of mutual codon randomization between all *ina* genes (see text).

witnessed by biases in codon usage). *Codon* randomization ($R_c$) is calculated as $R_c = R_n^3$. Table 2 calculates $R_n$ and $R_c$ values from measurements of third-base substitution frequencies. From the $R_c$ values, it is possible to calculate the significance of conservation between pairwise combinations of proteins. However, it is more desirable to calculate a single significance figure for conservation among all the proteins. What is the most objective way to estimate total mutual randomization of all the proteins? I have chosen to sum the $R_c$ values for the protein pairs InaZ/InaW, InaW/InaX, and InaX/IceE, because these pairings link all four proteins with the minimum values; InaV and InaA are too little randomized relative to InaZ and IceE, respectively, to be worth including. The $R_c$ values sum to 1.53; therefore, if an acidic residue (for example) has been conserved in all genes, the probability $p$ of this having happened by chance is estimated as $p = (4/61)^{1.53}$. Significance $s$ is given by $s = 1 - p$. It may be worth restating that this approach was designed to err on the side of caution by overestimating the potential for randomization and thus underestimating the significance of its absence. Using this approach, Table 3 calculates the significance of the various types of residue conservation among the six Ina proteins of known sequence. It is apparent that the conservation of all residue classifications except "hydrophilic" and "hydrophobic" are significant at the 95% confidence level.

## Ina Proteins

### Identification

When membrane proteins were extracted from *E. coli* strains carrying cloned *ina* genes and subjected to sodium dodecyl sulfate–polyacrylamide gel electrophoresis (SDS-PAGE), Ina proteins were seen as novel bands (Corotto et al., 1986; Wolber et al., 1986; Lindow et al., 1989). Strains expressing *inaW* gave a band of mobility corresponding to a molecular weight of 180 kDa, and the expression of *inaZ* and *iceC* gave rise to bands estimated at 153 and 150 kDa, respectively. This simple approach worked because there is a relatively low background of high molecular weight species among *E. coli* membrane proteins. This was not true in the pseudomonads, where *inaW*, *inaZ*, and *iceC* occur naturally, and the identification of Ina proteins in their native hosts required the use of antibodies.

Antibodies have been raised against the putative InaW protein (the apparently 180-kDa protein observed in *E. coli* expressing *inaW*) (Deininger et al., 1988), against the protein product of a *lacZ-iceC* gene fusion (Lindow et al., 1989), and against synthetic peptides designed to mimic the repetitive sequences of Ina proteins (Mueller et al., 1990; Ruggles et al., 1991). As far as they have been compared, the various antibodies have all shown broadly similar specificities. The bands identified by SDS-PAGE (at 180, 153, and 150 kDa) were seen, upon probing of Western blots, to be the main reactive bands in the respective *E. coli* clones and to have counterparts of identical mobility in preparations of membrane protein from the native Ice$^+$ species. Because the antipeptide antibodies were directed against *predicted* structural components, they provide compelling evidence for the correct identification of the Ina proteins. Further evidence is provided by the absence of these bands in strains of *P. syringae* and *P. fluorescens* that have been deprived of their *ina* genes by means of directed mutagenesis (Deininger et al., 1988). Antibodies have also been used in attempts to probe the biochemical nature of the nucleating site (Ruggles et al., 1991; see also Chapter 4).

The estimates of 180 and 153 kDa do not correspond very well to the predicted molecular weights of InaW and InaZ (which are 120 and 119 kDa, respectively). This discrepancy would be explained by posttranslational modification of Ina proteins. However, posttranslational modification would have to be improbably similar in *E. coli* and *Pseudomonas* to account for the apparently identical molecular weights of an Ina protein produced in both. Moreover, in vitro transcription and translation yields proteins of electrophoretic mobilities indistinguishable from those seen in vivo (G. Mueller and G. Warren, *unpublished*); it seems very unlikely that identical posttranslational modification could have occurred in the in vitro system. Thus the anomalous gel behavior of the Ina proteins is more plausibly attributable to their possessing anomalous chemical properties, e.g., abnormally low affinities for SDS.

## Purification of Protein

Purification of Ina proteins was achieved from *E. coli* strains expressing *ina* genes from strong promoters (Wolber et al., 1986). Under these circumstances, inclusion bodies are liable to form and may have been the source of the material that was purified. The InaZ protein was found by protein sequencing to have an unprocessed N terminus, but this is not particularly significant, given its possible origin in inclusion bodies.

Ice nucleation activity was undetectable in the initial preparations of Ina protein (which had been denatured during purification), but it became detectable when a nondenaturing purification was developed (Wolber et al., 1986; Watanabe et al., 1993). Ice nucleation sites (active at –13°C) were present at a mean frequency of one per 30,000 InaZ protein molecules; the most active fraction contained one nucleation site per approximately 17,000 InaZ molecules (i.e., one per $2 \times 10^9$ Da of protein). It seems likely that most of the InaZ protein in this preparation is inactive, and the active nucleation sites are actually smaller: first, because the mean mass estimate differs by four orders of magnitude from the in vivo estimate of 150 kDa

**Technical note:** For Western blotting of Ina proteins, the standard protocols must be modified to obtain quantitative transfer (Deininger et al., 1988). The methanol content of the transfer buffer is reduced, and SDS (not normally present) is included in it. The rationale for these alterations is to counteract the consequences of the Ina proteins' relatively low affinity for SDS. See also Baertlein et al. (1992) for an example implementing the modified protocol.

(Govindarajan and Lindow, 1988; see also Chapter 4), and secondly, because ice nucleation sites can be formed in cells containing many fewer InaZ protein molecules in toto (Southworth et al., 1988). Therefore we may conclude that active ice nucleation sites are a minor component of such a protein preparation. This means that many types of analysis on the InaZ protein preparation (e.g., structural determinations) would be of dubious value in ascertaining the nature of the nucleating site.

### Heterogeneity of Protein Distribution

What is the basis of the variation in nucleation thresholds within bacterial populations? Mueller et al. (1990) examined the distribution of Ina proteins within and between *E. coli* cells by immunofluorescence microscopy. They found apparent aggregates of Ina protein that varied considerably between cells in their sizes and frequencies. This is consistent with the inference of Watanabe et al. (1990) that subpopulations of cells within a clone differ in their rates of assembly and degradation of nucleating sites.

Mueller et al. (1990) saw by microscopy that the sizes of Ina protein aggregates varied both with expression level (the aggregates were larger when more protein was present) and with growth temperature (the aggregates were larger toward lower growth temperatures, at which ice nucleation sites with higher thresholds are produced). Both trends are consistent with the hypothesis that the quantity of Ina protein in an aggregate determines the threshold temperature of the ice nucleation site. Size-dependence of threshold temperature was observed by Govindarajan and Lindow (1988) and is also implied by theory (Burke and Lindow, 1990; Fletcher, 1958; see also Chapter 4). However, it should be remembered that the microscopically observed aggregates were in a range of sizes considerably larger than the nucleation site sizes inferred by observation or predicted by theory.

The low-temperature growth conditions most conducive to high ice nucleation activities caused irregular distortions of the rod-like shapes of *E. coli* cells when Ina proteins were being produced at high levels. In the distorted rods, Ina protein aggregates were associated with the cell wall on the inner sides of the bends. This association may be a manifestation of the presence of Ina proteins in the outer membrane (Lindow et al., 1989; see also Chapter 4).

## Genetic Manipulations

### Expression Levels and Expression Kinetics

If we assume Ina protein to be the major constituent of the ice nucleation site, it follows that ice nucleation sites with higher nucleation thresholds will contain more molecules of Ina protein. It would be expected, therefore, that the sites with higher thresholds (and greater amounts of Ina protein) would show a concomitantly greater dependence on the supply of Ina protein.

The prediction was tested in a study where a variety of promoters were employed to vary the expression level of *inaZ* over a 3,000-fold range in *E. coli* (Southworth et al., 1988). Results were contrary to expectation. The frequency of nucleation sites of all temperature thresholds showed similar relationships to the concentration of InaZ protein—only the constant of proportionality varied with threshold temperature (Fig. 4). At lower concentrations in *E. coli*, frequency varied as the second power (i.e., the square) of the protein's concentration; toward higher concentrations, it varied as the third power until the frequency asymptote was approached. Watanabe et al. (1990) also found a third-power relationship between the protein's concentration and nucleation site frequency in the nonequilibrium situation of protein decay following inhibition of protein synthesis. In two Ice$^{-}$ *P. syringae* strains, expression of an introduced *inaZ* gene that varied over a 400-fold (Southworth et al., 1988) or a 40-fold (Lindgren et al., 1989) range also showed a second to third power relationship to the frequency of ice nucleation sites, although the data in these cases were not sufficient to rule out a power relationship that varied for nucleation sites with different threshold temperatures.

The power relationship between the intracellular concentration of InaZ protein and the frequency of ice nucleation sites suggests that the formation of every type of nucleation site is limited by the same type of molecular event; the second to third power dependence suggests that the event might be a bi- or trimolecular aggregation of Ina protein molecules. Why should the formation of ice nucleation sites large and small occur in proportion to the rate at which a protein dimer or trimer is formed? It seems necessary to postulate that the *threshold* of a potential nucleating site is not normally limited by protein availability but by a stochastic process (such as random termination of the process of aggregation); it is still possible, and consistent with the data, that the *number* of potential nucleating sites could be limited by protein availability.

It has been proposed that the *effective* size of an ice nucleation site may be only a portion of the aggregated InaZ protein (Warren and Wolber, 1987), owing to the need for coherent alignment of lattice-matching units. If this is accepted, a simple model can reconcile the observations on nucleation site size (Govindarajan and Lindow, 1988), distribution of ice nucleation protein within cells (Mueller et al., 1990), and the dose-dependency discussed in this section. It is proposed that after the rate-limiting step of di- or trimerization, all available Ina protein is incorporated

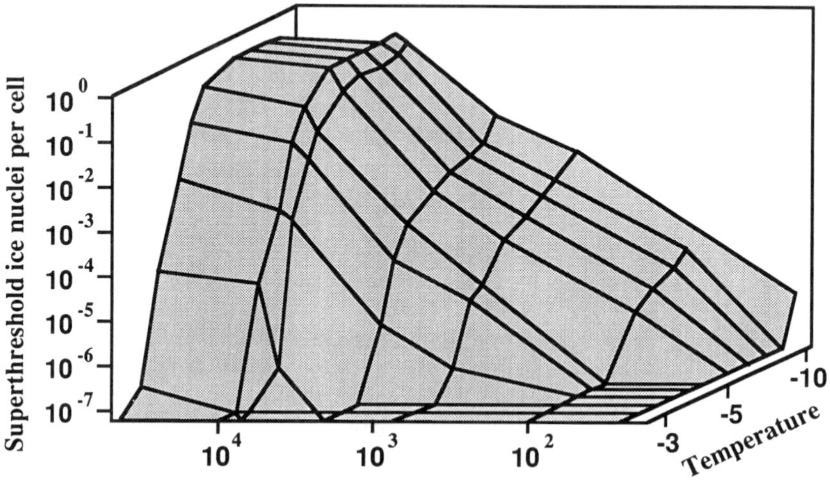


**Figure 4.** The relationship between ice nucleation frequency (y axis) and InaZ protein concentration (x axis) in *E. coli*, using logarithmic scales for each. The third dimension (z axis) represents the degree of supercooling at which activity was measured. Where nucleation was undetectable, limit points were plotted.

into larger aggregates; within these aggregates, the largest region able to coherently lattice-match to ice (the effective size of the nucleation site) is stochastically determined. The boundaries of the largest coherent unit might be set by stochastic events such as the choice between alternative orientations when the proteins aggregate. It is noteworthy that a simulation of stochastic disruption of a large template

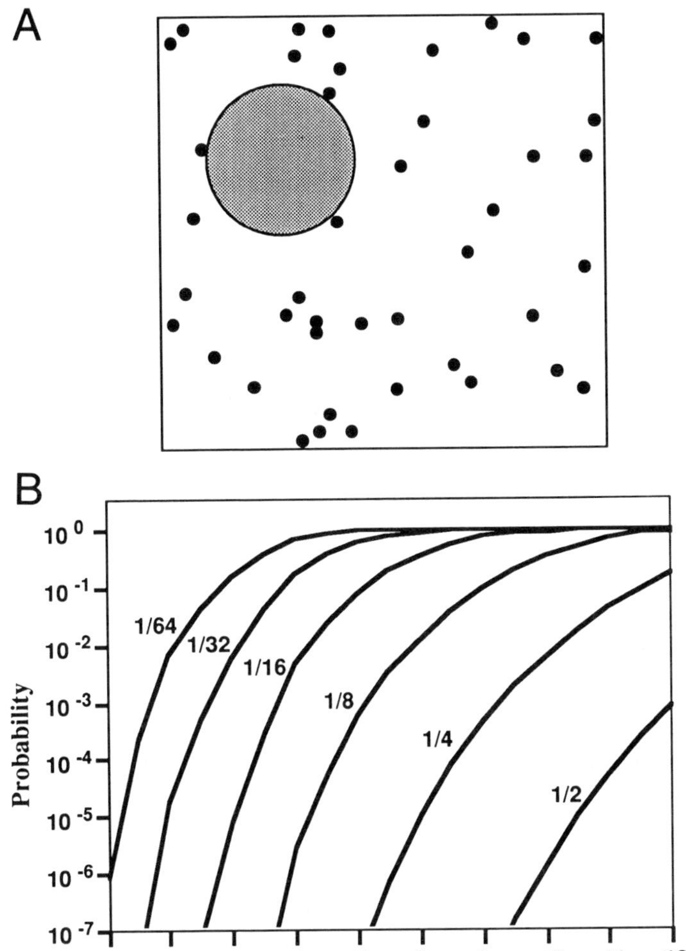


**Figure 5**. Modeling of the frequency distribution of nucleation thresholds, based on stochastic disruption of large templates. **A**, Assumptions of the model: the template is disrupted at random sites (represented by dots), and the effective nucleation site is the largest disk that does not include a disruption point (shaded circle). **B**, The frequency of occurrence (per 1,000 $nm^2$ of template) of effective sites large enough to nucleate at various supercooling thresholds. Curves are presented for six values of the frequency of template disruption, from one disruption per 64 $nm^2$ to one per 2 $nm^2$. For the frequency calculations, the disk areas of effective nucleation sites were derived from the radii given by Fletcher (1970), with the parameter *m* equal to 1; they vary from 22 $nm^2$ for a site with threshold at −12°C to 890 $nm^2$ for a −3°C threshold.

predicts a relationship between nucleation site frequency and temperature that resembles the real thing (Fig. 5).

The kinetics of assembly of ice nucleating sites have been studied by Watanabe et al. (1990), who used a *tac-inaZ* fusion to permit induction of InaZ protein synthesis. They found that the minimum time for appearance of new ice nucleating sites of all temperature thresholds was between 2.5 and 5 minutes, although the accumulation of Type 1 nucleating sites to their maximum level was slower than for Types 2 and 3. Proteolytic degradation products appeared with the same kinetics as InaZ protein itself, indicating that the onset of degradation is very rapid.

## Mutations in Coding Sequences

Gurian-Sherman et al. (1993) took a genetic approach toward understanding the differences between nucleating sites of Types 1, 2, and 3. After hydroxylamine mutagenesis of a cloned *iceE* gene in vitro, they isolated a range of *iceE* mutants. The majority of these showed a deficiency that was restricted to Type 1 nucleating sites, or was greater for Type 1 than for Type 3. (They named this phenotype "warm temperature ice nucleus deficient" or WIND.) A minority were completely deficient; mutations with a reverse-WIND phenotype (deficiency in Type 3 without deficiency in Type 1 nucleating sites), or with equal deficiency of all types of nucleating sites, occurred but were rare. This distribution of phenotypes is quite striking: it indicates that the formation of Type 1 nucleating sites is far more sensitive to substitutions of single amino acids (the results of point mutations induced by hydroxylamine) than is the formation of Type 3 sites. Thus it would not be tenable to suggest that the relatively infrequent formation of Type 1 sites is merely an accidental consequence of the production of Type 3 sites; selection for Type 3 alone would not prevent the accumulation of WIND mutations.

Twenty-two WIND mutations were mapped by fragment-substitution experiments, and two were sequenced. All mapped to the repetitive portion of *iceE*. The repetitive domain of an Ina protein is believed to function as a lattice-matching subunit (Green and Warren, 1985; Kajava and Lindow, 1993; see also Chapter 6). How could alteration of such a subunit affect the frequency of Type 1 nucleating sites but not the frequency of Type 3, when it is a component of both? It is interesting to interpret this in terms of the different possibilities for the origin of the distribution of temperature thresholds. If the distribution is governed by the extent of aggregation, then the WIND mutant proteins might be impaired in their ability to aggregate. On the other hand, if the distribution is governed by the frequency of template disruption, and the WIND mutant proteins make templates that are more prone to such disruption, the frequency of large effective nucleating sites will be reduced much more than that of small nucleating sites, without necessarily affecting the extent of protein aggregation. These alternatives should be distinguishable by experiment.

Among the mapped WIND mutations, an apparent gradient of severity was noted: the average deficiency of mutations within a region was worst for the region closest to the protein's C-terminus. However, deficiency of mutations in each region varied over a considerable range, and it is not clear if the observed trend is statistically significant.

When comparing the structures of the *ina* genes in Figure 1, it is apparent that periodicity is more strictly conserved than is the length of the repeated DNA. Green et al. (1988) generated deletions and duplications of repetitive DNA in order to compare the phenotypes of those that retained 48-residue periodicity and those that did not. Deletions disruptive to 48-residue periodicity had a WIND phenotype; possible explanations and the tests that could distinguish between them are the same as for the WIND point mutants discussed above. Mutations that retained 48-residue periodicity in Block 2 (ranging from a 3 × 48-codon deletion to an 8 × 48 duplication) had very little effect on the frequency of nucleation sites at any temperature threshold. This is entirely consistent with what was concluded about requirements for functionality based on the comparison of naturally occurring *ina* genes.

Deletions of nonrepetitive C-terminal sequences have deleterious effects on all types of nucleation sites—their phenotypes are quite distinct from those of the WIND mutants. The region that is essential for full activity corresponds closely to what can now be recognized as the consensus. Deletions extending further upstream have rapidly declining activities (Green et al., 1988), and the highly acidic region downstream of the sequence VEADIP could not be substituted by a highly acidic random peptide.

The nonrepetitive N-terminal sequences can be completely deleted with only partial loss of ice nucleation activity (Green et al., 1988). Type 1 nucleation sites are completely eliminated, and Type 2 sites almost so; Type 3 sites are depressed up to three orders of magnitude. One deletion (removing the first 145 N-terminal residues) retained enough Type 3 activity to qualify as a WIND mutant.

Transposons were at one time used as mutagens to define the extent of *ina* genes in cloned DNA. They revealed that insertion mutations within the repetitive region of an *ina* gene can appear leaky because they give rise to revertants at fairly high frequency. Some revertants were characterized (Corotto et al., 1986) and found to consist of deletions internal to the *ina* gene's repetitive sequences, lacking any remnant of transposon sequences. These were probably produced by recombination between repeats flanking the transposon.

## Concluding Comments

The results of mutagenesis are consistent with the implications of comparing sequences of the naturally occurring *ina* genes. The periodicity is crucial for function; the exact length of the periodic domain is not. The highly conserved nonrepetitive portions are required; the nonconserved portions are dispensible. However, the *products* of these genes do not easily yield their secrets. The Ina proteins' association with membranes, their strange dose/activity relationship, and their reluctance to retain activity during purification are bothersome attributes for the purposes of practical investigation. They are also enigmatic traits that will make the understanding of Ina proteins' function the more satisfying when it is achieved.

## Acknowledgment

I wish to thank David Pridmore for his generosity in providing the sequence of *inaV*.

## Literature Cited

Abe, K., Watabe, S., Emori, Y., Watanabe, M., and Arai, S. 1989. An ice nucleation active gene of *Erwinia ananas*: Sequence similarity to those of *Pseudomonas* species and regions required for ice nucleation activity. FEBS Lett. 258:297-300.

Arai, S., Abe, K., Watabe, S., Emori, Y., and Watanabe, M. 1989. Molecular cloning of an ice nucleation gene from *Erwinia ananas* and its expression in *Escherichia coli*. FEMS Microbiol. Lett. 61:53-56.

Baertlein, D. A., Lindow, S. E., Panopoulos, N. J., Lee, S. P., Mindrinos, M. N., and Chen, T. H. H. 1992. Expression of a bacterial ice nucleation gene in plants. Plant Physiol. 100:1730-1736.

Burke, M. J., and Lindow, S. E. 1990. Surface properties and size of the ice nucleation site in ice nucleation active bacteria: Theoretical considerations. Cryobiology 27:80-84.

Corotto, L. V., Wolber, P. K., and Warren, G. J. 1986. Ice nucleation activity of *Pseudomonas fluorescens*: Mutagenesis, complementation analysis and identification of a gene product. EMBO J. 5:231-236.

Deininger, C. A., Mueller, G. M., and Wolber, P. K. 1988. Immunological characterization of ice nucleation proteins from *Pseudomonas syringae*, *Pseudomonas fluorescens*, and *Erwinia herbicola*. J. Bacteriol. 170:669-675.

Eberhard, W. G. 1989. Why do bacterial plasmids carry some genes and not others? Plasmid 21:167-174.

Edwards, A. R., VanDenBussche, R.A., Wichman, H.A., and Orser, C. S. 1994. Unusual pattern of bacterial ice nucleation gene evolution. Mol. Biol. Evol. 11:911-920.

Fletcher, N. Y. 1958. Size effect in heterogeneous nucleation. J. Chem. Phys. 29:572-576.

Fletcher, N. H. 1970. The Chemical Physics of Ice. Cambridge University Press, Cambridge.

Govindarajan, A. G., and Lindow, S. E. 1988. Size of bacterial ice-nucleation sites measured in situ by radiation inactivation analysis. Proc. Natl. Acad. Sci. USA 85:1334-1338.

Green, R., and Warren, G. 1985. Physical and functional repetition in a bacterial ice nucleation gene. Nature 317:645-648.

Green, R. L., Corotto, L. V., and Warren, G. J. 1988. Deletion mutagenesis of the ice nucleation gene from *Pseudomonas syringae* S203. Mol. Gen. Genet. 215:165-172.

Gurian-Sherman, D., Lindow, S. E., and Panopoulos, N. J. 1993. Isolation and characterization of hydroxylamine-induced mutations in the *Erwinia herbicola* ice nucleation gene that selectively reduce warm temperature ice nucleation activity. Mol. Microbiol. 9:383-391.

Hasegawa, Y., Sakai, N., Yoshitome, H., Kawate, S., Obata, H., and Tokuyama, T. 1990. Cloning of bacterial ice nucleation genes of *Pseudomonas viridiflava* in *Escherichia coli*. J. Ferment. Bioeng. 70:143-146.

Hottinger, H., Niederberger, P., Pridmore, D., and Staeger-Roose, U. 1991. Additives for promoting ice nucleation in ingestible biological materials. European patent no. 0 424 771 A1.

Kajava, A., and Lindow, S. E. 1993. A model of the three-dimensional structure of ice nucleation proteins. J. Mol. Biol. 232:709-717.

Lindgren, P. B., Frederick, R., Govindarajan, A. G., Panopoulos, N. J., Staskawicz, B. J., and Lindow, S. E. 1989. An ice nucleation reporter gene system: Identification of inducible pathogenicity genes in *Pseudomonas syringae* pv. *phaseolicola*. EMBO J. 8:1291-1301.

Lindow, S. E. 1987. Competitive exclusion of epiphytic bacteria by Ice$^-$ mutants of *Pseudomonas syringae*. Appl. Environ. Microbiol. 53:2520-2527.

Lindow, S. E., Lahue, E., Govindarajan, A. G., Panopoulos, N. J., and Gies, D. 1989. Localization of ice nucleation activity and the *iceC* gene product in *Pseudomonas syringae* and *Escherichia coli*. Mol. Plant-Microbe Interact. 2:262-272.

Mueller, G. M., Wolber, P. K., and Warren, G. J. 1990. Clustering of ice nucleation protein correlates with ice nucleation activity. Cryobiology 27:416-422.

Orser, C., Staskawicz, B. J., Loper, J., Panopoulos, N. J., Dahlbeck, D., Lindow, S. E., and Schroth, M. N. 1983. Cloning of genes involved in bacterial ice nucleation and fluorescent pigment/siderophore production. Pages 353-361 in: Molecular Genetics of the Bacteria-Plant Interaction. A. Puhler, ed. Springer-Verlag, Berlin.

Orser, C., Staskawicz, B. J., Panopoulos, N. J., Dahlbeck, D., and Lindow, S. E. 1985. Cloning and expression of bacterial ice nucleation genes in *Escherichia coli*. J. Bacteriol. 164:359-366.

Ruggles, J. A., Nemecek-Marshall, M., and Fall, R. 1991. Antibody-induced shift of bacterial ice nucleation freezing spectra. Cryo Lett. 12:245-252.

Southworth, M. W., Wolber, P. K., and Warren, G. J. 1988. Nonlinear relationship between concentration and activity of a bacterial ice nucleation protein. J. Biol. Chem. 263:15211-15216.

Warren, G., and Corotto, L. 1989. The consensus sequence of ice nucleation proteins from *Erwinia herbicola, Pseudomonas fluorescens* and *Pseudomonas syringae*. Gene 85:239-242.

Warren, G., Corotto, L., and Wolber, P. 1986. Conserved repeats in diverged ice nucleation structural genes from two species of *Pseudomonas*. Nucl. Acids Res. 14:8047-8060.

Warren, G. J., Lindemann, J., Suslow, T. V., and Green, R. L. 1987. Ice nucleation deficient bacteria as frost protection agents. Pages 215-227 in: Applications of Biotechnology to Agricultural Chemistry. H. M. LeBaron, R. O. Mumma, R. C. Honeycutt, and J. H. Duesing, eds. American Chemical Society, Washington, DC.

Warren, G. J., and Wolber, P. K. 1987. Heterogeneous ice nucleation by bacteria. Cryo Lett. 8:204-215.

Watanabe, N. M., Southworth, M. W., Warren, G. J., and Wolber, P. K. 1990. Rates of assembly and degradation of bacterial ice nuclei. Mol. Microbiol. 4:1871-1880.

Watanabe, S., Abe, K., Hirata, A., Emori, Y., Watanabe, M. and Arai, S. 1993. Large-scale production and purification of an *Erwinia ananas* ice nucleation protein and evaluation of its ice nucleation activity. Biosci. Biotechnol. Biochem. 57:603-606.

Wolber, P. K. 1993. Bacterial ice nucleation. Adv. Microbiol. Physiol. 34:203-237.

Wolber, P. K., Deininger, C. A., Southworth, M. W., Vandekerckhove, J., Van Montagu, M., and Warren, G. J. 1986. Identification and purification of a bacterial ice-nucleation protein. Proc. Natl. Acad. Sci. USA 83:7256-7260.

Zhao, J., and Orser, C. S. 1990. Conserved repetition in the ice nucleation gene inaX from Xanthomonas campestris pv. translucens. Mol. Gen. Genet. 223:163-166.

CHAPTER 6

# Molecular Modeling of the Three-Dimensional Structure of Bacterial Ina Proteins

**Andrey V. Kajava**

A single protein located on the bacterial outer membrane is essential for the ice nucleation activity of each Ice[+] bacterial strain (Wolber et al., 1986; Warren et al., 1987; Lindow et al., 1989; see also Chapter 4). Genes encoding ice nucleation-active (Ina) proteins have been sequenced from five bacterial species, and all encode proteins of approximately 120 kDa with similar primary structures (Green and Warren, 1985; Warren et al., 1986; Abe et al., 1989; Warren and Corotto, 1989; Zhao and Orser, 1990; see also Chapter 5). However, it is not obvious from the amino acid sequences of these proteins how they interact with water. Knowledge of the three-dimensional (3-D) protein structure would probably provide insight into these interactions. To engineer the ice nucleation proteins for improved activity in practical applications, it will also be necessary to know their 3-D structure.

Direct experimental approaches, such as X-ray structural analysis or nuclear magnetic resonance (NMR) studies of Ina proteins, are problematic. X-ray determination is hampered because 1) ice nucleation proteins function and, consequently, have an appropriate structure only in the presence of membranes (Maki et al., 1974; Phelps et al., 1986; Wolber et al., 1986; Govindarajan and Lindow, 1988a), and it is difficult to crystallize such large membrane-associated proteins; and 2) as a rule, direct X-ray determination is applied easily only to globular proteins, whereas, judging from amino acid sequence, much of a typical bacterial Ina protein is likely to be nonglobular. NMR studies also hold little promise of success, because it is difficult to obtain enough protein for NMR, and the proteins are too large for interpretation of NMR spectra. These constraints increase the interest of theoretical approaches for determining the 3-D structure of Ina proteins.

## Principles Underlying Modeling of the 3-D Structure of Ina Proteins

No generally successful method has yet been found to predict the tertiary structures of proteins. However, correct prediction of the 3-D structure of the ice nucleation proteins is plausible because of two factors.

First, the primary structures of these proteins are regular. The presumptive translation products of five ice nucleation genes each consist mostly (approximately 80%) of an internally repetitive domain (Warren and Corotto, 1989; see also Chapter 5). It seems plausible to infer that the repetitive portion of the protein forms that part of the template structure which contacts ice, since the regular alignment of identical molecules requires structural repetition. The regular portion of the primary structure repeats at highest fidelity with a period of 48 residues. Each 48-residue unit can be further subdivided into three 16-residue medium-fidelity repeats and, finally, into six 8-residue low-fidelity repeats. This regularity imposes a major constraint on the structural modeling of the ice nucleation proteins. It follows from known 3-D structures of proteins that regular primary structure, as a rule, corresponds in space to a regular fibrillar structure (Fraser and MacRae, 1973). So the 3-D structure of the ice nucleation proteins should be regular. A model that can incorporate the necessity of all three orders of repetition (8-, 16-, and 48-residue) would be preferable.

The second factor that makes theoretical predictions plausible derives from the *function* of an Ina protein. General considerations and theoretical calculations (Fletcher, 1970) indicate that a protein structure which promotes ice formation

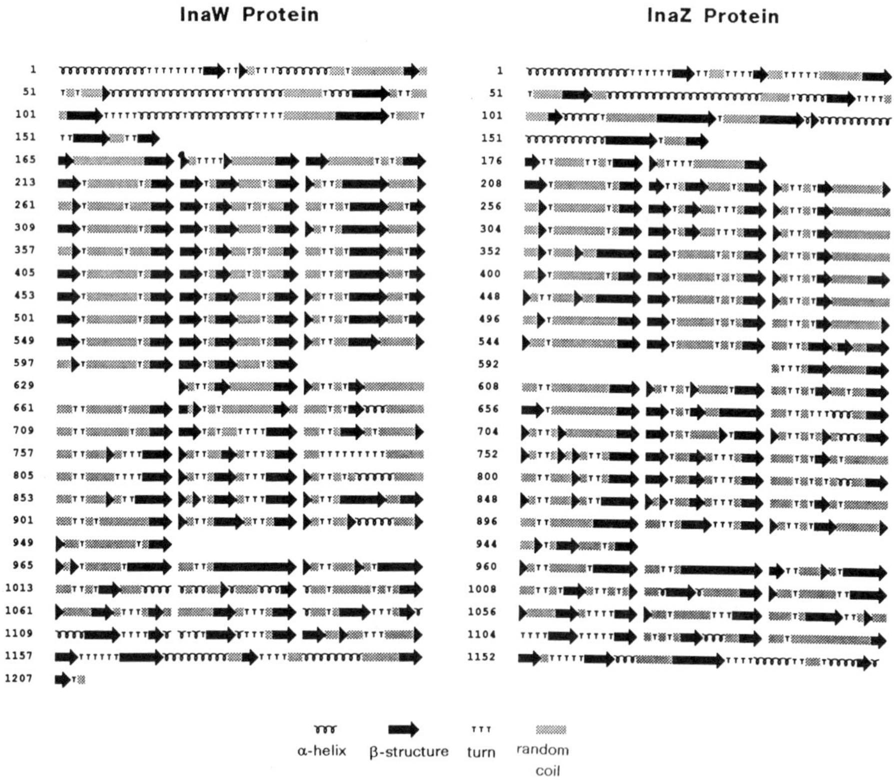


**Figure 1.** Predicted secondary structures of the InaW and InaZ proteins. The layout of the predictions is isometric with that of the amino acid sequences (see Chapter 5). Reprinted with permission from Warren et al. (1986).

must order water molecules on its surface in an ice-like manner. The arrangement of donor-acceptors of hydrogen bonds in such a protein must be complementary to the arrangement of donor-acceptors on ice. However, this is not sufficient for effective ice nucleation, since there exist protein structures (the "antifreezes") with ice-like sites consisting of three to five compactly arranged donor-acceptors of hydrogen bonds that do not initiate but rather inhibit ice formation (Yang et al., 1988). Other physical conditions must be satisfied for a protein molecule to act as an ice nucleator: the ice-like site of the molecule must be larger than 100 Å (10 nm), the radius known roughly as the size of the critical nucleus (Fletcher, 1970). Also, the more rigid the ice-like protein structure, the more effective it is expected to be in nucleating ice.

Even in the absence of experimental data on Ina protein structure, these two constraints (following from regularity and function) severely restrict the range of plausible models and hence facilitate molecular modeling. Moreover, several common principles of 3-D protein structures also constrain the modeling of protein structures: 1) the absence of significant covalent and steric tensions, 2) close packing, and 3) involvement of all donor-acceptors of hydrogen bonds in bonding with each other or with water molecules. This so called "prohibition on dehydration" was drawn from the observation that there are practically no unpaired donor-acceptors in the nonpolar interiors of known protein structures. For examination of the hydrogen bonding, the following criterion could be used: the A–H· · · B bond is formed when the distance AB is between 2.6 and 3.2 Å and the angle HAB is less than 30° (Venkachatalam, 1968).

## Secondary Structure Prediction

In nearly all proteins, the local folding of the chain leads to the formation of α-helices or β-sheets, and these assemble to give the molecules their 3-D structures. It can be assumed that the ice nucleation protein is also an aggregate of secondary structure elements. When the algorithm of Garnier et al. (1978) to predict protein secondary structure from the amino acid sequence was applied to the ice nucleation proteins (Warren et al., 1986), their nonrepetitive N- and C-terminal domains were predicted to consist of both α-helices and β-strands typical of globular protein arrangement (Fig. 1). The correct prediction of the tertiary structure of these domains is rather difficult due to the irregular amino acid sequences. It will be more appropriate to obtain direct experimental evidence of their 3-D structures. In contrast, the repetitive portions were predicted to consist of alternating β-strands and random-coil regions (Fig. 1). Other algorithms (Chou and Fasman, 1974; Ptitsyn and Finkelstein, 1983) display similar results. It should be noted, however, that the algorithms for predicting secondary structure were worked out with globular proteins; they should be applied with caution to large, repetitive amino acid sequences that probably assume a nonglobular structure. Regular primary structures require a special stereochemical analysis of their packing in all possible regular conformations (α-helix, β-structure, and chains in a polyproline conformation). Stereochemical analysis has revealed that the ice nucleation proteins are very unlikely to self-aggregate into a complex consisting of α-helices because of the absence of the periodicity $(H–X–X–H–X–X–X)_n$ (where H is a nonpolar residue and X is any hydrophilic residue) which is typical for the α-helical fibrillar proteins (Fraser and MacRae, 1973). It was also impossible to unite polypeptide chains to form a polyproline helix because of the absence of collagen-like periodicity $(Gly–X–X)_n$ in the sequence. The β conformation was found to be the only regular conformation of the ice nucleation proteins that allowed the assembly of different repetitive 3-D structures, since the β-strands can interact with each other by interpeptide hydrogen bonds, largely independent of the amino acid sequence.

Thus, on the basis of all these predictions, one can conclude that the repetitive portion of the ice nucleation proteins most probably consists of β-strands.

## Models of the 3-D Structure of Ina Proteins

Three attempts have been made to predict the 3-D structure of the ice nucleation proteins (Warren et al., 1986; Mizuno, 1989; Kajava and Lindow, 1993).

Warren et al. (1986) proposed two structural models that have triangular and hexagonal shapes (Fig. 2). One of the models (Fig. 2B) represents an up-and-down β-sheet consisting of 3 × 8 residue β-strands. This β-structure is much longer in the hydrogen bond direction than in the β-strand one. The β-sheet is folded in a triangular prism. The other model (Fig. 2C) is formed by right- and left-handed helical folds united by the antiparallel β-structure interactions into a hexagonal double helix. These arrangements of the protein were suggested by the authors because 1) they are consistent with the results of secondary structure prediction, which suggest that mainly β-strands appear in this polypeptide; 2) it was possible to incorporate the three levels of structural repetition into the models; and 3) protein folding of the models resembles the space group P6-3/mmc of ice Ih (defined below).

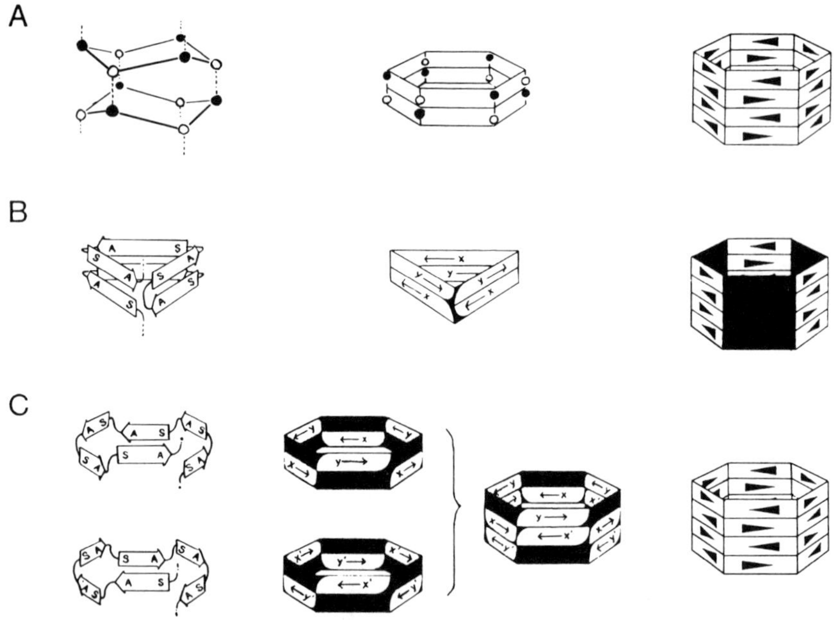


**Figure 2.** Structural models suggested by Warren et al. (1986) in which the ice nucleation protein displays a symmetry related to that of ice (not to scale). **A,** Ice; **B,** 48-residue unit of triangular model; **C,** 48-residue units of both chains of antiparallel double-helix model. Their individual structures, the symmetries of these structures, and their extended symmetries are shown at left, center, and right. Reprinted with permission from Warren et al. (1986).

The model, however, was presented only schematically, without any details at the molecular level. The construction of their molecular structures reveals some discrepancies between the models and principles of protein structures. For example, the β-structures of the models are flat, whereas β-structures of proteins always form a right-handed turn (Chothia, 1973). The triangular and hexagonal prisms, both constructed from the twisted β-strands, must be left-handed along the prism axis, and this must distort their symmetry relative to that of ice. Another disadvantage of these arrangements is the lack of close packing of the protein structures inside the prisms. There are no such "open" forms among experimentally determined protein structures.

The first model developed at the atomic level was suggested by Mizuno (1989) (Fig. 3). The model was deduced from the assumption that the 3-D structure of the ice nucleation proteins must be a helix, with each octapeptide repeat assuming an identical conformation. (The model neglects the two higher orders of periodicity [16 and 48 residues] that are found in the Ina proteins). This assumption facilitated the generation of modeled structures and their energy calculation as well as the search for ice-like regions on the protein surface.

A detailed analysis of ice-like sites requires determination of the ice crystals that should be taken for the comparison with the protein templates. The most favorable ice crystals, with an optimal tetrahedral orientation of hydrogen bonds of water molecules, can exist in two forms: ice Ih (hexagonal) and ice Ic (cubic). Ice Ih is the most common structure. At the same time ice Ic could act as an intermediate in the nucleation of ice Ih, because 1) ice Ic can be irreversibly transformed into ice Ih when freezing occurs near 0°C (Eisenberg and Kauzmann, 1969); and 2) ice Ih and

**Figure 3.** Skeletal axial projection of the turn of a helix suggested by Mizuno (1989) for the ice nucleation proteins. The helix twist and height per octapeptide are 60.6° and 1.41 Å, respectively. The inside surface of the helix consists of nonpolar leucine side chains. Contours around the atoms of the inside surface demonstrate van der Waals volumes. A water molecule (W) in the center of the nonpolar cavity shows its inability to form all necessary hydrogen bonds.

ice Ic have a similar surface, and ice Ic formed during nucleation could permit the subsequent growth of ice Ih. An X-ray diffraction study of ice structures formed in proteinaceous gels showed that either hexagonal or cubic ice I could become the predominant phase, depending on cooling conditions (Dowell et al., 1962). Therefore, planes both of ice Ih and ice Ic, densely populated by hydrogen-bonding sites, were considered.

As a result of Mizuno's calculations, two helices formed by repetition of identical octapeptide conformations were selected. The helices have internal and external surfaces approximated by a hexagonal shape. One of the helices (Fig. 3) has an epitaxial fit to ice Ih, the other, a fit to ice Ic. Although, according to Mizuno (1989), the external surface of these two helices have ice-like sites, the protein structures themselves are unusual (irregular conformations of the octapeptide segments, intersegmental interactions, and the helical architecture as a whole). As a rule, a calculated minimum energy structure is highly dependent on initial conformations and input constraints. The condition that the ice protein chain form a helix has led to structures with a hexagonal shape, similar to the models of Warren et al. (1986). Thus, Mizuno's models have a similar disadvantage concerning the cavities inside the structure. A stereochemical analysis reveals the inability of water molecules (Fig. 3) inside these nonpolar cavities to form all the necessary hydrogen bonds; i.e., some hydrogen bonding groups would remain dehydrated.

These shortcomings of previous models encouraged Kajava and Lindow (1993) to undertake another attempt to predict the 3-D structure of ice nucleation proteins. The goal of the new model was to demonstrate an ice-like site consisting of β-strands—the theoretically predicted elements of an Ina protein's secondary structure. Figure 4 shows that the antiparallel β-strands, oriented in a such a manner that they contact each other via side chains, have donors and acceptors in an arrange-

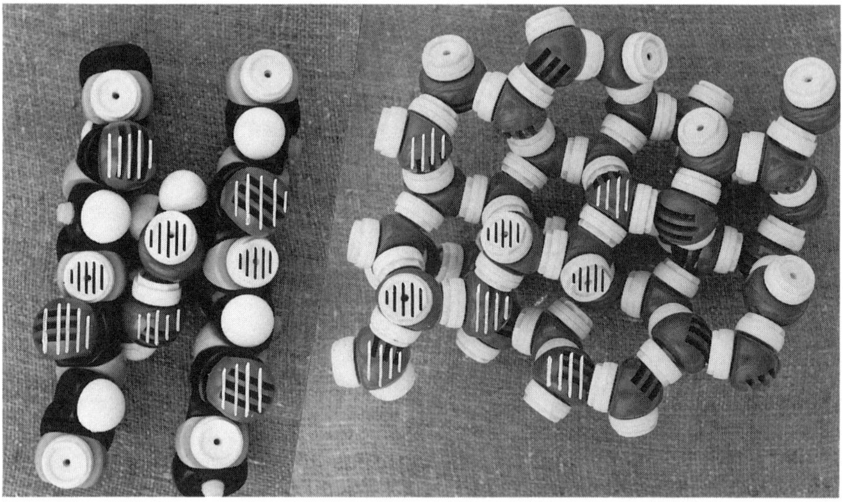

**Figure 4.** Space-filling models of two antiparallel β-strands having an ice-like arrangement of the donor-acceptors of hydrogen bonds (left) and a fragment of ice Ic (right). The β-strands contact each other by serine and threonine side chains. Donor-acceptors of the β-strands and of the ice, which form similar patterns, are hatched. The ice-like pattern can be extended by its repetitions along the chain and in the direction perpendicular to the chains.

ment similar to that in ice Ic. To create this ice-like pattern, the β-strands should have particular side chains, namely, side chains of serine and threonine (in some cases also aspartic and glutamic acids, asparagine and glutamine). The amino acid sequences of ice nucleation proteins have anomalously large contents of serine and threonine residues, consistent with the possibility of their adopting the modeled structure.

Single chains in β-conformations are highly flexible. To fix their conformation, the β-strands should participate in the formation of interpeptide hydrogen bonds (forming a β-structure). The repetitive domain of an ice nucleation protein, however, has an extremely low proportion of hydrophobic residues (about 30%), and a stereochemical analysis indicates that a polypeptide chain with such a distribution of residues cannot be involved in a structure consisting of extensive (each more than two β-strands) β-layers (Fig. 5, left) because of the dehydration of hydrophilic residues in the zone of interlayer contact. In this case, a broad, rigid ice-like template, which at the same time satisfies principles of protein structure modeling, can be constructed from two-stranded β-structures stacked with each other (Fig. 5, right).

Another constraint for modeling was that the β-strands that take part in this arrangement belong to the same polypeptide chain. Since adjacent β-strands along the chain are most likely to interact with each other, inferred from the 3-D structure of other proteins, the most probable two-stranded β-structure is a β-hairpin. The arrangement of stacked β-hairpins shown in Figure 6 was chosen as the most probable for the following reasons:

1) This folding pattern corresponds to the 8-, 16-, and 48-residue periodicities of the primary structure.
2) Conservative glycine residues, which are the best residues for bending of the chain, are located in every turn of the suggested structure. Highly conservative serine and threonine residues occur only in the middle of these β-strands and can effectively participate in the ice-like template as in Figure 4.
3) There is no glycine in the "glycine position" in every sixth octapeptide; this octapeptide thus does not produce a sharp bend.

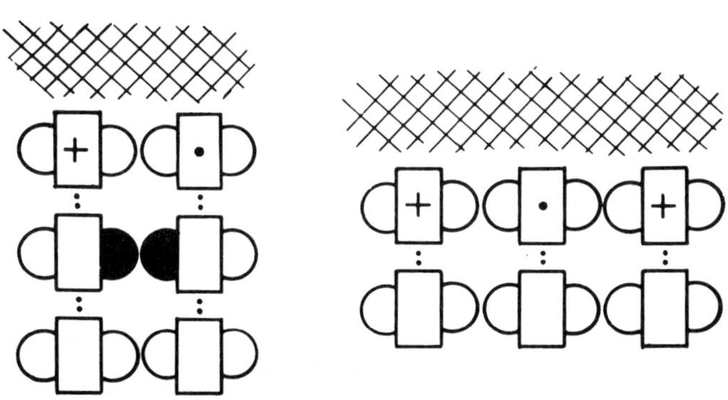

**Figure 5.** Schematic representation of possible β-layer arrangements viewed along the chains. Hatched regions denote ordered water near ice-like template. Crosses and points on the chains show opposite β-strand orientations needed for formation of the ice-like surface. Black half-circles mean side chains which are dehydrated between β-layers. Reprinted with permission from Kajava and Lindow (1993).

4) The lengths of the β-strands in this model coincide with the average lengths of β-strands (six to eight residues) which occur in known protein structures (Sternberg and Thornton, 1977).
5) Analysis of the primary structures reveals that the 48-residue units have four highly conservative octapeptides and two octapeptides with a low conservation placed along the chain as shown in Figure 6. One can expect that repeats with the greatest fidelity will participate in the formation of the functionally important ice-like surface. The suggested folding pattern, unlike all others, has only the highly conserved residues on the suggested ice-like surface (Fig. 6). The other side of the units might interact via hydrogen and ionic bonds with the polar heads of membrane lipids which, in turn, could induce and support such a flat protein structure (see "Influence of the Bacterial Membrane on Protein Structure" later in this chapter).

Inspection of known protein structures reveals that some, for example, type 1 module of fibronectin (Baron et al., 1990) and N-terminal domain of CD4 (Ryu et

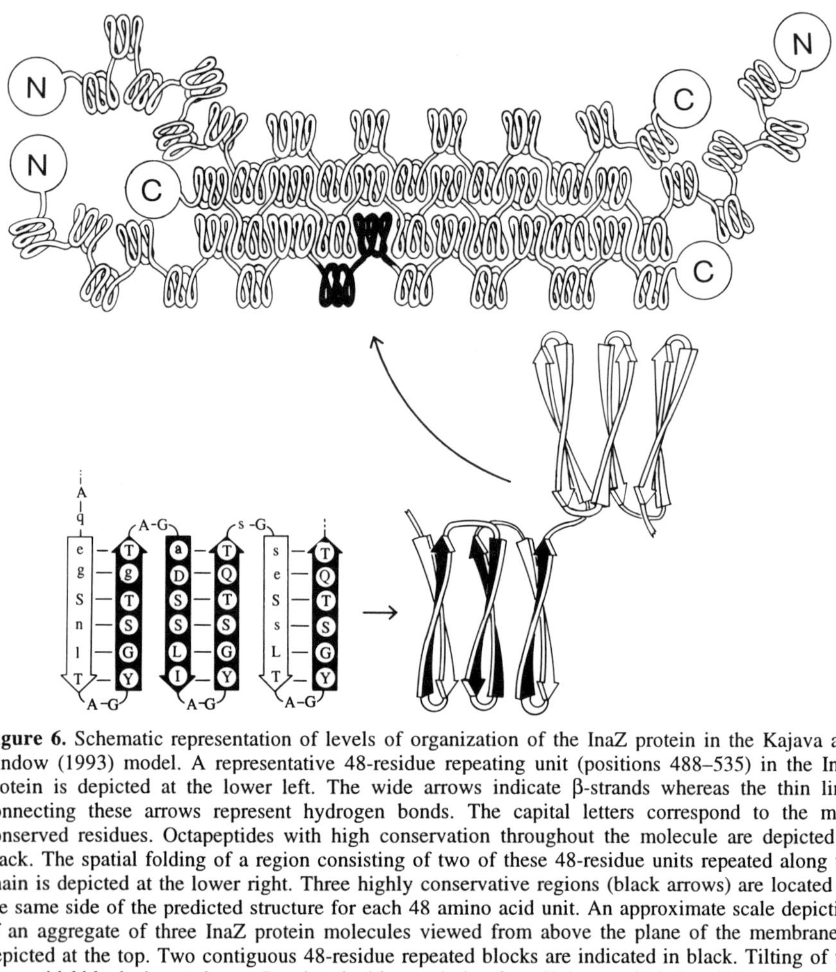


**Figure 6.** Schematic representation of levels of organization of the InaZ protein in the Kajava and Lindow (1993) model. A representative 48-residue repeating unit (positions 488–535) in the InaZ protein is depicted at the lower left. The wide arrows indicate β-strands whereas the thin lines connecting these arrows represent hydrogen bonds. The capital letters correspond to the most conserved residues. Octapeptides with high conservation throughout the molecule are depicted in black. The spatial folding of a region consisting of two of these 48-residue units repeated along the chain is depicted at the lower right. Three highly conservative regions (black arrows) are located on the same side of the predicted structure for each 48 amino acid unit. An approximate scale depiction of an aggregate of three InaZ protein molecules viewed from above the plane of the membrane is depicted at the top. Two contiguous 48-residue repeated blocks are indicated in black. Tilting of the trapezoidal blocks is not shown. Reprinted with permission from Kajava and Lindow (1993).

al., 1990), have stacked β-hairpins similar to those proposed above.

The folding pattern of a 48-residue unit has a mirror topology if compared with a neighboring unit along the chain (Fig. 6). Modeling of the molecular structure revealed that the conformation of β-hairpins in adjacent units could be the same and similar to that in other known proteins; i.e., the β-strands in a slightly right-handed twist and their β-turns constructed with $\alpha_L\alpha_L$- or $\beta\alpha_L$-conformations

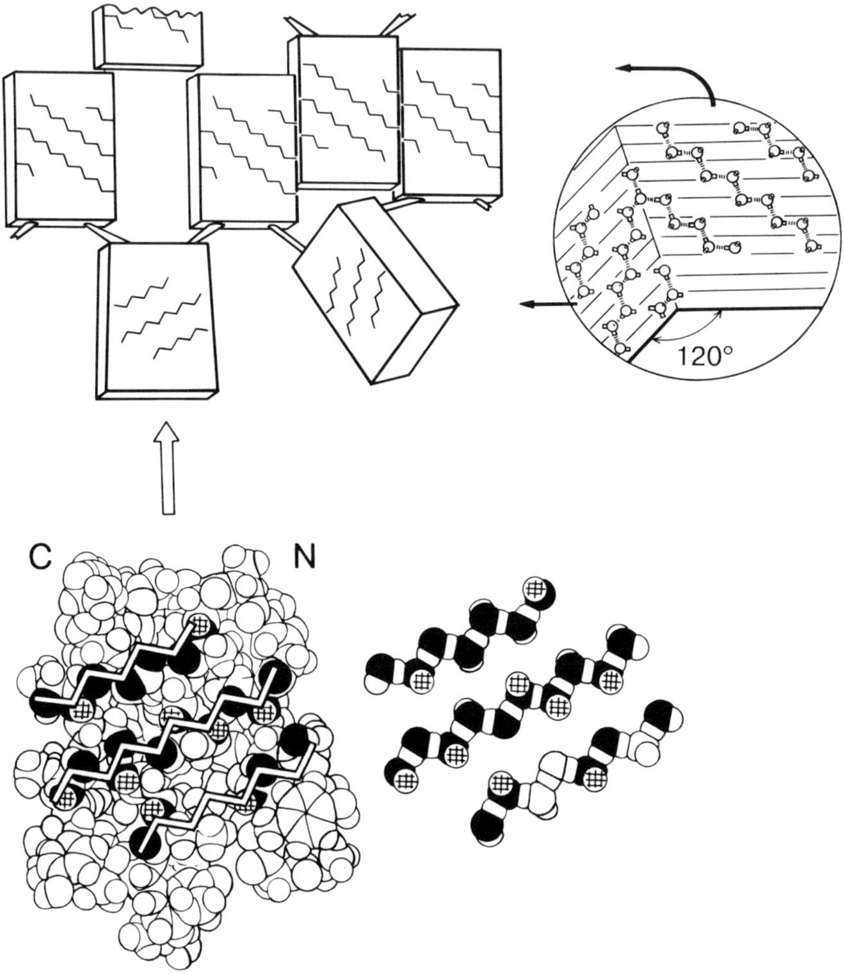


**Figure 7.** Space-filling models of a 48-amino acid portion of the proposed three-dimensional structure of ice nucleation proteins which has a trapezoidal structure (bottom left) and the (110) surface of ice Ic (bottom right). Hatched atoms and those depicted in black in both the Ina protein and in ice Ic represent corresponding donor-acceptors of hydrogen bonds. The zigzagged lines superimposed on the protein structure (bottom left) represent chains of water molecules in the lattice of ice Ic to facilitate viewing. The top of the figure shows tilting of the 48-amino acid trapezoidal blocks of an ice protein upon interdigitation of the rectangular 48 amino acid blocks from two different ice protein molecules when viewed from above the plane of the membrane. The zigzag lines on the protein surface (top left) show the orientation of the ice-like pattern. An enlarged representation of the two ice crystal planes corresponding to the two ice-like protein planes comprised of rectangular and trapezoidal blocks is depicted in the circle.

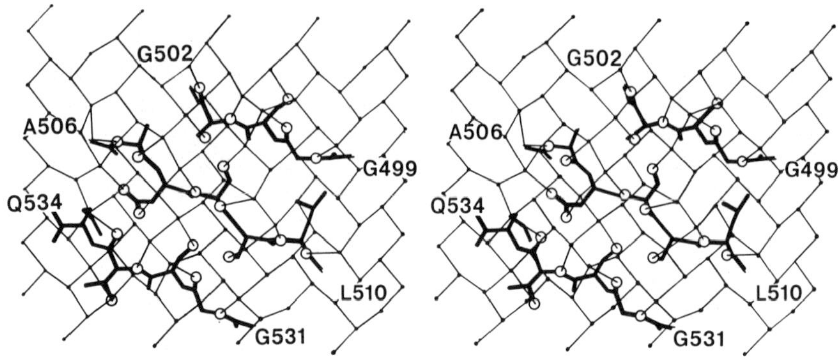


**Figure 8.** Stereo view of ice crystal over ice-like site of InaZ protein. Thin lines represent hydrogen bonds of the ice crystal; thick lines represent polypeptide chains. Black dots on the hydrogen bond network represent water molecules. Open circles denote protein donor-acceptors of hydrogen bonds, which form a pattern similar to that on the (110) plane of ice Ic. Letters and numbers denote identity and position, respectively, of the residues within InaZ protein. Reprinted with permission from Kajava and Lindow (1993).

(Efimov, 1986; Wilmot and Thornton, 1988) ($\alpha$, $\alpha_L$, $\beta$ mean conformations of a residue from a corresponding region of the Ramachandran plot). The structures of the units are closely packed and have no forbidden van der Waals contacts or dehydrated donor-acceptors of hydrogen bonds. Both adjacent units use the NH and CO groups of the backbone and serine, threonine, and aspartic acid side chains to organize the ice-like template (Fig. 7). A lattice-match of the ice-like protein surface with the (110) face of ice Ic is represented in Figure 8. A good fit of the van der Waals surfaces of the protein and the ice crystal should additionally decrease the interfacial energy.

The differences of the polypeptide conformation of the folds from quasi-mirror symmetry are localized in the crossovers between one group of β-hairpins and the next. Because of this difference, type 1 units have a rectangular shape and type 2 units have a trapezoidal shape (Figs. 6 and 7). The structural model with the two quasi-mirror-symmetrical units alternating along the chain leads to the expectation of an additional 96-residue periodicity. However, an analysis of the primary structure of ice nucleation proteins does not support this consequence of the model.

If it is assumed that the N- and C-terminal domains of ice nucleation proteins are globular and that the central portion consists of "rectangular" and "trapezoidal" structures alternating with each other, the whole structure of the InaZ protein would appear schematically as in Figure 6. There is an interruption of the 48-residue regularity in the InaZ protein, where the sequence contains 16 additional residues. In this case, an ice-nucleation block might have an additional (fourth) β-hairpin (Fig. 6). It is worth noting that InaW protein has a similar +16 interruption, IceE and InaX proteins have no such interruption, and the 48-residue regularity of InaA protein is interrupted three times.

The rectangular blocks are oriented to one side and the trapezoidal blocks to the other side relative to the central axis of the regular domain.

## Aggregation of the Ina Proteins

Ice nucleation activity in a population of bacterial cells occurs at temperatures from −12°C to −2°C, corresponding to nucleant masses that range from about 1 to

50 ice proteins, respectively (Govindarajan and Lindow, 1988b). Ina proteins must therefore be able to aggregate in an unlimited fashion with additive effects. Little is known about the general shape of the aggregate. The general shape for the aggregate was estimated by calculations of the dependence of nucleation temperature/molecular mass for various shapes of the nucleator using the heterogeneous ice nucleation theory (Burke and Lindow, 1990). The best fit between the measured and calculated data was obtained for a flat, disk-like nucleator with a diameter depending on the number of molecules in the aggregate. This hypothetical nucleator generates ice embryos only on its upper surface and would appear as a hydrated patch on the membrane surface. Patches (1,000–2,000 Å in diameter) that *might* represent nucleation sites have been observed on negatively stained Ice$^+$ bacteria using an electron microscope (Welch and Speidel, 1989).

The mechanism of the adjustment of separate molecules into an aggregate with a large continuous ice-like site is not evident in the prism models (Warren et al., 1986; Mizuno, 1989). Kajava and Lindow (1993) proposed that Ina proteins form a flat aggregate on the surface of the membrane by virtue of interdigitation of 48-residue units (Fig. 6). Their modeling has revealed that a rectangular block can be

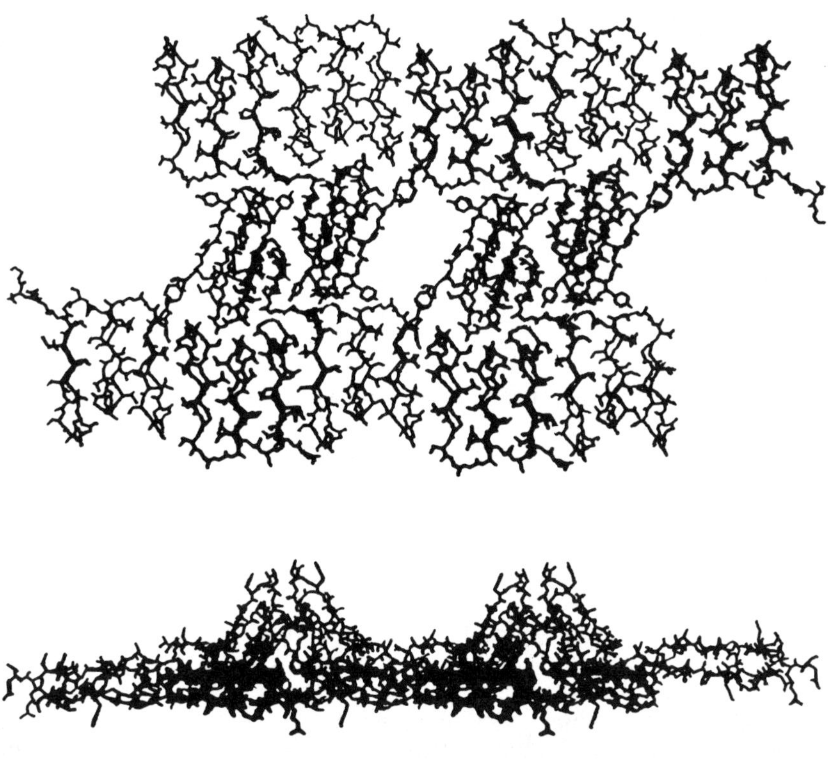

**Figure 9.** Two projections of Kajava and Lindow's skeletal model of fragments of four InaZ proteins when viewed from above the plane of the membrane (top) and along the plane of the membrane (bottom). The structure, consisting of 14 48-residue units, reflects its refinement after energy minimization. One interdigitated band (central band on the upper projection and protruding from the flat structure on the lower projection) is composed of tilted trapezoidal blocks while other bands are composed of rectangular blocks.

docked well between other rectangular blocks and cannot be close-packed between trapezoidal blocks. Interdigitated rectangular blocks in the zone of intermolecular contacts have the same arrangement of the conservative serine, threonine, and glutamine residues as that within the 48-residue units, indicating that rectangular ice-like templates can merge into a continuous ice-like band. Because of a preference for "rectangle-rectangle" packing, the proteins interact in an antiparallel orientation; their specific staggering is dictated by the distribution of 48- and 64-residue units (Fig. 6). Modeling has demonstrated that an interdigitated arrangement of trapezoidal blocks cannot lie in the same plane because of structural constraints. Molecular modeling has also revealed that the space between adjacent rectangular blocks along the chain is inadequate for insertion of the rectangular block from another molecule when both rectangular and trapezoidal blocks lie in the same plane. Broadening of this interblock space occurs when trapezoid units are tilted relative to the plane of rectangle blocks (Figs. 7 and 9). Kajava and Lindow (1993) therefore proposed a quaternary structure of ice proteins with alternating flat "rectangular" bands and wavy "trapezoidal" bands. The plane of the trapezoidal blocks is tilted 120° with respect to the plane of the rectangular blocks (Figs. 7 and 9). This arrangement permits the efficient docking of different molecules and maintains an ice-like surface that is common to one ice crystal over the whole aggregate even though each band has a different orientation of ice-like templates. The aggregate would have nucleation sites more than 250 Å in length and 25*n* Å in width (where *n* is the number of molecules in the aggregate). This is consistent with the sizes of nucleation sites as predicted by Fletcher (1970).

The proposed quaternary structure with an ice-like surface that is common to one ice crystal can be constructed only from the units consisting of three β-hairpins because the tilting of the trapezoid blocks depends on the number of β-hairpins in the unit. Moreover, units with an even number of β-hairpins have such an arrangement of interunit connections that broadening of the space between rectangular units does not lead to the tilting of the trapezoidal blocks. Thus, within the frame of this model, the threefold periodicity in ice nucleation proteins is related to the continuity of the ice-like template and is unrelated to the space groups of ice I as it was in the previous models (Warren et al., 1986; Mizuno, 1989).

## Influence of the Bacterial Membrane on Protein Structure

Experimental data (Lindow et al., 1989), and the known function and hydrophilicity of the ice nucleation protein, imply that it is located on the surface of the bacterial membrane rather than buried in this membrane. It is not clear how the flat membrane surface might facilitate formation of the structure with a three- or six-fold symmetry (Warren et al., 1986; Mizuno, 1989). At the same time, existence of the flat quaternary structure of ice nucleation proteins (Kajava and Lindow, 1993) cannot be imagined without support by the membrane. This structure suggests possible pathways for folding of the molecule. Judging from their primary structures, the aperiodic N- and C-terminal domains might self-assemble in solution into globules and then participate in anchoring to membranes and the aggregation of Ina protein molecules. The repetitive portion of the protein, because of a lack of nonpolar residues, cannot form a hydrophobic core and hence probably has no fixed 3-D structure in solution. Although small toxin molecules also consist of three β-hairpins (e.g., Almassy et al., 1983) and have ratios of hydrophilic to hydrophobic residues similar to those of the repetitive domains of ice nucleation proteins, they maintain their native structure in solution by means of four to five disulfide bridges (which are absent in ice nucleation proteins). The repetitive part of ice nucleation proteins therefore needs assistance to obtain a fixed structure, and the surface of the bacterial outer membrane may provide such support. The flat structure described above appears as a result of self-association of the repetitive part of Ina proteins on the membrane. The 48-residue block nearest the N- or C-terminal domain may fold first because of stabilization from the neighboring globule. The appearance of the first block might determine the order of alternation of rectangular and trapezoidal blocks along the molecule. The preferred interdigitation of rectangular units suggests that the association of two antiparallel molecules of InaZ protein is the first step in aggregate assembly. This bimolecular interaction provides a possible basis for the approximately second-order dependence of nucleation activity on protein concentration (Southworth et al., 1988; see also Chapter 5). In the final step of self-assembly, interdigitation of rectangular blocks may lead to a tilting of the trapezoidal blocks, thus permitting further interdigitation of the "trapezoidal" sides.

Thus, due to the significant constraints on molecular modeling it was possible to suggest the detailed structural models for the bacterial ice nucleation proteins. It should be mentioned that the results of deletions, insertions (Green and Warren, 1985; Green et al., 1988) and single substitutions of amino acids (Gurian-Sherman et al., 1993) in the repetitive part of the protein were also taken into consideration during the modeling. However plausible a structural model may be, it needs additional experimental support. The growing biological and commercial importance of ice nucleation proteins hopefully will lead to the appearance of new experimental evidence which will support, refute, or demand modification of the suggested models.

## Literature Cited


Abe, K., Watabe, S., Emori, Y., Watanabe, Y., and Arai, S. 1989. An ice nucleation active gene of *Erwinia ananas*: Sequence similarity to those of *Pseudomonas* species and regions required for ice nucleation activity. FEBS Lett. 258: 297-300.

Almassy, R. J., Fontecilla-Camps, C. J., Suddath, F. L., and Bugg, C. E. 1983. Structure of variant-3 scorpion neurotoxin from *Centruroides sculpturarus* Ewing, refined at 1.8 Å resolution. J. Mol. Biol. 170:497-527.

Baron, M., Norman, D., Willis, A., and Campbell, I. D. 1990. Structure of the fibronectin type 1 module. Nature 345:642-646.

Burke, M. J., and Lindow, S. E. 1990. Surface properties and size of the ice nucleation site in ice nucleation active bacteria: Theoretical considerations. Cryobiology 27:80-84.

Chou, P. Y., and Fasman, G. D. 1974. Prediction of protein conformation. Biochemistry 13:222-245.

Chothia, C. 1973. Conformation of twisted β-pleated sheets in proteins. J. Mol. Biol. 75:295-302.

Dowell, L. G., Moline, S. W., and Rinfret, A. P. 1962. A low-temperature X-ray diffraction study of ice structures formed in aqueous gelatin gels. Biochem. Biophys. Acta 59:158-167.

Efimov, A. V. 1986. Standard conformations of a polypeptide chain in irregular regions of proteins. Mol. Biol. (USSR) 20:208-216.

Eisenberg, D., and Kauzmann, W. 1969. The Structure and Properties of Water. Oxford University Press, New York.

Fletcher, N. H. 1970. Pages 97-103 in: The Chemical Physics of Ice. Cambridge University Press, Cambridge.

Fraser R. D., and MacRae, T. P. 1973. Conformation in Fibrous Proteins and Related Synthetic Polypeptides. Academic Press, New York.

Garnier, J., Osguthorpe, D. J., and Robson, B. 1978. Analysis of accuracy and implications of simple methods for predicting secondary structure of globular proteins. J. Mol. Biol. 120:97-120.

Govindarajan, A. G., and Lindow, S. E. 1988a. Phospholipid requirement for expression of ice nuclei in *Pseudomonas syringae*. J. Biol. Chem. 263:9333-9338.

Govindarajan, A. G., and Lindow, S. E. 1988b. Size of bacterial ice-nucleation sites measured in situ by radiation inactivation analysis. Proc. Natl. Acad. Sci. USA 85:1334-1338.

Green, R. L., Corotto, L. V, and Warren, G. J. 1988. Deletion mutagenesis of the ice nucleation gene from *Pseudomonas syringae* S203. Mol. Gen. Genet. 215:165-172.

Green, R. L., and Warren, G. J. 1985. Physical and functional repetition in a bacterial ice nucleation gene. Nature 317:645-648.

Gurian-Sherman, D., Panopoulos, N. J., and Lindow, S. E. 1993. Isolation and characterization of hydroxylamine-induced mutations in the *Erwinia herbicola* ice nucleation gene that selectively reduce warm temperature ice nucleation activity. Mol. Microbiol. 9:383-391.

Kajava, A. V., and Lindow, S. E. 1993. A model of the three-dimensional structure of ice nucleation proteins. J. Mol. Biol. 232:709-717.

Lindow, S. E., Lahue, E., Govindarajan, A. G., Panopoulos, N. J., and Gies, D. 1989. Localization of ice nucleation activity and the *iceC* gene product in *Pseudomonas syringae* and *Escherichia coli*. Mol. Plant-Microbe Interact. 2:262-272.

Maki, L. R., Galyan, E. L., Chang-Chien, M., and Caldwell, D. R. 1974. Ice nucleation induced by *Pseudomonas syringae*. Appl. Microbiol. 28:456-459.

Mizuno, H. 1989. Prediction of the conformation of ice nucleation protein by conformational energy calculation. Proteins 5:47-65.

Phelps, P., Giddings, T. H., Prochoda, M., and Fall, R. 1986. Release of cell-free ice nuclei by *Erwinia herbicola*. J. Bacteriol. 167:496-502.

Ptitsyn, O. B., and Finkelstein, A. V. 1983. Theory of protein secondary structure and algorithm of its prediction. Biopolymers 22:15-25.

Ryu, S.-E., Kwong, P. D., Truneh, A., Porter, T. G., Arthos, J., Rosenberg, M., Dai, X., Xuong, N.-H., Axel, R., Sweet, R. W., and Hendrickson, W. A. 1990. Crystal structure of an HIV-binding recombinant fragment of human CD4. Nature 348:419-426.

Southworth, M. W., Wolber, P. K., and Warren, G. J. 1988. Nonlinear relationship between concentration and activity of a bacterial ice nucleation protein. J. Biol. Chem. 263:15211-15216.

Sternberg, M. J. E., and Thornton, J. M. 1977. Conformation of proteins—Analysis of β-pleated sheets. J. Mol. Biol. 110:285-296.

Venkachatalam, C. M. 1968. Stereochemical criteria for polypeptides and proteins. V. Conformation of a system of 3 linked peptide units. Biopolymers 6:1425-1436.

Warren, G., and Corotto, L. 1989. The consensus sequence of ice nucleation proteins from *Erwinia herbicola*, *Pseudomonas fluorescens*, and *Pseudomonas syringae*. Gene 85:239-243.

Warren, G., Corotto, L., and Wolber, P. 1986. Conserved repeats in diverged ice nucleation structural genes from two species of *Pseudomonas*. Nucleic Acids Res. 14:8047-8060.

Warren, G. J., Lindemann, J., Suslow, T. V., and Green, R. L. 1987. Ice nucleation deficient bacteria as frost protection agents. Pages 215-227 in: Applications of Biotechnology to Agricultural Chemistry. H. M. LeBaron, R. O. Mumma, R. C. Honeycutt, and J. H. Duesing, eds. American Chemical Society, Washington, DC.

Welch, J. F., and Speidel, H. K. 1989. Visualization of potential bacterial ice nucleation sites. Cryo Lett. 10:309-314.

Wilmot, C. M., and Thornton, J. M. 1988. Analysis and prediction of the different types of β-turns in proteins. J. Mol. Biol. 203:221-232.

Wolber, P. K., Deininger, C. A., Southworth, M. W., Vandekerckhove, J., van Montagu, M., and Warren, G. J. 1986. Identification and purification of a bacterial ice-nucleation protein. Proc. Natl. Acad. Sci. USA 83:7256-7260.

Yang, D. S. C., Sax, M., Chakrabartty, A., and Hew, C. L. 1988. Crystal structure of an antifreeze polypeptide and its mechanistic implications. Nature 333:232-237.

Zhao, J., and Orser, C. S. 1990. Conserved repetition in the ice nucleation gene *inaX* from *Xanthomonas campestris* pv. *translucens*. Mol. Gen. Genet. 223:163-166.

CHAPTER 7

# Freezing Tolerance in Plants: An Overview

**T. H. H. Chen, M. J. Burke, and L. V. Gusta**

Food and fiber production, the most important global industry, interconnects nations through trade, primarily nations that have a surplus with those that have the financial ability to purchase. Adverse climatic and meteorological conditions, e.g., frost or drought, which cause losses in both yield and quality, can have a dramatic effect on the economics of a country. Even in a large country such as Canada, crop losses due to an abiotic stress have both a regional and national impact. For example, an unseasonal frost in mid-August of 1992 in western Canada resulted in over a billion dollar loss in crop production. Not only yield but also the quality of the crop was reduced. Wheat grain was devalued from high-quality bread flour to feed wheat. This not only resulted in a major loss to the producer but also affected the grain handling services, railroads, international trade, millers, farm sales, etc.

The growing season in the temperate zone is generally defined by temperature and the length of the frost-free period. Even though global warming may result in an increase in temperature, the risk of an unseasonal frost may also increase due to earlier seeding and earlier emergence. Fruit trees may also flower earlier and be injured by frosts.

Because frost plays a major role in plant distribution and crop yield and quality, the study of frost protection and avoidance is now receiving intensive interest both at the industrial and research levels. Frost injury is primarily due to the formation of ice in plant tissues at subzero temperatures. Methods of preventing frost injury have received considerable attention since the recent discovery of biological ice nucleators. In the absence of nucleators, water will not crystallize when cooled several degrees below zero. Biological ice nucleators, such as bacteria or fungi, are thought to be the main extrinsic agents that initiate freezing at temperatures close to zero. Recent evidence suggests that intrinsic nucleators, which also may be present in plants, initiate freezing at warm subzero temperatures. In contrast, certain flower buds, e.g., those of azalea, and the wood of the hardwoods lack both intrinsic and extrinsic nucleators. Azalea flower buds will supercool from −15 to −20°C; the xylem ray parenchyma cells of American elm will supercool to −45°C in midwinter. This overview and next three chapters provide insight into the recent progress made in understanding how plants either avoid ice formation or tolerate its presence.

## Plant Cold Hardiness

Plants in nature face several types of freezing stress including unseasonal frosts (usually in the growing season), and low-temperature extremes (usually when the plant is dormant and not actively growing). Depending on the minimum temperature, plants may be partially damaged or killed, with resultant reduced yield and quality or complete crop failure. During active growth in the spring and summer, certain crops are killed at the moment of ice formation, e.g., cucumbers (–2 to –3°C), whereas cereal grains can tolerate the presence of ice to temperatures as low as –9°C. Some winter crops can acclimate to autumn temperatures as low as –30°C. The most hardy plants, such as woody perennials, cannot tolerate –3°C during active growth. When fully acclimated, however, these plants tolerate temperatures as low as –196°C (Guy et al., 1986).

In tender plants, freezing injury occurs when ice forms in the tissue, regardless of the initiation temperature of freezing. The presence of ice results in either mechanical damage and/or dehydration injury to the tissue. If ice does not form at subzero temperatures, water in the cells is said to be supercooled, and no apparent injury occurs. Thus for tender plants to survive freezing temperature, they must somehow avoid or escape ice formation. On the other hand, hardy plants tolerate ice formation in their tissue if ice is excluded from the cytoplasm. The ability of hardy plants to survive freezing is dependent on many factors, including site of ice nucleation, ice nucleation temperature, rate of cooling during crystallization, rate of ice growth, minimum temperature of exposure, duration of exposure to freezing conditions, etc. (Levitt, 1980).

The inherent ability of temperate plants to acclimate to cold and their rate of acclimation are the two major factors that limit low-temperature survival. Cold acclimation is thought to be a complex genetic trait induced by low temperature and results in both morphological and molecular changes. Recent advances in the field of plant cryobiology and the identification and characterization of genes associated with cold acclimation have greatly enhanced our understanding of plant freezing resistance.

## Freezing Injury in Plants

During the growing season, radiation frosts on clear, windless nights are the most common type of freezing stress. Large leaves oriented parallel to the sky rapidly lose heat to the open sky through black body radiation and can cool to temperatures substantially below ambient (Chandler, 1958). Advective cooling, whereby leaf and air temperatures drop at similar rates, is caused by the inflow of cold air and can also occur during the growing season. As mentioned earlier, certain tender plant tissues, such as in cucumbers and tomatoes, are damaged at the moment of ice formation. Therefore, in order for these plants to escape unseasonable frost, ice nucleation must be avoided by lowering the freezing point of the tissue water or by maximizing the extent and duration of supercooling.

Perennial plants that survive harsh winter conditions develop freezing tolerance in the autumn. The temperature and the site of ice nucleation have a profound effect on the ultimate level of freezing tolerance. Generally, the more a plant supercools before it freezes, the more injury it will sustain. Siminovitch and Scarth (1938) showed that the probability of lethal intracellular freezing increases in hardy cells allowed to supercool substantially before freezing. They demonstrated that initiation of ice formation close to 0°C reduces the probability of injury. Olien (1964) further demonstrated that supercooling promotes nonequilibrium freezing, which is injurious to the tissues. Rapid nonequilibrium freezing that follows significant supercooling leads to a large Gibbs free energy for ice formation, which provides the energy for destructive work on tissues at the ice-liquid interface. Winter wheat crown tissue nucleated with ice at –3°C is less hardy than at just below 0°C (Gusta and Fowler, 1977).

Rajashekar et al. (1983) determined that leaves of *Solanum acaule* nucleated at –1°C survived a controlled freeze test to –7°C. If the leaves were nucleated at –3°C, however, they were killed at that temperature. Similar findings have also been reported for flower buds of *Prunus* spp. (Andrews et al., 1986; Gross et al., 1984). These studies demonstrated that in most plant species if freezing initiates close to 0°C, the plant is able to survive a colder temperature than if extensive supercooling is allowed to occur.

Once nucleation occurs in a plant, ice quickly moves into the extracellular space and propagates through large vessels with a high water content (Levitt, 1980). From the vessels, ice spreads throughout the extracellular spaces (Ashworth, 1990; Chandler, 1958; Levitt, 1980) and continues to grow until it reaches parts of the plant containing no water or warmer regions. It must be noted that the temperature of plant parts can vary by 5°C during a radiative frost. Measurements of rates of ice growth in woody stem tissues revealed rates of ice propagation as high as 60 to 74 cm/min, i.e., comparable to the rate of ice growth in U-tubes filled with pure water (Sakai and Larcher, 1987). In field conditions, freezing initiates from a few ice nucleation sites or ice seeding sites and then spreads quickly throughout the plant through vessels. In hardy plants, this is probably an effective way of preventing supercooling and reducing the risk of intracellular ice formation.

In partially frozen tissues, ice crystallization occurs extracellularly, and ice grows through movement of water vapor, films, and channels in and around cell walls and the extracellular spaces (Olien, 1967). When the freezing rate is sufficiently slow to allow water to diffuse to ice loci extracellularly, the volume of cytoplasm gradually decreases, concentrating the cell sap, which in turn depresses the freezing point of the intracellular water (Burke et al., 1976). Equilibrium is reached when the chemical potential of the cell water equals the chemical potential of the extracellular ice. The ice nucleation temperature (supercooling point) of the intracellular fluids also decreases upon cell dehydration. At temperatures close to 0°C, cells are in equilibrium with extracellular ice and are unlikely to undergo intracellular nucleation. Extracellular freezing results in freeze dehydration, cell volume reduction, and concentration of cell solutes including salts, which, if too extensive, may injure the membranes. Under rapid freezing conditions, such as when extensive supercooling precedes nucleation, water diffusion to the extracellular ice does not proceed quickly enough to concentrate the cell sap. Such rapidly cooled cells often freeze intracellularly due to ice seeding from extracellular ice or from intracellular ice nucleation. The ability of the protoplasm to tolerate the strain exerted by extracellular ice formation determines the freezing tolerance of a plant. Freezing tolerance or tolerance to extracellular freezing is therefore a form of avoidance, i.e., avoidance of intracellular ice formation.

Levitt (1980) classified freezing injury as follows: 1) primary direct injury due to intracellular freezing; 2) secondary freeze-induced dehydration injury due to extracellular freezing, and 3) injury due to other tertiary freeze-induced stresses. In his early work, Levitt described the principal mechanism of frost hardiness, i.e., tolerance of extracellular freezing and extracellular ice and avoidance of intracellular ice (Levitt and Scarth, 1936). Although this work was done in the 1930s, it is still relevant today. Levitt (1980) describes four "moments of freezing injury" or points in the freeze-thaw cycle where plant cells are possibly injured: 1) during ice growth as the cells dehydrate, 2) at the point where a low-temperature limit has been reached and a temperature-dependent structural transition has occurred, 3) during thawing as ice melts and cells rehydrate, and 4) following thawing when biochemical dysfunctions induced in the frozen state take effect. Although extracellular freezing is necessary for the survival of hardy plants, freeze-induced dehydration results in injury when tissues are cooled below a critical temperature or if they are held frozen for a prolonged period of time at a temperature slightly warmer than the killing temperature. This may result in a form of drought stress, a physical impairment of cell membranes, reduction of cellular volume below a critical size, or concentration of the cell sap to result in the denaturation of proteins.

Early workers believed the principal cause of freezing injury was physical damage caused by ice crystals. The volume increase associated with the freezing of water was thought to rupture plant cells, thus destroying their cellular structure. This early explanation was rejected when it was observed that cells were collapsed (not expanded) during freezing and did not rupture while in a frozen state.

It is generally accepted that intracellular freezing always results in cell death, probably due to mechanical destruction of membrane systems associated with ice growth in the protoplasm. Some cells survive intracellular freezing if freezing is so rapid that aqueous glasses and very fine ice crystals are formed. In these cases, thawing must also be very rapid to prevent the very small ice crystals from growing during the thawing process to a size that results in injury. The extremely rapid cooling and thawing rates used in some cryopreservation procedures (above 1,000°C per minute) are good examples of this.

The nature of frost-induced damage to the plasma membrane is still open to question. Steponkus et al. (1990) stressed the importance of specific lipid components in the plasma membrane that may prevent or reduce injury during freezing. Possible causes of membrane injury during freezing may be the activation of phospholipase D (Yoshida and Sakai, 1974), a lamellar-to-hexagonal II phase transition (Gordon-Kamm and Steponkus, 1984), and irreversible endocytotic vesiculation of the plasma membrane (Steponkus et al., 1982). Palta's group, however, emphasizes the importance of injury to membrane ATPases caused directly or indirectly by the freeze-thaw process (Iswari and Palta, 1989).

## Metastable Water and Plant Hardiness

Supercooling discussed throughout this book is only one of the several metastable states of water important in plant stress responses. The following are the four principal metastable states for plant stress responses:

1. *Supercooled (sometimes called undercooled) tissue and cell solutions where the solutions are between the equilibrium freezing point and the homogeneous ice nucleation temperature (usually between –1 and –41 °C).* This metastable condition is terminated by heterogeneous or homogeneous ice nucleation and subsequent ice growth (Burke et al., 1976)

2. *Supersaturated cell solutions where the concentration of one solute (usually a sugar) is above the crystallization concentration.* An example occurs during the dehydration of seeds when cell solutions pass through the eutectic point and reach sugar concentrations greater than the sugar crystallization point. The metastable condition is terminated upon the formation of the first sugar crystal, which then grows and precipitates the sugar from solution. Supersaturation is the first critical step in glass formation in the cell solutions of seeds (Leopold et al., 1992).
3. *Tension stressed (hydrostatic) cell and tissue solutions resulting from stretching, cohesion, and tensile strength of the solutions.* For tall trees, it has long been proposed that water in xylem vessels has a tensile strength sufficient to pull water to the tops of the trees. Such tensile strength leads to hydrostatic tensions of up to 0.5 MPa and allows a sufficient free energy lowering to pull water up to the top of the tree. It has also been proposed that water in deep supercooled ray parenchyma cells is retained in the cells due to hydrostatic tensions as large as 40 MPa. The metastable condition is terminated when a cavitation occurs, causing a water vapor bubble to form and releasing the tension (George and Burke, 1977).
4. *Glassed cell solutions.* This is a unique metastable condition that cannot be terminated below the glass transition temperature. Aqueous glasses are extremely viscous solutions with the viscosity brought about by high solute (sugar) concentrations at a sufficiently low temperature. Glasses have been reported to form at temperatures as high as 30°C in seeds (Williams and Leopold, 1989; Bruni and Leopold, 1991, 1992). In extremely cold hardy woody stems, glasses are reported to form below –20°C (Hirsh et al., 1985). Glasses in such systems may be supercooled, supersaturated, and under hydrostatic tension; however, they are not subject to ice nucleation (heterogeneous or homogeneous), solute crystallization, or water vapor cavitation so long as the solution remains below the glass transition temperature. Indeed the glass is extremely stable.

Clearly, glass formation can be a highly desirable metastable state for conditions of low temperature and extreme desiccation. Glasses in living tissues are usually associated with high sugar concentrations; sucrose and oligosaccharides (raffinose and stachyose) are common in plants, and trehalose is common in animals and fungi (Leopold et al., 1992). Glasses are formed from sugars that protect macromolecules and membranes and function as important storage materials (Crowe et al., 1984a,b). Glasses fill space and stop further cell collapse or changes in solute concentration, pH, or dehydration. They are extremely viscous and prevent all physical and chemical reactions that require molecular diffusion including ice nucleation, solute crystal nucleation, and vapor cavitation (Burke, 1986).

For glasses to form in most plant systems, the solutions must usually be dehydrated to a point where a sugar is supersaturated, and sometimes such solutions are supercooled and/or under hydrostatic tension. An example would be in the dehydration of a corn embryo at room temperature where sucrose becomes supersaturated. Koster (1991) suggested that a low concentration of raffinose in these seeds effectively poisons the initiation of sucrose crystal formation and therefore prevents precipitation of sucrose from solution. Hirsh has demonstrated that twig cells of *Populus* spp. dehydrate when frozen slowly and concentrate proteins and sugars (predominant sugars are sucrose, raffinose, and stachyose) until a sugar-protein-water glass transition is reached at about –20°C (Hirsh, 1987; Hirsh et al., 1985).

Hirsh (1987) provides further evidence for two additional glass transitions at lower temperatures in twigs of *Populus*. Hydrophilic proteins, as first described by Siminovitch and Briggs (1953) and discussed later in this chapter, may also play a key role in glass formation. Hydrophilic proteins at sufficiently high concentrations may add to the solution viscosity; they would fill space and might also prevent crystallization of critical solutes.

Many trees use deep supercooling of cells in woody tissues to survive to their homogeneous ice nucleation temperature, usually about –41°C. Principal cells of this type are ray parenchyma and pith cells in the xylem. At low temperatures, these supercooling cells do not collapse and appear fully turgid (Malone and Ashworth, 1991). Most interesting in Malone and Ashworth's work is that the very hardy species with ray parenchyma and pith cells that survive –196°C also do not collapse and appear turgid at low temperatures. One might speculate that these very hardy species have glass formation as outlined by Hirsh (1987) and that the glasses occur in ray parenchyma and pith cells due to high sugar concentrations. Only slight dehydration and cell collapse might lead to a glass transition, which protects the cells from further dehydration, ice nucleation (to 0°K), vapor cavitation, and solute crystallization.

Indeed, metastable water is very important for our considerations of plant survival of extreme conditions. Most of the metastable solutions are subject to termination by ice nucleation, vapor cavitation, and/or solute crystallization. However, if the solution glasses, it becomes resistant to these occurrences.

## Changes During Plant Cold Acclimation

Although it is generally thought that actively growing plants have a limited ability to tolerate ice formation in their tissues, many plants tolerate –5 to –10°C during the growing season. For example, winter rye (*Secale cereale* 'Puma') growing in the field in mid-June at day temperatures of 30°C can tolerate –10°C. Plants grown in an environmental chamber maintained at 30°C, however, are killed at –3°C (L. V. Gusta, *unpublished*). The reason for such a difference in freezing tolerance is currently unknown. The $LT_{50}$ (temperature at which 50% of cells were killed) of some *Solanum* species growing at warm temperatures can be as low as –5 to –6°C (Rajashekar et al., 1983). Upon exposure to hardening conditions, some of these species can survive –8 to –12°C.

Chen et al. (1976) studied the freezing characteristics of hardy and nonhardy *Solanum* species and concluded that tissues of hardy species tolerate more frost-induced dehydration than do tissues of nonhardy species. *Solanum* species have been divided into five hardiness categories: Group 1, frost resistant and able to cold harden; Group 2, frost resistant and unable to cold harden; Group 3, frost sensitive and able to cold harden; Group 4, frost sensitive and unable to cold harden; and Group 5, chilling sensitive (Chen and Li, 1980b). Recently, Stone et al. (1993) performed an interspecific cross between a Group 1 species (*Solanum commersonii*), and a Group 4 species (*S. cardiophyllum*). In the segregating generations, frost tolerance and ability to cold harden were separately inherited, implying that these two traits were controlled by different loci, not closely linked.

Low growing temperatures are not the only stimulus inducing freezing tolerance. Short days (Chen and Li, 1978), dehydration or drought (Chen and Li, 1978; Cloutier and Andrews, 1984; Siminovitch and Cloutier, 1982), and abscisic acid

(ABA) treatment (Chen et al., 1983; Chen and Gusta, 1983; Keith and McKersie, 1986; Lang et al., 1989) also induce freezing tolerance. It is unknown if the various external stimuli induce freezing tolerance via a common metabolic pathway or by alternative pathways.

### Extracellular Polysaccharides

Olien et al. (1965) identified a group of compounds termed freezing inhibitors that are dispersed in the liquid water of the intercellular spaces. The inhibitors are large polysaccharide polymers, generally with a molecular weight of several hundred thousand Daltons. The interaction of these polymers with ice is variable; some bind or are captured in the growing ice crystals with little effect on crystal growth rate or crystal structure. Others form a cohesive film at the ice-liquid interface and slow the rate of ice crystal growth during freezing. These effective inhibitors influence the path of ice growth through the plant and the subsequent size and shape of ice crystals.

In suspension-cultured pear cells, cold acclimation is accompanied by changes in soluble extracellular polysaccharides (Wallner et al., 1986). Specifically, there is an increase in the release of a relatively small neutral polysaccharide into the culture medium, a decrease in production of a large molecular weight pectic polysaccharide, and a deposition of callose at the cell surface (Wallner et al., 1986). The localization of callose at the plasma membrane–cell wall interface in pear cells may stabilize this region from the freezing stress.

### Cell Wall

There is growing evidence that the cell wall plays an important role in freezing injury and resistance (Bartolo and Wallner, 1986; Griffith et al., 1985; Olien and Smith, 1977; Paroschy et al., 1980; Rajashekar and Burke, 1982). Negative turgor measurements (Rajarshekar and Burke, 1982) have prompted some researchers to suggest that enhanced rigidity of cell walls increases the cold hardiness of plants by decreasing the severity of cell volume reduction during extracellular freezing.

Cell wall augmentation has often been observed as a plant response to low temperature. For example, pea epicotyl cell wall weight increases by 40% during cold acclimation (Weiser et al., 1990). A hardy potato species, *S. acaule,* has a thicker cell wall than a frost-sensitive species, *S. tuberosum* (Chen et al., 1977). The epidermal and mesotome sheath cell walls of hardened Puma rye leaves are significantly thicker than those of nonhardened leaves (Griffith et al., 1985). Ultrastructural studies of bromegrass suspension culture cells indicated that ABA-treated cells have markedly thicker cell walls than do untreated cells (Tanino et al., 1991). In cold-acclimating pea seedlings, cell wall weight not only increases but also shows compositional changes: arabinosyl content increases by 100%, cell wall glycosyl residues and cellulose increase by approximately 20%, and hydroxyproline content increases by 80% (Weiser et al., 1990).

Most biochemical studies on the cold acclimation of cells have concentrated on the soluble rather than the insoluble dry matter fractions (Levitt, 1980). Carbohydrates account for a significant proportion of the dry matter accumulation during cold acclimation (Chen and Li, 1980a; Levitt, 1980). Although carbohydrate reserves are important as an energy source for winter survival (Levitt, 1980), structural carbohydrates may also be closely associated with hardiness. In bromegrass cells, Tanino et al. (1990) observed that ABA causes a significant accumulation of dry matter, particularly in the 85% ethanol-insoluble fraction, of which cell walls represent the most abundant component. From $^{14}$C-sucrose labeling studies, it was demonstrated that a high percentage of the exogenous sucrose is incorporated into the cell wall fraction. Weiser et al. (1990) observed that increased freezing tolerance is associated with elevated levels of extensin, a cell wall glycoprotein that cross-links cellulose microfibrils, thereby adding rigidity and strength to cell walls. Northern blot analysis indicated that the level of specific extensin transcripts increases during cold acclimation (Weiser et al., 1990). Perhaps an increase in the rigidity of the cell wall increases the resistance of the cell to collapse during freezing-induced dehydration.

It has been postulated that membrane damage is the result of physical tearing that occurs when the cell wall separates from the plasma membrane during freezing (Tanino et al., 1991). A greater cell wall-to-plasma membrane adhesion has been suggested to increase the resistance to cell collapse and dehydration stress (Bartolo and Wallner, 1986). The number of membrane attachments to the cell wall increases during cold acclimation in alfalfa cell cultures (Johnson-Flanagan and Singh, 1986). The changes observed in cell wall and membrane composition during cold acclimation may alter the attachment characteristics and therefore may serve to increase plasma membrane-to-cell wall adhesion.

### Plasma Membrane

Since the plasma membrane is considered the primary site of freezing injury, any freezing tolerance mechanism should account for protection of the membrane from freezing stress. During cold acclimation, changes to lipid (Steponkus et al., 1988; Uemura and Yoshida, 1984; Yoshida and Uemura, 1984) and protein (Uemura and Yoshida, 1984; Yoshida and Uemura, 1984) constituents, as well as altered biophysical properties (Steponkus et al., 1990; Yoshida, 1984a,b) of membranes have been reported. Sugawara and Sakai (1978) observed that the number of inner membrane particles of the plasma membrane primarily on the fracture surface facing the cell wall (face E), was markedly reduced in hardened Jerusalem artichoke tuber calli but was restored to initial levels following dehardening. Similar results were also observed in *S. acaule* (Toivio-Kinnucan et al., 1981) and wheat (Pearce, 1985) during hardening and dehardening.

Lipid analysis of purified plasma membrane fractions have demonstrated either dramatic or slight changes in the lipid composition of the plasma membrane during the development of cold hardiness (Uemura and Yoshida, 1984; Yoshida and Uemura, 1984). In purified plasma membranes of cold-acclimated winter rye seedlings, Uemura and Yoshida (1984) reported 1) slight changes in the degree of fatty acid unsaturation and proportions of phospholipids; 2) a small change in sterol composition consisting of increased beta-sitosterol and decreased campestrol and stigmasterol levels; 3) an increase in the phospholipid-to-protein ratio; and 4) a decline in the sterol-to-phospholipid ratio.

Steponkus et al. (1990) also demonstrated the following changes in the properties of plasma membrane: 1) the cryobiological behavior of hardy and nonhardy rye are very different; 2) fusion of liposomes prepared from hardy plants to nonhardy rye protoplasts increased the freezing tolerance of the nonhardy protoplasts (Steponkus et al., 1988). These studies provided evidence that changes in the lipid composition of the plasma membrane are related to cold hardiness.

Yoshida (1984b) noted significant changes in the plasma membrane protein pattern of mulberry bark cells during acclimation to cold. An altered glycoprotein fraction coincided with the development of freezing tolerance. Uemura and Yoshida (1984) observed several proteins present in the plasma membrane of hardened winter rye seedlings that were not present in nonhardened tissue. In winter canola plants, Johnson-Flanagan and Singh (1987) reported the identification of a membrane protein associated with the induction of freezing tolerance.

## Cytoplasmic Changes

### Sugars

All major classes of organic compounds located in the cytoplasm, of which sugars are the most abundant, have been implicated in the hardening response of plants (Levitt, 1980; Sakai and Larcher, 1987). Major changes in total osmotic potential, which accompany seasonal changes in freezing tolerance, are related to changes in the concentrations of sugars and polyhydric alcohols (Sakai and Larcher, 1987). In herbaceous and woody plants, soluble carbohydrates increase from fall to winter and decrease in the spring during dehardening (Sakai and Larcher, 1987). Changes in sugar content during artificial hardening follow a very similar response pattern (Sakai and Larcher, 1987).

In addition to quantitative changes in the sugar content of plant cells, qualitative changes also occur during cold hardening. For example, Parker (1959) observed that levels of raffinose and stachyose increased markedly in the bark and leaves of six conifer species during late fall; concentrations of sucrose and sometimes glucose and fructose also increased. The proportion of the various sugars differed in 18 woody plant species, but the accumulation of a specific sugar was not consistently correlated with hardiness (Sakai, 1962). The type of carbohydrate that accumulates during hardening is dependent on a species-specific pattern of carbohydrate metabolism. In several species, polyhydric alcohols, such as sorbitol or mannitol, account for approximately 40% of the total soluble carbohydrate content and may therefore contribute to hardening (Ichiki and Yamaya, 1982; Raese et al., 1977). Sugars accumulate in the chloroplasts of acclimating cabbage, spinach (Krause et al., 1978), and wheat (Trunova and Zvereva, 1974).

Sugar feeding trials have been used to demonstrate the role of sugars in frost hardiness. For example, the maximum hardiness potential of winter cereals and callus tissues of woody plants could not be attained unless, in addition to exposure to low temperature, the plants and tissues were fed sucrose (Sakai, 1962; Ogolevets, 1976). In the dark, *Chlorella* spp. hardened only marginally, but the addition of glucose caused a dramatic increase in frost resistance (Hatano, 1978).

There is ample evidence of a close correlation between the accumulation of soluble carbohydrates and freezing tolerance. Suggested possibilities as to how sugars are involved in protection against freezing injury include the following:

1. *Osmotic effect.* Sugars decrease crystallization of water and therefore reduce freeze-induced dehydration.
2. *Metabolic effect.* Metabolization of sugars in the cytoplasm during acclimation produces other protective substances or metabolic energy.
3. *Cryoprotective effect.* Sugars may protect cellular constituents including membranes during freeze/thaw cycle (Sakai and Larcher, 1987).
4. *Glass effect.* High sugar concentrations may stop all biochemical and most physical activity (dehydration) when they form a solid glass (Bruni and Leo-

pold, 1991; Williams and Leopold, 1989; Hirsh et al., 1985).

It is also possible that specific interactions between sugars and cell structures protect or ameliorate the deleterious effects of freezing.

**Proteins**

Protein accumulation during cold acclimation has been frequently observed in numerous herbaceous plant species (Chen and Li, 1980a; Guy, 1990; Johnson-Flanagan and Singh, 1988; Lee and Chen, 1993a) under both natural and artificial conditions. In most instances, there is an increase in both total protein content (Levitt, 1980) and in specific polypeptide species (Gilmour et al., 1988; Guy et al., 1985; Guy and Haskell, 1987; Lang et al., 1989; Lee et al., 1990; Lee et al., 1992; Perras and Sarhan, 1989; Robertson et al., 1987). The requirement for protein synthesis during cold acclimation has been demonstrated in wheat (Trunova, 1982), winter canola (Kacperska-Palacz et al., 1977), *Chlorella* (Hatano, 1978), and *Solanum* species (Chen et al., 1983). A high correlation between the increase in protein content and cold hardiness was observed in all of these species. If applied prior to cold acclimation, cycloheximide, a protein synthesis inhibitor, suppressed the hardening process (Chen *et al.*, 1983), thus indicating the need for protein synthesis.

Guy (1990) proposed that changes during cold acclimation involve both the induction of freezing tolerance and metabolic adjustment to constraints imposed by low temperature. This may involve structural proteins as well as enzymes. The implication is that protein changes can include either the modification of existing polypeptides to generate different isoforms, or the production or enhanced production of new polypeptides with unique properties, or the creation of a new metabolic pathway. On the basis of these assumptions, it is not surprising to observe changes to many enzymes during cold acclimation (Guy, 1990; Levitt, 1980).

Aside from proteins with catalytic or structural properties, proteins may have direct cryoprotective roles. For example, Volger and Heber (1975) isolated a protein fraction from frost-hardy spinach leaves that protected membranes in vitro from freezing injury. Recently, Hincha et al. (1989, 1990) identified protein fractions from cold-acclimated, frost-hardy cabbage and spinach leaves that protected non-hardy spinach thylakoids against a freeze-thaw stress. On a per molecule basis, the protective proteins are about 20,000 to 40,000 times more effective than sucrose in preventing freeze-thaw rupture to isolated thylakoid membranes (Hincha et al., 1989). A correlation between hydrophilic proteins and hardiness in black locust was observed by Siminovitch and Briggs (1953). It was concluded that hydrophilic proteins improve the water-binding capacity of the cytoplasmic fluid and therefore reduce the freezing stress. Recently, many cold-induced mRNAs encoding hydrophilic proteins have been cloned, suggesting that these types of proteins may protect the cytoplasm or membranes from freeze-induced dehydration stress. Some of the cold-regulated genes of *Arabidopsis* encode for heat-stable proteins (Lin et al., 1990). Although the exact role of these heat-stable proteins is not known, they have a random coil structure and may represent a unique group of proteins that somehow reduce a freezing stress.

Recently, several cDNA clones from *S. commersonii* cell cultures have been isolated during the induction of cold hardiness by low temperature or ABA treatment (Zhu et al, 1993). Three of the cDNA clones have high homology to tobacco osmotin, which accumulates in tobacco cells undergoing gradual osmotic adjustment to NaCl stress (Singh et al., 1989).

In ABA-hardened bromegrass cell cultures, Lee and Chen (1993b) isolated an ABA-responsive cDNA clone that shares 82% nucleic acid sequence similarity to barley dehydrin (Close et al., 1989). In barley, dehydrin mRNA accumulates during the programmed dehydration phase of embryo development and is also induced by a desiccation stress or ABA treatment. As discussed above, freezing is also a form of desiccation stress. The enhanced expression of osmotin- or dehydrin-like genes during the development of freezing tolerance in plant cells indicates a requirement for tolerance to the desiccation stress manifested by the freezing stress.

## Ice Nucleation and Freezing Tolerance in Plants

Ice formation involves the initial ice nucleation and the consequent crystallization of water molecules. In contrast to the common belief that aqueous solutions freeze at the melting point of the solid phase (ice), cell solutions rarely freeze at their melting point. Liquid water, supercooled several degrees below its freezing point, will freeze only if spontaneous ice formation occurs (usually only below –38°C) or if an extrinsic ice nucleator is present that acts as a catalyst for the liquid-solid phase transition. There are two general types of ice nuclei in biological systems: homogeneous and heterogeneous. In homogeneous ice nucleation, the nuclei form spontaneously in the liquid (usually below –38°C). In heterogeneous ice nucleation, nucleation occurs as a result of extrinsic nuclei, such as ice. Potent biological heterogeneous nucleators are bacteria, fungi, plants, and insects (see other chapters in the book). Extrinsic ice nuclei are the major cause of ice formation in tender growing plants (see Chapter 2).

It is generally believed that most plant cells do not contain intrinsic ice nucleators; however, there are exceptions (see Chapter 8). Krog et al. (1979) reported flower tissues of the Afro-alpine plant *Lobelia telekii* contain a slightly viscous fluid in the central part of the inflorescence. Freezing of this central fluid occurs near 0°C and results in a slow and steady release of heat, which maintains the plant's temperature at about 0°C during a frost. Analysis of the chemical composition of this central fluid revealed that the nucleating agents are carbohydrate, probably high molecular weight polysaccharides. Thus, this plant has evolved a unique way of surviving a frost in the alpine zone of Mount Kenya.

Various studies have provided evidence for effective ice nucleators within plant tissues., Andrews et al. (1986) measured the ice nucleation temperatures of peach and sweet cherry flowers, fruits, and stems. The mean ice nucleation temperature of flowers was between –4 and –6°C, while those of 0.5-cm stem segments were between –6 and –9°C. The stem tissue contained a lower concentration of ice nucleators active above –5°C than did the floral tissue. Homogenization of the tissue reduced the nucleation temperature, which suggests that tissue structural integrity is necessary for the manifestation of optimal ice nucleation activity.

In other studies, woody tissue of *Prunus* spp. nucleated at –2°C, and the nucleator was considered to be *not* of bacterial origin for the following reasons: First, peach stems nucleated at –2°C even though INA bacteria could not be detected. Second, suppression of INA bacteria in orchards with either bactericides or antagonistic bacteria failed to prevent extensive supercooling (Proebsting and Gross, 1988). Third, based on seasonal monitoring of activity, bacteria ice nuclei did not remain active for extended periods in nature (Cody et al., 1987). Ashworth and Davis (1984) conducted the following studies on peach to establish that the ice nuclei were of plant origin. Autoclaved shoots were supercooled to approximately –4°C, which was 1.5 to 2.0°C lower than the ice nucleation temperature of untreated peach shoots. Soaking 5-cm peach stem sections in water for 4 hr lowered the main ice nucleation temperature to below –4°C; however, ice nucleation activity was fully restored by air-drying woody stem sections for a few hours. The ice nuclei in woody tissues were inactivated between 40 and 50°C but unaffected by treatment with bacterial ice nucleation inhibitors (i.e., NaOCl, tartaric acid, triton XQS-20), sulfhydryl reagents (i.e., *p*-hydroxymercuribenzoate and iodine), and pronase. Ice nuclei could not be dislodged from stems by sonication. They were shown to be distributed uniformly in bud and internodal stem tissue. The ice nucleation activity in outer and inner stem tissues was also indistinguishable. The development of ice nuclei in immature peach and sweet cherry stem did not occur until midsummer, was complete by August, and was unaffected by seasonal changes and additional growth.

## Molecular Biology of Cold Hardiness Development

Genetic aspects of cold hardiness have been reviewed (Limin and Fowler, 1983; Thomashow, 1990) and will not be reiterated here. The complex genetic basis of cold tolerance has traditionally been studied using techniques of quantitative analysis (Eunus et al., 1962) supported by cytogenetic evidence. Monosomic analysis and substitution lines have been used in wheat to locate cold tolerance genes on 15 of the 21 chromosomes (Cahalan and Law, 1979; Sutka, 1981; Sutka and Veisz, 1988). Similar studies in other species indicate that cold hardiness is controlled by many genes.

### Molecular Biology of Plant Cold Acclimation

In the past few years, rapid progress has been made in the characterization of molecular changes associated with frost tolerance. In all the systems studied, the induction of freezing tolerance is generally accompanied by the appearance of new mRNAs and new polypeptides. cDNAs of cold-regulated genes have been cloned from *Arabidopsis thaliana* (Gilmour et al., 1992; Hajela et al., 1990; Kurkela and Franck, 1990, 1992; Nordin et al., 1991), *Bromus inermis* (Lee and Chen, 1993b), *Brassica napus* (Weretilnyk et al., 1993), *Hordeum vulgare* (Dunn et al., 1990; Hughes et al., 1992; Cattivelli and Bartels, 1990), *Medicago* spp. (Luo, et al., 1992; Mohapatra et al., 1989), *S. commersonii* (Zhu et al., 1993), and *Triticum aestivum* (Guo et al., 1992; Houde et al., 1992). For a thorough review on molecular biology of cold acclimation, the reader is directed to reviews by Guy (1990), Lee and Chen (1993a), and Thomashow (1990).

The nucleotide sequences for some of these cDNAs have been published. The search of the GenBank or EMBL data bases has identified homology with reported sequences of known DNAs and proteins (Table 1). The expression of some of these genes is also induced by ABA, a plant hormone known for its ability to induce freezing tolerance in plants (Chen et al., 1983; Chen and Gusta, 1983).

It appears that plants have evolved sophisticated mechanisms that provide protection to all major freezing-sensitive sites of plant cells. There are few examples in which the putative functions of the gene products have been proposed (Table 2). The putative functions of these cold-regulated proteins include 1) cryoprotective properties (Lin and Thomashow, 1992), 2) altered lipid metabolism (Hughes et al.,

**Table 1.** cDNAs of abscisic acid (ABA) or cold-regulated genes associated with cold acclimation

| Plant species | cDNA clone or gene | Up-regulated[a] LT | ABA | Drought | Identity/homology[b] (%) | Reference |
|---|---|---|---|---|---|---|
| *Arabidopsis thaliana* | pHH 7.2 (*cor* 47) | + | + | + | . . . | Hajela et al., 1990 |
| | pHH 28 (*cor* 160) | + | + | + | . . . | Thomashow et al., 1993 |
| | pHH 29 (*cor* 6.6) | + | + | + | *kin* 1 | Thomashow et al., 1993 |
| | pHH 67 (*cor* 15) | + | + | + | . . . | Thomashow et al., 1993 |
| | *kin* 1 | + | + | + | Antifreeze protein of winter flounder (28% a.a.) | Kurkela and Franck, 1990 |
| | *kin* 2 | + | + | + | *kin* 1 (95%) | Kurkela and Franck, 1992 |
| *Bromus inermis* (bromegrass) | pBGA 55 | – | + | ? | Barley dehydrin (55% a.a.) | Lee and Chen, 1993 |
| | pBGA 56 | + | + | ? | Wheat germin (45% a.a.) | |
| | pBGA 61 | + | + | ? | Barley aldose reductase (89% a.a.) | |
| | pBGA 72 | – | + | ? | Rabbit phosphoglucomutase (69% a.a.) | |
| | pBGA 85 | + | + | ? | Human cathepsin D (53% a.a.) | |
| | pBGA 86 | – | + | ? | Barley embryo globulin (76% Nt) | |
| *Hordeum vulgare* (barley) | BLT 14 | + | ? | ? | . . . | Dunn et al., 1990 |
| | blt 4 | + | + | + | Maize phospholipid transfer proteins (42% a.a.) | Hughes et al., 1992 |
| | pT 59 | + | ? | ? | . . . | Cattivelli and Bartels, 1990 |
| | pA 086 | + | ? | ? | . . . | |
| *Medicago* spp. (alfalfa) | pSM 2075 | + | + | + | *Chenopodium* light-induced glycine-rich protein (60% a.a.), maize ABA inducible gene (42.3% a.a.), bean glycine-rich cell wall protein (49.5% Nt), barley dehydrin (30% Nt) | Luo et al., 1991 |
| | pUM 90-1 | + | + | + | pSM 2075 | Luo et al., 1992 |
| | pUM 90-2 | + | + | + | pSM 2075 | |
| | pUM 91-4 | + | + | + | pSM 2075 | |
| *Solanum commersonii* (potato) | pA 13 | + | + | + | Tobacco osmotin, tomato NP 24 protein | Zhu et al., 1993 |
| | pA 35 | + | + | + | Maize λ-amylase/trypsin inhibitor | |
| | pA 81 | + | + | + | Tobacco pathogenesis-related protein, thaumatin (89–52% a.a.) | |
| *Spinacia oleracea* (spinach) | CAP 85 | + | ? | + | Barley dehydrin | Neven et al., 1993 |
| *Triticum aestivum* (wheat) | pWG1 (*cor* 39) | + | + | + | Arabidopsis *cor* 47 | Guo et al., 1992 |
| | Wcs 120 | + | – | – | Barley dehydrin and RAB | Houde et al., 1992 |

[a] LT = low temperature, + = up-regulated, – = not affected by the treatment, and ? = not reported.
[b] a.a. = amino acid, and Nt = nucleotide.

**Table 2.** Putative functions of polypeptides encoded by cold-regulated genes

| Plant | Cold-regulated gene | Identity | Putative function | Reference |
|---|---|---|---|---|
| Alfalfa | pSM 2075 | Dehydrin | Desiccation tolerance | Luo et al., 1991 |
| *Arabidopsis* | *cor* 15 | . . . | Cryoprotection | Lin and Thomashow, 1992 |
| Barley | Blt 4 | Phospholipid transfer proteins | Lipid metabolism | Hughes et al., 1992 |
| Bromegrass | pBGA 55 | Dehydrin | Desiccation tolerance | Lee and Chen, 1993 |
| | pBGA 61 | Aldose reductase | Sugar metabolism | Lee and Chen, 1993 |
| | pBGA 72 | Phosphoglucomutase | Sugar metabolism | Lee and Chen, 1993 |
| Potato | PA 13 | Osmotin | Desiccation tolerance | Zhu et al., 1993 |
| Spinach | Hsp 70 | Heat shock proteins | Minimizing protein denaturation | Neven et al., 1992 |

1992), 3) protein protection (Neven et al., 1992), 4) desiccation tolerance (Luo et al., 1991; Lee and Chen, 1993b; Zhu et al., 1993), and 5) sugar metabolism (Lee and Chen, 1993b). The preliminary results from molecular biology studies thus support the findings of the earlier physiological studies. With the further isolation and characterization of cold-regulated genes, it is hoped that the mechanisms of cold acclimation and freezing tolerance in plants will be elucidated.

## Cryoprotective Proteins

Heat-stable polypeptide fractions prepared from cold-acclimated spinach and cabbage prevent freeze-induced rupture of isolated thylakoid membranes (Heber, 1968; Volger and Heber, 1975; Hincha et al., 1989, 1990). Thomashow and associates (Hajela et al., 1990; Lin et al., 1990) isolated cDNA clones for four *Arabidopsis* cold-regulated genes that encode hydrophilic, boiling-stable polypeptides. One of the *Arabidopsis* cold-regulated peptides encoded by *cor*15 protects the cold-labile enzyme, lactate dehydrogenase, against freeze-inactivation (Lin and Thomashow, 1992). On a per molecule basis, *cor*15 was about 1 million times more effective than sucrose and 100 to 1,000 times more effective than other proteins including bovine serum albumin, a recognized cryoprotectant and protein stabilizer. Although the in vitro assay data are encouraging, there is no direct evidence that these polypeptides contribute significantly to freezing tolerance of cold-acclimated plants in vivo.

Exposure of spinach seedlings to 5°C leads to an increase in the synthesis of 70-kDa proteins that also are present in leaf tissues grown at 20°C (Neven et al., 1992). Protein sequence analyses and immunological cross-reactivity indicate that this group of proteins belongs to the 70-kDa heat-shock protein family, of which several are chaperonins. A mechanism that avoids aggregation of proteins and favors folding and renaturation of proteins at low temperature would be advantageous in restoring the proper protein conformation and function upon thawing and returning to favorable temperatures. Neven et al. (1992) proposed that elevated synthesis of several CAP79 isoforms represents an adjustment of cellular metabolism to biophysical constraints experienced during cold acclimation. A biophysical constraint may be that chilling and freezing conditions favor protein denaturation.

## Antifreeze Proteins

Many freezing-tolerant insects produce antifreeze proteins that modify extracellular ice crystals (Duman et al., 1991). Griffith et al. (1992) hypothesized that frost-tolerant plants may also produce similar proteins in the apoplast. These proteins are able to modify the normal growth of ice crystals and depress the freezing temperature of water noncolligatively. Although plant antifreeze genes are not available, Cutler et al. (1989) demonstrated that antifreeze protein isolated from winter flounder acted as ice nucleation inhibitor when vacuum infiltrated into leaves.

Georges et al. (1990) constructed a synthetic gene coding for the winter flounder antifreeze protein that was electroporated into protoplasts of corn. The production of fusion protein was detected in the protoplasts; however, the amount of the fusion protein produced and the antifreeze activity of the fusion protein were not clear. Hightower et al. (1991) produced transgenic tobacco and tomato plants that express genes encoding antifreeze proteins. The antifreeze gene was expressed at high steady-state mRNA levels in leaves from transformed plants. A full-length protein was synthesized in transgenic plants (0.1% of total soluble leaf protein), which was active in inhibiting ice recrystallization.

### Ice Nucleation Proteins

INA bacteria can catalyze ice formation at temperatures as warm as –2°C (see Chapter 2). Genes encoding ice nucleation proteins in different strains of *Pseudomonas syringae* have been cloned and structurally characterized (see Chapter 4). Since most plant cells do not contain intrinsic ice nuclei that are active close to 0°C, it would be desirable to introduce the genetic determinants of ice nucleation proteins into plant cells to prevent extensive supercooling. As stated earlier, if a plant supercools to –3 to –5°C and then freezes, ice growth will be too rapid to allow equilibrium freezing to occur (Olien, 1964). Introduction of an ice nucleation gene that produces an intrinsic nucleator active close to 0°C would prevent extensive supercooling, and less injurious equilibrium freezing would occur.

Baertlein et al. (1992) reported the stable introduction of an ice nucleation gene (*inaZ*) from *Pseudomonas syringae* pv. *syringae* into *Nicotiana tabacum*, a freezing-sensitive plant species, and *S. commersonii*, a freezing-tolerant species. Transgenic plants of both species show an increase in ice nucleation activity over untransformed controls. The most active transformant contains over 100 ice nuclei/mg of tissue at –10.5°C. Nontransformed plants supercool to –12°C in contrast to the –4°C for the *inaZ* transformants. The levels of ice nucleation activity measured at –9°C were comparable to that of naturally occurring *P. syringae* on a per cell basis. However, there is a much lower number of warm temperature ice nuclei (i.e., above –5°C) in transgenic plants than in INA bacteria. Fewer than 1 in 100,000 plant cells contain ice nuclei active at –5°C, whereas about 1 in 100 bacterial *P. syringae* cells contain ice nuclei active at this temperature.

## Conclusions and Future Research

Of the many mechanisms that have evolved to enable plants to survive freezing stress, tolerance to freezing is probably the most common. The ability to acclimate to cold in response to low, nonfreezing temperatures is essential for many plant species to overwinter under natural conditions. Cold acclimation involves a series of changes that results in increased levels of freezing tolerance. Given the complexity of changes required for achieving maximal freezing tolerance, many genes are likely to be involved in the overall cold acclimation process. With the isolation of more cold-regulated genes associated with cold hardiness and a better understanding of their regulation and function, we may be able to decipher the process of cold acclimation. Only then can we attempt to genetically manipulate the freezing tolerance of crop plants.

## Literature Cited

Andrews, P. K., E. L. Proebsting, Jr., and D. C. Gross. 1986. Ice nucleation and supercooling in freezing-sensitive peach and sweet cherry tissues. J. Am. Soc. Hortic. Sci. 111:232-236.

Ashworth, E. N. 1990. Formation and spread of ice in plant tissues. Hortic. Rev. 12: 215-254.

Ashworth, E. N., and G. A. Davis. 1984. Ice nucleation within peach trees. J. Am. Soc. Hortic. Sci. 109:198-201.

Baertlein, D. A., S. E. Lindow, N. J. Panopoulos, S. P. Lee, M. N. Mindrinos, and T. H. H. Chen. 1992. Expression of a bacterial ice nucleation gene in plants. Plant Physiol. 100:1730-1736.

Bartolo, M. E., and S. J. Wallner. 1986. Cold hardiness and cellulose resistance induced by wounding.

(Abstr.) Plant Physiol. 80:122.
Bruni, F., and A. C. Leopold. 1991. Glass transitions in soybean seed. Relevance to anhydrous biology. Plant Physiol. 96: 660-663.
Bruni, F., and A. C. Leopold. 1992. Pools of water in anhydrobiotic organisms: A thermally stimulated depolarization current study. Biophys. J. 63: 663-672.
Burke, M. J. 1986. The glassy state and survival of anhydrous biological systems. Pages 358-363 in Membranes Metabolism and Dry Organisms. A.C. Leopold, ed.. Cornell Press, Ithaca, NY.
Burke, M. J., L. A. Gusta, H. A. Quamme, C. J. Weiser, and P. H. Li. 1976. Freezing and injury to plants. Annu. Rev. Plant Physiol. 27:507-528.
Cahalan, C., and C. N. Law. 1979. The genetic control of cold resistance and vernalization requirement in wheat. Heredity 42:125-132.
Cattivelli, L., and D. Bartels. 1990. Molecular cloning and characterization of cold-regulated genes in barley. Plant Physiol. 93:1504-1510.
Chandler, W. H. 1958. Cold resistance in horticultural plants: A review. Proc. Am. Soc. Hortic. Sci. 64:552-572.
Chen, H. H., and P. H. Li. 1980a. Biochemical changes in tuber-bearing *Solanum* species in relation to frost hardiness during cold acclimation. Plant Physiol. 66:414-421.
Chen, H. H., and P. H. Li. 1980b. Characteristics of cold acclimation and deacclimation in tuber-bearing *Solanum* species. Plant Physiol. 65:1146-1148.
Chen, H. H., P. H. Li, and M. L. Brenner. 1983. Involvement of abscisic acid in potato acclimation. Plant Physiol. 71:362-365.
Chen, H. H., and P. H. Li. 1978. Interaction of low temperature, water stress and short days in the induction of stem frost hardiness in red osier dogwood. Plant Physiol. 62:833-835.
Chen, P., M. J. Burke, P. H. Li. 1976. The frost hardiness of several *Solanum* species in relation to the freezing of water, melting point depression and tissue water content. Bot. Gaz. 137:313-317.
Chen, P. M., P. H. Li, and W. P. Cunningham. 1977. Ultrastructural differences in leaf cells of some *Solanum* species in relation to their frost resistance. Bot. Gaz. 138:376-285.
Chen, T. H. H., and L. V. Gusta. 1983. Abscisic acid induced freezing resistance in cultured plant cells. Plant Physiol. 73:71-75.
Close, T. J., A. A. Kortt, and P. M. Chandler. 1989. A cDNA-based comparison of dehydration-induced proteins (dehydrins) in barley and corn. Plant Mol. Biol. 13:95-108.
Cloutier, Y., and C. J., Andrews. 1984. Efficiency of cold hardiness induction by desiccation stress in four winter cereals. Plant Physiol. 78:595-598.
Cody, Y. S., D. C. Gross, E. L. Proebsting, Jr., and R. A. Spotts. 1987. Suppression of ice nucleation-active *Pseudomonas syringae* by antagonistic bacteria in fruit tree orchards and evaluation of frost control. Phytopathology 77:1036-1044.
Crowe, J. H., M. A. Whittam, D. Chapman and L. M. Crowe. 1984a. Interactions of phospholipid monolayers with carbohydrates. Biochem. Biophys. Acta 769:151-159.
Crowe, L. M., R. Mouradian, J. H. Crowe, S. A. Jackson and C. Wormersly. 1984b. Effect of carbohydrates on membrane salinity at low water activities. Biochem. Biophys. Acta 769:141-150.
Cutler, A. J, M. Saleem, E. Kendall, L. V. Gusta, F. Georges, and G. L. Fletcher. 1989. The winter flounder antifreeze protein improves the cold hardiness of plant tissues. J. Plant Physiol. 135:351-354.
Dunn, M. A., M. A. Hughes, R. S. Pearce, and P. L. Jack. 1990. Molecular characterization of a barley gene induced by cold treatment. J. Exp. Bot. 41(232):1405-1413.
Duman, J. G., L. Xu, L. G. Neven, D. Tursman, D. W. Wu. 1991. Hemolymph proteins involved in insect subzero-temperature tolerance: Ice nucleators and antifreeze proteins. Pages 94-127 in: Insects at Low Temperature. R. E. Lee, Jr. and D. L. Denlinger, eds. Chapman and Hall, New York.
Eunus, A. M., L. P. V. Johnson, and R. Askel. 1962. Inheritance of winter hardiness in a eighteen-parent diallel cross of barley. Can. J. Genet. Cytol. 4:356-376.
George, M. F., and M. J. Burke. 1977. Cold hardiness and deep supercooling in xylem of Shagbark Hickory. Plant Physiol. 59:319-325.
Georges, F, M. Saleem, and A. J. Cutler. 1990. Design and cloning of a synthetic gene for the flounder antifreeze protein and its expression in plant cells. Gene 91:159-165.
Gilmour, S. J., R. K. Hajela, and M. F. Thomashow. 1988. Cold acclimation in *Arabidopsis thaliana*. Plant Physiol. 87:745-750.
Gilmour, S. J., N. N. Artus, M. F. Thomashow. 1992. cDNA sequence analysis and expression of two cold-regulated genes of *Arabidopsis thaliana*. Plant Mol. Biol. 18:13-21.
Gordon-Kamm, W. J., P. L. Steponkus. 1984. Lamellar-to-hexagonal II phase transitions in the plasma membrane of isolated protoplasts after freezing-induced dehydration. Proc. Natl. Acad. Sci. USA.

81:6373-6377.

Griffith, M., N. P. A. Huner, K. E. Espelie, and P. E. Kolattukudy. 1985. Lipid polymers accumulate in the epidermis and mesotome sheath cell walls during low temperature development of winter rye leaves. Protoplasma 125:53-64.

Griffith, M., P. Ala, D. S. C. Yang, W. C. Hon, and B. A. Moffatt. 1992. Antifreeze protein produced endogenously in winter rye leaves. Plant Physiol. 100:593-596.

Gross, D. C., E. L. Proebsting, Jr., and P. K., Andrews. 1984. The effects of INA bacteria on the temperatures of ice nucleation and freeze injury of *Prunus* flower buds at various stages of development. J. Am. Soc. Hortic. Sci. 109:375-380.

Guo, W., R. W. Ward, and M. F. Thomashow. 1992. Characterization of a cold-regulated wheat gene related to *Arabidopsis* cor 47. Plant Physiol. 100:915-922.

Gusta, L. V., and D. B. Fowler. 1977. Factors affecting the cold survival of winter cereals. Can. J. Plant Sci. 57:213-219.

Guy, C. L., K. J. Niemi, A. Fennell, and J. V. Carter. 1986. Survival of *Cornus sericea* L. stem cortical cells following immersion in liquid helium. Plant Cell Environ. 9:447-450.

Guy, C. L. 1990. Cold acclimation and freezing stress tolerance: Role of protein metabolism. Annu. Rev. Plant Physiol. Plant Mol. Biol. 41:187-223.

Guy, C. L., K. J. Niemi, and R. Brambl. 1985. Altered gene expression during cold acclimation of spinach. Proc. Natl. Acad. Sci. USA 82:3673-3677.

Guy, C. L., and D. Haskell. 1987. Induction of freezing tolerance in spinach is associated with the synthesis of cold acclimation induced proteins. Plant Physiol. 84:872-878.

Hajela, R. K., D. P. Horvath, S. J. Gilmour, and M. F. Thomashow. 1990. Molecular cloning and expression of *cor* (cold-regulated) Hagenes in *Arabidopsis thaliana*. Plant Physiol. 93:1246-1252.

Hatano, S. 1978. Studies on frost hardiness in *Chlorella ellipsoidea*: Effects on antimetabolites, surfactants, hormones, and sugars on the hardening process in the light and dark. Pages 175-196 in: Plant Cold Hardiness and Freezing Stress. P. H. Li and A. Sakai, eds. Academic Press, New York.

Heber, U. 1968. Freezing injury in relation to loss of enzyme activities and protection against freezing. Cryobiology 5:188-201.

Hincha, D. K., U. Herber, and J. M. Schmitt. 1989. Freezing ruptures thylakoid membranes in leaves, and rupture can be prevented *in vitro* by cryoprotective proteins. Plant Physiol. Biochem. 27:795-801.

Hincha, D. K., U. Heber, and J. M. Schmitt. 1990. Proteins from frost-hardy leaves protect thylakoids against mechanical freeze-thaw damage *in vitro*. Planta 180:416-419.

Hightower, R., C Baden, E. Penzes, P. Lund, and P. Dunsmuir. 1991. Expression of antifreeze proteins in transgenic plants. Plant Mol. Biol. 17:1013-1021.

Hirsh, A. G. 1987. Vitrification in plants as a natural form of cryoprotection. Cryobiology 24:214-228.

Hirsh, A. G., R. J. Williams, and H. T. Meryman. 1985. A novel method of natural cryopreservation: Intracellular glass formation in deeply frozen *Populus*. Plant Physiol. 79:41-56.

Houde, M., J. Danyluk, J. F. Laliberte, E. Rassart, R. S. Dhindsa, and F. Sarhan. 1992. Cloning, characterization, and expression of a cDNA encoding a 50-kilodaton protein specifically induced by cold acclimation in wheat. Plant Physiol. 99:1381-1387.

Hughes, M. A., M. A. Dunn, R. S. Pearce, A. J. White, and L. Zhang. 1992. An abscisic-acid-responsive, low temperature barley gene has homology with a maize phospholipid transfer protein. Plant Cell Environ. 15:861-865.

Ichiki, S., and H. Yamaya. 1982. Sorbitol in tracheal sap of dormant apple (*Malus domestica* Borkh.) shoots as related to cold hardiness. Pages181-187 in: Plant Cold Hardiness and Freezing Stress, Vol. II. P. H. Li and A. Sakai, eds. Academic Press, New York.

Iswari, S., and J. P. Palta. 1989. Plasma membrane ATPase as a site of functional alteration during cold acclimation and freezing injury. Pages 123-138 in: Low Temperature Stress Physiology in Crops. P. H. Li, ed. CRC Press, Boca Raton, FL.

Johnson-Flanagan, A. M., and J. Singh. 1986. Membrane deletion during plasmolysis in hardened and non-hardened plant cells. Plant Cell Environ. 9:299-305.

Johnson-Flanagan, A. M., and J. Singh. 1987. Alteration of gene expression during the induction of freezing tolerance in *Brassica napus* suspension cultures. Plant Physiol. 85:699-705.

Johnson-Flanagan, A. M., and J. Singh. 1988. Protein synthesis and freezing tolerance in plant cells. Crit. Rev. Plant Sci. 7:279-302.

Kacperska-Palacz, A., E. Dlugokecka, J. Breitenwald, and B. Wcislinska. 1977. Physiological mechanisms of frost tolerance: Possible role of protein in plant adaptation to cold. Biol. Plant. 10:10-17.

Keith, C. N., and B. D. McKersie. 1986. The effect of abscisic acid in the freezing tolerance of callus cultures of *Lotus corniculatus* L. Plant Physiol. 80:766-770.

Koster, K. L. 1991. Glass formation and desiccation tolerance in seeds. Plant Physiol. 96:302-304.

Krause, G. H., R. J. Klosson, and V. Troster. 1978. On the mechanism of freezing injury and cold acclimation of spinach leaves. Pages 58-76 in: Plant Cold Hardiness and Freezing Stress. Vol. II. P. H. Li and A. Sakai, eds. Academic Press, New York.

Krog, J. O., K. E. Zachariassen, B. Larsen, and O. Smidsrod. 1979. Thermal buffering in Afro-alpine plants due to nucleating agent- induced water freezing. Nature 282:300-301.

Kurkela, S., and M. Franck. 1990. Cloning and characterization of a cold- and ABA-inducible *Arabidopsis* gene. Plant Mol. Biol. 15:137-144, 1990.

Kurkela, S., and M. Borg-Franck. 1992. Structure and expression of *kin* 2, one of two cold- and ABA-induced genes of *Arabidopsis thaliana*. Plant Mol. Biol. 19:689-692.

Lang, V., P. Heino, and E. T. Palva. 1989. Low temperature acclimation and treatment with exogenous abscisic acid induce common polypeptides in *Arabidopsis thaliana* L. Heynh. Theor. Appl. Genet. 77:729-734.

Lee, S. P., B. Zhu, T. H. H. Chen, and P. H. Li. 1992. Induction of freezing tolerance in potato (*Solanum commersonii*) suspension cultured cells. Physiol. Plant. 84:41-48.

Lee, S. P., T. H. H. Chen, and L. H. Fuchigami. 1990. Changes in the translatable RNA population during abscisic acid induced freezing tolerance in bromegrass suspension culture. Plant Cell Physiol. 32:45-56.

Lee, S. P., and T. H. H. Chen. 1993a. Molecular biology of plant cold hardiness development. Pages 1-29 in: Advances in Plant Cold Hardiness. P. H. Li and L. Christersson, eds. CRC Press, Boca Raton, FL.

Lee, S. P., and T. H. H. Chen. 1993b. Molecular cloning of abscisic acid-responsive mRNA's expressed during the induction of freezing tolerance in bromegrass (*Bromus inermis* Leyss) suspension culture. Plant Physiol. 101:1089-1096.

Leopold, A. C., F. Bruni and R. J. Williams. 1992. Water in dry organisms. Pages 161-169 in: Water and Life. G. Somero, ed.. Springer-Verlag. Berlin.

Levitt, J. 1980. Responses of plants to environmental stresses. In: Chilling, Freezing and High Temperature Stresses, Vol. 1. Academic Press, New York.

Levitt, J., and G. W. Scarth. 1936. Frost-hardening studies with living cells. I. Osmotic and bound water changes in relation to frost resistance and the seasonal cycle. Can. J. Res. Sect. C. 14:267-305.

Limin, A. E., and D. B. Fowler. 1983. Genetics of cold hardiness. Pages 139-207 in: New Frontiers in Winter Wheat Production. D. B. Fowler, L. V. Gusta, A. E. Slinkard, and B. A. Hobins, eds. Univ. Saskatchewan, Saskatoon.

Lin, C., W. W. Guo, E. Everson, and M. F. Thomashow. 1990. Cold acclimation in *Arabidopsis* and wheat. Plant Physiol. 94:1078-1083.

Lin, C., and M. F. Thomashow. 1992 A cold-regulated *Arabidopsis* gene encodes a polypeptide having potent cryoprotective activity. Biochem. Biophys. Res. Commun. 183:1103-1108.

Luo, M., J. H. Liu, S. Mohapatra, R. D. Hill, and S. S. Mohapatra. 1992. Characterization of a gene family encoding abscisic acid- and environmental stress-inducible proteins of alfalfa. J. Biol. Chem. 267:15367-15374.

Luo, M., L. Lin, R. D. Hill and S. S. Mohapatra. 1991. Primary structure of an environmental stress and abscisic acid-inducible alfalfa protein. Plant Mol. Biol. 17:1267-1269.

Malone, S. R., and E. N. Ashowrth. 1991. Freezing stress response in woody tissues observed using low-temperatgure scanning electron microscopy and freeze substitution techniques. Plant Physiol. 95:871-881.

Mohapatra, S. S., L. Wolfraim, R. J. Poole, and R. S. Dhindsa. 1989. Molecular cloning and relationship to freezing tolerance of cold-acclimation-specific genes of alfalfa. Plant Physiol. 89:375-380.

Neven, L. G., D.W. Haskell, C.L. Guy, N. Denslow, P.A. Klein, L.G. Green, and A. Silverman. 1992. Association of 70-kilodalton heat-shock cognate proteins with acclimation to cold. Plant Physiol. 99:1362-1369.

Nordin, K., P. Heino, E. T. Palva. 1991. Separate signal pathways regulate the expression of a low temperature-induced gene in *Arabidopsis thaliana*. Plant Mol. Biol. 16:1061-1071.

Ogolevets, I. V. 1976. Study of hardening of isolated callus tissue of trees with different frost resistance. Fiziol. Rast. 23:139-145.

Olien, C. R. 1964. Freezing processes in the crown of "Hudson" barley, *Hordeum vulgare* (L.) 'Hudson'. Crop Sci. 4:91-95.

Olien, C. R. 1965. Interference of cereal polymers and related compounds with freezing. Cryobiology 2:47-54.

Olien, C. R. 1967. Freezing stresses and survival. Annu. Rev. Plant Physiol. 18:387-408.

Olien, C. R., and M. N. Smith. 1977. Ice adhesions in relation to freeze stress. Plant Physiol. 60:499-

503.
Parker, J. 1959. Seasonal variations in sugars of conifers with some observations on cold resistance. Forest Sci. 5:56-63.
Paroschy, J. H., A. G. Meiering, R. L. Peterson, G. Hostetter, and A. Neff. 1980. Mechanical winter injury in grapevine trunks. Am. J. Enol. Vitic. 31:227-232.
Pearce, R. S. 1985. A freezing-fracture study of the cell membranes of wheat adapted to extra cellular freezing and to growth at low temperatures. J. Exp. Bot. 36:369-381.
Perras, M., and F. Sarhan. 1989. Synthesis of freezing tolerance proteins in leaves, crown, and roots during cold acclimation of wheat. Plant Physiol. 89:577-585.
Proebsting, E. L. Jr., and D. C. Gross. 1988. Field evaluations of frost injury to deciduous fruit trees as influenced by ice nucleation-active *Pseudomonas syringae*. J. Am. Soc. Hortic. Sci. 113:498-506.
Raese, J. I., M. V. Williams, and H. D. Billingsley. 1977. Sorbitol and other carbohydrates in dormant apple shoots as influenced by controlled temperatures. Cryobiology 14:373-378.
Rajashekar, C., and M. J. Burke. 1982. Liquid water during slow freezing based on cell water relations and limited experimental testing. Pages 211-220 in: Plant Cold Hardiness and Freezing Stress, vol. 2. P. H. Li and A. Sakai, eds. Academic Press, New York.
Rajashekar, C., P. H. Li, and J. V. Carter. 1983. Frost injury and heterogeneous ice nucleation in leaves of tuber-bearing *Solanum* species. Plant Physiol. 71:749-755.
Robertson, A. J., L. V. Gusta, M. J. T. Reaney, and M. Ishikawa. 1987. Protein synthesis in bromegrass (*Bromus inermis* Leyss) cultured cells during the induction of frost tolerance by abscisic acid or low temperature. Plant Physiol. 84:1331-1336.
Sakai, A. 1962. Studies on the frost hardiness of woody plants. 1. The causal relation between sugar content and frost hardiness. Contrib. Inst. Low Temp. Sci. B 11:1-40.
Sakai, A., and W. Larcher. 1987. Frost Survival of Plants. Berlin, Springer-Verlag. 321 Pages
Siminovitch, D., and D. R. Briggs. 1953. Studies on the chemistry of the living bark of the black locust in relation to its hardiness. III. The validity of plasmolysis and desiccation tests for determining the frost hardiness of bark tissues. Plant Physiol. 28:15-34.
Siminovitch, D., and Y. Cloutier. 1982. Twenty-four-hour induction of freezing and drought tolerance in plumules of winter rye seedlings by desiccation stress at room temperature in the dark. Plant Physiol. 69:250-255.
Siminovitch, D., and G. W. Scarth. 1938. A study of the mechanism of frost injury to plants. Can. J. Res. C16:467-481.
Singh, N. K., D. E. Nelson, D. Kuhn, P. M. Hasegawa, and R. A. Bressan. 1989. Molecular cloning of osmotin and regulation of its expression by ABA and adaptation to low water potential. Plant Physiol. 90:1096-1101.
Steponkus, P. L., M. F. Dowgert, R. Y. Evans, and W. Gordon-Kamm. 1982. Cryobiology of isolated protoplasts. Pages 459-474 in: P. H. Li and A. Sakai, eds. Plant Cold Hardiness and Freezing Stress, Vol 2. Academic Press, New York.
Steponkus, P. L., M. Uemura, R. A. Balsamo, T. Arvinte, and D. V. Lynch. 1988. Transformation of the cryo-behavior of rye protoplasts by modification of the plasma membrane lipid composition. Proc. Natl. Acad. Sci. USA 85:9026-9030.
Steponkus, P. L., D. V. Lynch, and M. Uemura. 1990. The influence of cold acclimation on the lipid composition and cryo-behavior of the plasma membrane of isolated rye protoplasts. Phil. Trans. R. Soc. Lond. B. 326:571-583.
Stone, J. M., J. P. Palta, J. B. Bamberg, L. S. Weiss, and J. F. Harbage. 1993. Inheritance of freezing tolerance in tuber-bearing *Solanum* species: Evidence for independent genetic control of nonacclimated freezing tolerance and cold acclimation capacity. Proc. Natl. Acad. Sci. USA 90:7869-7873.
Sugawara, Y., and A. Sakai. 1978. Cold acclimation of callus cultures of Jerusalem artichoke. Pages 197-210 in: Plant Cold Hardiness and Freezing Stress, vol. 1. P. H. Li and A. Sakai, eds. Academic Press, New York.
Sutka, J., and O. Veisz. 1988. Reversal of dominance in a gene on chromosome 5A controlling frost resistance in wheat. Genome 30:313-317.
Sutka, J. 1981. Genetic studies of frost resistance in wheat. Theor. Appl. Genet. 59:145-152.
Tanino, K. K., T. H. H. Chen, L. H. Fuchigami, and C. J. Weiser. 1991. Abscisic acid-induced cellular alterations during the induction of freezing tolerance in bromegrass cells. J. Plant Physiol. 137:619-624.
Tanino, K. K., T. H. H. Chen, L. H. Fuchigami, and C. J. Weiser. 1990. Metabolic alterations associated with abscisic acid induced frost hardiness in bromegrass suspension culture cells. Plant Cell Physiol. 31:505-511.
Thomashow, M. F., S. J. Gilmour, and C. Lin. 1993. Cold-regulated genes of *Arabidopsis thaliana*.

Pages 31-44 In: Advances in Plant Cold Hardiness. P. H. Li and L. Christersson, eds. CRC Press, Boca Raton, FL.

Thomashow, M. F. 1990. Molecular genetics of cold acclimation in higher plants. Adv. Genet. 28:99-131.

Toivio-Kinnucan, M. A., H. H. Chen, P. H. Li, and C. Stushnoff. 1981. Plasma membrane alterations in callus tissues of tuber-bearing *Solanum* species during cold acclimation. Plant Physiol. 67:478-483.

Trunova, I. I. 1982. Mechanism of winter wheat hardening at low temperature. Pages 41-54 in: Plant Cold Hardiness and Freezing Stress, vol. 2. P. H. Li and A. Sakai, eds. Academic Press, New York.

Trunova, I. I., and C. N. Zrereva. 1974. Protective effect of sugars during freezing of chloroplast suspensions. Fiziol. Rast. 21:1000-1006.

Uemura, M., and S. Yoshida. 1984. Involvement of plasma membrane alterations in cold acclimation of winter rye seedlings (*Secale cereale* L. cv. Puma). Plant Physiol. 75:818-826.

Volger, H. G., and U. Heber. 1975. Cryoprotective leaf proteins. Biochem. Biophys. Acta. 412:335-349.

Wallner, S. J., M. T. Wu, and S. J., anderson-Krengel. 1986. Changes in extracellular polysaccharides during cold acclimation of cultured pear cells. J. Am. Soc. Hortic. Sci. 111:769-773.

Weretilnyk, E., W. Orr, T. C. White, B. Lu, and J. Singh. 1993. Characterization of three related low-temperature-regulated cDNAs from winter *Brassica napus*. Plant Physiol. 101:171-177.

Weiser, R. L., S. J. Wallner, and J. W. Waddell. 1990. Cell wall and extensin mRNA changes during cold acclimation of pea seedlings. Plant Physiol. 93:1021-1026.

Williams, R. J., and A. C. Leopold. 1989. The glassy state in corn embryos. Plant Physiol. 89: 977-981.

Yoshida, S., and A. Sakai. 1974. Phospholipid degradation in frozen plant cells associated with freezing injury. Plant Physiol. 53:509-511.

Yoshida, S., and M. Uemura. 1984. Protein and lipid composition of isolated plasma membranes from orchard grass (*Dactylis glomerata* L.) and changes during cold acclimation. Plant Physiol. 75:31-37.

Yoshida, S. 1984a. Studies on freezing injury of plant cells. 1. Relation between thermotropic properties of isolated plasma membrane vesicles and freezing injury. Plant Physiol. 75:38-42.

Yoshida, S. 1984b. Chemical and biophysical changes in the plasma membrane during cold acclimation of mulberry bark cells (*Morus bombycis* Koidz. cv. Goroji). Plant Physiol. 76:257-265.

Zhu, B, T. H. H. Chen, and P. H. Li. 1993. Expression of an ABA-responsive osmotin-like gene during the induction of freezing tolerance in *Solanum commersonii*. Plant Mol. Biol. 21:729-735.

CHAPTER 8

# Ice Nucleation Activity Associated with Plants and Fungi

**Edward N. Ashworth and Thomas L. Kieft**

While the majority of attention regarding biological ice nucleation has been focused on a few species of epiphytic bacteria, the ability to initiate freezing at relatively warm temperatures has evolved in many eukaryotic organisms, as well. Examples of organisms that generate ice nucleation sites have been reported among plants (Ashworth et al., 1985a; Gross et al., 1988), fungi (Kieft, 1988), and animals (Duman et al., 1984; Storey and Storey, 1985). This chapter deals with the first two of these kingdoms of life: the plants and the fungi. In the plants, a major research question concerns the relative importance of ice nucleation sites produced by epiphytic ice nucleation-active bacteria and intrinsic nucleation sites synthesized by the plants themselves. Another important question is the degree to which plant tissue actually supercools in nature. The first part of this chapter reviews the field and laboratory methods of quantifying ice nucleation activity (INA), particularly with respect to plant tissue. This is followed by a discussion of the distribution of INA in plant tissue and of the relative importance of bacterial and plant INA. The last section presents information on two groups of fungi that have recently been discovered to produce intrinsic warm-temperature ice nucleation sites. These comprise several species of lichen fungi and a few species of the free-living Ascomycete *Fusarium*. Discussion of the possible selective advantages and biotechnological applications of both plant and fungal INA are presented.

## Formation of Ice on and in Plant and Fungal Tissues

### Temperature Range Over Which Freezing Is Initiated

Although water can be supercooled to −38°C, the homogeneous ice nucleation temperature; the extent of supercooling in plant and fungal tissues is limited. Freezing of these tissues in nature is generally initiated at warmer temperatures and occurs by heterogeneous ice nucleation. Attempts to characterize the ice nucleation sites associated with these tissues invariably begin with experiments to determine the temperature range over which freezing is initiated. Published reports indicate that freezing in plants and fungi is initiated over a wide range of temperatures (−1 to −10°C) (Salt and Kaku, 1967; Marcellos and Single, 1979; Cary and Mayland, 1970; Krog et al., 1979; Yelenosky, 1983; Ashworth et al., 1985b; Anderson and Ashworth, 1985; Kieft and Ahmadjian, 1989; among others). However, these differences are likely a better reflection of measuring methods than of differences between species (Ashworth, 1992).

**Field Determinations**

The technique that should provide the most realistic assessment of when freezing is initiated in plants and fungi involves direct measurements on intact organisms during a frost episode in nature. Such measurements utilize remote data-logging equipment to monitor the temperature of plant tissues and an inert reference in the field. Freezing can be detected because the accompanying release of the heat of fusion raises tissue temperature relative to the temperature of the inert reference (Fig. 1). Such direct measurements of ice formation on both woody (Ashworth and Davis, 1984, 1986; Ashworth et al., 1985b) and herbaceous species (Anderson and Ashworth, 1985; Ashworth et al., 1985c) growing in the field demonstrate that freezing is initiated just below 0°C (Table 1). These observations indicate that effective ice nucleation sites are commonly associated with a variety of species.

Although direct field measurements allow determinations to be made on intact plants and in natural environments, this has not been the method of choice for most investigators. Measuring tissue temperature in the field is more troublesome and requires more specialized equipment than is needed for laboratory or growth chamber measurements. In addition, the environmental conditions in the field are both variable and unpredictable. Therefore, each determination will be made under unique conditions and cannot be readily replicated. Scheduling experiments is also difficult and depends upon changing weather conditions. This can be particularly frustrating if measurements are to be made on tissues at specific stages of development, such as developing flower buds of perennial fruit crops. Researchers investigating ice formation in frost-tender species have the additional disadvantage

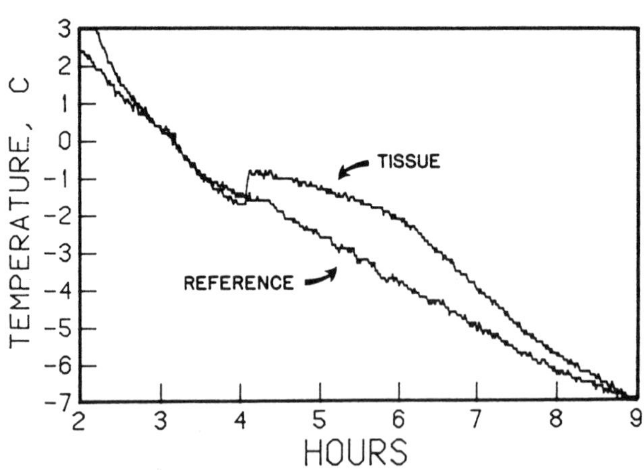


**Figure 1.** Thermal analysis of an intact nectarine tree growing in the orchard. The plot is a comparison of ambient reference and scaffold limb temperature. A distinct exotherm was observed near −2°C due to the freezing of water within the plant tissue. Measurements began at dusk with the elapsed time indicated on the x-axis (Reprinted, with permission, from Ashworth et al., 1985b).

**Table 1.** Temperature at which freezing was initiated in plants under field conditions[a]

| Species | Mean ice initiation temperature (°C) |
|---|---|
| *Prunus persica* | −1.6 |
| *Malus domestica* | −1.3 |
| *Fagus sylvatica* | −1.6 |
| *Cornus florida* | −1.8 |
| *Ilex crenata* | −1.6 |
| *Juniperus chinensis* | −1.3 |
| *Pyrus communis* | −2.1 |
| *Rubus* spp. | −1.6 |
| *Pinus strobus* | −1.2 |
| *Taxus cuspidata* | −2.0 |
| *Lycoperiscon esculentum* | −2.0 |
| *Zea mays* | −2.5 |
| *Glycine max* | −2.7 |
| *Phaseolus vulgaris* | −2.7 |
| *Gossypium hirsutum* | −2.5 |

[a] Data compiled from Anderson and Ashworth (1985), Ashworth et al. (1985b), and Ashworth and Davis (1986).

that their experimental material will not survive to allow for additional determinations within the same growing season. Despite these inconveniences, there is no substitute for field measurement to provide an accurate description of what occurs in nature.

Thermal imaging has recently been demonstrated to be a novel and useful means to observe and follow plant freezing remotely (Le Grice et al., 1993). This technology enables the thermal image to be recorded or videotaped over time to show nucleation points and ice propagation in the foliage. The images can be color enhanced to provide a means of accurately determining the temperature of individual leaves. Since the images are recorded by infrared, the freezing events can be recorded in both the daylight and at night.

### Controlled Environment and Laboratory Determinations

To conduct investigations beyond the descriptive phase requires alternative approaches. Studies to determine if ice nucleation temperature can be manipulated, or to investigate the nature of the ice nucleation sites in plant and fungal tissues, require a more convenient and more rigorously controlled experimental protocol. This has led investigators to use either controlled environment equipment or other laboratory procedures. Experiments have been conducted using both portable cooling equipment to freeze whole plants or plant parts in the field (Gartman et al., 1992) and controlled environment chambers to expose potted plants or excised tissues to subzero temperatures (Cary and Mayland, 1970; Rajashekar et al., 1983; Yelenosky, 1983; Ashworth et al., 1985a; among others). This equipment enables investigators to expose plants to freezing conditions at their convenience. Usually temperatures are uniform throughout the chamber. Cooling rates can be controlled, and plants can be exposed to reproducible conditions in successive experiments. Depending upon the size of the equipment, intact plants or large specimens can be examined. The principal disadvantage of using such equipment is that specimen tissues are primarily cooled by advection. Therefore, most controlled environment equipment cannot mimic the conditions experienced during a radiation frost.

A variety of laboratory procedures have been used to measure when ice formation is initiated in plants. Measurements have been conducted using excised tissues (i.e., leaves, flowers, stem pieces, leaf disks) (Salt and Kaku, 1967; Marcellos and Single, 1979; Lindow et al., 1982; Proebsting et al., 1982; Rajashekar et al., 1983; Ashworth and Davis, 1984; among others), or tissue homogenates (Rajashekar et al., 1983; Ashworth and Davis, 1984; Ashworth et al., 1985c). Generally, either thermocouples or similar devices are used to measure the release of the heat of fusion upon freezing, or samples are placed in water and freezing of the surrounding solution is monitored. These laboratory procedures have the advantage of precise cooling rates, uniform temperatures, and reproducible conditions. They also facilitate analysis of replicate samples and enable investigators to estimate the concentration of ice nucleation-active sites within a tissue. The main disadvantages of these techniques are the inability to simulate radiation frost conditions and the necessity of extrapolating data obtained from freezing small tissue specimens to predict the behavior of intact organisms.

Published data obtained from experiments using controlled environment chambers or laboratory procedures often vary. While some studies have measured freezing initiated at –2 to –3°C (Ashworth and Davis, 1984; Anderson and Ashworth, 1985; Ashworth et al., 1985a,c), others have found that tissues supercool to –5 to –10°C before freezing (Marcellos and Single, 1976, 1979; Rajashekar et al., 1983). These latter reports suggest that ice nucleation-active sites effective at temperatures above –5°C are not commonly associated with plant tissues.

## Factors Affecting Ice Formation in Plant and Fungal Tissues

There would be no compelling reason to review the different techniques used to measure ice nucleation if each of the protocols mentioned gave similar results. Unfortunately, this does not appear to be the case, and published values for comparable specimens range from –1 to –10°C. While some of this variability could reflect differences among species, much of the variation likely reflects differences in experimental protocol. If investigators are to attempt to identify the source of INA associated with plant and fungal tissues, then an awareness of factors which affect ice nucleation is critical.

### Specimen Size

It has been well documented that small specimens of plant and fungal tissue will supercool to lower temperatures than either larger specimens or intact organisms. This effect has been observed in a number of species including peach (Ashworth and Davis, 1984; Ashworth et al., 1985a,c); *Buxus* (Kaku, 1971); conifers (Kaku and Salt, 1968); tomato (Anderson and Ashworth, 1985); maize, cotton, bean, soybean (Ashworth et al., 1985c); and lichens (Ashworth and Kieft, 1992). An example of this phenomenon is presented in Table 2. The INA of bean (*Phaseolus vulgaris* L.) was measured by various techniques that utilized different amounts of tissue. In one instance, tissue was homogenized in distilled water, and 10-µl suspensions of homogenized tissue were used for the assay. An analysis of 100 replicate samples noted that the first drop that froze had supercooled to –9.8°C, and the median nucleation temperature was –10.5°C. From these data, one might conclude that bean seedlings did not contain ice nucleation-active sites effective above –9°C. However, this conclusion would be incorrect. Bean seedlings exposed to subzero temperatures, either in a controlled environment chamber or in the field, super-

**Table 2.** Effect of assay technique and tissue mass on the temperature at which ice nucleation occurred in bean[a]

| Specimen | Mean tissue mass | Temperature of nucleation (°C) First | Median |
|---|---|---|---|
| 2% homogenate, 10 μl | 0.2 mg | −9.8 | −10.5 |
| Leaf discs, 6.5 mm OD | 4.5 mg | −7.5 | −11.0 |
| 2% homogenate, 1 ml | 20 mg | −7.0 | −8.0 |
| Seedlings in lab | 2.7 g | −1.3 | −3.9 |
| Seedlings in field | 2.7 g | −1.3 | −2.7 |

[a] Data from Ashworth et al. (1985c).

**Table 3.** Influence of both specimen size and inoculation with ice nucleation-active ($Ina^+$) bacteria on ice nucleation within tomato plant tissues[a,b]

| Mean tissue mass (g) | Mean temperature of nucleation (°C) $Ina^+$ bacteria | Control |
|---|---|---|
| 0.1 | −3 | −8 |
| 1 | −3 | −6 |
| 10 | −2 | −4 |
| 100 | −2 | −2 |

[a] Data recalculated from Anderson and Ashworth (1985).
[b] Inoculated tissues were sprayed with a solution of $10^8$ *Pseudomonas syringae* cells per milliliter.

cooled very little before freezing (Table 2). These results demonstrate that using small tissue specimens to assay INA and subsequently extrapolating from these results to predict what might occur with intact plants under field conditions will often be inappropriate and misleading.

There are several possible explanations for the above phenomenon. One possibility is that ice nucleation-active sites are infrequent and not uniformly distributed within tissues. Therefore, smaller specimens will be less likely to contain an ice nucleation site active at warm temperatures and more likely to supercool. Specimen size would have no effect if ice nucleation sites were abundant and evenly distributed throughout the tissue. An example of this was noted in experiments conducted with tomato plants (Table 3). Plants inoculated with $Ina^+$ bacteria froze between −2 and −3°C regardless of the specimen size. While inoculated plants contained numerous ice nucleation sites, control plants that had not been inoculated contained few sites, and a logarithmic relationship between specimen size and the temperature of ice nucleation was observed (Anderson and Ashworth, 1985).

An alternative explanation for why specimen size influences the extent of supercooling is based on the premise that ice nucleation occurs as a function of probability. Therefore, the larger the sample, the greater the probability that a nucleation event will occur. A third possibility incorporates aspects of both previous explanations. Larger specimens would be more likely to contain effective ice nucleation sites, and there would be a greater probability that one of these sites would catalyze ice nucleation. This last possibility would appear most consistent with a modified singular hypothesis of ice nucleation (Vali and Stansbury, 1966). This hypothesis states that heterogeneous ice nucleation sites have a characteristic temperature range at which ice formation is catalyzed, but a stochastic element influences whether an ice nucleus of critical size is formed.

Irrespective of which of the above explanations accounts for the increased supercooling of small specimens relative to larger specimens, this phenomenon has important implications. One is that the ice nucleation sites within plants and lichens that catalyze ice formation at warmer temperatures are either present in low concentration or not very active. The second is that because of the low concentration and activity, assaying these sites and investigating their nature is more difficult.

### Cooling Rate

Temperatures generally decline gradually (<1–2°C/hour) during a natural freeze-stress (Levitt, 1980; Ashworth et al., 1985b; Steffens et al., 1989). However, comparable cooling rates are not always used when determining the temperatures at which freezing is initiated in plants and fungi. Justification for using faster cooling rates is often based on the premise that ice nucleation is strictly a function of temperature. This may not be entirely justified. Several studies have indicated that cooling rate can influence ice nucleation in plant tissues (Cary and Mayland, 1970; Yelenosky, 1983; Anderson and Ashworth, 1985; and Ashworth et al., 1985a,c). Therefore, although temperature has a greater effect on ice nucleation than length of low-temperature exposure, cooling rate can have a significant effect. Experiments that utilize rapid cooling rates have the potential to overestimate the extent of supercooling.

### Moisture on Tissue Surfaces

Water often condenses on tissue surfaces as ice or dew. External ice may inoculate the freezing of internal moisture (Cary and Mayland, 1970; Single and Marcellos, 1981), and many researchers apply moisture to tissue surfaces prior to conducting laboratory freezing experiments in an attempt to simulate a hoar frost. In a variety of species, the application of moisture to tissue surfaces reduces the extent of supercooling compared with dry controls (Modlibowska, 1962; Cary and Mayland, 1970; Thomas and Barber, 1974; Yelenosky, 1983; Ashworth et al., 1985c; Cary and Lindow, 1986). The observation that moistened tissues freeze at warmer temperatures than do comparable dry specimens suggests that ice nucleation sites on tissue surfaces are more effective than internal sites. Presumably these external nucleation sites are only effective when sufficient water is present to permit nucleation and the subsequent spread of ice crystals into internal tissues. Whether these external ice nucleation sites are a component of the plant tissue, a foreign component deposited from the atmosphere, epiphytic Ina$^+$ bacteria, or airborne ice crystals has not been established and could vary depending upon plant species and environmental conditions.

## Location of Ice Crystals On and In Plant and Fungal Tissues

### Formation of Ice on Plant and Fungal Surfaces

Plant and fungal tissues frequently exhibit external ice crystals during frost episodes. These crystals can appear as either dendritic ice crystals or frozen dew. In the first case, ice probably formed by direct deposition from the vapor phase; whereas in the latter, water first condensed on the tissue surface as a liquid and subsequently froze. Both ice deposition and water condensation are enhanced when tissue temperature drops below air temperature. Tissues exposed to clear night skies and calm conditions will radiate heat energy and become cooler than the surrounding air. Hoar frost will form on objects that cool below freezing. Water will condense as dew when the surface temperature drops below the dew point. If the temperature continues to decline and goes below freezing, dew can subsequently freeze.

Both direct ice deposition and water condensation involve a phase transition and require nucleation. On the basis of the environmental conditions at which they occur in the field, it is likely that each event is catalyzed by heterogeneous nucleation. Ice formation could be catalyzed by a constitutive ice nucleation-active component associated with the tissue, a material or microorganism deposited as an aerosol on the tissue surface prior to the frost episode, or by ice crystals deposited from the atmosphere. Whatever catalyzes ice formation must be widely dispersed, since the distribution of both dew and hoar frost can be extensive on exposed surfaces.

**Formation of Ice Within Tissues**

When ice does form within plant and fungal tissues, it generally forms extracellularly. Examinations of frozen tissues have revealed that ice forms in the intercellular spaces, and that cells contract during freezing (Wiegand, 1906; Siminovitch and Scarth, 1938; Levitt, 1980; Pearce, 1988; Pearce and Ashworth, 1992). It is believed that intracellular ice formation rarely, if ever, occurs in the field (Levitt, 1980). When intracellular ice does form, it is likely seeded by extracellular ice and results from membrane failure (Morris and McGrath, 1981; Dowgert and Steponkus, 1983).

It is not clear how extracellular ice is initiated within plant and fungal tissues. Several possible scenarios can be envisioned, and include seeding from external ice, nucleation of liquid water in the apoplast, or direct deposition or condensation from the vapor phase. The mechanism could vary among species and under different environmental conditions.

Ice formation within tissues may be seeded by external ice. Ice crystals readily form on tissue surfaces close to 0°C, and it has been demonstrated that inoculation of freezing on leaf surfaces can induce internal freezing (Kitaura, 1967a,b; Cary and Mayland, 1970). In addition, plants tend to supercool to lower temperatures before freezing if they are freeze-stressed in a growth chamber or laboratory environment where external ice does not form (Cary and Mayland, 1970; Yelenosky, 1983; Ashworth et al., 1985c; Cary and Lindow, 1986). This effect can be diminished by application of water to leaf surfaces prior to the freeze-stress. This treatment decreases the extent of supercooling and promotes freezing at warmer temperatures (Modlibowska, 1962; Cary and Mayland, 1970; Thomas and Barber, 1974; Yelenosky, 1983; Ashworth et al., 1985c). These observations suggest that tissue freezing is catalyzed by external ice nucleation-active sites. It is not clear, however, how ice on the tissue surface could propagate and seed intercellular ice formation. Plant cuticles are an effective barrier to water movement, and would likely prevent ice propagation as well. Instead, ice may propagate through the stomata. Even when closed, stomata may provide openings large enough for ice propagation. However, the greatest barrier to propagation through the stomata may be the requirement that a continuous film of liquid water be present between the external ice and internal tissue water. There is no evidence that such a film of water exists, and this seems particularly unlikely during conditions when ice is directly deposited on the tissue surface from the vapor phase. Therefore, despite the observations that ice readily forms on plant surfaces, there is no direct evidence that external ice seeds the formation of internal ice crystals during a frost episode in the field.

External ice formation is not a prerequisite for ice formation within a tissue. Ice can form internally in the absence of external ice, and freezing may be initiated internally by the nucleation of water within the apoplast. Several studies have reported that ice formation was initiated near the xylem, and ice subsequently spread through the vascular system (Salt and Kaku, 1967; Kaku, 1971). These observations suggest that an ice nucleation-active component is associated with tracheids and xylem vessels. Unfortunately, these observations may not be relevant to what occurs in the field. The investigators used a cryomicroscope and examined thin tissue specimens mounted in water or oil and cooled by conduction. These tissues supercooled more than 7 to 10°C before freezing, so neither the temperature at which freezing was initiated nor the conditions in which the tissues were cooled resembled those experienced by tissues in the field.

Ice may form within plant tissues via direct deposition from the vapor phase or by the freezing of condensed water. At night with the stomata closed, the humidity within the intercellular spaces should approach 100%. When the temperature drops below 0°C, the internal gas phase becomes saturated with respect to water and supersaturated with respect to ice (Rogers and Yau, 1989). If suitable ice nucleation sites are present on cell surfaces within the intercellular spaces, direct deposition of ice from the vapor phase or condensation of moisture may occur.

**Distribution of Ice Within Plant Tissues and Its Potential Relationship to the Distribution of Ice Nucleation Sites**

Unfortunately, our understanding of how ice forms within plants and fungi is limited. Techniques are not available that permit an examination of ice nucleation, crystal growth, and subsequent spread of ice within complex tissues under conditions representative of natural frost episodes. As an alternative, researchers have examined the distribution of ice within frozen tissues and tried to deduce where freezing was initiated and how ice spread. These studies demonstrated that ice crystals are not uniformly distributed. For example, large ice crystals form within the scales of overwintering buds (Wiegand, 1906; Quamme, 1978; Sakai, 1982; Ishikawa and Sakai, 1982, 1985; Ashworth et al., 1989; Ashworth, 1990), in the bark cortex of woody perennials (Wiegand, 1906; Ashworth et al., 1988; Malone and Ashworth, 1991), and just below the epidermis of leaf tissues (Nath and Fisher, 1971; Pearce and Ashworth, 1992).

The nonuniform distribution of ice within tissues may reflect an uneven distribution of ice nucleation sites. If true, this may provide plants a means of controlling the location of ice. It may also provide a clue regarding the nature of the ice nucleation sites. However, the distribution of ice could also reflect differences in both localized cooling rates and patterns of ice propagation. For example, the location of ice within cereal leaves varied depending upon the conditions at which freezing occurred. Pearce (1988) observed ice crystals associated with the vascular bundles of laboratory freeze-stressed leaves; whereas an examination of wheat leaves freeze-stressed in the field revealed that ice was primarily associated with the substomatal cavity, and formed near the epidermis (Pearce and Ashworth, 1992). It was proposed that the former pattern in which ice was associated with the vascular tissue may have resulted from seeding ice formation in the laboratory and the subsequent propagation of ice through the vascular tissue. Therefore, although examining the distribution of ice crystals within tissues may help identify the location of nucleation sites, such inferences may be misleading. Experimental conditions must be carefully evaluated so that observations reflect what occurs in nature.

## Ice Nucleation Activity Associated with Plants

Much of the INA associated with plant tissues appears to be of bacterial origin. Several species of bacteria exhibit INA, and two species—*Pseudomonas syringae* and *Erwinia herbicola*—are widely dispersed epiphytes that colonize a number of different plants (Lindow 1982, 1983a,b; and Chapters 2, 3, and 13 in this book). Lindow and co-workers reported that strains of $Ina^+$ bacteria were effective in catalyzing heterogeneous ice nucleation at temperatures associated with frost injury in plants (Arny et al., 1976; Lindow et al., 1978a,b, 1982a; Hirano et al., 1985). In addition, it has recently been shown that some fungal species also exhibit INA (Kieft, 1988; Pouleur et al., 1992; Tsumuki et al., 1992). Therefore, it is likely that microorganisms associated with plant tissues are a major source of ice nucleation sites.

Whether plants themselves contain intrinsic INA over this same temperature range has been debated. Several reports have indicated that in the absence of sufficient populations of $Ina^+$ bacteria, plant tissues would supercool to −8 to −10°C before freezing (Arny et al., 1976; Lindow et al., 1978b, 1982a; Anderson et al., 1982; Lindow, 1983a; Hirano et al., 1985; Marshall, 1988). Such observations have led to the proposal that controlling $Ina^+$ bacteria would provide a method of frost protection. In contrast, other investigators have noted that plant tissues often supercool less than 2 to 3°C even when $Ina^+$ bacteria populations are low or undetectable (Ashworth and Davis, 1984; Gross et al., 1984; Ashworth et al., 1985, Anderson and Ashworth, 1985; Andrews et al., 1986; Anderson et al., 1987; Cody et al., 1987; Proebsting and Gross, 1988; Constantinidou et al., 1991, 1992). The application of surface disinfectants, bactericides, chemicals inhibitory to bacterial ice nucleation, and antagonistic organisms, which lower the population of $Ina^+$ bacteria on the tissue, fail in some cases to alter either the temperature at which freezing was initiated or the sensitivity of the tissue to frost (Ashworth and Davis, 1984; Cody et al., 1987; Proebsting and Gross, 1988). Such evidence suggests that some plants contain a nonbacterial component active in catalyzing ice formation. Although apparently present in low concentrations, this intrinsic component may limit the extent of supercooling even in the absence of $Ina^+$ bacteria.

The above debate hinges on determining whether $Ina^+$ bacteria are present in sufficient populations on plants to catalyze ice formation. While apparently straightforward, answering this question has not been as easy as might be anticipated. This requires determining the population of $Ina^+$ bacteria on a tissue and the minimum population required to initiate ice formation. Ice nucleation-active bacteria are not uniformly distributed among leaves within a crop canopy, and can vary 500-fold (Hirano et al., 1982). This complicates estimating the population on tissues within a plant, or within a canopy. Estimating the minimum population necessary for ice nucleation is also difficult, since cells within a population of $Ina^+$ bacteria vary in their effectiveness of catalyzing ice formation (Maki et al., 1974). Therefore, the ice nucleation temperature is a function of both the population of $Ina^+$ bacteria and the proportion of cells within that population which are active ice nucleating agents. This latter component is called the nucleation frequency and refers to the number of ice nuclei detected within a population of bacteria (Lindow, 1982, 1983b; Lindow et al., 1982a,b). Reported values for bacterial ice nucleation frequencies range from one cell in ten, to one in $10^8$ cells. Nucleation frequency values are apparently a function of species, strain, growth temperature, media composition, and substrate consistency (Maki et al., 1974; Lindow et al., 1982b, 1983b; Hirano et al., 1985; Rogers et al., 1987; O'Brien and Lindow, 1988). Unfortunately the nucleation frequency of a bacterial strain in culture does not correlate with the nucleation frequency observed on plant tissues (O'Brien and Lindow, 1988). This complicates determining the relationship between the population of Ina$^+$ bacteria and the temperature at which freezing is initiated in plant tissues.

Maki and co-workers (1974) reported that dead *P. syringae* cells retained INA, and Lindow (1982, 1983a,b) suggested that dead bacteria may contribute significantly to the INA associated with plant tissues. The contribution of dead Ina$^+$ bacteria to INA is difficult to assess, since the number of dead bacteria cannot be readily measured. Likewise, one cannot assume that all plant-associated INA that can not be ascribed to living Ina$^+$ bacteria reflect the activity of dead bacteria.

In the absence of detectable Ina$^+$ bacteria, it is tempting to speculate that plants contain intrinsic INA. However, the inability to detect Ina$^+$ bacteria does not prove their absence. Bacteriological techniques have a limit of detection, and values below the detection limit should not be automatically equated to zero. Since ice nucleation frequency values as high as one cell in 10 have been reported, it is not possible to exclude Ina$^+$ bacteria as a source of INA associated with plant and fungal tissues. Likewise, it is also difficult to eliminate dead bacteria as a potential ice nucleation site. However, we must not assume that Ina$^+$ bacteria are the sole source of biological INA associated with plant and fungal tissues. Direct and indirect evidence indicates the presence of intrinsic plant components that exhibit INA.

## Woody Stem Tissues of *Prunus* Species

Several reports have failed to observe a direct relationship between the population of Ina$^+$ bacteria and the temperature at which freezing was initiated in *Prunus* stem tissues (Ashworth and Davis, 1984; Gross et al., 1984; Ashworth et al., 1985; Andrews et al., 1986; Cody et al., 1987; Proebsting and Gross, 1988). Ice formation was initiated between –2 and –3°C in these tissues both during laboratory experiments (Ashworth and Davis, 1984; Ashworth et al., 1985a; Anderson and Ashworth, 1987) and during natural freeze episodes in the field (Ashworth et al., 1985b). The temperature at which freezing was initiated was similar regardless of whether or not Ina$^+$ bacteria could be detected (Gross et al., 1984; Ashworth et al., 1985a, Andrews et al., 1986; Proebsting and Gross, 1988). Treatment of tissues with surface disinfestants, bactericides, chemicals that inhibit bacterial ice nucleation, or antagonistic bacteria did not affect either the temperature at which freezing was initiated or the frost-sensitivity (Ashworth and Davis, 1984; Cody et al., 1987; Proebsting and Gross, 1988). The combined indirect evidence suggested that these tissues contained an intrinsic component that could catalyze ice nucleation in the absence of Ina$^+$ bacteria.

The intrinsic component within *Prunus* stem tissues has been neither identified nor well characterized. Mature stem tissues supercool less than younger tissues, suggesting that this component develops during the maturation of the shoot (Ashworth et al., 1985; Gross et al., 1988). The ice nucleation temperature of mature shoots was consistent throughout the fall and winter (Ashworth, 1985a), and also did not change as bud development progressed (Ashworth and Davis, 1984; Gross et al., 1984). This suggested that INA is a constitutive feature of mature woody tissue. Properties of the intrinsic ice nucleation-active site(s) are different from those of the active protein associated with $Ina^+$ bacteria. Gross and co-workers (1988) observed that INA associated with *Prunus* stems was insensitive to pronase, *p*-hydroxymercuribenzoate, and iodine. All three of these treatments inactivate the bacterial ice nucleating protein. They also noted that INA of *Prunus* stems was reduced by soaking the tissue in water, a treatment that did not affect activity associated with $Ina^+$ bacteria. The intrinsic ice nucleating component of *Prunus* stems exhibited differences in heat stability and sensitivity to several compounds which inhibit bacterial INA (Ashworth and Davis, 1984; Gross et al., 1988). Homogenizing stem tissues also reduced INA which suggests that either structural activity is necessary for ice nucleation or that a compound which interferes with ice nucleation is released upon homogenization.

### Leaf Tissues of *Veronica, Pisum, Vicia,* and *Buxus*

Kaku (1973) observed that leaf tissues of *Veronica persica* Poir., *Pisum sativum* L., *Vicia faba* L., *V. sativa* L., and *Buxus microphylla* Siebold & Zucc. freeze at warmer temperatures than stem tissues. In *Veronica* leaves, the most effective ice nucleation sites were associated with the leaf blade. In contrast, the more effective nucleation sites of *Buxus* were associated with the midrib rather than the blade, petiole, or stem. Homogenizing either *Veronica* or *Buxus* leaves reduced the INA. Following homogenization, the residual insoluble materials were more effective at catalyzing ice formation than the supernatant (Kaku, 1973). This suggested that the ice nucleation sites were associated with structural components of the leaf. In an earlier study, Kaku (1971) observed that mature *Buxus* leaves supercooled less than younger leaves, and speculated that this was related to maturation of the vascular system.

### Inflorescence of *Lobelia telekii*

*Lobelia telekii* grows in the alpine zone on Mount Kenya and is exposed to frost episodes throughout the year (Krog et al., 1979). The inflorescence of this plant has a hollow central cylinder that contains a slightly viscous fluid. During a frost episode, ice formation is apparently initiated within the central cylinder. As ice forms within this cylinder, the release of the heat of fusion maintains inflorescence temperature near 0°C and provides a measure of frost protection. In nature, these plants supercool little before freezing. An analysis of the fluid within the hollow cylinder of the inflorescence revealed that it contained an ice nucleating agent active at –4.6°C (Krog et al., 1979; Zachariassen and Hammel, 1988). Identification of the ice nucleation-active component within *Lobelia* has not been reported. Nucleation activity is not inactivated by 100°C treatment, and the fluid is primarily composed of a carbohydrate polymer (Krog et al., 1979). Nucleation assays with diluted samples indicate that nucleation sites are present in low concentration (Zachariassen and Hammel, 1988).

### Leaves of *Secale cereale*

Griffith and co-workers concluded that epiphytic $Ina^+$ bacteria could not account for the INA associated with rye (*Secale cereale* L.) leaves and that an intrinsic ice nucleation-active component initiated freezing in this tissue (Brush et al., 1994). The temperature threshold for nucleation of mesophyll cell suspensions was –7°C, and it was estimated that one such nucleation site was present per $10^5$ cells. Treatment of cell suspensions with either heat, urea, pronase E, or sulfhydryl reagents reduced INA, and thus suggested that the ice nucleation site had a proteinaceous component. Ice nucleation activity could also be reduced by treating cell suspensions with phospholipase C, boric acid, and periodic acid. The combined evidence suggests that the intrinsic ice nucleation-active components within rye leaves are complexes of proteins, phospholipids, and carbohydrates.

In other studies, Griffith and co-workers reported that proteins within the apoplast of rye leaves exhibit both antifreeze and INA (Griffith et al., 1992, 1993; Marentes et al., 1993). Both quantitative and qualitative changes in apoplast proteins were observed during cold acclimation, and one or more of the isolated proteins apparently had INA (Marentes et al., 1993). Apoplast extracts from acclimated rye leaves had a threshold ice nucleation temperature of –9°C compared with –13°C for extracts from nonacclimated leaves. Trypsin and boiling treatments reduced INA of these extracts. The extraction of apoplast proteins did not change the ice nucleation temperature of rye leaves (Marentes et al., 1993). Whether this limited supercooling results from incomplete extraction of apoplast protein(s) or another nonextractable component is not known. However, extraction of apoplast proteins did reduce the freeze tolerance of rye leaves, suggesting that these proteins play an important role in low-temperature survival.

### Citrus Tissues

Based upon the differential sensitivity of ice nucleation-active populations and ice nucleation temperature to various chemical treatments, it was concluded that both bacterial and nonbacterial ice nucleation sites were present on citrus (Constantinidou et al., 1991; Constantinidou and Menkissoglu, 1992). The nonbacterial ice nucleating agent was sensitive to protease, guanidine HCl, and copper sulfate and so was likely to contain proteinaceous components. The nonbacterial ice nucleating agent did not appear to be associated with fungi or other microbes; however, until characterized further, it cannot be assumed to be an intrinsic component of the plant.

As noted in the previous sections, several independent research groups have reported that a significant portion of the INA associated with plant tissues does not appear to be of bacterial origin. These observations have led to proposals that some plants contain intrinsic ice nucleation-active components. In most cases, the evidence for endogenous plant INA was based upon the inability to detect $Ina^+$ bacteria, or the differential sensitivity of ice nucleation sites to various treatments. However, the isolation of proteins (Marentes et al., 1993) and complex carbohydrates (Krog et al., 1979) with INA provides direct evidence for intrinsic ice nucleation sites within plant tissues.

## Selective Advantages of Ice Nucleation Activity Associated with Plant Tissues

Ice usually forms in the intercellular spaces within plant tissues and causes cells to dehydrate. The presence of ice nucleation-active components within the apoplast would influence both the temperature when freezing is initiated and the location of extracellular ice. Both of these factors can influence plant survival. For example, several studies have noted that excessive supercooling led to increased injury when ice formation was eventually initiated (Simonovitch and Scarth, 1938; Gusta and Fowler, 1977; Lindstrom and Carter, 1983; Rajashekar et al., 1983). Presumably the increased damage results from rapid rates of ice crystal growth when supercooled water eventually freezes. Experiments by Pruppacher (1967a,b) demonstrated that the rate of ice crystal growth increases exponentially with a linear increase in supercooling. Therefore, minimizing supercooling would reduce the rate of ice formation and cellular dehydration.

The location of ice within tissues also likely affects freeze-survival. Ice crystals are not uniformly distributed in many tissues. This has been particularly well documented in overwintering flower buds of many perennial plant species (Wiegand, 1906; Quamme, 1978; Sakai, 1979, 1982; Ishikawa and Sakai, 1982, 1985; Ashworth et al., 1988; Ashworth, 1990) where ice forms preferentially in the bud scales and subtending axis tissue and not within the developing floral tissues. Ishikawa and Sakai (1982) described this pattern as extraorgan freezing and suggested that it may be an adaptation to prevent ice formation within tender tissues. Whether the distribution of ice in plant tissues reflects the distribution of ice nucleation sites or some other characteristic of the tissue is not known.

It would of course be a disadvantage for freezing to be initiated in species and tissues that are freeze-sensitive. In these cases, the promotion of supercooling would enhance survival and extend the growing season. Yet there are few examples of species in which the entire plant supercools to low temperatures before freezing. It is perhaps surprising that a species has not adapted to both avoid colonization by Ina$^+$ bacteria and eliminate significant intrinsic ice nucleation sites.

## Potential of Manipulating INA Associated with Plant Tissues

Strategies to reduce the concentration and effectiveness of ice nucleation sites associated with frost-sensitive plants have been proposed and evaluated by Lindow and co-workers (Lindow 1983a,b; Lindow et al., 1983; Lindow and Connell, 1984; and Chapter 13 in this book). The approach has focused on controlling populations of Ina$^+$ bacteria, or inactivating the ice nucleation site. In many species, success may also require altering the properties of intrinsic ice nucleation sites. However, until the existence, role, and identification of such sites is established, we are unable to evaluate the potential for this approach.

It has also been proposed that frost tolerance of some species could be improved by promoting ice nucleation (Rajashekar et al., 1983). Supercooling potentially could be reduced by either seeding tissues with artificial ice nucleating agents (Rajashekar et al., 1983), promoting colonization by ice nucleation-active microorganisms (Lindow, 1982) or by transferring genes coding for an ice nucleation-active protein to a host plant (Baertlein et al., 1992). Whether plants in the field supercool enough to warrant this approach has not been established. However, a variation of these approaches may have utility in facilitating the cryopreservation of plant germ plasm. Controlling the extent of supercooling and the location of ice formation within cultured plant tissues may enhance survival.

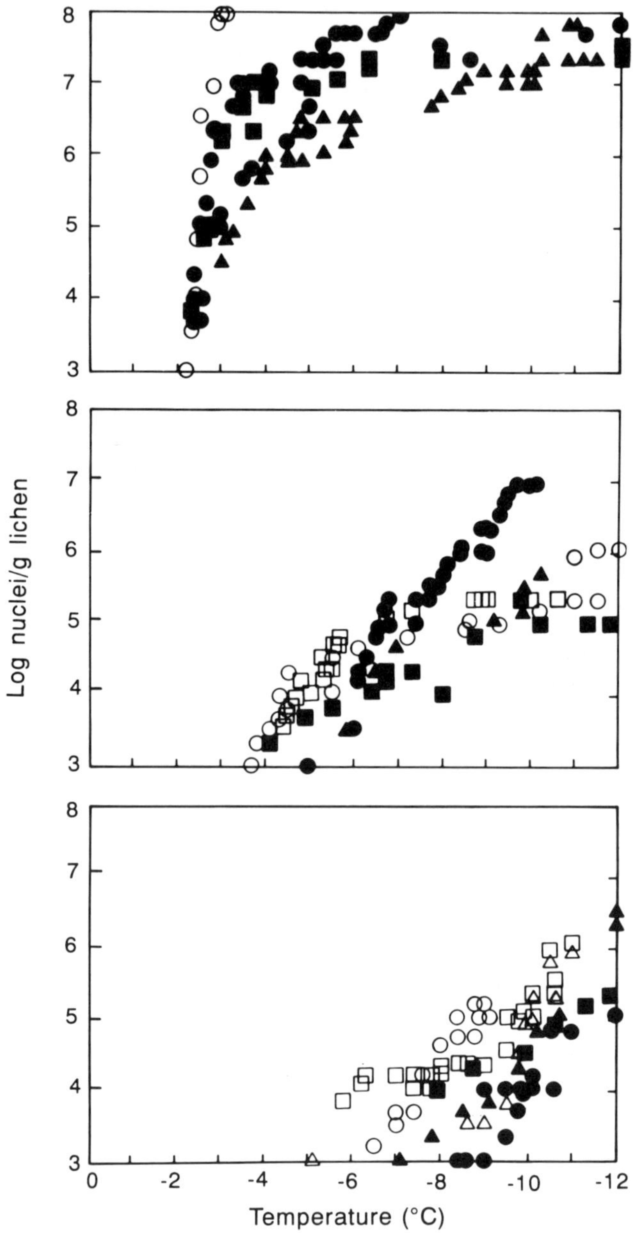


**Figure 2.** Cumulative ice nucleation spectra of lichens (Kieft, 1988). **A,** Lichens showing ice nucleation activity at temperatures above –3°C: *Rhizoplaca chrysoleuca* (■), *Xanthoparmelia* sp. (●), *Xanthoria elegans* (▲), all from near Jemez Springs, NM, and *Rhizoplaca chrysoleuca* (○) from the Magdalena Mountains, NM. **B,** Lichens with intermediate ice nucleation activity: *Platismatia* sp. (●), *Acarospora* sp., from near Jemez Springs, NM (■), *Acarospora* sp., Platoro, CO (▲), *Acarospora* sp. from near Socorro, NM (○), *Leptogium* sp. (□). (C) Lichens with low levels of ice nucleation activity: *Psora decipiens* (●), *Xanthoparmelia* sp., Magdalena Mountains (■), *Xanthoparemlia* sp. from near Socorro, NM (▲), *Letharia* sp. (○), *Peltigera* sp. (□), *Usnea* sp (△).

## Ice Nucleation Activity Associated with Fungi

Warm-temperature INA has been found within members of the fungal kingdom as well as in the plant kingdom. In some species, the fungal ice nucleation sites are as potent as those produced by the most active bacterial ice nucleators. Kieft (1988) reported a wide range of cumulative ice nucleation spectra for a variety of homogenized lichens. Threshold temperatures for INA (i.e., the warmest temperature at which INA is detected) ranged from less than –8°C for the least active lichens to –2.3°C for the most active nucleators (Fig. 2). One of the lichen partnerships with the highest threshold nucleation temperatures was *Rhizoplaca chrysoleuca*. As part of the same study, Kieft reported that an axenic culture of the lichen fungus *R. chrysoleuca* initiated ice formation in supercooled water at temperatures as warm as –2°C and that the cumulative number of nucleation sites active at –5°C or warmer was approximately $10^7$ per gram dry weight of fungus (Fig. 3). More recently, Pouleur et al. (1992) and Tsumuki et al. (1992) reported INA in species of the free-living fungus *Fusarium*. To date, reports of warm-temperature INA (i.e., maximum threshold temperatures more than –5°C or Type 1 INA as described by Phelps et al. (1986)) in fungi, have been limited to lichen fungi and to the genus *Fusarium*. The fungal ice nucleation sites have a number of similarities to, but also differences from, the more thoroughly studied bacterial nucleators. These differences and similarities have implications for the evolution of biological ice nucleation and may shed light on the intrinsic properties of biological ice nucleation sites. The unique properties of fungal ice nucleators also suggest a number of potential applications.

### Lichens and Lichen Fungi

Lichens are symbiotic associations of fungi with algae and/or cyanobacteria. The association is generally an obligate one for the fungus (the mycobiont); however, many of the photosynthetic partners (the photobionts) can also be found in natural environments without the mycobionts. The lichen association is designated by the taxonomic name of the fungus. Nearly all lichen fungi are Ascomycetes; the photobionts comprise a few genera of green algae, with the genus *Trebouxia* being the most common, plus some species of cyanobacteria, e.g., *Nostoc* (Ahmadjian, 1967). Cyanobacteria can be the sole photobiont or in some cases, they are a third symbiont. The advantage of the symbiosis to the fungus is fairly obvious: the photobiont releases photosynthate in the form of sugars or sugar alcohols for heterotrophic metabolism by the fungus. When cyanobacteria are present, they confer the added advantage of nitrogen fixation. The fungi, in return, provide a protective structure for the photobiont and may also add to the nutrient budget of the partnership by solubilizing nutrients from the mineral substrata through the action of carbonic and organic acids. The fungal structures and physiology may also function in tolerance of environmental extremes (especially desiccation) experienced by many lichens. Epilithic lichens occur as pioneer organisms on mineral surfaces and are thus important agents in the initial stages of pedogenesis. Other lichens are epiphytic, i.e., the woody structures of perennial plants form the substrata for these species. Lichens are famous for their ability to survive harsh environmental conditions that include desiccation, heat, and freezing. Lichens are dependent on the atmosphere for moisture, either as precipitation or as water vapor. Many lichens can remain viable in a quiescent state while severely desiccated for prolonged periods;

desert lichens can remain viable for longer than a year at moisture contents below 1% by weight (Rundel, 1988). When moisture becomes available, photosynthesis commences within minutes. Many species of lichens survive prolonged exposure to temperatures in excess of 70°C (Kappen, 1973); and lichens have been shown to survive freezing to at least –50°C (Nash et al. 1987a,b). Ice nucleation activity in lichens may be an adaptation for increased uptake of atmospheric water vapor.

The original survey of INA in lichens (Kieft, 1988; Fig. 2) showed a number of trends. Overall, the threshold temperatures for freezing were higher for the epilithic lichens than for the epiphytic lichens. Also, the one lichen with a cyanobacterial photobiont, *Peltigera* sp., was a relatively poor nucleator. Extensive efforts to iso-

**Table 4.** Ice nucleation activities (INA) detected in pure cultures of free-living fungi, lichen fungi (mycobionts), and lichen algae and cyanobacteria (photobionts)

| Organism | Nuclei per gram at −5°C | Warmest INA detected (°C) | Reference |
|---|---|---|---|
| Free-living fungi | | | |
| *Fusarium avenaceum* | $1.0 \times 10^9$ | −2.5 | Pouleur et al., 1992 |
| *F. acuminatum* | ND[a] | ≥−5.0 | Pouleur et al., 1992 |
| *Fusarium* sp. from insect larval gut | ND | −5.5 | Tsumuki et al., 1992 |
| Lichen mycobionts | | | |
| *Acarospora fuscata* | 0 | −9.1 | Kieft and Ahmadjian, 1989 |
| Antarctic endolithic lichen fungus | 0 | −5.9 | Kieft and Ahmadjian, 1989 |
| *Cladonia bellidiflora* | 0 | −8.3 | Kieft and Ahmadjian, 1989 |
| *C. boryi* | $8.3 \times 10^3$ | −5.0 | Kieft and Ahmadjian, 1989 |
| *C. cariosa* | 0 | −6.3 | Kieft and Ahmadjian, 1989 |
| *C. chlorophaea* 1429 | 0 | −10.0 | Kieft and Ahmadjian, 1989 |
| *C. cristatella* 113 | 0 | −6.3 | Kieft and Ahmadjian, 1989 |
| *C. pleurota* | 0 | −9.1 | Kieft and Ahmadjian, 1989 |
| *C. rangiferina* | $5.7 \times 10^3$ | −5.0 | Kieft and Ahmadjian, 1989 |
| *C. subcariosa* | 0 | −6.2 | Kieft and Ahmadjian, 1989 |
| *Lecanora dispersa* | $2.0 \times 10^4$ | −4.2 | Kieft and Ahmadjian, 1989 |
| *Pertusaria flavicans* | $3.1 \times 10^4$ | −4.1 | Kieft and Ahmadjian, 1989 |
| *Phaeographis leucophelia* | 0 | −8.3 | Kieft and Ahmadjian, 1989 |
| *Rhizoplaca chrysoleuca* | $1.3 \times 10^8$ | −1.9 | Kieft, 1988; Kieft and Ahmadjian, 1989 |
| Lichen photobionts | | | |
| *Coccomyxa peltigera* var. *variolosa* | 0 | −8.3 | Kieft and Ahmadjian, 1989 |
| *Nostoc* | | | |
| from *Peltigera canina* | 0 | −9.1 | Kieft and Ahmadjian, 1989 |
| from *P. rufescens* | 0 | −7.5 | Kieft and Ahmadjian, 1989 |
| from *P. spuria* | 0 | −9.1 | Kieft and Ahmadjian, 1989 |
| *Trebouxia erici* from *Cladonia cristatella* | 0 | −9.2 | Kieft and Ahmadjian, 1989 |
| *Trebouxia incrustata* from *Lecanora dispersa* | 0 | −9.1 | Kieft and Ahmadjian, 1989 |
| *Trebouxia* | | | |
| from *Acarospora fuscata* | 0 | −5.1 | Kieft and Ahmadjian, 1989 |
| from *Aspicilia calcarea* | 0 | −6.3 | Kieft and Ahmadjian, 1989 |
| from *Cladonia coniocraea* | 0 | −11.3 | Kieft and Ahmadjian, 1989 |
| from *Lecidea tumida* | 0 | −6.2 | Kieft and Ahmadjian, 1989 |
| from *Parmelia tinctorum* | 0 | −16.0 | Kieft and Ahmadjian, 1989 |
| from *Parmeliopsis hyperopta* | 0 | −6.3 | Kieft and Ahmadjian, 1989 |
| from *Stereocaulon saxatile* | 0 | −8.8 | Kieft and Ahmadjian, 1989 |

[a]No data.

late bacteria with INA from the lichens were unsuccessful, suggesting that the lichens themselves might be the source of the nucleation sites. The finding of warm-temperature INA in a pure culture of the mycobiont *R. chrysoleuca* supported the hypothesis that lichen mycobionts synthesize intrinsic ice nucleation sites. This finding led to a survey of a greater number of lichen symbionts for INA. Many of the individual symbionts of lichens have been cultured axenically in vitro (Ahmadjian et al., 1980). Kieft and Ahmadjian tested many of Ahmadjian's pure cultures of symbionts and reported that warm-temperature ice nucleation, when it occurred, was restricted to the mycobionts; threshold temperatures in the photobionts were all below –5°C (Kieft and Ahmadjian, 1989) (Table 4). The INA data for the mycobionts also corroborated the results for the homogenized lichens (Kieft, 1988). For instance, pure cultures of the fungus *R. chrysoleuca* and the closely related *Lecanora dispersa* were among those with the warmest threshold temperatures.

Lichens were first tested for INA (Kieft, 1988) using the droplet-freezing assay (Vali, 1971). More recently, Ashworth and Kieft (1992) tested INA in whole lichen thalli by thermal analysis. With thermocouple probes attached to lichen thalli, freezing was detected by the release of the latent heat of fusion (the exotherm) as the temperature of the lichen was slowly decreased. By this method of analysis, it was found that the range of threshold temperatures for freezing in lichens is much narrower than that detected by the droplet-freezing assay. Freezing was detected at temperatures above –5°C in all lichens tested, even epiphytic lichens like *Usnea* that had appeared to be poor nucleators in the droplet-freezing assay. This discrepancy in results from the two methods can be explained by the differences in con-

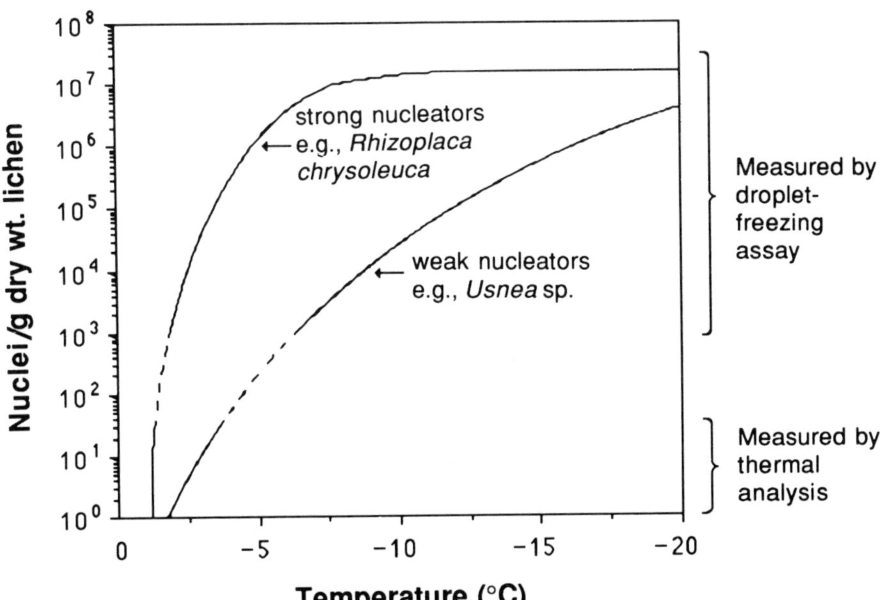


**Figure 3.** Idealized cumulative ice nucleation spectra for a strong fungal ice nucleator, e.g., *Rhizoplaca chrysoleuca,* and a weak fungal ice nucleator, e.g., *Usnea* sp., showing the ranges of data collected by the droplet-freezing assay and thermal analysis methods.

centrations of warm-temperature ice nucleation sites among the various species of lichens and by the differences in sensitivities of the two tests (Fig. 3). The thermal analysis method can detect a single ice nucleus in a relatively large sample, whereas the lower limit of detection for the droplet-freezing assay is about $10^3$ nucleation sites per gram. It appears that some lichens previously thought not to have warm-temperature ice nucleation sites actually possess them, albeit in low numbers. When freezing is initiated at a single site, ice formation is propagated throughout the lichen thallus. Thus, INA appears to be even more widespread among lichens than was previously known.

## Free-living Fungi: *Fusarium* spp.

In the years since Schnell and Vali (1972) first reported the presence of warm-temperature INA in decaying leaf litter, many free-living fungi have been screened for INA, especially as part of tests for INA among epiphytic microorganisms and microbes from the guts of insects (Lindow et al., 1978a; Tsumuki, 1992; Pouleur et al., 1992; S. Lindow, *personal communication*). However, it has only very recently been demonstrated that the nonlichenized fungi identified that have warm-temperature INA are all members of the genus *Fusarium*. Pouleur et al. (1992) screened 145 strains comprising 20 genera of fungi for their ability to nucleate ice at –5°C. Of these, 16 strains of *Fusarium avenaceum* and 5 strains of *Fusarium accuminatum* were positive. All others were negative at this temperature, including 12 other species of *Fusarium*. The threshold temperature for nucleation in *F. avenaceum* was –2.5°C, and the number of nucleation sites per gram of fungus was approximately $10^9$ (Table 4). INA could be detected in macroconidia that were physically separated from the mycelium (Pouleur et al., 1992). *F. avenaceum* and *F. accuminatum* are pathogens of plants, causing root rot in a variety of plants, including alfalfa, especially in cold climates (Pouleur et al., 1992). Tsumuki et al. (1992) screened bacteria from the guts of insect larvae (the rice stem borer *Chilo suppressalis*) for INA at –10°C. Of the approximately 600 colonies tested (75% bacterial, 25% fungal), only one fungus, an unidentified strain of *Fusarium,* was positive. The threshold temperature of freezing for this strain was –5.5°C (Table 4). As part of the same study, it was shown that inoculation of plants and insect larvae with this *Fusarium* strain raised the temperature of freezing by several degrees.

As noted by Pouleur et al. (1992), the fungi that display warm-temperature INA are phylogenetically related. Nearly all lichenized fungi are members of the Ascomycetes; *Fusarium* spp. are in the form-class Deuteromycetes (the Fungi Imperfecti) but would belong to the Ascomycetes if they had known perfect stages. This suggests a possible common evolutionary origin for the ice nucleation trait within these phylogenetically related fungi.

## Characteristics of Fungal Ice Nucleation Sites

Currently, the lichen ice nucleators are not nearly as well characterized as their bacterial counterparts. As with the bacterial nucleators, characterization of fungal ice nucleation sites has been hampered by the difficulties of isolating them in large numbers. Warm-temperature ice nucleation sites are generally produced in low numbers, usually fewer than one per cell. Consequently, characterization of fungal ice nucleation sites has thus far been limited to indirect means. Activities have been measured after various physical and chemical treatments to gain insights regarding their structure and function. In lichens, *R. chrysoleuca* has been the source of nucleators for study. A few similar studies have also been performed on *F. avenaceum*.

As demonstrated for bacteria, the lichen fungal ice nucleation sites appear to be proteinaceous. For example, lichen INA is reduced following treatment with the proteases pronase E and papain (Kieft and Ruscetti, 1990). Further support for the proteinaceous nature of lichen ice nuclei was shown in their inactivation by the denaturing agents guanidine hydrochloride and urea. The effect of these denaturants also suggests that, as with the bacterial ice nucleation proteins, warm-temperature ice nucleation sites may form by the aggregation of many smaller protein subunits. Theories of physical chemistry indicate that warm-temperature INA requires molecules of large size (Fletcher, 1958; Rogers et al., 1987; Burke and Lindow, 1990). Govindarajan and Lindow (1988a) showed by gamma irradiation inactivation analysis that a direct relationship exists between the number of aggregated bacterial ice nucleation proteins and the temperature of freezing. Using the same approach, Kieft and Ruscetti (1992) found a similar pattern in *R. chrysoleuca* (Fig. 4).

The first clue that fungal ice nucleators differed from those of bacteria appeared when it was demonstrated that lichen ice nucleation sites are relatively temperature-stable. Fungal ice nuclei maintained activity even after exposure to temperatures of 60°C. As stated earlier, epilithic lichens are frequently subjected to these temperatures in warm desert and semiarid environments, so it is not surprising to find temperature stability in substances produced by these organisms. Bacterial nucleators are inactivated relatively quickly at 40°C (Maki and Willoughby, 1978).

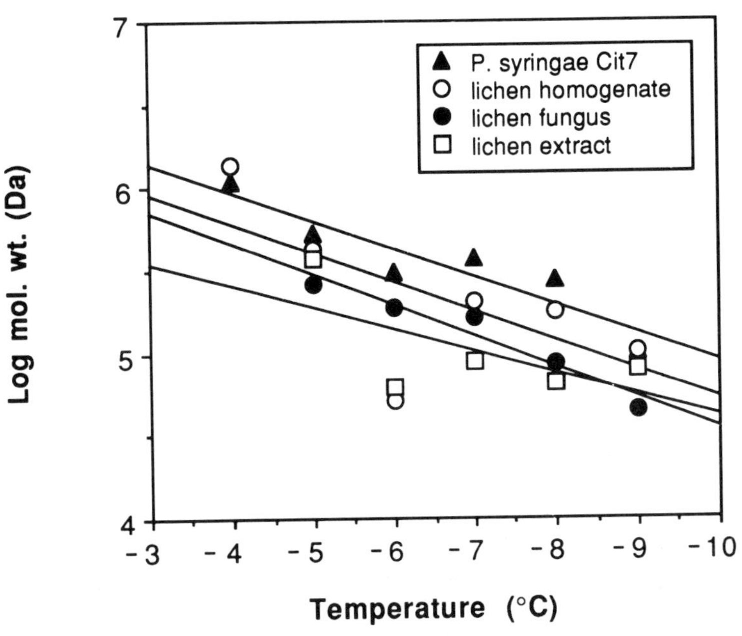


**Figure 4.** Molecular masses of the ice nucleation sites in *Pseudomonas syringae* strain Cit7 bacterial cells; field-collected, homogenized *Rhizoplaca chrysoleuca* lichen; cells of the lichen fungus *R. chrysoleuca* cultured axenically; and a crude extract of the lichen *R. chrysoleuca*, all determined by gamma-irradiation-inactivation analysis (Redrawn from Kieft and Ruscetti, 1992).

The lichen nucleation sites also differ from those of bacteria in being active over an extraordinarily broad pH range. Threshold temperatures in an extract of ice nucleation sites from *R. chrysoleuca* were unchanged over a pH range from 2 to 11 (Kieft and Ruscetti, 1990), a property that might be explained by a high proportion of uncharged amino acid residues in the ice nucleation protein. Bacterial INA is diminished below pH 4 (Kozloff et al., 1983).

The greatest difference between the lichen and bacterial ice nucleation relates to the involvement of membrane lipids. Bacterial ice nucleation proteins occur in the outer membrane of gram-negative bacteria (Lindow et al., 1989), and INA requires the presence of lipid material (Govindarajan and Lindow, 1988b). The nucleation sites of the lichen *R. chrysoleuca* differ in that they can still nucleate ice at warm temperatures in a crude extract completely delipidated with a nonpolar solvent such as chloroform (Kieft and Ruscetti, 1990). The crude extract can also be filtered through a 0.22-μm pore size filter, and activity is retained, indicating the presence of cellfree nuclei (Kieft and Ruscetti, 1990).

The *Fusarium* ice nucleators have properties in common with the lichen ice nucleation sites. These include activity over a broad pH range (pH 1 to 13) and temperature stability to 60°C (Pouleur et al., 1992). The ice nucleation-active *Fusarium* spp. also appear to generate cellfree nucleation sites, since filtrates from the fungi that pass through a 0.22-μm filter retain warm-temperature activity (Pouleur et al., 1992). These similarities between INA in *Fusarium* and lichen fungi also suggest a common genetic origin.

At the present time, the structure of the fungal ice nucleation protein(s) is unknown, and it is also not known whether other molecules, e.g., carbohydrates, inorganic cofactors, etc., are involved in the nucleation process. Similarly, the genetic basis of fungal INA is largely unexplored. In one preliminary study, C. Orser (*personal communication*) found no sequence homology between a bacterial INA gene probe and DNA from a lichen fungus. INA in lichens may have resulted from convergent evolution, i.e., the gene(s) coding for INA in fungi may have arisen independently from those in bacteria. However, the fungal ice nucleators may share intrinsic properties required for INA with those in other taxonomic groups. These may include large molecular size and proteins with repeating sequences of hydrophilic amino acids.

## Selective Advantages of Fungal INA

Although the selective advantage of lichen ice nuclei has not been firmly established, a number of explanations have been proposed. The most plausible has to do with the moisture relations of lichens. As stated above, lichens are dependent on the atmosphere for water and they are also extremely freeze-tolerant. Many lichens occur in habitats that undergo daily freeze-thaw cycles. Because the vapor pressure of water is lower over ice than over supercooled water, deposition freezing (freezing of water vapor on ice surfaces) on lichens could be enhanced by warm-temperature INA. Ice nucleation-active lichens may gain extra moisture through deposition freezing at night and then absorb the moisture as water when it melts the following day. This mechanism of moisture uptake would be particularly beneficial under conditions in which night temperatures drop below 0°C but not low enough for nonbiological ice nuclei to be active. As anecdotal evidence in support of this theory, the warmest threshold temperatures and the highest densities of ice nucleation sites in field-collected *R. chrysoleuca* are found in the mountains of New Mexico during the fall and spring, when nighttime low temperatures are typically approximately –5°C and conditions are generally dry. During midwinter and midsummer at the same sites, when nighttime low temperatures are below –5°C and above 0°C, respectively, and precipitation is more common, *R. chrysoleuca* samples are less active as ice nucleators. A second possible explanation for lichen INA relates to freeze tolerance. When freezing is initiated at warmer temperatures, the rate of ice crystal growth and cellular dehydration is slower than when freezing initiates at colder temperatures. Therefore, promoting ice nucleation may enhance freeze tolerance. These two hypotheses are not mutually exclusive; INA may benefit lichens by enhancing both moisture uptake and freeze tolerance.

The selective advantage of INA in *Fusarium* may be similar to that proposed for bacteria, i.e., the ability to initiate ice formation at relatively warm temperature appears to make these fungi conditional pathogens of plants. Infection of alfalfa roots by *Fusarium* spp. is enhanced by freezing (Richard et al., 1982, 1985; Pouleur et al., 1992). As with ice nucleating bacteria on leaf surfaces, fungi associated with plant roots may damage plant tissue by initiating freezing of supercooled water at relatively warm temperatures, thereby gaining access to plant nutrients and also facilitating invasion of plant root tissue. On the other hand, pathogenic fungi, including *Fusarium* spp., are normally not dependent on wounds or other openings for entry into plants as are pathogenic bacteria like *P. syringae*. This suggests that INA in *F. acuminatum* and *F. avenaceum* may not be essential for invasion of plant tissue.

Fungal ice nucleators may contribute, along with other biogenic ice nucleators, to heterogeneous ice nucleation in the atmosphere, thereby influencing hydrologic cycles. They may contribute to INA on the surface of plants, as well. For instance, lichens may function as ice nucleators on the surface of woody plants. Fungal nucleators may also initiate freezing at warm temperatures in insects (Duman and Montgomery, 1991; Tsumuki et al., 1992).

## Potential Applications of Fungal Ice Nucleators

The potential applications of ice nuclei have been reviewed elsewhere and include uses in snowmaking, cloud seeding, frozen food additives, and as signal transducers in immunological tests (Warren, 1987; Margaritas and Bassi, 1991). To date, these developments have been pursued exclusively with bacterial ice nucleation proteins; however, fungal ice nucleators may also gain favor. The unique properties of fungal ice nucleators (temperature stability, activity over a broad pH range, activity in the absence of lipids, etc.) may make these substances especially attractive for biotechnological applications requiring warm-temperature ice nucleation. For instance, the temperature stability would simplify storage and lengthen the shelf-life of ice nucleation products. The fact that lichen nucleators do not require lipid material for activity may make them especially suitable for the food industry and for immunological applications. Using lichen nuclei would avoid the use of bacterial outer membranes (which contain bacterial endotoxin). A purified preparation of lichen nuclei might also pose fewer nonspecific binding problems than lipid-associated bacterial ice nucleation-active proteins when used in immunological assays. Unfortunately, knowledge of the genetics of the fungal systems lags far behind that of the bacteria. Additionally, because lichens grow very slowly both in nature and in the laboratory, the commercial use of lichen nuclei would require that the gene(s) coding for this trait be cloned and expressed in a faster-growing organism. Unfortunately, the genetic studies of lichen fungi are only in the beginning stages (Ahmadjian, 1991). Because *Fusarium* is much more amenable than the lichens to growth in the laboratory, progress in fungal INA studies may proceed more rapidly with this organism.

## Concluding Remarks

Extracellular freezing in plant and fungal cells is initiated by heterogeneous ice nucleation and generally begins following only a few degrees of supercooling. Although much of the INA associated with plant and fungal tissues can often be ascribed to $Ina^+$ bacteria, there is significant evidence that plant and fungal tissues also contain intrinsic ice nucleation sites. Unfortunately, the ice nucleation-active components within these tissues have been only partially characterized. Progress in identifying these intrinsic ice nucleation-active sites has been hampered by their apparent low concentration within tissues and the difficulty of studying ice formation in tissues under conditions similar to those in nature.

## Literature Cited


Ahmadjian, V. 1967. The Lichen Symbiosis. Blaisdell, Waltham, MA.

Ahmadjian, V. 1991. Molecular biology of lichens: A look to the future. Symbiosis 11:249-254.

Ahmadjian, V., Russell, L. A., and Hildreth, K. C. 1980. Artificial reestablishment of lichens. I. morphological interactions between the photobionts of different lichens and the mycobionts *Cladonia cristatella* and *Lecanora chrysoleuca*. Mycologia 78:73-89.

Anderson, J. A., and Ashworth., E. N. 1985. Ice nucleation in tomato plants. J. Am. Soc. Hort. Sci. 110:291-296.

Anderson, J. A., Ashworth, E. N., and Davis, G. A. 1987. Nonbacterial ice nucleation in peach shoots. J. Am. Soc. Hort. Sci. 112:215-218.

Anderson, J. A., Buchanan, D. W., Stall, R. E., and Hall, C. B. 1982. Frost injury of tender plants increased by *Pseudomonas syringae* van Hall. J. Am. Soc. Hort. Sci. 107:123-125.

Andrews, P. K., Proebsting, E. L., Jr., and Gross, D. C. 1986. Ice nucleation and supercooling in freeze-sensitive peach and sweet cherry tissues. J. Am. Soc. Hort. Sci. 111:232-236.

Arny, D. C., Lindow, S. E., and Upper, C. D. 1976. Frost sensitivity of *Zea mays* increased by application of *Pseudomonas syringae*. Nature 262:282-284.

Ashworth, E. N. 1990. The formation and distribution of ice within forsythia flower buds. Plant Physiol. 92:718-725.

Ashworth, E. N. 1992. Formation and spread of ice in plant tissues. Hort. Rev. 13:215-255.

Ashworth, E. N., Anderson, J. A., and Davis, G. A. 1985a. Properties of ice nuclei associated with peach trees. J. Am. Soc. Hort. Sci. 110:287-291.

Ashworth, E. N., Anderson, J. A., Davis, G. A., and Lightner, G. W. 1985b. Ice formation in *Prunus persica* under field conditions. J. Am. Soc. Hort. Sci. 110:322-324.

Ashworth, E. N., and Davis, G. A. 1984. Ice nucleation within peach trees. J. Am. Soc. Hort. Sci. 109:198-201.

Ashworth, E. N., and Davis, G. A. 1986. Ice formation in woody plants under field conditions. HortScience 21:1233-1234.

Ashworth, E. N., Davis, G. A., and Anderson, J. A. 1985c. Factors affecting ice nucleation in plant tissues. Plant Physiol. 79:1033-1037.

Ashworth, E. N., Davis, G. A., and Wisniewski, M. E. 1989. The formation and distribution of ice within dormant and deacclimated peach flower buds. Plant Cell Environ. 12:521-528.

Ashworth, E. N., Echlin, P., Pearce, R. S., and Hayes, T. L. 1988. Ice formation and tissue response in apple twigs. Plant Cell Environ. 11:703-710

Ashworth, E. N., and Kieft, T. L. 1992. Measurement of ice nucleation in lichens using thermal analysis. Cryobiology 29:400-406.

Baertlein, D. A., Lindow, S. E., Panapoulos, N. J., Lee, S. P., Mindrinos, M. N., and Chen, T. H. H. 1992. Expression of a bacterial ice nucleation gene in plants. Plant Physiol. 100:1730-1736.

Brush, R. A., Griffith, M., and Mlynarz, A. 1994. Characterization and quantification of intrinsic ice nucleators in winter rye (*Secale cereale*) leaves. Plant Physiol. 104:725-735.

Burke, M. J., and Lindow, S. E. 1990. Surface properties and size of the ice nucleation site in ice nucleation-active bacteria: Theoretical considerations. Cryobiology 26:80-84.

Cary, J. W., and Lindow, S. E. 1986. The effect of leaf water variables on ice nucleating *Pseudomonas syringae* in beans. HortScience 21:1417-1418.

Cary, J. W., and Mayland, H. F. 1970. Factors influencing freezing of supercooled water in tender plants. Agron. J. 62:715-719.

Cody, Y. S., Gross, D. C., Proebsting, E. L., Jr., and Spotts, R. A. 1987. Suppression of ice nucleation-active *Pseudomonas syringae* by antagonistic bacteria in fruit tree orchards and evaluations of frost control. Phytopathology 77:1036-1044.

Constantinidou, H. A., Menkissoglu, O., and Stergiadou, H. C. 1991. The role of ice nucleation-active bacteria in supercoolng of citrus tissues. Physiologia Plantarum 81:548-554.

Constantinidou, H. A., and Menkissoglu, O. 1992. Characteristics and importance of heterogeneous ice nuclei associated with citrus fruits. J. Exp. Bot. 43:585-591.

Dowgert, M. F., and Steponkus, P. L. 1983. Effect of cold acclimation on intracellular ice formation in isolated protoplasts. Plant Physiol. 72:978-988.

Duman, J. G., and Montgomery, S. L. 1991. Subzero temperature adaptations in arthropods from the summit of Mauna Kea, Hawawii. J. Exp. Zool. 259:409-412.

Duman, J. G., Morris, J. P., and Castellino, F. J. 1984. Purification and composition of an ice nucleating protein from queens of the hornet, *Vespula maculata*. J. Comp. Physiol. 154:79-83.

Fletcher, N. H. 1958. Size effects in heterogeneous ice nucleation. J. Chem. Phys. 29:572-576.

Gartman, T. L., Ewers, F. W., and Davis, S. D. 1992. Differential sensitivities of five co-occurring species of chaparral shrubs to below freezing air temperatures. Bull. Ecol. Soc. Am. 73:184.

Govindarajan, A. G., and Lindow, S. E. 1988a. Size of bacterial ice nucleation sites measured in situ by radiation inactivation analysis. Proc. Natl. Acad. Sci. USA 85:1334-1338.

Govindarajan, A. G., and Lindow, S. E. 1988b. Phospholipid requirement for expression of ice nuclei in *Pseudomonas syringae* and in vitro. J. Biol. Chem. 263:9333-9338.

Griffith, M., Ala, P., Yang, D. S. C., Hon, W-C., and Moffatt, B. A. 1992. Antifreeze protein produced endogenously in winter rye leaves. Plant Physiol. 100:593-596.

Griffith, M., Marentes, E., Ala, P., and Yang, D. S. C. 1993. The role of ice-binding proteins in frost tolerance of winter rye. Pages 177-186 in: Advances in Plant Cold Hardiness, P. H. Li and L. Christersson, eds. CRC Press, Boca Raton, FL.

Gross, D. C., Proebsting, E. L., Jr., and Andrews, P. K. 1984. The effects of ice nucleation-active bacteria on temperatures of ice nucleation and freeze injury of *Prunus* flower buds at various stages of development. J. Am. Soc. Hort. Sci. 109:375-380.

Gross, D. C., Proebsting, E. L., Jr., and MacCrindle-Zimmerman, H. 1988. Development, distribution, and characteristics of intrinsic, nonbacterial ice nuclei in *Prunus* wood. Plant Physiol. 88:915-922.

Gusta, L. V., and Fowler, D. B. 1977. Factors affecting the cold survival of winter cereals. Can. J. Plant Sci. 57:213-219.

Hirano, S. S., Baker, L. S., and Upper, C. D. 1985. Ice nucleation temperature of individual leaves in relation to population sizes of ice nucleation-active bacteria and frost injury. Plant Physiol. 77:259-265.

Hirano, S. S., Nordheim, E. V., Arny, D. C., and Upper, C. D. 1982. Lognormal distribution of epiphytic bacterial populations on leaf surfaces. Appl. Environ. Microbiol. 44:695-700.

Ishikawa, M., and Sakai, A. 1982. Characteristics of freezing avoidance in comparison with freezing tolerance: a demonstration of extraorgan freezing. Pages 325-340 in: Plant Cold Hardiness and Freezing Stress, vol. 2. P. H. Li, A. Sakai, eds. Academic Press, New York.

Ishikawa, M., and Sakai, A. 1985. Extraorgan freezing in wintering flower buds of *Cornus officinalis* Sieb et Zucc. Plant Cell Environ. 8:333-338.

Kaku, S. 1971. Changes in supercooling and freezing processes accompanying leaf maturation in *Buxus*. Plant Cell Physiol. 12:147-155.

Kaku, S. 1973. High ice nucleating ability in plant leaves. Plant Cell Physiol. 14:1035-1038.

Kaku, S., and Salt, R. W. 1968. Relationship between freezing temperature and length of conifer needles. Can. J. Bot. 46:1211-1213.

Kappen, L. 1973. Response to extreme environments. Pages 311-380 in: The Lichens. V. Ahmadjian and M. E. Hale, eds. Academic Press, New York.

Kieft, T. L. 1988. Ice nucleation activity in lichens. App. Environ. Microbiol. 54:1678-1681.

Kieft, T. L., and Ahmadjian, V. 1989. Biological ice nucleation activity in lichen mycobionts and photobionts. Lichenologist 21:355-362.

Kieft, T. L. , and Ruscetti, T. 1990. Characterization of biological ice nuclei from a lichen. J. Bacteriol. 172:3519-3523.

Kieft, T. L., and Ruscetti, T. 1992. Molecular sizes of lichen ice nucleation sites determined by gamma radiation inactivation analysis. Cryobiology 29:407-413.

Kitaura, K. 1967a. Freezing and injury of mulberry trees by late spring frost. Bull. Seric. Exp. Stn. 22:202-323.

Kitaura, K. 1967b. Supercooling and ice formation in mulberry trees. p. 143-156 in: Cellular Injury and Resistance in Freezing Organisms. E. Asahina, ed. Inst. Low Temp. Sci. Hokkaido Univ. Sapporo, Japan.

Kozloff, L. M., Schofield, M. A., and Lute, M. 1983. Ice nucleation activity of *Pseudomonas syringae* and *Erwinia herbicola*. J. Bacteriol. 153:222-231.

Krog, J. O., Zachariassen, K. E., Larsen, B., and Smidsrod, O. 1979. Thermal buffering in Afro-alpine plants due to nucleating agent-induced water freezing. Nature 282:300-301.

Le Grice, P., Fuller, M. P., and Campbell, A. 1993. An investigation of the potential use of thermal imaging technology in the study of frost damage to sensitive crops. Page 4 in: Proc. Int. Conf. on Biological Ice Nucleation, 6th. University of Wyoming, Laramie.

Levitt, J. 1980. Responses of Plants to Environmental Stresses. 2nd ed. Vol. 1 Academic Press, New York.

Lindow, S. E. 1982. Epiphytic ice nucleation-active bacteria. Pages 335-362 in: Phytopathogenic Prokaryotes, vol. 1. M. S. Mount and G. H. Lacy, eds. Academic Press, New York.

Lindow, S. E. 1983a. Methods of preventing frost injury caused by epiphytic ice-nucleation-active bacteria. Plant Dis. 67:327-333.

Lindow, S. E. 1983b. The role of bacterial ice nucleation in frost injury to plants. Annu. Rev. Phytopathol. 21:363-384.

Lindow, S. E., Arny, D. C., and Upper, C. D. 1978a. Distribution of ice nucleation-active bacteria on plants in nature. Appl. Environ. Microbiol. 36:831-838.

Lindow, S. E., Arny, D. C., and Upper, C. D. 1978b. *Erwinia herbicola*: A bacterial ice nucleus active in increasing frost injury to corn. Phytopathology 68:523-527.

Lindow, S. E., Arny, D. C., and Upper, C. D. 1982a. Bacterial ice nucleation: A factor in frost injury to plants. Plant Physiol. 70:1084-1089.

Lindow, S. E., Arny, D. C., and Upper, C. D. 1983. Biological control of frost injury: An isolate of *Erwinia herbicola* antagonistic to ice nucleation bacteria. Phytopathology 73:1097-1106.

Lindow, S. E., and Connell, J. H. 1984. Reduction of frost injury to almond by control of ice nucleation-active bacteria. J. Am. Soc. Hort. Sci. 109:48-53.

Lindow, S. E., Hirano, S. S., Barchet, W. R., Arny, D. C., and Upper, C. D. 1982. Relationship between ice nucleation frequency of bacteria and frost injury. Plant Physiol. 70:1090-1093.

Lindow, S. E., Lahue, E., Govindarajan, S. E., Panapoulos, N. J., and Gies, D. 1989. Localization of ice nucleation activity and the iceC gene product in *Pseudomonas syringae* and *Escherichia coli*. Mol. Plant-Microbe Interact. 2:262-272.

Lindstrom, O. M., and Carter, J. V. 1983. Assessment of freezing injury of cold-hardened undercooled leaves of *Solanum commersonii*. Cryo Lett. 4:361-370.

Maki, L. R., and Willoughby, K. J. 1978. Bacteria as biogenic sources of freezing nuclei. J. Appl. Meteorol. 17:1049-1053.

Maki, L. R., Galyan, E. L., Chang-Chien, M., and Caldwell, D. R. 1974. Ice nucleation induced by *Pseudomonas syringae*. Appl. Microbiol. 28:456-459.

Malone, S. R., and Ashworth, E. N. 1991. Freezing stress response in woody tissues observed using low-temperature scanning electron microscopy and freeze substitution techniques. Plant Physiol. 95:871-881.

Marcellos, H., and Single, W. V. 1976. Ice nucleation on wheat. Agric. Meteorol. 16:125-129.

Marcellos, H., and Single, W. V. 1979. Supercooling and heterogeneous nucleation of freezing in tissues of tender plants. Cryobiology 16:74-77.

Marentes, E., Griffith, M., Mlynarz, A., and Brush, R. A. 1993. Proteins accumulate in the apoplast of winter rye leaves during cold acclimation. Physiol. Plant. 87:499-507.

Margaritas, A., and Bassi, A. S. 1991. Principals and biotechnological applications of bacterial ice nucleation. Crit. Rev. Biotechnol. 11:277-295.

Marshall, D. 1988. A relationship between ice-nucleation-active bacteria, freeze damage, and genotype in oats. Phytopathology 78:952-957.

Modlibowska, I. 1962. Some factors affecting supercooling of fruit blossoms. J. Hort. Sci. 37:249-261.

Morris, G. J., and McGrath, J. J. 1981. Intracellular ice nucleation and gas bubble formation in *Spirogyra*. Cryo Lett. 2:341-352.

Nash, T. H., Kappen, L., Losch, R., Larson, D. W., and Mathes-Sears, U. 1987a. Cold resistance of lichens with *Trentepohlia* or *Trebouxia* photobionts from the North American West Coast. Flora 179:241-251.

Nash, T. H., Kappen, L., Losch, R., Mathes-Sears, U., and Larson, D. W 1987b. Cold resistance of lichens. Progress and Problems in Lichenology in the Eighties. Bibl. Lichenol. 25:313-317.

Nath, J., and Fisher, T. C. 1971. Anatomical study of freezing injury in hardy and nonhardy alfalfa varieties treated with cytosine and guanine. Cryobiology 8:420-430.

O'Brien, R. D., and Lindow, S. E. 1988. Effect of plant species and environmental conditions on ice nucleation activity of *Pseudomonas syringae* on leaves. Appl. Environ. Microbiol. 54:2281-2286.

Pearce, R. S. 1988. Extracellular ice and cell shape in frost-stressed cereal leaves: A low-temperature scanning-electron-microscopy study. Planta 175:313-324.

Pearce, R. S., and Ashworth, E. N. 1992. Localization of ice and cell shape in leaves of overwintering wheat during frost stress in the field. Planta 188:324-331.

Phelps, P., Giddings, T. H., Prochoda, M., and Fall, R.. 1986. Release of cell-free nuclei by *Erwinia herbicola*. J. Bacteriol. 167:496-502.

Pouleur, S., Richard, C., Martin, J.-G., and Antoun, H. 1992. Ice nucleation activity in *Fusarium acuminatum* and *Fusarium avenaceum*. Appl. Environ. Microbiol. 58:2960-2964.

Proebsting, E. L., Jr., Andrews, P. K., and Gross, D. C. 1982. Supercooling young developing fruit and floral buds in deciduous orchards. HortScience 17:67-68.

Proebsting, E. L., Jr., and Gross, D. C. 1988. Field evaluations of frost injury to deciduous fruit trees as influenced by ice nucleation-active *Pseudomonas syringae*. J. Am. Soc. Hort. Sci. 103:57-61.

Pruppacher, H. R. 1967a. Interpretation of experimentally determined growth rates of ice crystals in supercooled water. J. Chem. Phys. 47:1807-1813.

Pruppacher, H. R. 1967b. Some relations between the structure of the ice-solution interface and the free growth rate of ice crystals in supercooled aqueous solutions. J. Colloid Interface Sci. 25:285-294.

Quamme, H. A. 1978. Mechanism of supercooling in overwintering peach flower buds. J. Am. Soc. Hort. Sci. 103:57-61.

Rajashekar, C. B., Li, P. H., and Carter, J. V. 1983. Frost injury and heterogeneous ice nucleation in leaves of tuber-bearing *Solanum* species. Plant Physiol. 71:749-755.

Richard, C., Michaud, R., Willemot, C., Bernier-Cardou, M., and Gagnon, C. 1985. Effect of frost on *Fusarium* root rot of alfalfa and possibility of double trait selection. Pages 209-211 in: Ecology and Management of Soilborne Plant Pathogens. C. A. Parker, A. D. Rovira, K. J. Moore, P. T. W. Wong, and J. F. Kollmorgen, eds. American Phytopathological Society, St. Paul, MN.

Richard, C., Willemot, C., Michaud, R., Bernier-Cardou, M., and Gagnon, C. 1982. Low-temperature interactions in *Fusarium* wilt and root rot of alfalfa. Phytopathology 72:293-297.

Rogers, J. S., Stall, R. E., and Burke, M. J. 1987. Low temperature conditioning of the ice nucleation-active bacterium *Erwinia herbicola*. Cryobiology 24:270-279.

Rogers, R. R., and Yau, M. K. 1989. A Short Course in Cloud Physics. Pergamon Press, New York.

Rundel, P. W. 1988. Water relations. Pages 17-36 in: CRC Handbook of Lichenology, vol. 2. M. Galun, ed. CRC Press, Boca Raton, FL.

Sakai, A. 1979. Freezing avoidance mechanism of primordial shoots of conifer buds. Plant Cell Physiol. 20:1381-1390.

Sakai, A. 1982. Extraorgan freezing of primordial shoots of winter buds of conifer. Pages 199-209 in: Plant Cold Hardiness and Freezing Stress, vol 2. P. H. Li, A. Sakai, eds. Academic Press, New York.

Salt, R. W., and Kaku, S. 1967. Ice nucleation and propagation in spruce needles. Can. J. Bot. 45:1335-1346.

Schnell, R. C., and Vali, G. 1972. Atmospheric ice nuclei from decomposing vegetation. Nature (London) 236:163-165.

Siminovitch, D., and Scarth, G. W. 1938. A study of the mechanism of frost injury to plants. Can. J. Res. C 16:467-481.

Single, W. V., and Marcellos, H. 1981. Ice formation and freezing injury in actively growing cereals. Pages 17-33 in: Analysis and Improvement of Plant Cold Hardiness. C. R. Olein and M. N. Smith, eds. CRC Press, Boca Raton, FL.

Steffens, K. L., Arora, R., and Palta, J. P. 1989. Relative sensitivity of photosynthesis and respiration to freeze-thaw stress in herbaceous species. Plant Physiol. 89:1372-1379.

Storey, J. M., and Storey, K. B. 1985. Triggering of cryoprotectant synthesis by the initiation of ice nucleation in the freeze tolerant frog, *Rana sylvatica*. J. Comp. Physiol. B 156:191-195.

Thomas, D. A., and Barber, H. N. 1974. Studies on leaf characteristics of a cline of *Eucalyptus urnigera* from Mount Wellington, Tasmania. I. Water repellency and the freezing of leaves. Aust. J.

Bot. 22:501-512.

Tsumuki, H., Haruyoshi, K., Maeda, T., and Okamoto, T. 1992. An ice-nucleating active fungus isolated from the gut of the rice stem borer, *Chilo suppressalis* Walker (Lepidoptera: Pyraldae). J. Insect Physiol. 38:119-125.

Vali, G. 1971. Quantitative evaluation of experimental results on the heterogeneous freezing nucleation of supercooled liquids. J. Atmos. Sci. 28:402-409.

Vali, G., and Stansbury, E. J. 1966. Time-dependent characteristics of the heterogeneous nucleation of ice. Can. J. Phys. 44:477-502.

Warren, G. J. 1987. Bacterial ice nucleation: Molecular biology and applications. Biotechnol. & Gen. Eng. Rev. 5:107-135.

Wiegand, K. M. 1906. Some studies regarding the biology of buds and twigs in winter. Bot. Gaz. 41:373-424.

Yelenosky, G. 1983. Ice nucleation active (INA) agents in freezing of young citrus trees. J. Am. Soc. Hort. Sci. 108:1030-1034.

Zachariassen, K. E., and Hammel, H. T. 1988. The effect of ice-nucleating agents on ice-nucleating activity. Cryobiology 25:143-147.

CHAPTER 9

# Deep Supercooling in Woody Plants and the Role of Cell Wall Structure

**Michael Wisniewski**

Deep supercooling of xylem parenchyma cells is a mechanism of freeze avoidance and can be defined as the ability of a population of cells to retain cellular water in a liquid phase at low, subfreezing temperatures by remaining free from internal, heterogeneous ice nuclei and isolated from the nucleating effect of extracellular ice (Burke, 1979). Supercooled water is in a metastable condition and will form intracellular ice in response to a heterogeneous nucleation event or when the homogeneous nucleation temperature of water (–40°C) is reached (Rassmussen and Mackenzie, 1972). Additionally, colligative properties of the cell sap may further depress the freezing point beyond the homogeneous nucleation point by several more degrees (George and Burke, 1977; Rassmussen and Mackenzie, 1972). When freezing does occur, intracellular ice is formed, which results in lysis of the cell. Deep supercooling has also been referred to as deep undercooling. The terms are synonymous, and although the latter term may represent more appropriate terminology, the term deep supercooling has found widespread and common use in the literature. Due to the intrinsic temperature constraints for deep supercooling, which limit its effectiveness as a strategy for freeze protection, it has been implicated in determining species distribution on both a latitudinal and elevational gradient (George et al., 1974; Becwar et al., 1981).

Of the many aspects of biological ice nucleation and cold hardiness of plants, deep supercooling is perhaps the most enigmatic. The ability of some plants to maintain symplastic water in an unfrozen condition and without movement of the water into the apoplast is a remarkable adaptation that has not failed to impress both biophysicists and plant physiologists. Although the ability of woody plant tissues to avoid freezing by deep supercooling was first suggested in the 1960s (Tumanov and Krasavtsev, 1962; Tumanov et al., 1969) and 1970s (Krasavtsev, 1970; Quamme et al., 1972, 1973), the mechanism that allows small domains of water to avoid freezing, despite the presence of extracellular ice, remains little understood (Franks, 1985). This is partly due to the fact that the properties which allow deep supercooling to occur apparently rely on the structural organization of the tissue or organ. This feature has made it very difficult to manipulate plant material in a way to discover the fundamental mechanism and/or properties that allow deep supercooling to occur.

Despite this limitation, evidence indicates that the structure of the cell wall plays an integral role in determining whether or not a species will exhibit deep supercooling. The purpose of this chapter is to provide an overview of the topic of deep supercooling, to discuss the role of cell wall structure in regulating the ability of plants to exhibit this trait, and to pose some possible areas of future research. The review and discussion are limited to deep supercooling of xylem tissues of woody plants (see also Ashworth, 1993; Wisniewski et al., 1993). Although deep supercooling of overwintering floral buds also occurs, the response of these tissues is unique, and the reader should refer to Chapter 10 of this volume for a detailed elucidation of this topic.

## Characterization of Freezing Events in Xylem Tissues

The freezing response of xylem tissues (or other material) can be monitored using differential thermal analysis (DTA) as outlined by Quamme et al. (1972) and since modified to take advantage of recent advances in digital data-acquisition equipment and microcomputers (see references in Wisniewski et al., 1990). Thermocouples are used to detect the latent heat released by water in the samples as it undergoes a liquid to solid phase change. Concomitantly, melting curves can also be obtained with this technique. Sample temperatures are compared to a reference (freeze-dried tissue) undergoing the same rate of cooling. This produces a flat baseline on the thermogram until the water in the sample undergoes a phase change, resulting in a difference in temperature between the sample and the reference. The sample/reference differential is recorded as a peak on the thermogram and represents an exotherm during freezing or an endotherm during melting. Differential scanning calorimetry and nuclear magnetic resonance have also been used to obtain invaluable information about deep supercooling in plant tissues (George and Burke, 1977; Quamme et al., 1982).

Using DTA, two exotherms have been observed during the controlled freezing of xylem tissues of many temperate hardwood and softwood trees (Quamme et al., 1972; George et al., 1974; Quamme, 1976; George and Burke, 1977; Becwar et al., 1981; Ashworth et al., 1983; Rajashekar and Reid, 1989). Examples of thermograms are illustrated in Figure 1. The initial "high-temperature exotherm" (HTE) represents the freezing of bulk water within the lumens of tracheary elements and extracellular spaces, whereas the subsequent, "low-temperature exotherm" (LTE), which often occurs near –40°C, represents the freezing of water within living cells (Fig. 1). Evidence for the existence of a large fraction of deep supercooled water has also been obtained from studies utilizing nuclear magnetic resonance in which a sharp change in the freezing curve is observed within the same temperature range of the LTE (George and Burke, 1977; Quamme et al., 1982). Providing further evidence for deep supercooling of xylem tissue, Ashworth et al. (1988) and Malone and Ashworth (1991) used low-temperature- or conventional scanning electron microscopy of freeze-substituted specimens to observe the response of ray cells to freezing. They noted, in xylem tissues from several species of trees which exhibit deep supercooling, that the cell walls of ray cells remained rigid during freezing. There was no evidence of cytorrhysis as had been observed in bark tissues, which exhibited extracellular freezing and a concomitant loss of cellular water to sites of extracellular ice.

In contrast to thermograms obtained from species that exhibit deep supercooling, only a single HTE is exhibited in species, such as *Salix babylonica*, that lack the ability to deep supercool (Fig. 1). In these species, survival of living xylem tissues depends upon mechanisms associated with freeze tolerance rather than freeze avoidance (see Chapter 7).

Although xylem tissues that deep supercool are often referred to as having a single Gaussian-shaped LTE, it should be noted that in some species, the shape of the LTE, as well as the number, can be quite variable and complex (Figs. 1 and 2). The basis for the complex shape of the exotherms, which is often accentuated during periods of acclimation and deacclimation (Ashworth, 1993), is not understood. It appears the xylem tissue in some species does not freeze in a homogeneous manner but instead, as indicated by the complex shape of the LTE, responds as a heterogeneous population of cells that freeze over a wide temperature range. In apple (*Malus pumila*), Ketchie and Kammereck (1987) observed at least three exotherms (Fig. 2) and indicated that they represent the freezing of different tissues of the stem (i.e., pith, primary xylem, secondary xylem). The shape of the LTE and the temperature at which it is initiated is also subject to seasonal changes (Quamme et al., 1972; George and Burke, 1977; Ashworth et al., 1983; Ketchie and Kammereck, 1987; Wisniewski and Ashworth, 1986; Arora et al., 1992) and in general follows a pattern similar to those of other processes involved in cold acclimation: increasing in the fall, reaching a maximum in midwinter, and then decreasing in the spring.

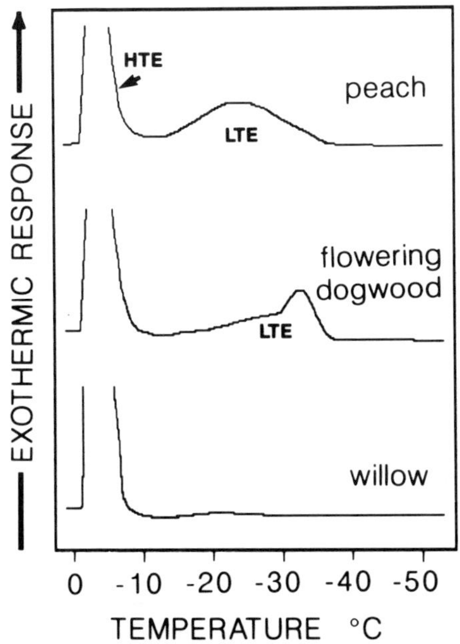


**Figure 1**. Freezing response of debarked, internodal twig sections of peach (*Prunus persica*), flowering dogwood (*Cornus florida*), and willow (*Salix babylonica*). The HTE (high-temperature exotherm) represents the freezing of intercellular water and water within the lumen of nonliving tracheary elements. The LTE (low-temperature exotherm) represents the freezing of intracellular, deep supercooled water, within living ray cells. Note the absence of any LTE in willow, which lacks the ability to deep supercool.

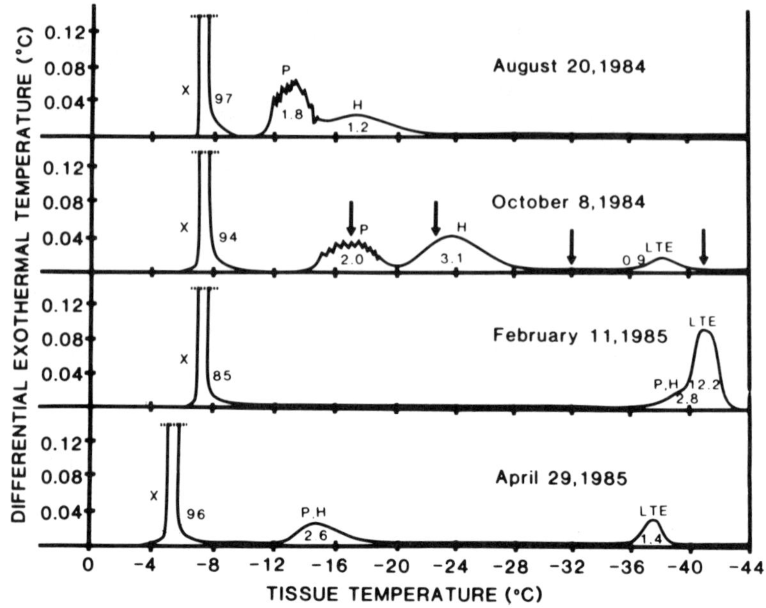


**Figure 2.** Typical exotherm patterns observed in basal internodes of current-season lateral shoots of apple cv. Golden Delicious prior to vegetative maturity (August), during acclimation (October), at maximum cold resistance (February), and during deacclimation (April). X represents a high-temperature exotherm whereas P, H, and LTE all represent low-temperature exotherms. LTE in this study correlated with injury only at the stage of maximum cold resistance. The number associated with the exotherm represents proportion of total tissue water, in percent. (Reprinted, by permission, from Ketchie and Kammereck, 1987.)

## Deep Supercooling and Tissue Injury

The initial research on deep supercooling of xylem tissues established a correlation between the temperature range of the LTE and tissue injury (Quamme et al., 1972; George et al., 1974; George and Burke, 1977; Hong et al., 1980). This was a significant discovery in regards to the biology and survival of woody plants at freezing temperatures and has led to many practical applications in evaluating woody plants for cold hardiness (Stushnoff, 1972; Quamme, 1976; Quamme and Stushnoff, 1983). In particular, DTA of stem tissues has played an important role in evaluating germ plasm of deciduous fruit and nut trees for cold hardiness (Quamme, 1976; Rajashekar et al., 1982; Ashworth et al., 1983; Rajashekar and Reid, 1989).

It is believed that the correlation of the LTE with tissue injury is due to the formation of intracellular ice within living cells, which results in lysis of the cell (Burke and Stushnoff, 1979). Hong and Sucoff (1980), in a study of eight species of woody plants, observed a linear relationship between the number of dead xylem ray cells and the amount of supercooled water that had frozen. This work confirmed and extended earlier observations made by George and Burke (1977) on shagbark hickory (*Carya ovata*). Using differential scanning calorimetry, the latter authors demonstrated that the bell-shaped LTE is actually composed of many independent freezing events (Fig. 3) and, based on low-temperature microscopy, indicated that the intact ray was the unit of freezing. Hong and Sucoff (1980) later

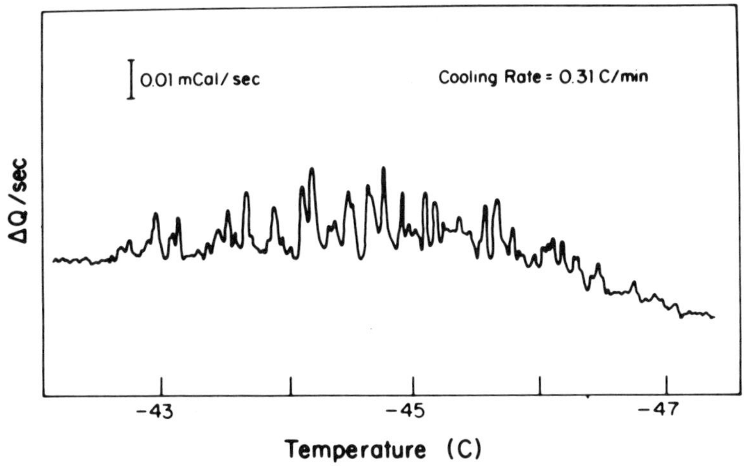


**Figure 3.** Exotherm fine structure as revealed by calorimetric analysis. A large peak roughly equals 100 ray parenchyma cells freezing simultaneously. (Reprinted, by permission, from George and Burke, 1977.)

reported individual or small groups of cells were the units of freezing and that ice spread very slowly, if at all, from the initial freezing units. Wisniewski and Ashworth (1985), using electron microscopy in combination with freeze-fixation, also confirmed thc units of freezing to be individual or small groups of cells within a ray.

It appears likely that cells that freeze near the homogeneous ice nucleation temperature of water do so due to spontaneous intracellular nucleation. Nevertheless, the cause of ice nucleation is by no means clear when cells freeze over a very broad temperature range, as indicated by an irregularly shaped, broad LTE. As Ashworth (1993) has correctly indicated, freezing at these temperatures could be due to the presence of an internal, heterogeneous ice-nucleating agent or due to external seeding by extracellular ice. These two alternatives have important implications regarding the mechanism of deep supercooling. In the first, seasonal changes in the temperature at which ice is initiated would be due to the activity of an intracellular ice-nucleating agent, whereas the latter would be due to some structural modification in the barrier that prevents extracellular ice from propagating, contacting the ray parenchyma cell, and either puncturing the plasma membrane or acting as a seeding agent. As Pitt (1992) indicated, the molecular basis of seeding is not well understood. Either ice is able to grow through aqueous channels in the plasma membrane (Mazur, 1965, 1977), or the extracellular ice interacts with the plasma membrane in such a manner as to cause intracellular nucleation on the membrane surface (Toner et al., 1990). The reader is referred to Pitt (1992) for a detailed discussion on the thermodynamics of intracellular ice formation.

In situations where the LTE is quite broad or exhibits several peaks, it is also unclear whether or not the entire freezing curve is associated with tissue injury. Ketchie and Kammereck (1987) observed multiple LTEs in apple and found that only the peaks at the warmer temperatures (peaks P and H, Fig. 2) were associated with injury. The peak that occurred near the temperature of the homogeneous nucleation point of water was only associated with injury at midwinter, when all peaks became superimposed, and occurred in the range of −35 to −40°C.

## Deep Supercooling and the Distribution of Woody Plants

An important extension that developed from the recognition of deep supercooling of xylem as a freeze-avoidance mechanism was its implication for defining the distribution of woody plants (George et al., 1974; Becwar et al., 1981). If the observed LTEs were due to simple supercooling, then the limits of this freeze avoidance should occur in a temperature range near the homogeneous nucleation point of water (–38°C). George et al. (1974), in a survey of 49 species of woody plants native to North America, found a good correlation between the temperature at which an LTE occurred and the minimum temperatures at the boundary of the northern range of that species. Based on this correlation, they proposed that deep supercooling played a major role in defining the northern limit of the distribution of many native species of hardwoods. In addition to defining the northern latitudinal limits of a species, Becwar et al. (1981) later extended this hypothesis to the distribution of woody plants on an elevational gradient. Again, regarding the distribution of a species, there was a good correlation between the extent of deep supercooling and the minimum temperature occurring at the elevational limit of the species. Comprehensive lists of species that exhibit deep supercooling and those that do not have been published (George et al., 1974, 1982; Becwar et al., 1981)

In attempting to associate anatomical features of xylem tissues with the ability to deep supercool, George et al. (1974) reported that large LTEs, and deep supercooling in general, predominate in species with ring porous xylem. It should be noted that species with ring porous xylem have very wide diameter vessel elements that are very efficient in conducting water compared with species with diffuse porous xylem (Zimmerman, 1983). These large-diameter vessel elements, however, are also subject to cavitation during freezing, resulting from the separation of diffused gases from water contained in vessel elements. This occurs during the initial freezing of bulk water, recorded as an HTE. Because of the large size of the air bubble that forms in vessel elements of ring porous xylem, the air cannot be redissolved when the ice melts at the return of warm temperatures, and therefore the xylem element remains nonfunctional in regard to water transport (Zimmerman, 1983; Ellmore and Ewers, 1985). Whether or not this characteristic is related to deep supercooling in a causal manner is not known but bears further consideration (Wisniewski et al., 1987b).

Gusta et al. (1983) reported several exceptions to the –40°C isotherm limit proposed by George et al. (1974). A major finding of this report was that, in a number of woody species, LTEs could occur at much lower temperatures than had been previously reported, allowing these species to survive in areas where winter temperatures of –45°C are not uncommon. Furthermore, in some species (e.g., red ash, *Fraxinus pennsylvanica*), LTEs observable in early winter completely disappeared during prolonged exposure of the tissue to subzero temperatures approaching –38°C. Based on their observations, the authors concluded that, in some species, xylem ray cells were able to gradually lose water to extracellular ice, resulting in a four- to fivefold concentration of the dissolved cell solutes (i.e., an increase from 0.5–1.0 osM to 4.0–5.0 osM). This intriguing data indicated that not only was cell wall structure an important component of the mechanism of deep supercooling, but that properties of the plasma membrane were also a critical element. Earlier work by Hong and Sucoff (1982) reported that rapid shifts in deep supercooling to lower temperatures could occur in apple during a 10-h exposure to –5°C. Dehydration of the living cells, however, was not responsible for this shift, and no causal mechanism was suggested. More recently, Wisniewski et al. (1991c) noted that contact of the plasma membrane with the cell wall was essential for maximum supercooling to occur. Again, the causal basis for this observation was not known, but collectively these reports indicate that the relationship between cell wall structure, the plasma membrane, and the ability to deep supercool needs a great deal of further investigation.

## Cell Wall Structure and Deep Supercooling

In order for deep supercooling to occur, a tissue must exhibit several features (George and Burke, 1977): 1) cells must be free of heterogeneous nucleating substances active at "warm" subzero temperatures; 2) a barrier must be present that excludes the growth of ice crystals into a cell; concomitantly 3) a barrier to water movement must exist that prevents a "rapid" loss of cellular water to extracellular ice in the presence of a strong vapor pressure gradient; and 4) cell walls must have sufficient tensile strength to counteract the negative hydrostatic pressures that result from the large vapor pressure deficit.

Current opinion holds that physical properties of the apoplast (i.e., cell wall structure) rather than the symplast (i.e., protoplast) largely account for the ability to deep supercool (George and Burke, 1977; George, 1983; Ashworth and Abeles, 1984). In this regard, porosity of the cell wall would play an essential role, although as noted by Wisniewski et al. (1987a), permeability should also be considered. George and Burke (1977) found that although the xylem of shagbark hickory exhibited an LTE around –40°C in midwinter, there was no substantial barrier to water exchange between the xylem ray parenchyma cells and neighboring vessel elements. They suggested the water potential of xylem ray cells at subfreezing temperatures was reduced by some means such that ice outside the ray cells and cellular water could be in thermodynamic equilibrium. They proposed a simple surface chemical effect, the "ink bottle" pore effect, to explain their observations. As illustrated in Figure 4, the ink bottle represents the water contained within a group of ray cells. The pore represents cell wall microcapillaries of ray cells containing vapor-phase water that is in equilibrium with ice present in neighboring vessel elements. The stability of the liquid to desiccation depends on the radius of curvature of the meniscus (liquid-vapor) according to the Kelvin equation. As illustrated, a unit of water separated from its external environment by a pore with a diameter of 100 Å would, due to the highly curved meniscus at point A, be in equilibrium at 80% relative humidity. Any further drop in relative humidity, however, would cause evaporation, resulting in a lowering of the meniscus to point B and subsequent evaporation of the remaining water. Utilizing this concept to explain the isolation and desiccation resistance of water within deep supercooled tissues, the water potential in supercooled cells would have to be lowered to about –400 bars in order for the cells to be in vapor equilibrium with ice at –40°C. How the cells could withstand so great a hydrostatic tension remains an open question.

Subsequent work by Ashworth and Abeles (1984), using controlled-pore-size glass particles and polycarbonate filters, also documented that pore size could have an effect on the temperature at which water within the pores would freeze. They found water within the pores supercooled and that the extent of supercooling increased as the diameter of the pores decreased. Within different sized pores, the observed melting point of water (Fig. 5) agreed very closely with predicted values generated using equations developed by Mazur (1965) and Homshaw (1980), although glass particles with a pore size of less than 7.5 nm (75 Å) could not be obtained. The freezing point of water within these smallest pores occurred at −10°C. As George and Burke (1977) had used the pore effect to explain observations related to desiccation resistance, Ashworth and Abeles (1984) proposed that ice could not propagate through a cell wall microcapillary until the temperature dropped below the melting point of the water within the pore, which in turn would be determined

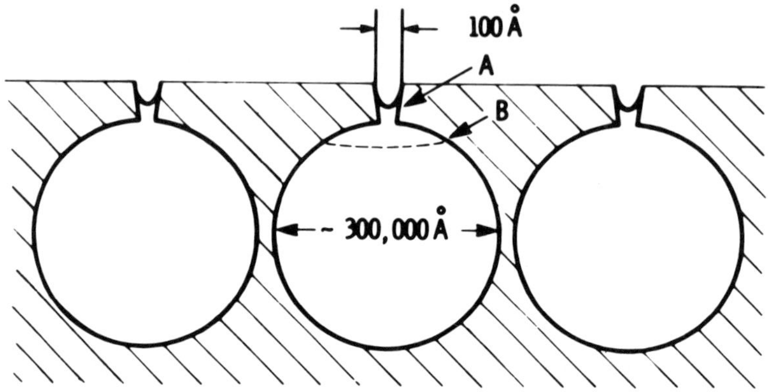


**Figure 4.** "Ink bottle" pores having volumes approximately equal to 100 ray parenchyma cells. The relative humidity calculated for the meniscus at point A is 80% and at point B is slightly below 100%. Such pores would not dehydrate significantly until the relative humidity was dropped to below 80%. (Reprinted, by permission, from George and Burke, 1977.)

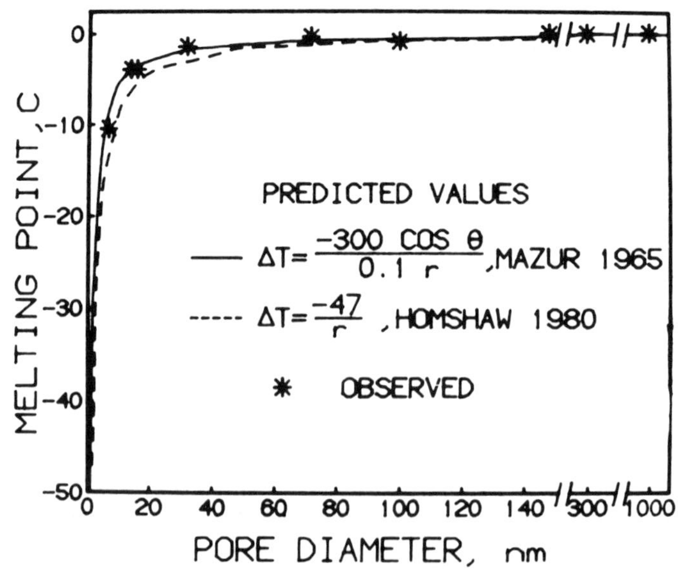


**Figure 5.** Relationship between pore diameter and melting point. Experimental values obtained using controlled pore glass and polycarbonate filters were compared to values predicted by equations developed by Mazur (1965) and Homshaw (1980). (Reprinted, by permission, from Ashworth and Abeles. 1984.)

by the diameter of the pore. There is a large discrepancy in the effect of pore size on freezing point depression in these two reports, which may be explained by the fact that water movement is assumed to occur in the vapor phase in George and Burke (1977), while the system of Ashworth and Abeles (1984) utilized a continuous column of water. An important corollary of both hypotheses, however, is that freezing or evaporative loss of water contained in the deep supercooled cells would be determined by the diameter of the largest size pores.

The data obtained from theoretical calculations and model systems clearly indicate that pore size should have a profound effect on the movement and freezing point of water within supercooled tissues. How these constraints on pore size relate to actual cell wall structure and tissue anatomy, however, is not clear. These hypotheses do not indicate whether the entire primary and/or secondary wall or just specific sites within the cell wall need exhibit restricted pore size. It is also unclear whether all cells or just living cells of xylem tissues need possess this structure. Furthermore, hypotheses on cell wall structure must also account for seasonal changes in the extent of deep supercooling, as well as differences in wood structure between species that exhibit deep supercooling and those that do not. Over the past several years, Wisniewski and co-workers (Wisniewski et al., 1987a,b, 1991a,b; Wisniewski and Davis, 1989) have focused on trying to elucidate the aspects of cell wall structure that promote deep supercooling and regulate seasonal variation in the extent of deep supercooling. Their data, as presented in the following sections, support a major role for the structure of the pit membrane of xylem parenchyma and further indicate that pectins may determine the porosity and/or permeability of this region of the cell wall.

## Characterization of Cell Wall Porosity and Permeability

Lanthanum nitrate has been extensively used as a free-space marker at the ultrastructural level in both plant and animal tissues, and is believed to penetrate spaces as small as 20 Å (see citations in Wisniewski, 1987a). It is administered as an ionic solution, and the lanthanum is precipitated when tissue is immersed in a fixative with a neutral to basic pH. The lanthanum is seen as black, electron-dense deposits when tissue sections are observed with a transmission electron microscope.

Utilizing this apoplastic tracer, the distribution of lanthanum deposits in stem tissues of several species of woody plants was characterized (Wisniewski et al., 1987a,b). Intergeneric and intrageneric comparisons were made between species that exhibit deep supercooling (*Prunus persica, Cornus florida, Betula lenta*) and those that do not (*Salix babylonica, Betula papyrifera*). The distribution of lanthanum was generally the same in all species. Although some differences were noted, they were not associated with the freezing response (i.e., supercooling vs. nonsupercooling) of the tissue. Whereas the primary cell walls of cortical cells were freely permeable to the lanthanum solution, both the primary and secondary cell walls of xylem parenchyma cells were impermeable (Fig. 6). This indicates either an extremely small pore size or low accessibility (i.e., porous but low permeability). In contrast, the pit membrane and amorphous (protective) layer of xylem parenchyma cells were very permeable. The absence of lanthanum deposits in secondary cell walls of ray cells, in all species examined, indicates that if pore size in the cell wall is integral to defining the freezing response of xylem tissue, then the structure and permeability of the pit membrane (and associated amorphous layer) rather than the secondary cell wall, should play an integral role. This is due to the fact that the widest capillaries limit the ability of a cell to retain water against a strong vapor-pressure gradient and prevent the intrusive growth of external ice crystals.

Evidence of pore diameters in the range of 5 to 170 nm (50–1,700 Å) for pit membranes and 10 to 130 nm (100–1300 Å) for xylem cell walls of hardwoods have been reported (Siau, 1984). However, as noted by Preston (1974), the term porosity can only be used loosely in that the path through the cell wall represents a very tortuous series of canals, angular in cross-section and mainly aligned parallel to the microfibrils. Whereas the "pore" system in pit membranes may come close to representing true capillaries, the "pore" system in the secondary cell wall must be more complex due to the laminate construction of the cell wall with opposing microfibril orientations in the $S_1$, $S_2$, and $S_3$ layers. Transition lamellae have also been reported (Vian et al., 1986), which would further complicate the organization of any pore system in the secondary cell wall. This feature, along with the demon-

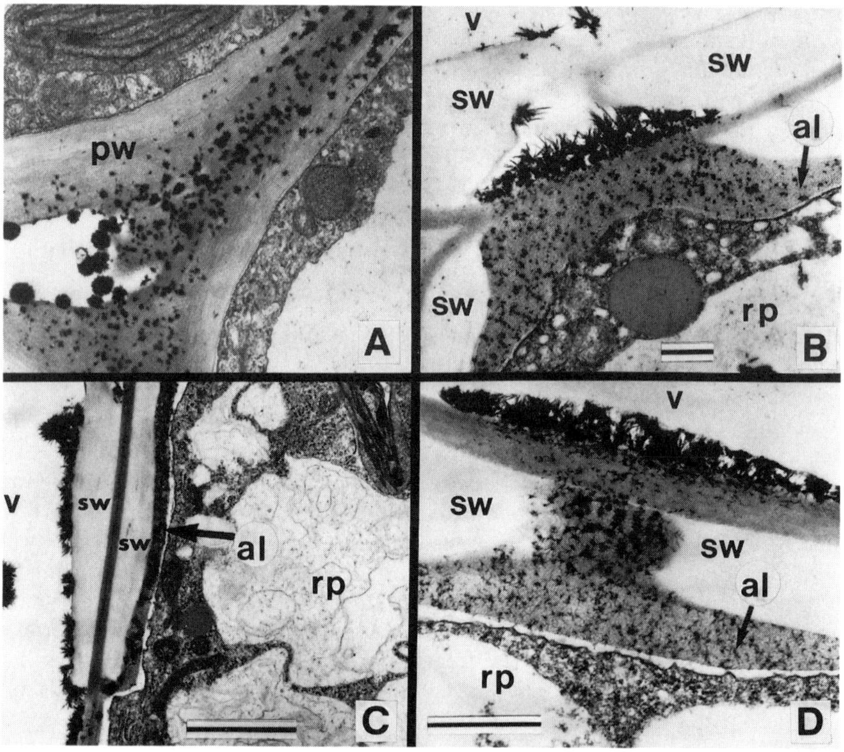


**Figure 6.** Distribution of lanthanum (black, crystalline deposits) in **A,** bark and **B–D,** ray cells. Note distribution of lanthanum within the primary cell walls (pw) of bark cells. In xylem ray cells, lanthanum was absent from the secondary cell walls (sw) but present in large amounts in the pit membrane and underlying amorphous layer (al). (See Figure 7 for diagrammatic illustration of xylem ray parenchyma cell and cell wall layers composing the pit membrane.) *Prunus persica* (A and B), *Salix babylonica* (C), *Cornus florida* (D). v = vessel element, rp = xylem ray parenchyma cell. (Reprinted, by permission, from Wisniewski et al., 1987a.)

strated permeability of the pit membrane to lanthanum, adds credibility to the idea that differences in the structure of pit membranes may be responsible for determining whether or not a cell or tissue will deep supercool.

The studies utilizing lanthanum as an apoplastic tracer clearly revealed that the pit membrane was the most permeable portion of the xylem ray cell. Therefore, if differences in cell wall structure exist between species that supercool and those that do not, and which directly impact on freezing response, they should be present in the pit membrane portion of the cell wall. Based on the hypotheses of George and Burke (1977) and Ashworth and Abeles (1984), pores in the size range of 60–100 Å or less would be expected to have an impact on freezing response. Lanthanum can penetrate voids as small as 20 Å. Therefore, apoplastic tracers of a larger size would be necessary to conduct exclusion studies to reveal more details about the pore structure of the pit membrane in supercooling vs. non-supercooling species. Research on the movement of solutions through xylem tissues has been reported in the wood technology literature (Siau, 1984) and supports the premise that, although cell wall capillaries exist in secondary cell walls, the major pathway of both longitudinal and lateral movement of solutions is via the pit system into the lumens of cells and then into the cell wall (Kininmonth, 1971, 1972; Murmanis and Chudnoff, 1979).

## Pit Membrane Structure and Deep Supercooling

Although the structure of the pit membrane of ray cells of hardwoods can be variable depending on the type of cell it interfaces (Esau, 1977), the structure of a typical pit membrane of a ray cell bordering a vessel element (such ray cells are designated as contact cells) is diagrammatically represented in Figure 7. As seen with an electron microscope, it is composed of an outer layer of electron-dense material (black cap), a middle layer of primary cell wall derived from both the ray cell and adjoining vessel element, and an inner layer (amorphous or protective layer) that lines the inside of the secondary cell wall but is thickened in the vicinity of the pit membrane. The function of this inner wall layer is not fully understood;

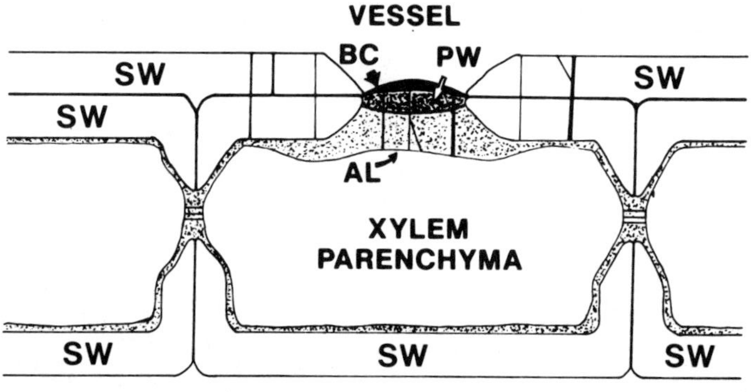


**Figure 7.** Diagrammatic representation of a xylem ray cell (transverse section) adjacent to a vessel element. The pit membrane consists of three layers: an outermost black cap (BC); a primary wall (PW), derived from both the ray cell and vessel element, and an amorphous layer (AL), which lies interior to the secondary cell wall (SW) of the ray cell. The channels are meant to diagrammatically illustrate cell wall microcapillaries. (Reprinted, by permission, from Wisniewski et al., 1991b.)

hence, ambiguous terminology has arisen to describe it (Schaffer and Wisniewski, 1989). It should be noted that only a single pit membrane is illustrated; in reality, a ray cell will have many of these pits interfacing with its neighboring cells.

Wisniewski et al. (1991a) found that when stem segments of peach and flowering dogwood were treated with macerase (an enzyme mixture rich in pectinase),

**Table 1.** Compounds or treatments that modify the structure of the pit membrane of xylem parenchyma[a]

| Compound or treatment | Portion of pit membrane affected[b] | Principal effect on low-temperature exotherm |
|---|---|---|
| Pectinase | BC, PW, AL | Flattening |
| Cellulase | BC | Shift to warmer temperature |
| Hemicellulase | BC | Shift to warmer temperature |
| Oxalic acid | BC, PW, AL | Flattening and slight shift to warmer temperature |
| EGTA[c] | BC, PW | Shift to warmer temperature |
| Water soak (3–10 days) | BC, PW, AL | Flattening and slight shift to warmer temperature |
| Sodium phosphate buffer | BC (slight) | Shift to warmer temperature |

[a]Source: Wisniewski et al. (1991); used by permission.
[b]BC = black cap or torus-like covering on vessel side of pit membrane; PW = primary wall portion of pit membrane; and AL = amorphous layer of pit membrane.
[c]Ethyleneglycol-bis ($\beta$-amino-ethyl ether)-$N,N'$-tetraacetic acid.

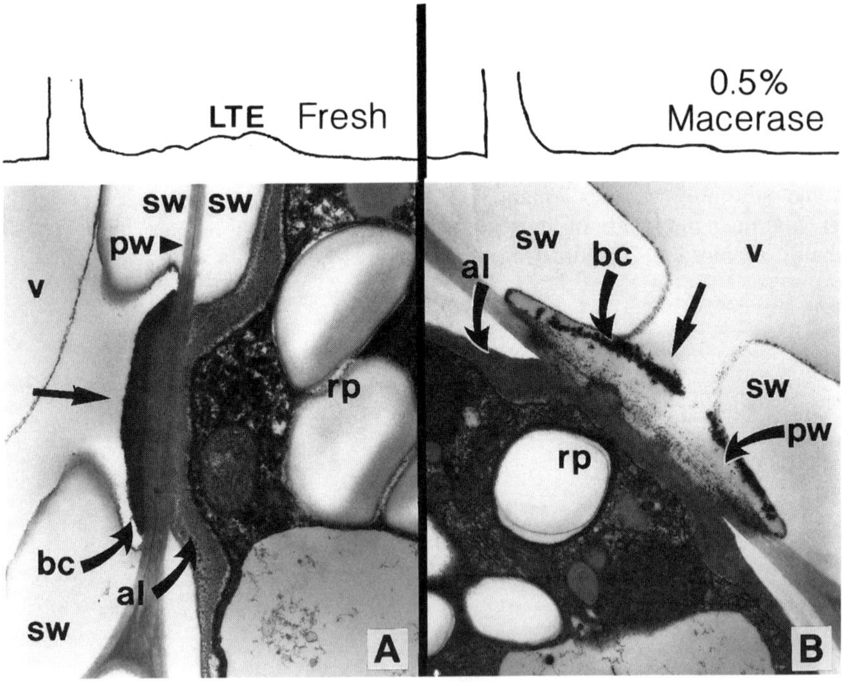


**Figure 8.** Effect of macerase on freezing response and pit membrane stucture of peach. **A,** Control tissue. Note LTE (low-temperature exotherm) and intact structure of the pit membrane (large arrow). **B,** Macerase-treated tissue. Note flattening of LTE and disruption of pit membrane structure (large arrow). v = vessel element, sw = secondary cell wall, pw = primary cell wall, bc = black cap, al = amorphous layer, rp = xylem ray parenchyma cell.

the enzyme had a profound effect on both pit membrane structure and the ability to deep supercool. In contrast, treatments with cellulase, hemicellulase, or phosphate buffer had relatively minor effects on the character of the LTE and pit membrane structure (Table 1). As illustrated in Figure 8, the macerase treatment resulted in a flattening (i.e., disappearance) of the LTE. This was associated with an almost complete digestion of the outer two layers of the pit membrane and a partial digestion of the inner amorphous layer. The other enzyme and buffer treatments appeared to affect only the outer layer (black cap) and mainly resulted in a shift of the LTE to warmer temperatures. No alterations were observed in the structure of the secondary cell wall or primary cell wall (outside of the pit membrane). Based on these observations, it was concluded that treatments (cellulase, hemicellulase, buffer) that resulted in only a slight modification in pit membrane structure (via partial dissolution of cell wall polysaccharides) altered the "barrier" properties of the cell wall. This caused a shift of the LTE to warmer temperatures. In comparison, the pectinase treatment, which substantially altered the structure of the pit membrane, nearly eliminated this "barrier" property, thus eliminating the ability of the tissue to deep supercool. These observations supported a previous report (Wisniewski and Davis, 1989) in which self-induced alterations in pit membrane structure in peach were accompanied by a dramatic shift of the LTE to warmer

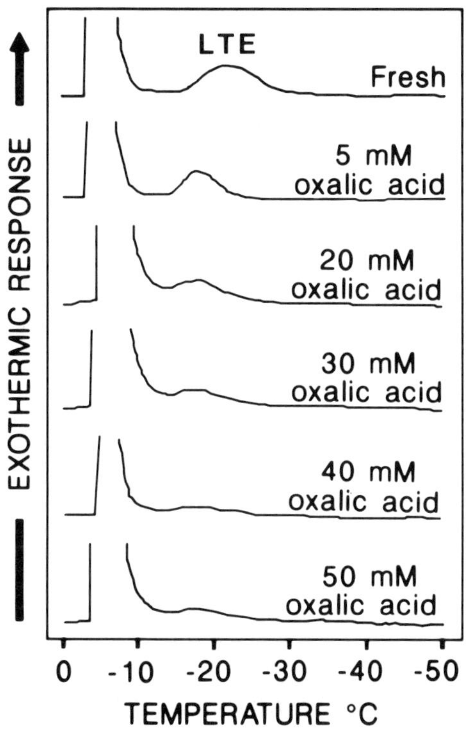


**Figure 9.** Differential thermal analysis profiles of 'Loring' peach xylem tissue before and after exposure to 5.0–50.0 mM oxalic acid for 24 h. Note the gradual flattening of the low-temperature exotherm (LTE) with increasing concentration of oxalic acid. (Reprinted, by permission, from Wisniewski et al., 1991b.)

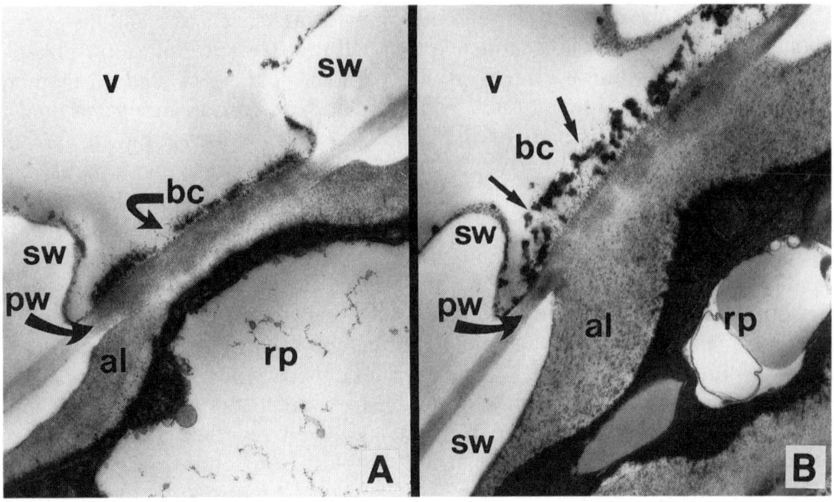


**Figure 10.** Pit membrane of ray cells of peach (*Prunus persica*) after exposure to oxalic acid for 24 h. **A,** Low concentration (5 mM) resulted in loosening of the black cap (bc, curved arrow). **B,** High concentration (50 mM) resulted in severe degradation of the black cap (bc, arrows) and underlying layers of the pit membrane. v = vessel element, sw = secondary cell wall, bc = black cap, pw = primary cell wall, rp = xylem ray parenchyma cell, al = amorphous layer. (Reprinted, by permission, from Wisniewski et al., 1991b.)

temperatures. The alterations were brought about by prolonged soaking of twigs in water. Both the alterations and the shift in the LTE were prevented by soaking the twigs in cycloheximide.

Subsequent work with calcium chelating agents (Wisniewski et al., 1991b) provided further evidence that the pit membrane of xylem ray cells is rich in pectin. A summary of the effects of compounds or treatments which modify the structure of the pit membrane and cause an alteration of the LTE profile is presented in Table 1. Besides the pectinase treatment, oxalic acid in particular had a significant effect on both the character of the LTE and the structure of the pit membrane (Figs. 9 and 10, respectively).

More recently, utilizing monoclonal antibodies (Knox et al., 1990) to esterified (JIM 7) and nonesterified (JIM 5) pectin, Wisniewski and Davis (*unpublished*) observed that in pit membranes of peach ray cells, nonesterified epitopes of pectin were confined to the outer layer of the pit membrane, whereas esterified epitopes predominated in the middle (primary wall) and inner layer (amorphous layer) (Fig. 11). Labeling of the pit membrane by either antibody (JIM 5 and JIM 7) was absent when tissues were first treated with pectinase. The amorphous layer was also abundant in epitopes of arabinogalactan-rich glycoprotein recognized by the monoclonal antibody JIM 14 developed by Knox et al. (1989, 1991). An increase in extensin, a cell wall glycoprotein, in epicotyl tissue of pea (*Pisum sativum*) during cold acclimation has been previously reported (Weiser et al., 1990).

Collectively, the data on cell wall structure and deep supercooling indicate that the type, amount , and degree of cross-linking of pectin within the pit membrane may determine the size of the pores and/or their permeability to water (some pectin polymers may increase hydrophobicity and therefore decrease permeability). The

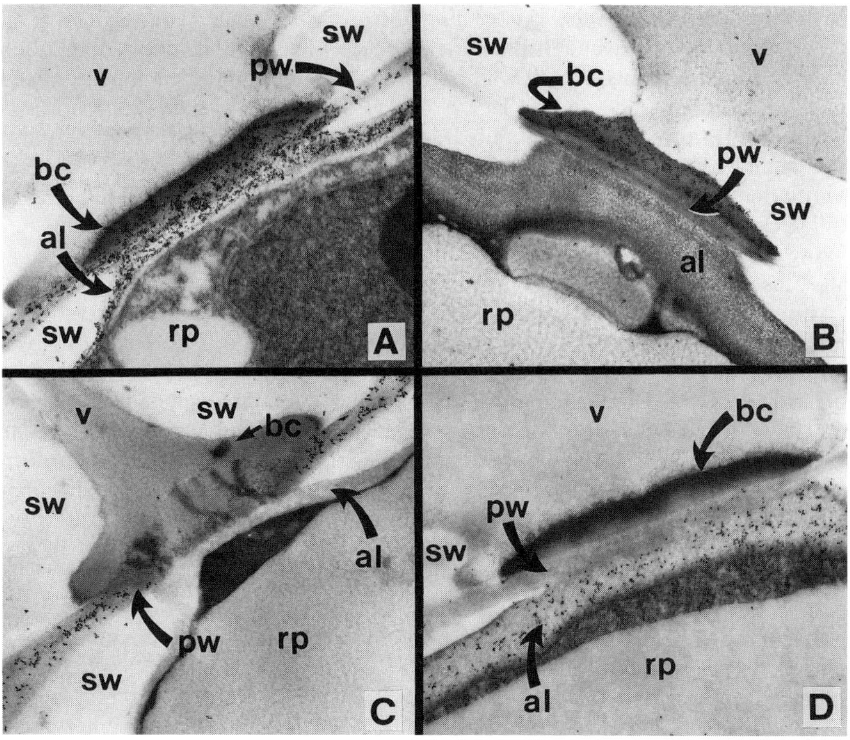


**Figure 11.** Distribution of monoclonal antibodies to esterified or nonesterified epitopes of pectin (antibodies termed JIM 7 and JIM 5) and to epitopes of arabinogalactan-rich glycoprotein (JIM 14) in pit membrane of ray cells of peach (*Prunus persica*). **A,** Distribution of esterified epitopes of pectin recognized by JIM 7. **B,** Distribution of epitopes of nonesterified pectin recognized by JIM 5. **C,** Drastic reduction of labeling with JIM 7 after treatment with pectinase. **D,** Distribution of epitopes recognized by JIM 14. v = vessel element, rp = xylem ray parenchyma, sw = secondary cell wall, bc = black cap, pw = primary cell wall, al = amorphous layer. (See Figure 7 for diagrammatic illustration of pit membrane structure.)

structure of the pit membrane, along with the tensile strength imparted by the secondary cell wall, could account for the ability to deep supercool. Baron-Epel et al. (1988) also suggested that pectins determine pore size in cell walls of soybean suspension cells. Pectin-mediated regulation of porosity/permeability of the pit membrane is an attractive hypothesis because it provides a plausible basis to explain the seasonal shifts that occur in the extent of deep supercooling. Loosening of cell wall structure (i.e., increased porosity/permeability) could occur in the spring via disruption of metal ion bridges (primarily $Ca^{2+}$) or breakage of covalent bonds by intrinsic pectinases. Such modifications do occur, for example, during tylose formation when the cell wall within the pit membrane loosens considerably to allow for growth and extension of the living ray cell into a neighboring vessel element (Beckman, 1971).

## Summary and Future Research

As previously stated, deep supercooling of plant tissues is perhaps one of the more enigmatic aspects of biological ice nucleation. Because the full expression of this trait is dependent on the existence of intact tissue, it has been difficult to discern the fundamental mechanism and/or properties that regulate deep supercooling. Although the importance of cell wall porosity has been implied via the use of theoretical calculations (George and Burke, 1977) and model systems (Ashworth and Abeles, 1984), how these constraints translate to actual cell wall structure and tissue anatomy is not well understood. As presented, the combined evidence indicates an important role for pit membrane structure in regulating deep supercooling and implicates pectin and the interaction of pectin with other cell wall constituents as playing a role in defining the porosity and/or permeability of this region of the cell wall. This evidence, however, is only correlative or inferential at this stage.

Although pectin-mediated regulation of deep supercooling is an attractive hypothesis because it can account for many of the observations presented, there are many questions that must be resolved before a definitive knowledge of the underlying mechanism of deep supercooling of xylem tissues is developed. Perhaps uppermost is the fact that components of cell wall structure that specifically dictate freezing response and differ between species that supercool and those that do not have yet to be documented. This is true in regards to both demonstrated porosity (Wisniewski et al., 1987a,b) and appearance during exposure to subzero temperatures (Ashworth et al., 1988, Malone and Ashworth, 1991). Furthermore, in apple (Ketchie and Kammereck, 1987), peach (Wisniewski and Davis, 1989), and some other species (Ashworth, 1993), it is clear that the xylem tissue does not exhibit a homogeneous freezing response. This is evidenced by multiple LTEs in apple and bimodal peaks in peach. How the complex freezing response of these tissues is regulated is not understood.

The work of Gusta et al. (1983) also indicates that, despite exhibiting deep supercooling, some species also have the ability to slowly dehydrate at very low temperatures (–30 to –40°C) resulting in a complete disappearance of the LTE. The viability of these tissues after exposure to cold temperatures would be dependent on variables associated with deep supercooling as well as additional cellular properties associated with cold acclimation. How these mechanisms of cold adaptation and ice nucleation are regulated is, again, not understood.

Due to the limits imposed on the cold hardiness of temperate fruit trees (and other economically important woody plants) by the temperature constraints associated with deep supercooling of xylem tissue and floral buds (see Chapter 10), selection to eliminate this trait in favor of other mechanisms of cold hardiness has been proposed (Burke and Stushnoff, 1979). The inheritance and genetic regulation of deep supercooling, however, have not been investigated. Although there is a distinct seasonality in the expression of deep supercooling of xylem tissues, which implies a controlled sequence of active processes, the biochemical and structural changes responsible for this seasonal regulation remain to be elucidated. As scientific evidence for the control and regulation of deep supercooling becomes available, determination of whether or not this trait can be manipulated using traditional methods of plant breeding or modern techniques of molecular biology will be forthcoming. It is hoped that a better understanding of deep supercooling of xylem tissues will directly or indirectly result in new strategies having the potential to prevent freeze injury in economically important horticultural plants.

## Literature Cited


Arora, R., Wisniewski, M., and Scorza, R. 1992. Cold acclimation in genetically related (sibling) deciduous and evergreen peach. Plant Physiol. 99:1562-1568.

Ashworth, E. N. 1993. Deep supercooling in woody plant tissues. Pages 204-213 in: Advances in Plant Cold Hardiness. P. H. Li and L. Christersson, eds. CRS Press, Boca Raton, FL.

Ashworth, E. N., and Abeles, F. B. 1984. Freezing behavior of water in small pores and the possible role in the freezing of plant tissues. Plant Physiol. 76:201-204.

Ashworth, E. N., Echlin, P., Pearce, R. S., and Hayes, T. L. 1988. Ice formation and tissue response in apple twigs. Plant Cell Environ. 11:703-710.

Ashworth, E. N., Rowse, D. J., and Billmeyer, L. A. 1983. The freezing of water in woody tissues of apricot and peach and the relationship to freezing injury. J. Am. Soc. Hort. Sci. 108:299-303.

Baron-Epel, O., Gharyal, P. K., and Schindler, M. 1988. Pectins as mediators of wall porosity in soybean cells. Planta 175:389-395.

Beckman, C. H. 1971. The plasticizing of plant cell walls and tylose formation—A model. Physiol. Plant Pathol. 1:1-10.

Becwar, M. R., Rajashekar, C. Hansen-Bristow, K. J., and Burke, M. J. 1981. Deep supercooling of tissue water and winter hardiness limitations in timberline flora. Plant Physiol. 68:111-114.

Burke, M. J. 1979. Discussion. Water in plants: The phenomenon of frost survival. Pages 259-281 in: Comparative Mechanisms of Cold Adaptations. L. S. Underwood, L. L. Tieszen, A. B. Callahan, and G. E. Folk, eds. Academic Press, New York.

Burke, M. J., and Stushnoff, C. 1979. Frost hardiness: A discussion of possible molecular causes of injury with particular reference to deep supercooling of water. Pages 199-225 in: Stress Physiology of Crop Plants. H. Mussell and R. C. Staples, eds. Wiley, New York.

Ellmore, G. S., and Ewers, F. W. 1985. Hydraulic conductivity in trunk xylem of elm, *Ulmus americana.* IAWA Bull. New Ser. 6:303-307.

Esau, K. 1977. Anatomy of Seed Plants. 2nd ed. John Wiley, New York.

Franks, F. 1985. Biophysics and Biochemistry at Low Temperatures. Cambridge University Press, London.

George, M. F. 1983. Freezing avoidance by deep supercooling in woody plant xylem: Preliminary data on the importance of cell wall porosity. Pages 84-95 in: Current Topics in Plant Biochemistry and Physiology. D. D. Randall, D. G. Blevins, R. L. Larson, and B. J. Rapp, eds. University of Missouri Press, Columbia, MO.

George, M., Becwar, R., and Burke, M. J. 1982. Freezing avoidance by deep undercooling of tissue water in winter-hardy plants. Cryobiology 19:628-639.

George, M. F., and Burke, M. J. 1977. Cold hardiness and deep supercooling in xylem of shagbark hickory. Plant Physiol. 59:319-325.

George, M. F., Burke, M. J., Pellet, H. M., and Johnson, A. G. 1974. Low temperature exotherms and woody plant distribution. HortScience 9:519-522.

Gusta, L. V., Tyler, M. J., and Chen, T. H. 1983. Deep undercooling in woody taxa growing north of the –40 C isotherm. Plant Physiol. 72:122-128.

Homshaw, L. G. 1980. Freezing and melting temperature hysteresis of water in porous materials: Application to the study of pore form. J. Soil Sci. 31:399-414.

Hong, S., and Sucoff, E. 1980. Units of freezing of deep supercooled water in woody xylem. Plant Physiol. 66:40-45.

Hong, S., and Sucoff, E. 1982. Rapid increase in deep supercooling of xylem parenchyma. Plant Physiol. 69:697-700.

Hong, S., Sucoff, E., and Lee-Stadelmann, O. 1980. Effect of freezing of deep supercooled water on the viability of ray cells. Bot. Gaz. 141:464-468.

Ketchie, D. O., and Kammereck, R. 1987. Seasonal variation of cold resistance in *Malus* woody tissue as determined by differential thermal analysis and viability tests. Can. J. Bot. 65:2640-2645.

Kininmonth, J. A. 1971. Permeability and fine structure of certain hardwoods and effects of drying. I. Transverse permeability of wood to microfiltered water. Holzforschung 25:127-133.

Kininmonth, J. A. 1972. Permeability and fine structure of certain hardwoods and effects of drying. II. Differences in fine structure of *Nothofagus fusca* sapwood and heartwood. Holzforschung 26:32-38.

Knox, J. P., Day, S., and Roberts, K. 1989. A set of cell surface glycoproteins forms an early marker of cell position but not cell type in the root apical meristem of *Daucus carota* L. Development 106:47-56.

Knox, J. P., Linstead, P. J., King, J., Cooper, C., and Roberts, K. 1990. Pectin esterification is spatially regulated both within cell walls and between developing tissues of root apices. Planta 181:512-521.

Knox, J. P., Linstead, P. J., Peart, J., Cooper, C., and Roberts, K. 1991. Developmentally regulated epitopes of cell surface arabinogalactan proteins and their relation to root tissue pattern formation. Plant J. 1:317-326.

Krasavtsev, O. A. 1970. Rate of water outflow from cells of frost resistant plants at negative temperatures. Sov. Plant Physiol. 17:417-422.

Malone, S. R., and Ashworth, E. N. 1991. Freezing stress response in woody tissues observed using low-temperature scanning electron microscopy and freeze substitution techniques. Plant Physiol. 95:871-881.

Mazur, P. 1965. The role of cell membranes in the freezing of yeast and other single cells. Ann. NY Acad. Sci. 125:658-676.

Mazur, P. 1977. The role of intracellular freezing in the death of cells cooled at supraoptimal rates. Cryobiology 14:251-272.

Murmanis, L., and Chudnoff, M. 1979. Lateral flow in beech and birch as revealed by the electron microscope. Wood Sci. Technol. 13:79-87.

Pitt, R. E. 1992. Thermodynamics and intracellular ice formation. Pages 63-99 in: Advances in Low-Temperature Biology, vol. 1. P. L. Steponkus, ed. JAI Press, London.

Preston, R. D. 1974. The Physical Biology of Plant Cell Walls. Chapman and Hall, London.

Quamme, H. A. 1976. Relationship of the low temperature exotherm to apple and pear production in North America. Can. J. Plant Sci. 56:493-500.

Quamme, H. A., Chen, P. M., and Gusta, L. V. 1982. Relationship of deep supercooling and dehydration resistance to freezing injury in dormant stem tissues of 'Strakrimson Delicious' apple and 'Siberian C' peach. J. Am. Soc. Hort. Sci. 107:299-304.

Quamme, H. A., and Stushnoff, C. 1983. Resistance to environmental stress. Pages 242-266 in: Methods in Fruit Breeding, J. N. Moore and J. Janick, eds. Purdue University Press, Lafayette, IN.

Quamme, H. A., Stushnoff, C., and Weiser, C. J. 1972. The relationship of exotherms to cold injury in apple stem tissues. J. Am. Soc. Hort. Sci. 97:608-613.

Quamme, H. A., Weiser, C. J., and Stushnoff, C. 1973. The mechanism of freezing injury in xylem of winter apple twigs. Plant Physiol. 51:273-277.

Rajashekar, C., and Reid, W. 1989. Deep supercooling in stem and bud tissues of pecan. HortScience 24:348-350.

Rajashekar, C. Westwood, M. N., and Burke, M. J. 1982. Deep supercooling and cold hardiness in the genus *Pyrus*. J. Am. Soc. Hort. Sci. 107:968-972.

Rasmussen, D. H., and Mackenzie, A. P. 1972. Effect of solute on ice-solution interfacial free energy: calculation from measured homeogeneous nucleation temperature. Pages 126-145 in: Water Structure at the Water Polymer Interface. H. H. G. Jellnek, ed. Plenum Press, New York.

Schaffer, K., and Wisniewski, M. 1989. Development of the amorphous layer (protective layer) in xylem parenchyma of cv. Golden Delicious apple, cv. Loring peach, and willow. Am. J. Bot. 76:1569-1582.

Siau, J. F. 1984. Transport Processes in Wood. Springer-Verlag, New York.

Stushnoff, C. 1972. Breeding and selection methods for cold hardiness in deciduous fruit crops. HortScience 7:10-13.

Toner, M., Cravalho, E. G., and Kavel, M. 1990. Thermodynamics and kinetics of intracellular ice formation during freezing of biological cells. J. Appl. Phys. 67:1582-1593.

Tumanov, I. I., and Krasavtsev, O. A. 1962. Hardening of northern woody plants by temperatures below zero. Sov. Plant Phsyiol. 9:474-482.

Tumanov, I. I., Krasavtsev, O. A., and Trunova, T. I. 1969. Investigation of the ice formation processes in plants by measuring heat evolution. Sov. Plant Physiol. 17:754-760.

Vian, B. Reis, D. Mosiniak, M., and Roland, J. C. 1986. The glucoroxylans and the helicoidal shift in celluose microfibrils in linden wood: Cytochemistry *in mure* and on isolated molecules. Protoplasma 131:185-189.

Weiser, R. L., Wallner, S. J., and Wadell, J. W. 1990. Cell wall and extensin mRNA changes during cold acclimation of pea seedlings. Plant Physiol. 93:1021-1026.

Wisniewski, M., Arora, R., and Davis, G. 1991c. Role of the protoplast in deep supercooling of xylem tissue. HortScience 26:727.

Wisniewski, M., and Ashworth, E. N. 1985. Changes in the ultrastructure of xylem parenchyma cells of peach (*Prunus persica*) and red oak (*Quercus rubra*) in response to a freezing stress. Am. J. Bot. 72:1364-1376.

Wisniewski, M., and Ashworth, E. N. 1986. Seasonal variation in deep supercooling and dehydrative resistance. HortScience 21:503-505.

Wisniewski, M., and Davis, G. 1989. Evidence for the involvement of a specific cell-wall layer in regulation of deep supercooling of xylem parenchyma. Plant Physiol. 91:151-156.

Wisniewski, M., Davis, G., and Arora, R. 1991b. Effect of macerase, oxalic acid, and EGTA on deep supercooling and pit membrane structure of xylem parenchyma of peach. Plant Physiol. 96:1354-1359.

Wisniewski, M., Davis, G., and Arora, R. 1993. The role of pit membrane structure in deep supercooling of xylem parenchyma. Pages 215-228 in: Advances in Plant Cold Hardiness. P. H. Li and L. Christersson, eds. CRC Press, Boca Raton, FL.

Wisniewski, M., Davis, G., and Schaffer, K. 1991a. Mediation of deep supercooling in *Prunus* and *Cornus* via enzymatic modifications in cell wall structure. Planta 184:254-260.

Wisniewski, M., Lightner, G., Davis, G., and Schiavone, M. 1990. System configuration for microcomputer-controlled, low-temperature, differential thermal analysis. Comp. Elec. Agric. 5:223-230.

Wisniewski, M., Ashworth, E. N., and Schaffer, K. 1987a. The use of lanthanum to characterize cell wall permeability in relation to deep supercooling and extracellular freezing in woody plants. I. Intergeneric comparisons between *Prunus, Cornus,* and *Salix.* Protoplasma 139:105-116.

Wisniewski, M., Ashworth, E. N., and Schaffer, K. 1987b. The use of lanthanum to characterize cell wall permeability in relation to deep supercooling and extracellular freezing in woody plants. II. Intrageneric comparisons between *Betula lenta* and *Betula papyrifera.* Protoplasma 141:160-168.

Zimmerman, M. H. 1983. Xylem Structure and the Ascent of Sap. Springer Verlag, Berlin, Heidelberg, New York.

CHAPTER 10

# Deep Supercooling in Buds of Woody Plants

**H. A. Quamme**

## Introduction

Ice formation at certain preferred sites of freezing within overwintering leaf and flower buds was first observed in woody plants by Weigand (1906). Dorsey (1934) suggested that in dormant peach flower buds, water withdrawal to preferred sites of freezing within the bud scales and bud axis protects the flower by lowering its freezing point. The release of latent heat of fusion associated with the freezing of supercooled water was first detected in cherry flower buds by Tumanov et al. (1969) using heat calorimetry, but the full significance of this finding to flower bud survival was not understood until thermal analysis studies performed on *Rhododendron mollis* and *R. mollis* × *R. roseum* flower buds by Graham (1971). Graham was able to show that water in the flowers of the azalea flower buds supercools and freezes at the temperature that corresponds to the killing point. The maintenance of supercooled water in certain tissues below the freezing point of surrounding tissue has since been termed deep supercooling (Burke et al., 1976). Since Graham's report (1971), deep supercooling has been observed in the flower buds of a number of species, vegetative buds of conifers, and mixed buds of grape (Table 1), and the mechanism of supercooling has been elucidated.

## Expression of Deep Supercooling in Buds

Thermal analysis of dormant buds of most woody plants usually reveals a high-temperature exotherm (HTE) just below the freezing point (–5 to –10°C) which results from extracellular ice formation. The HTE may be followed by one or more low-temperature exotherms (LTEs) that arise from the freezing of supercooled water within the flower(s) or the shoot primordium of the bud. Typical thermal profiles are shown for peach, *R. japonicum*, *Abies firma*, and grape (Fig. 1).

The expression of supercooling in dormant buds cooled at rates that correspond to environmental rates (<5°C/h) varies with species and anatomical structure. The flower buds of forsythia (Ashworth, 1990; Ishikawa and Sakai, 1982) and peach

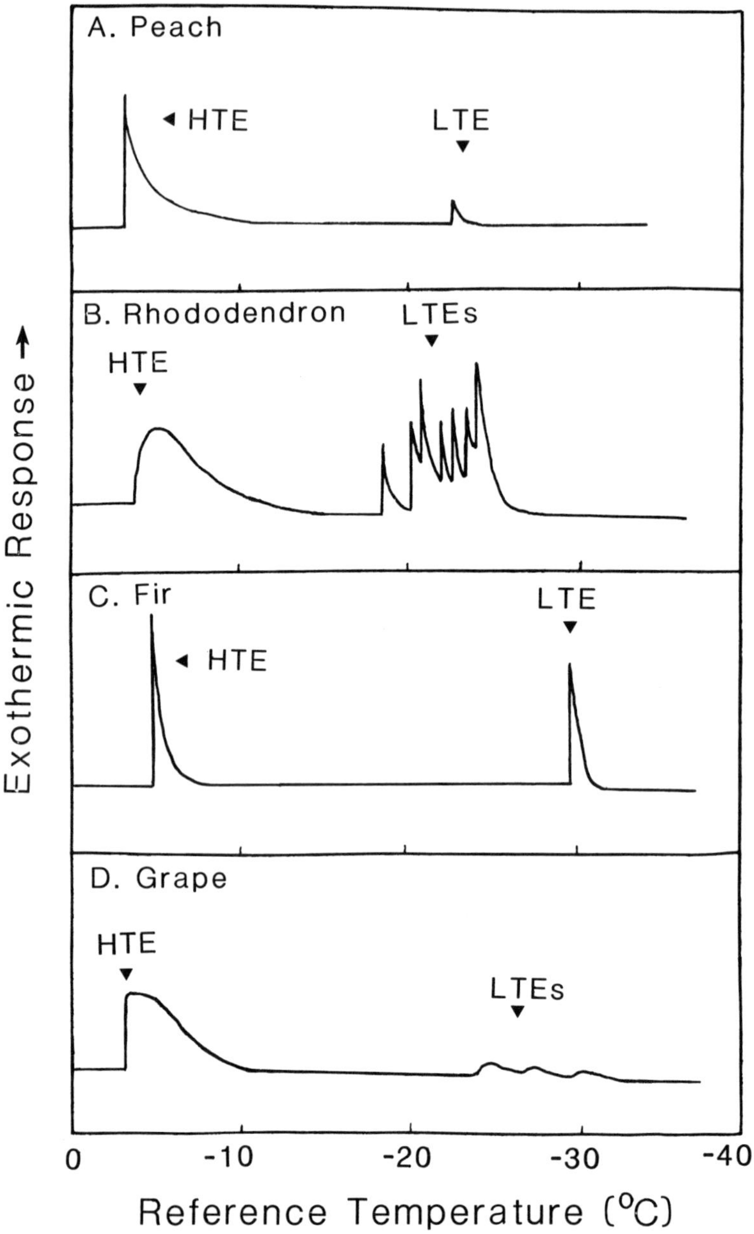


**Figure 1.** A differential thermal analysis profile of **A,** a peach flower bud; **B,** a *Rhododendron japonicum* flower bud (reproduced, by permission, from Ishikawa and Sakai, 1981); **C,** a vegetative bud of *Abies firma* (reproduced, by permission, from Sakai, 1978); and **D,** a mixed bud of grape. HTE = high-temperature exotherm; LTE = low-temperature exotherm.

**Table 1.** Occurrence of deep supercooling in buds of woody plant species

| Genus and species | Common name | Bud type[a] | Range in temperature of LTE (°C) | Reference |
|---|---|---|---|---|
| Species in which deep supercooling has been detected in flower buds | | | | |
| **Angiosperms** | | | | |
| *Andromeda polifia* | Bog rosemary | F | −18 to −22 | Ishikawa and Sakai, 1982 |
| *Arcterica nana* | | F | −18 to −25 | Ishikawa and Sakai, 1982 |
| *Cassiope lycopodioides* | | F | −18 to −25 | Ishikawa and Sakai, 1982 |
| *Cerciciphyllum japonica* | Katsura tree | F | −19 to −24 | Ishikawa and Sakai, 1982 |
| *Chamaedaphne calyculata* | | F | −21 to −26 | Ishikawa and Sakai, 1982 |
| *Cornus florida* | Flowering dogwood | F | −15 to −23 | Sakai, 1979a |
| *Forsythia* spp. | Forsythia | F | −23 | Nus et al., 1981 |
| *koreana* | | F | −12 to −20 | Ishikawa and Sakai, 1982 |
| *suspensa* | | F | −24 to −28 | Ishikawa and Sakai, 1982 |
| *viridissima* | | F | −10 to −25 | Ashworth et al., 1992 |
| *Ledum palustre* | Wild rosemary | F | −22 to −24 | Ishikawa and Sakai, 1982 |
| *Phyllodoce* sp. | | F | −12 to −21 | Ishikawa and Sakai, 1982 |
| *Pieris japonica* | | F | −10 to −23 | Ishikawa and Sakai, 1982 |
| *Prunus* | | | | |
| *americana* | Wild plum | F | <−10 | Burke and Stushnoff, 1978 |
| *armeniaca* | Apricot | F | −19 to −24 | Ashworth et al., 1981; Quamme, 1974 |
| *avium* | Sweet cherry | F | −12 to −32 | Andrews et al., 1983b; Quamme, 1974 |
| *besseyi* | Western sand cherry | F | −12 to −45 | Burke and Stushnoff, 1978; Quamme et al., 1982 |
| *besseyi* × *P. salicina* | Sapalta cherry | F | <−11 | Burke and Stushnoff, 1978 |
| *cerasus* | Sour cherry | F | −20 to −30 | Burke and Stushnoff, 1978 |
| *domestica* | European plum | F | −28 | Quamme, 1974 |
| *japonica* | Dwarf flowering cherry | F | <−11 | Burke and Stushnoff, 1978 |
| *nigra* × *P. tenella* | Muckle plum | F | <−11 | Burke and Stushnoff, 1978 |
| *pennsylvanica* | Pin cherry | F | <−13 | Burke and Stushnoff, 1978 |
| *persica* | Peach | F | −15 to −27 | Proebsting and Sakai, 1979; Quamme, 1974 |

(*continued on next page*)

[a] F = flower bud, V = vegetative bud, and M = mixed bud.

**Table 1.** (*continued*)

| Genus and species | Common name | Bud type[a] | Range in temperature of LTE (°C) | Reference |
|---|---|---|---|---|
| *Prunus* (continued) | | | | |
| *salicina* | Japanese plum | F | −24 | Quamme, 1974 |
| *salicina* × *munsoniana* | Redglow plum | F | <−7 | Burke and Stushnoff, 1978 |
| *salicina* × (*P. salicina* × *P. americana*) | Pipestone plum | F | <−17 | Burke and Stushnoff, 1978 |
| *Ribes* | | | | |
| *nigrum* | Black currant | F | −12 to −35 | Warmund et al., 1991 |
| *sativum* | Red currant | F | −10 to −31 | Warmund et al., 1991 |
| *Rhododendron* | | | | |
| *brachycarpum* | Rhododendron | F | −22 to −28 | Ishikawa and Sakai, 1982 |
| *dauricum* | Rhododendron | F | −27 to −31 | Ishikawa and Sakai, 1982 |
| *dilatatum* | Rhododendron | F | −16 to −25 | Ishikawa and Sakai, 1981 |
| *keiskei* | Rhododendron | F | −11 to −17 | Ishikawa and Sakai, 1981 |
| *japonica* | Rhododendron | F | −14 to −34 | Ishikawa and Sakai, 1982 |
| *obtusum* | Rhododendron | F | −19 to −29 | Ishikawa and Sakai, 1981 |
| *tschonoskii* | Rhododendron | F | −20 to −24 | Ishikawa and Sakai, 1981 |
| × *mollis* | Azalea | F | −16 to −22 | Graham and Mullin, 1976a,b |
| × *mollis* × *R. roseum* | Azalea | F | −17 to −38 | Graham and Mullin, 1976a,b |
| *kosterianum* | Azalea | F | −12 to −27 | George et al., 1974 |
| *Rubus* | | | | |
| *ideas* | Red raspberry | F | −19 to −37 | Warmund and George, 1990 |
| spp. | Black berry | F | −16 to −28 | Warmund et al., 1988 |
| *Tsusiophyllum tanakae* | | F | −20 to −24 | Ishikawa and Sakai, 1982 |
| *Ulmus* | | | | |
| *davidiana* | | F | −25 to −33 | Ishikawa and Sakai, 1982 |
| *pumila* | Manchurian elm | F | −21 to −27 | Ishikawa and Sakai, 1982 |
| *Vaccinium* | | | | |
| *corymbosum* | Blueberry | F | <−23 | Biermann et al., 1979 |
| *smallii* | Blueberry | F | −12 to −20 | Ishikawa and Sakai, 1981 |
| *vitis-idea* | | | −15 to −29 | Ishikawa and Sakai, 1981 |

| | | | | |
|---|---|---|---|---|
| *Vitis* | | | | |
| *riparia* | Grape | F | −27 to −42 | Pierquet et al., 1977 |
| *vinifera* | Grape | M | −22 to −25 | Andrews et al., 1984; Quamme, 1986; Wolf and Pool, 1987 |
| spp. | Grape | M | −22 to −28 | Andrews et al., 1984; Quamme, 1986; Wolf and Pool, 1987 |
| **Gymnosperms** | | | | |
| *Abies* | | | | |
| *balsama* | Fir | V | ≈−35 | Sakai, 1978, 1979 |
| *firma* | Fir | V | −30 | Sakai, 1978, 1979 |
| *homolepis* | Fir | V | −32 | Sakai, 1978, 1979 |
| *sachalinensis* | Fir | V | ≈−30 | Sakai, 1978, 1979 |
| *veitchii* | Fir | V | ≈−30 | Sakai, 1978, 1979 |
| *Larix leptolepis* | Larch | V | −30 | Sakai, 1978, 1979 |
| *Juniperus virginia* | Juniper | F | −40 | George, 1982 |
| Species in which deep supercooling of the buds is absent | | | | |
| **Angiosperms** | | | | |
| *Aeseulus turbenata* | Chestnut | ... | ... | Ishikawa and Sakai, 1982 |
| *Betula platyphylla* | Birch | ... | ... | Ishikawa and Sakai, 1982 |
| *Cornus stolonifera* | Red-osier dogwood | ... | ... | Ishikawa and Sakai, 1982 |
| *Gauteria* | | | | |
| *adinothix* | | ... | ... | Ishikawa and Sakai, 1982 |
| *migueliana* | | ... | ... | Ishikawa and Sakai, 1982 |
| *Malus domestica* | Apple | ... | ... | Quamme, 1976 |
| *Populus* spp. | Popular | ... | ... | Ishikawa and Sakai, 1982 |
| *Prunus* | | | | |
| *maaki* | | ... | ... | Burke and Stushnoff, 1978 |
| *padus* | European bird cherry | ... | ... | Burke and Stushnoff, 1978 |
| *virginiana* | Choke cherry | ... | ... | Burke and Stushnoff, 1978 |
| *Pyrus communis* | Pear | ... | ... | Quamme, 1976 |
| **Gymnosperms** | | | | |
| *Picea* spp. | Spruce | ... | ... | Sakai, 1978 |

(Proebsting and Sakai, 1979; Quamme, 1974) and male cone of juniper (George, 1982) produce a single LTE that corresponds to a single flower. A single LTE is also produced by the vegetative buds of fir that corresponds to freezing of the shoot primordium (Sakai, 1978, 1979b). In larch, a single shoot primordium produces several LTEs that correspond to leaf primordia (Sakai, 1978). Racemose flower buds of blueberry (Biermann et al., 1979), flowering dogwood (Sakai, 1979a), *Prunus* spp. (Burke and Stushnoff, 1978), and *Rhododendron* spp. (George et al., 1974a; Graham and Mullin, 1976a; Kaku et al., 1980; Ishikawa and Sakai, 1981) produce multiple exotherms. The number of LTEs produced in *Rhododendron* spp. corresponds to the number of florets (Graham and Mullin, 1976a; Kaku et al., 1980), but the number of LTEs in sweet cherry (Andrews et al., 1983b), black and red currant (Warmund et al., 1991), black raspberry (Warmund et al., 1988), and red raspberry (Warmund and George, 1990) is less than the number of florets. The exact cause for the discrepancy is unknown, but it may result from an absence of freezing in some of the florets, simultaneous freezing of florets, or an inability to sense LTE of all florets. The inflorescence of *Prunus serotina*, which is racemose, freezes as a unit and produces a single LTE (Kader and Proebsting, 1992). One or two broad LTEs are present on the differential thermal analysis (DTA) profiles of black and red currant flower buds in addition to sharp LTEs. The broad LTEs correspond to the freezing of water in the outer nonliving region of the cane periderm tissue attached to the flower bud, whereas the sharp LTEs correspond to the freezing of the flowers (Warmund et al., 1991).

Grape has a complex bud containing primary, secondary, and tertiary buds, each of which may contain an apical meristem and flowers. Up to three LTEs, thought to correspond to the freezing of each bud type, are detected below the HTE (Andrews et al., 1984; Quamme, 1986; Pierquet and Stushnoff, 1980; Wolf and Pool, 1987). The LTEs of each bud type occur at a progressively lower temperature and are progressively smaller in size. The size of the LTEs is probably related to bud size.

The LTE of many buds including those of many conifers (Sakai, 1978, 1979b), *Prunus* spp. (Ashworth, 1982; Burke and Stushnoff, 1978; Kader and Proebsting, 1992; Quamme, 1983; Quamme et al., 1982; Proebsting and Sakai, 1979), and *Rhododendron* spp. (Kaku et al., 1980), occurs at a warmer temperature with increasing cooling rate. The response is unlike that of the xylem of many of these same plant species, in which LTEs usually vary little with cooling rate (Burke and Stushnoff, 1978). The exception is grape, in which the LTEs of buds remain at constant temperature irrespective of cooling rate (Quamme, 1986).

Deep supercooling appears to be a true avoidance mechanism for low-temperature survival for flower buds of certain genera. Flowers of both peach (Quamme, 1983) and *Rhododendron* spp. (Graham and Mullin, 1976b; Kaku et al., 1981) remain supercooled throughout the winter. Excised flowers of *R. japonicum* (Ishikawa and Sakai, 1981) and peach (Quamme, *unpublished*) freeze and are killed at temperatures well above the LTE temperature of the intact flower bud when inoculated with ice. Thus, the floral tissue of these two genera is inherently nonhardy and survives only by supercooling.

Avoidance of freezing injury by supercooling appears to be transitory in some species and may be important to survival in these species only under certain environmental conditions. The LTEs of blueberry (Biermann et al., 1979), *Prunus besseyi*, and *Prunus pennsylvanica* (Burke and Stushnoff, 1978) buds shift rapidly to low temperature and eventually disappear when exposed to freezing temperatures above the LTE. In midwinter, LTEs are often absent in the buds of these plants. The flowers of these species survive the disappearance of LTEs.

## Freezing Pattern in Dormant Buds

Although few attempts have been made to observe intracellular ice in buds that exhibit deep supercooling, the supposition is that the freezing is intracellular because of the sudden occurrence of the LTE and the invariably lethal effects on the tissue. Ice crystals have been observed with a light microscope within the tissue subtending the flower in frozen *R. japonicum* (Ishikawa and Sakai, 1981), peach (Dorsey, 1934; Quamme, 1974), and plum (Dorsey and Strausbaugh, 1923), but not in the flower itself. At subzero temperatures, the shoot primordium of fir (Sakai, 1979b) and flower buds of peach (Quamme, 1978) appear greenish and pliable but suddenly turn white and stiff when nucleated with ice crystals. The presence of intercellular ice in peach flowers has been verified by rapidly cooling frozen flowers in supercooled liquid nitrogen, fracturing the flower, and examining the surfaces of the fractures in a cryostage with a scanning electron microscope (Quamme et al., *unpublished*).

Extracellular ice is absent in dormant peach flowers but is present in the peduncle and in the base of the peach flower during the late stages of deacclimation (Ashworth et al., 1989). It is also present in the peduncle and base of dormant forsythia flowers before deacclimation (Ashworth et al., 1992). The presence of intracellular ice was not determined in either deacclimated peach flowers or dormant forsythia, but the upper regions of these flowers are believed to deep supercool because LTEs are present.

The tissue subtending the supercooled flower in the flower bud, including the scales and flower bud axis, appears to freeze extracellularly. Ice accumulates in preferred sites at the base of the scales and in the flower bud axis just below the flower in peach (Dorsey, 1934; Quamme, 1978) and plum (Dorsey and Strausbaugh, 1923) and in the scales of *R. japonicum* (Ishikawa and Sakai, 1981). In vegetative buds of fir, ice accumulates in a cavity beneath the shoot primordium and in the scales (Sakai, 1979b). Ice crystals that form by extracellular freezing disrupt and produce voids in the tissue (Ashworth et al., 1989, 1992; Ishikawa and Sakai, 1981). Although cell damage occurs near the voids, bud survival is not reduced.

The initiation of freezing in woody plant buds appears to result from the propagation of ice from the shoot. Freezing within several species spreads throughout the tree from a few intrinsic nucleation sites (Anderson and Smith, 1989; Ashworth and Davis, 1984; Ashworth et al., 1985; Andrews et al., 1983b). In peach, the excised flowers and bud scales do not differ appreciably in their nucleation temperatures. Nucleation of the flower bud occurs at warmer temperatures as the amount of tissue attached to the bud increases, which implies that nucleation occurs outside the bud (Quamme and Gusta, 1987).

## Mechanism of Deep Supercooling in Dormant Buds

Melting points of water have not been determined in the buds of many plants with LTEs. However, it is presumed that the LTE is not a eutectic point, because the LTE temperature is usually much lower than the freezing point depression, which seldom exceeds –4°C in plants (Levitt, 1972). The melting point of the water associated with freezing of *R. kosterianum* (George et al., 1974a) and peach flowers (Quamme, 1978) is approximately –2°C, whereas the freezing point of the flowers is below –18°C. Nuclear magnetic resonance measurements made on *R. kosterianum* (George et al., 1974a) and peach (Rajashekar, 1989) confirm that injury to the flower follows the sudden freezing of a fraction of supercooled water. Ice formation in peach flowers (Quamme, 1978) and fir shoot primordia (Sakai, 1979b), probed with an ice crystal under a microscope in a cold chamber, appeared to spread from the point of inoculation. The presence of the inoculating ice crystal allows a lowered activation energy for the subsequent rapid ice crystal growth.

Although supercooling of buds has been studied in a number of species, it is not known how the supercooled state of the tissue is maintained or how freezing of supercooled tissue is initiated. One of the requirements for supercooling of tissue is a low nucleation temperature of the cellular water solution. It is known that in the absence of nucleators, pure water supercools to temperatures as low as –38°C (homogeneous nucleation point) (Fletcher, 1970). The presence of solutes further lowers the homogeneous nucleation point by a value that is proportional to the freezing point depression (Rasmussen and MacKenzie, 1972). Yeast cells in oil suspensions supercool to near the homogeneous nucleation point (Rasmussen et al., 1975). Thus, in yeast cells, nucleators appear to be either absent or to have low activity. In plant cells that supercool, nucleators may also be either absent or have low activity.

A second requirement for deep supercooling is the prevention of ice propagation into the supercooled flowers or apical meristem from the bud axis. Barriers to ice propagation may exist that involve the pore structure of the tissue. Melting point and vapor pressure are decreased in structures with fine pores by the effects of increased surface tension. Mazur (1965) and Homshaw (1980) developed equations that predict the decreases in the melting point of water with pore size. These equations show that melting point decreases from 2°C in pores of 20 nm diameter to below –40°C in pores of 1 nm diameter. Using glass particles with a range in internal pore size, Ashworth and Abeles (1984) demonstrated that melting point depression fits the equations of Mazur (1965) and Homshaw (1980). They also demonstrated that the presence of solutes further suppresses the melting point of water in pores (Ashworth and Abeles, 1984). In tissue that lacks extracellular spaces, ice propagation would be restricted and supercooling facilitated by the presence of water in cell wall microcapillaries. The presence of solutes in water of the cell wall microcapillaries would further restrict ice propagation.

The freezing behavior of water in small pores may explain the restricted ice propagation and its accumulation at certain preferred sites in buds. The flowers of both peach (Ashworth, 1982; Quamme et al., *unpublished*) and *R. japonicum* (Ishikawa and Sakai, 1981) and shoot primordia of fir (Sakai, 1979b) do not contain extracellular spaces that are present in the bud axis. Restricted pore size is probably an important factor preventing ice spread into the flower. The exclusion of azosulfamide (Ashworth, 1982) and acid fuchsin dye (Quamme, 1978) from the flowers in peach buds is an indication of the small pore size within the flower.

In winter buds of fir and larch, a dense zone of collenchyma cells is present at the base of the shoot primordium. This anatomical structure prevents ice propagation into the shoot primordium from the subtending tissue. Removal of the basal tissue from excised shoot primordium of fir raises the LTE from –25 to –15°C

(Sakai, 1978). Dormant peach flowers also have a basal zone of cells that lacks intercellular spaces (Quamme et al., *unpublished*). The flower can be induced to freeze by ice inoculation above the basal zone but not below it. Similar anatomical barriers that restrict the spread of ice are also present in wheat flowers (Marcellos and Single, 1976) and stems (Olien, 1967).

Loss of supercooling in peach during deacclimation is associated with the development of functional xylem (Ashworth, 1984). Vascular traces are present in the base of the forsythia flower, in which extracellular ice forms, but not in the petals, filaments, and pistil that supercool (Ashworth et al., 1992). The xylem of *Prunus virginiana* and *P. padus* flowers, which does not supercool, is more differentiated than that of *P. serotina*, which does supercool (Kader and Proebsting, 1992). Therefore, fully differentiated xylem may act as a conduit for ice propagation into the flower.

The possibility that antinucleating substances may restrict ice propagation in supercooled flowers was first proposed by George et al. (1974a) but to date has not been confirmed. Antifreeze proteins similar to those that exist in winter cereals (Griffith et al., 1992) may be present in flowers of woody plants, but this has yet to be determined.

Pore structure of the tissue also affects the extracellular freezing pattern. Ice first forms in the larger pores of structures having irregular porosity. Water moves from the pores to freeze at the interface of the ice crystal because of the difference in water potential. The result is continued growth of the ice crystal where it first formed. The ice crystal grows as long as the structural material expands to accommodate the volume change, until the structural forces balance the water potential, or until the structure breaks (Everett, 1961; Jackson and Chalmers, 1958).

Migration of water to preferred sites of freezing appears to be essential to the supercooling of peach flower buds. Disruption of the flower bud axis by cutting the flower bud just below the flower prevents supercooling (Ashworth, 1982; Quamme, 1978). Scale removal studies show that the first two proximal scales subtending the peach flower are necessary for supercooling to occur below −10°C (Quamme, *unpublished*). Rapid cooling (>15°C/h) of peach flower buds, which disrupts water migration to preferred sites, eliminates supercooling (Ashworth, 1982; Quamme, 1983).

Ice propagation through the tissue can be prevented by discontinuities in water flow towards growing ice crystals. Using a model based on the initial solute concentration of the flower parts and heat and water flow during freezing, Cary (1985) demonstrated the possibility of a discontinuity in the liquid phase that allows the ovaries of fully developed peach and plum flowers to supercool. Ice formation is restricted to the stem and receptacle because a dry layer forms in the receptacle during the initial stages of cooling. If a discontinuity in the liquid phase occurs, water can only migrate from the supercooled tissue to the surface of adjacent ice crystals through the vapor phase. Such a discontinuity may also occur in dormant winter buds. The water and osmotic potential of the peach flower are higher than that of the subtending tissue, which may promote development of a water discontinuity (Quamme and Gusta, 1987).

A possibility exists that a discontinuity in the liquid phase can occur at the plasma membrane. A lethal freeze (Ashworth, 1982; Quamme, 1978) and heat treatment (Ashworth, 1982), both of which disrupt the plasma membrane of the flower, raise the LTE temperature but do not eliminate the LTE. Nucleation of the peach flower can be induced at temperatures below −10°C by probing the flower through the scales with ice (Quamme, 1978). Presumably, once ice propagation is initiated in some flower cells, it can spread easily from cell to cell through the tissue. In deacclimating peach (Ashworth et al., 1992) and dormant forsythia flowers (Ashworth et al., 1989), the discontinuity appears to occur among different parts of the flower.

Although a discontinuity in the liquid phase at the base of the flower or the shoot primordium prevents the spread of ice, it does not prevent water migration. Gravimetric determinations of the water content of flowers of dogwood (Sakai, 1979a), peach (Quamme, 1983), *Rhododendron* spp. (Graham and Mullin, 1976a; Kaku et al., 1981) and shoot primordia of fir (Sakai, 1979b) confirm that water migrates from the flower or bud apical meristem during freezing. In buds of blueberry (Biermann et al., 1979), juniper (George, 1982), and several *Prunus* spp. (Burke and Stushnoff, 1978; Quamme et al., 1982), the size of the LTE, which is an indication of the amount of supercooled water, decreases with a slow cooling rate (>5°C/h) or increased length of storage time at subzero temperatures. Water equilibration between different parts of the peach flower bud, however, is not uniform (Quamme, 1983). The flower loses water more slowly than the bud axis at the flower base. The water content of the flower bud axis near the flower base drops quickly after freezing, whereas that of the flower drops slowly after prolonged storage at subzero temperatures.

If the water content of the flower decreases, then the LTE occurs at a colder temperature. This has been demonstrated in the following species: blueberry (Biermann et al., 1979), sweet cherry (Andrews and Proebsting, 1987), flowering dogwood (Ishikawa and Sakai, 1985), peach (Quamme, 1983), and *Rhododendron* spp. (Graham and Mullin, 1976b; Kaku et al., 1981; George et al., 1974a). The relationship between water content of the excised flower and LTE is linear in peach (Quamme, 1983), *R. mollis,* and *R. mollis* × *R. roseum* (Graham and Mullin, 1976b). It is not known exactly how dehydration decreases the temperature of the LTE, but Ishikawa and Sakai (1981) observed that after drying, the freezing point of sap expressed from flowers of *R. japonicum* decreased. Presumably, cell solutes were concentrated by drying, and this decreased the nucleation temperature of water within the pore structure.

Sakai (1979a) termed the segregation of ice within plant tissues "extraorgan" freezing. As a consequence of extraorgan freezing, water is withdrawn from one tissue to freeze in another where it is less injurious. Extraorgan freezing also occurs in seeds of *Celastrus arbriculatus* and lettuce (Ishikawa and Sakai, 1982).

The structure of buds appears to be adapted to accommodate ice at preferred sites. Following freezing and thawing, voids are produced in the tissue of dormant flower buds of peach (Ashworth et al., 1989) and forsythia (Ashworth et al., 1992). It is unknown if there is a predisposition for void formation at preferred sites of freezing, but cell walls may be more susceptible to separation and deformation at these sites than in surrounding tissue. In peach, the cells at preferred sites are relatively large and have thinner walls of freezing than the cells in surrounding tissue (Quamme, *unpublished*).

Although ice appears to propagate suddenly from a single point in bud tissue, it is unknown whether ice propagates from cell to cell along the cell wall or through the cytoplasm. Possibly, the plasmodesmata could allow ice propagation from cell to cell. In xylem ray parenchyma, single cells or small groups of cells freeze over a broad temperature range (Hong and Sucoff, 1982). The pit structure of xylem ray parenchyma appears to be an important constraint to water permeability and ice propagation. A "protective layer" with low porosity underlies the pit membrane in the xylem of species that deep supercool, and this may prevent cell-to-cell ice propagation (Wisniewski and Ashworth, 1986; Wisniewski et al., 1987). The sudden spread of ice that occurs in buds that deep supercool suggests that no such barrier to ice propagation is present among cells. The lack of a protective layer in the pit structure of buds would also explain their more rapid water loss and decline in the LTE temperature during cooling compared to the xylem.

Another difference in deep supercooling between buds and the xylem ray parenchyma is that the rays are interspersed among lignified elements such as fibers and vessels. Rigidity of the tissue may restrict a decrease in cell volume, which may explain the slow equilibration between supercooled water and ice within the xylem (George and Burke, 1977; Quamme et al., 1973). The supercooled water within buds of many species appears to equilibrate rapidly with ice in adjacent tissue, whereas that of the xylem does not (Biermann et al., 1979; Burke and Stushnoff, 1978). The deep supercooled cells of buds may have cell walls that are more elastic than those of xylem ray parenchyma cells.

The possibility also exists that flowers of some species, such as peach and *R. japonicum*, which are subject to ice nucleation from the surface, may be inoculated by freezing rain or dew or by hoarfrost. Ice is probably excluded from the surface of the flowers of these species by the bud scales. Ice forms within the scales surrounding the flower, but not at the surface of the bud scale. Thus, ice within the flower bud, which is capable of nucleating the flower, does not appear to be in contact with the flower surface (Monet and Bastard, 1980; Quamme, 1978; Ishikawa and Sakai, 1981).

## Acclimation Cycle of Buds That Deep Supercool

The seasonal changes in supercooling of buds have been extensively studied in peach (Quamme, 1974 and 1983), *Rhododendron* spp. (Graham and Mullin, 1976b; Kaku et al., 1981), and sweet cherry (Andrews and Proebsting, 1987). In these species, LTE temperatures generally decrease with the onset of winter and increase with bud development in the spring. Depending on stage of dormancy and air temperature, fluctuation in LTE temperature occurs throughout the winter until bud development.

Peach and cherry flower buds attain a minimum hardiness level in early autumn and remain at this level throughout the rest period if cold weather does not occur. Upon exposure to prolonged periods of subzero temperatures, the flower buds increase in hardiness. After the completion of rest, which coincides with microspore meiosis, there is a progressive loss in hardiness above the minimum level, and hardiness levels fluctuate more widely with air temperature (Proebsting, 1970). LTEs are present in flower buds of these species at all stages of dormancy until bud swell (Andrew and Proebsting, 1987; Quamme, 1983), but the nature of the changes in bud development that affect the expression of supercooling is unclear.

At bud swell, hardiness decreases rapidly, and LTEs occur at warmer temperatures and eventually disappear. Although the LTEs disappear, swollen flower buds can still tolerate some freezing. The disappearance of the LTEs coincides with vascular differentiation in the flower (Ashworth, 1984). After the petal tips emerge through the calyx, tolerance is dependent on the HTE temperature. The flowers tolerate almost no freezing after petal emergence (Andrews et al., 1982; Kaku et al., 1981).

Changes in LTE temperature of the flower buds of peach (Quamme, 1983), sweet cherry (Andrews et al., 1983b; Andrews and Proebsting, 1987), and several *Rhododendron* species (Graham and Mullin, 1976b; Kaku et al., 1981) during the acclimation cycle are significantly correlated with changes in minimum air temperature and water content of the flower bud. In peach, the water content of the whole flower bud and LTE temperature are dependent upon the air temperature (Quamme, 1983). Peach and cherry flower buds exhibit greater hardiness if the air temperature is below –2°C (approximately, the melting point) than if the air temperature is above this temperature. The exception is that buds do not lose hardiness at air temperatures above –2°C when they are at the minimum hardiness level during rest (Proebsting and Mills, 1972). Exposure to controlled temperatures confirms that the LTE temperature of sweet cherry flower decreases when the air temperature is below the melting point of the subtending tissue and increases when it is above this point (Andrews and Proebsting, 1987). In controlled environment chambers, the drop in the LTE is associated with a change in internal water balance, but in the orchard external water loss also occurs (Quamme, 1983). Johnston (1925), who first observed the inverse relationship between water content and flower bud hardiness in peach, determined that in the orchard, flower buds lose water when frozen and regain water from other parts of the tree during thawing. Ishikawa and Sakai (1981) also demonstrated that dehydration by wind enhances the supercooling ability of *R. japonicum* flower buds. Changes in flower bud hardiness, therefore, depend on changes in water distribution within the flower bud that affect supercooling. The flower bud can undergo changes in water distribution as a result of internal water migration during freezing or thawing, or as a result of dehydration of the bud and uptake of water from the other parts of the tree.

Stone fruit orchards are often protected from spring frost injury by wind machines and/or orchard heaters. Protection of flower buds below –20°C in midwinter is possible but requires accurate determination of flower bud hardiness. Unfortunately, the length of time required to measure flower bud hardiness makes protection programs impractical. To overcome this problem, Andrews et al. (1987) developed a model that predicts the hourly median LTE of cherry based on changes in air temperature and chill unit accumulation, starting with an initial LTE measurement. Their model, using initial and updated LTEs, successfully predicted daily changes in hardiness of sweet cherry flower buds during two winter periods. The predicted values are accurate enough to permit the practice of winter freeze protection of cherry orchards.

## Relationship of Deep Supercooling in Buds to Plant Distribution

The natural distribution of many woody plants is limited by the deep supercooling capacity of the xylem ray parenchyma (George et al., 1974b). The production area of many cultivated fruit crops is also limited by the supercooling capacity of the xylem (Quamme, 1976; Quamme et al., 1982). In these species, the temperature of the LTE closely coincides with the average annual minimum temperature at the most northern limit of natural distribution or cultivation. Species that are found in colder regions than those where deep supercooling species grow do not deep supercool and are capable of hardening to lower temperatures, e.g., red osier dogwood, poplar, and willow (George et al., 1974b).

The relationship of deep supercooling capacity of buds to species distribution is less clear. *Prunus maaki, P. padus, P. pennsylvanica,* and *P. virginiana*, which do not supercool, survive further north than *Prunus* species that do supercool (Burke and Stushnoff, 1978). Among species with buds that deep supercool, the northern limit of distribution of riparian grape coincides closely with the average annual winter minimum temperatures and the LTE temperatures of both buds and xylem (Pierquet et al., 1977). On the other hand, the LTE temperatures (–22 to –26°C) of primary buds of cultivated grape (Andrews et al., 1984; Quamme, 1986; Wolf and Pool, 1987) coincide approximately with winter average annual minimum temperatures found at the northern limit of cultivation (–21 to –24°C). This also appears to be true of sweet cherry, peach, and apricot, in which crop production is dependent on flower bud survival (Quamme et al., 1982). The buds of fir and larch deep supercool. These two species are neither as hardy nor as well distributed into cold climates as other conifer species that do not deep supercool (Sakai, 1978). Other species that deep supercool either do not have LTEs in buds, e.g., apple and pear (Quamme, 1976), or the LTEs of the bud occur at a lower temperature than that of the xylem LTE (Burke and Stushnoff, 1978).

It is unclear what adaptive advantage, if any, deep supercooling has for the plant. Not all species exhibit deep supercooling, and compared with species that do not supercool, the level of protection from freezing is limited. Perhaps supercooling confers some advantage to the plant other than winter survival. In the flowers of some species, e.g., apricot, peach, and sweet cherry, flower bud development occurs early in the spring. Speculation is that bud survival by deep supercooling may not require changes in cellular structure necessary to withstand the dehydration of extracellular freezing. This may allow cell division required for flower development to take place in late winter or early spring. In turn, early flowering may be required for early fruit ripening and/or large fruit size. Other *Prunus* spp. that do not deep supercool or have temporary deep supercooling produce fruit neither as large nor as early in the season as apricot, cherry, and peach. One can also speculate that selection pressure for these characteristics, and thus selection for high supercooling temperatures in these crops, is in part due to domestication and breeding. The improvement of large, early fruited cultivars of *Prunus* spp. by breeding may have been at the expense of cold hardiness (Quamme, 1991).

## Measurement of Bud Hardiness with Thermal Analysis

Thermal analysis is an ideal method of determining the hardiness levels of species that supercool, because the buds are killed at the moment of ice formation. Thermal analysis measurements are more easily automated and require little time and labor compared with other viability tests such as browning, regrowth, and conductivity. Sample size is small, because a definite killing temperature can be determined by the LTE, and it is not necessary to freeze buds over a range of test temperatures to obtain the killing temperature.

The LTE of buds can be detected directly with either thermocouples or thermistors (Quamme, 1974; Graham and Mullin, 1976a), or by DTA using single sensors (George et al., 1974a), or with a thermopile (Andrews et al., 1983a). Scanning differential calorimeters can also be used for measuring the LTE but are not practical for routine hardiness determinations.

The use of a thermocouple or thermistor is the simplest method to measure the temperature of the LTE of buds. Ice or silicone grease is used to ensure good thermal contact between the sensor and the bud (Quamme, 1991). The LTE can readily be detected on the time-temperature profile if the LTE is a large sharp, discrete event, but is difficult to detect if the LTE is small or spread over a wide temperature range. Nus et al. (1981) and George (1982) describe methods for amplifying the LTE in buds using DTA with thermistors to improve the sensitivity of detection.

Routine determination of bud hardiness often requires a large number of measurements of LTE. The number of LTE measurements may be increased by placing more than one bud on a single sensor (Proebsting and Sakai, 1979) or by connecting thermocouples in series to monitor more than one bud per channel (Quamme et al., 1975). Proebsting and Sakai (1979) found that the yield of LTEs in peach buds decreases as the number of buds placed on the thermocouple increases, and at most, 25 peach flower buds could be placed on a single channel at a 1°C/h cooling rate without appreciably affecting the yield of LTEs.

Thermopiles can be used to monitor several buds at a time and have the advantage that the signal is amplified. A thermoelectric module (Melcor Materials Electronic Products Corp., Trenton, N.J.), which was designed as a cooling plate, has been adapted to measure several buds simultaneously (Andrews et al., 1983a). A thermal conducting paste (Omegatherm Omega Engineering, Inc., Stamford, Conn.) or silicone grease improves the thermal contact between the buds and the plate. More LTEs are detected from a given number of peach flower buds with a thermoelectric module than with a thermocouple.

In measuring the hardiness level of buds with thermal analysis, the point of bud excision and cooling rate are important. In both peach (Proebsting and Sakai, 1979; Quamme, 1978) and grape (Quamme, 1986; Wolf and Pool, 1987), if the bud is cut too close to the shoot, the LTE occurs at a higher temperature than if it is not. Rapid cooling (>15°C/h) also raises the supercooling point of peach flower buds (Ashworth, 1982; Proebsting and Sakai, 1979; Quamme, 1983).

Close agreement has been found between the LTE measurements and bud injury as measured by browning evaluation. The distribution of LTEs of peach flower buds is very close to the standard skewed mortality curve for flower buds evaluated by the tissue browning method (Proebsting and Sakai, 1979). The LTE temperature values agree with temperatures at which 50% of the buds are killed for several peach (Quamme et al., 1975) and apricot cultivars (Ashworth et al., 1981). The mean LTE temperature of grape buds falls within 1°C of the temperature that kills 50% of the buds (Andrews et al., 1984; Quamme, 1986).

## Summary

The avoidance of freezing in buds by deep supercooling represents a unique survival adaptation in many woody plant species. In buds that deep supercool, ice is segregated in preferred sites of freezing. The mechanism of this segregation is not fully understood but appears to involve tissue structure, cell wall porosity, and differences in water potential. Seasonal changes in bud hardiness are related to changes in the supercooling point that can be predicted from changes in temperature exposure. The supercooling capacity of buds limits the northern distribution of several woody plant species. The adaptive value of the deep supercooling mechanism in buds is unclear, but measurement of the supercooling point by thermal analysis can be used to determine bud hardiness.

## Literature Cited

Anderson, J. A., and Smith, M. W. 1989. Ice propagation in peach shoots and flowers. HortScience 24:480-482.

Andrews, P. K., and Proebsting, E. L. 1987. Effects of temperature on deep-supercooling characteristics of dormant and deacclimating sweet cherry flower buds. J. Am. Soc. Hort. Sci. 112:334-340.

Andrews, P. K., Proebsting, E. L., and Gross, D. 1982. Supercooling young developing fruit and floral buds in deciduous orchards. HortScience 17:67-68.

Andrews, P. K., Proebsting, E. L., and Campbell, G. S. 1983a. An exotherm sensor for measuring the cold hardiness of deep supercooled flower buds by differential thermal analysis. HortScience 18:77-78.

Andrews, P. K., Proebsting, E. L., and Gross, D. C. 1983b. Differential thermal analysis and freezing injury of deacclimating peach and sweet cherry reproductive organs. J. Am. Soc. Hort. Sci. 108:755-759.

Andrews, P. K., Proebsting, E. L., Jr., and Sweet Lee, C. 1987. A conceptual model of the changes in deep supercooling of dormant sweet cherry flower buds. J. Am. Soc. Hort. Sci. 112:320-324.

Andrews, P. K., Sandridge, C. R., III, and Toyama, T. K. 1984. Deep supercooling of dormant deacclimating *Vitis* buds. J. Am. Enol. Vitic. 35:175-177.

Ashworth, E. N. 1982. Properties of peach flower buds which facilitate supercooling. Plant Physiol. 70:1475-1479.

Ashworth, E. N. 1984. Xylem development in *Prunus* flower buds and relationships to deep supercooling. Plant Physiol. 74:862-865.

Ashworth, E. N. 1990. The formation and distribution of ice within forsythia flower buds. Plant Physiol. 92:718-725.

Ashworth, E. N., and Abeles, F. B. 1984. Freezing behaviour of water in small pores and the possible role in the freezing of plant tissues. Plant Physiol. 76:201-204.

Ashworth, E. N., and Davis, G. A. 1984. Ice nucleation within peach trees. J. Am. Soc. Hort. Sci. 109:198-201.

Ashworth, E. N., Anderson, J. A., and Davis, G. A. 1985. Properties of ice nuclei associated with peach trees. J. Am. Soc. Hort. Sci. 110:287-291.

Ashworth, E. N., Davis, G. A., and Wisniewski, M. E. 1989. The formation and distribution of ice within dormant and deacclimated peach flower buds. Plant Cell Environ. 12:521-528.

Ashworth, E. N., Lightener, G. W., and Rowse, D. J. 1981. Evaluation of apricot flower bud hardiness using a computer-assisted method of thermal analysis. HortScience 16:754-756.

Ashworth, E. N., Willard, T. S., and Mallone, S. R. 1992. The relationship between vascular differentiation and distribution of ice within forsythia flower buds. Plant Cell Environ.. 15:607-612.

Biermann, J., Stushnoff, C., and Burke, M. J. 1979. Differential thermal analysis and freezing injury in cold hardy blueberry flower buds. J. Am. Soc. Hort. Sci. 104:444-449.

Burke, M. J., and Stushnoff, C. 1978. Frost hardiness: A discussion of possible molecular causes of injury with particular reference to deep supercooling of water. Pages 197-225 in: Stress Physiology in Crop Plants. H. Mussell and R. Staples, eds. Wiley & Sons, New York.

Burke, M. J., Gusta, L. V., Quamme, H. A., Weiser, C. J., and Li, P. H. 1976. Freezing and freezing injury in plants. Annu. Rev. Plant Physiol. 27:507-528.

Cary, J. W. 1985. Freeze survival in peach and prune flowers. Plant Sci. Lett. 37:265-271.

Dorsey, M. J. 1934. Ice formation in the fruit bud of peach. Proc. Am. Soc. Hort. Sci. 31:22-27.

Dorsey, M. J., and Strausbaugh, P. D. 1923. Winter injury to plum during dormancy. Bot. Gaz. 76:113-142.

Everett, D. H. 1961. The thermodynamics of frost damage to porous solids. Trans. Faraday Soc. 57:1541-1551.

Fletcher, N. H. 1970. The Chemical Physics of Ice. Cambridge University Press, Cambridge.

George, M. F. 1982. Freezing avoidance by supercooling of tissue water in vegetative and reproductive structures of *Juniperus virginiana*. Pages 367-377 in: Plant Cold Hardiness and Freezing Stress. Vol. 2. P. H. Li and A. Sakai, eds. Academic Press, New York.

George, M. F., and Burke, M. J. 1977. Cold hardiness and deep supercooling in xylem of shagbark hickory. Plant Physiol. 59:319-325.

George, M. F., Burke, M. J., and Weiser, C. J. 1974a. Supercooling in overwintering azalea flower buds. Plant Physiol. 54:29-35.

George, M. F., Burke, M. J., Pellett, H. M., and Johnson, A. G. 1974b. Low temperature exotherms and woody plant distribution. HortScience 9:519-522.

Graham, P. R. 1971. Cold injury and its determination in selected *Rhododendron* species. Master's thesis. University of Minnesota, St. Paul.

Graham, P. R., and Mullin, R. 1976a. The determination of lethal freezing temperatures in buds and stems of deciduous azalea by a freezing curve method. J. Am. Soc. Hort. Sci. 101:3-7.

Graham, P. R., and Mullin, R. 1976b. A study of flower bud hardiness in azalea. J. Am. Soc. Hort. Sci. 101:7-10.

Griffith, M., Ala, P., Yang, D. S. C., Hou, W., and Moffatt, B. A. 1992. Antifreeze protein produced endogenously in winter rye leaves. Plant Physiol. 100:593-596.

Homshaw, L. G. 1980. Freezing and melting temperature hysteresis of water in porous materials: Application to study of pore form. J. Soil Sci. 31:399-414.

Hong, S., and Sucoff, E. 1982. Rapid increase in deep supercooling of xylem parenchyma. Plant Physiol. 69:697-700.

Ishikawa, M., and Sakai, A. 1981. Freezing avoidance mechanisms by supercooling in some *Rhododendron* flower buds with reference to water relations. Plant Cell Physiol. 22:953-967.

Ishikawa, M., and Sakai, A. 1982. Characteristics of freezing avoidance in comparison with freezing tolerance: A demonstration of extraorgan freezing. Pages 325-340 in: Plant Cold Hardiness and Freezing Stress: Mechanisms and Crop Implications. Vol. 1. P. H. Li and A. Sakai, eds. Academic Press, New York.

Ishikawa, M., and Sakai, A. 1985. Extraorgan freezing in wintering flower buds of *Cornus officinalis* Sieb. at Zucc. Plant Cell Environ. 8:333-338.

Jackson, K. A., and Chalmers, B. 1958. Freezing of liquids in porous media with special reference to frost heave in soils. J. Appl. Phys. 29:1178-1181.

Johnston, E. S. 1925. Moisture relations of peach buds during the winter and spring. Md. Agric. Exp. Stn. Bull. 255:59-86.

Kader, S. A., and Proebsting, E. L. 1992. Freezing behavior of *Prunus*, subgenus *Padus*, flower buds. J. Am. Soc. Hort. Sci. 117:955-960.

Kaku, S., Tawaya, M., and Kunishige, M. 1980. Supercooling ability of *Rhododendron* flower buds in relation to cooling rate and cold hardiness. Plant Cell Physiol. 21:1205-1216.

Kaku, S., Tawaya, M., and Jeon, F. B. 1981. Supercooling ability, water content, and hardiness of *Rhododendron* flower buds during cold acclimation and deacclimation. Plant Cell Physiol. 22:1561-1569.

Levitt, J. 1972. The hardiness of plants. Academic Press, New York.

Marcellos, H., and Single, W. V. 1976. Ice nucleation on wheat. Agric. Meterorol. 16:125-129.

Mazur, P. 1965. The role of cell membranes in freezing of single cells. Ann. New York Acad. Sci. 125:658-676.

Monet, R., and Bastard, Y. 1980. Etude du mécanisme du gel sur des bourgeons floraux et fleurs du Pêcher, par analyse thermique. C. R. Acad. Sci. Paris, Ser. D 291:113-116.

Nus, N. L., Weigle, J. L., and Schoradle, J. J. 1981. Superimposed amplified exotherm differential analysis system. HortScience 16:753-754.

Olien, C. R. 1967. Freezing stresses and survival. Annu. Rev. Plant Physiol. 18:387-408.

Pierquet, P., and Stushnoff, C. 1980. Relationships of low temperature exotherm to cold injury in *Vitis riparia* Michx. Am. J. Enol. Vitic. 31:1-6.

Pierquet, P., Stushnoff, C., and Burke, M. J. 1977. Low temperature exotherms in stem and bud tissues of *Vitis riparia* Michx. J. Am. Soc. Hort. Sci. 97:608-613.

Proebsting, E. L., Jr., 1970. Relation of fall and winter temperatures to flower bud behavior and wood hardiness of deciduous fruit trees. HortScience 5:422-424.

Proebsting, E. L., Jr., and Mills, H. H. 1972. A comparison of hardiness responses in fruit buds of 'Bing' cherry and 'Elberta' peach. J. Am. Soc. Hort. Sci. 97:802-806.

Proebsting, E. L., Jr., and Sakai, A. 1979. Determining $T_{50}$ of peach flower buds with exotherm analysis. HortScience 14:597-598.

Quamme, H. A. 1974. An exothermic process involved in freezing injury to flower buds of several Prunus species. J. Am. Soc. Hort. Sci. 99:315-317.

Quamme, H. A. 1976. Relationship of the low temperature exotherm to apple and pear production in North America. Can. J. Plant Sci. 56:493-500.

Quamme, H. A. 1978. Mechanism of supercooling in overwintering peach flower buds. J. Am. Soc.

Hort. Sci. 103:57-61.

Quamme, H. A. 1983. Relationship of air temperature to water content and supercooling of overwintering peach flower buds. J. Am. Soc. Hort. Sci. 108:697-701.

Quamme, H. A. 1986. Use of thermal analyses to measure the freezing resistance of grape buds. Can. J. Plant Sci. 66:945-952.

Quamme, H. A. 1991. Application of thermal analysis to breeding fruit crops for increased cold hardiness. HortScience 26:513-517.

Quamme, H. A., and Gusta, L. V. 1987. Relationship of ice nucleation and water status to freezing patterns in dormant peach flower buds. HortScience 22:465-467.

Quamme, H. A., Layne, R. E. C., Jackson, H. O., and Spearman, G. A. 1975. An improved exotherm method for measuring cold hardiness of peach flower buds. HortScience 10:521-523.

Quamme, H. A., Layne, R. E. C., and Ronald, W. G. 1982. Relationship of supercooling to cold hardiness and northern distribution of several cultivated and native *Prunus* species and hybrids. Can. J. Plant Sci. 62:137-148.

Quamme, H. A., Weiser, C. J., and Stushnoff, C. 1973. The mechanism of freezing injury in xylem of winter apple twigs. Plant Physiol. 51:273-277.

Rajashekar, C. B. 1989. Supercooling characteristics of isolated peach flower bud primordia. Plant Physiol. 89:1031-1034.

Rasmussen, D. H., and MacKenzie, A. P. 1972. Effect of solute on ice-solution interfacial free energy, calculation from measured homogeneous nucleation temperatures. Pages 126-145 in: Water Structure at the Water Polymer Interface. H. H. G. Jellineck, ed. Pleneum Press, New York.

Rasmussen, D. H., Macauley, M. N., and MacKenzie, A. P. 1975. Supercooling and nucleation in ice in single cells. Cryobiol. 12:328-339.

Sakai, A. 1978. Low temperature exotherms of winter buds of hardy conifers. Plant Cell Physiol. 19:1439-1446.

Sakai, A. 1979a. Deep supercooling of winter flower buds of *Cornus florida*. L. HortScience 14:69-70.

Sakai, A. 1979b. Freezing avoidance mechanism of primordial shoots of conifer buds. Plant Cell Physiol. 20:1381.

Tumanov, I. I., Krasavtsev, O. A., and Trunova, T. I. 1969. Investigation of ice formation process in plants by measuring heat evolution. Soviet Plant Physiol. 16:754-760.

Warmund, M. R., and George, M. F. 1990. Freezing survival and supercooling in primary and secondary buds of *Rubus* spp. Can. J. Plant Sci. 70:893-904.

Warmund, M. R., George, M. F., and Cumbie, B. G. 1988. Supercooling in 'Darrow' blackberry buds. J. Am. Soc. Hort. Sci. 113:418-422.

Warmund, M. R., George, M., and Takeda, F. 1991. Supercooling in floral buds of 'Danka' black and 'Red Lake' red currants. J. Am. Soc. Hort. Sci. 116:1030-1034.

Weigand, K. M. 1906. Some studies regarding the biology of buds and twigs in winter. Bot. Gaz. 103:373-424.

Wisniewski, M., and Ashworth, E. N. 1986. A comparison of seasonal ultrastructure changes in stem tissues of peach (*Prunus persica*) that exhibit contrasting mechanisms of cold hardiness. Bot. Gaz. 147:407-417.

Wisniewski, M., Ashworth, E. N., and Schaffer, K. 1987. The use of lanthanum to characterize cell wall permeability in relation to deep supercooling and extracellular freezing in woody plants. I. Intergeneric comparisons between *Prunus, Cornus* and *Salix*. Protoplasma 139:105-116.

Wolf, T. K., and Pool, R. M. 1987. Factors affecting exotherm detection in the differential thermal analysis of grapevine dormant buds. J. Am. Soc. Hort. Sci. 112:520-525.

CHAPTER 11

# The Roles of Ice Nucleators in Cold Tolerant Invertebrates

**John G. Duman, T. Mark Olsen, King Lun Yeung, and Fred Jerva**

Ice nucleators initiate heterogeneous nucleation by organizing water molecules into embryo crystals of critical size at temperatures above those at which homogeneous nucleation would occur. In freeze-avoiding animals, the prevention of inoculative freezing and extension of supercooling abilities to temperatures beyond those likely to be experienced are of obvious importance. Consequently, elimination of ice nucleators, either over evolutionary or shorter (i.e., seasonal) time frames, and/or masking or inhibition of ice nucleation activity by antifreezes in winter becomes crucial. In contrast, in many freeze-tolerant animals, there appears to be a requirement that ice formation be initiated at "fairly high" subzero temperatures, thus apparently providing selection pressure for the evolution of extracellular ice nucleators for this purpose. However, nucleation temperatures of freeze-tolerant species vary widely, and many species seem not to have this requirement.

Thus, the absence or presence of ice nucleators is critical to the overwintering success of both freeze-avoiding and freeze-tolerant species. Freeze-avoiding species must eliminate or mask ice nucleators, whereas many freeze-tolerant species have apparently selected for potent ice nucleators. While this trend applies to vertebrates (see Chapter 12) as well as invertebrates, this review will concern the latter.

## Freeze-Avoiding Species

While temperatures above those at which freezing occurs may initiate lethal chill, coma, etc., freeze-avoiding organisms die if ice forms in their tissues. Therefore, adaptations that lower the temperature at which body fluids remain liquid confer significant advantages to these organisms. A number of processes may contribute to avoidance of freezing. Some of these are antifreeze production, ice nucleator removal, prevention of inoculative freezing, microhabitat selection, desiccation, and, of course, appropriate acclimatization processes to control the timing of these seasonal adaptations (For reviews see Sömme, 1982; Baust and

Rojas, 1985; Zachariassen, 1985; Bale, 1987; Block, 1990; Lee and Denlinger, 1991; Duman et al., 1991a). These adaptations are not mutually exclusive, and animals exhibit various combinations of these processes. This section will concentrate on the basic physiological mechanisms for promotion of supercooling, namely, removal of ice nucleators, antifreeze production to overcome ice nucleator activity, and the interactions of these two adaptations. The magnitude of the supercooling produced by antifreeze depends greatly on the presence or absence of ice nucleators and on the activity of ice nucleators if they are present. However, unless there are mechanisms in place to prevent inoculative freezing, seeding across the body surface may preclude the chance to supercool. Therefore, it is necessary to briefly discuss inoculative freezing.

## Inoculative Freezing

Prevention of inoculation from external ice by a physical surface barrier (i.e., the wax-coated cuticle of insects or the cornified epithelium of some vertebrates) is of paramount importance to a freeze-avoiding animal. The absence of such a barrier requires production of sufficient concentrations of antifreezes to lower the freezing point of the body fluids below environmental temperatures. This may be energetically costly, if not impossible, when microhabitat temperatures are very low. Antifreeze proteins, even the most active ones at high concentrations, are capable of depressing the hysteretic freezing point of water by only ~6°C (Duman et al., 1993). Also, since the molal freezing point constant for water is 1.86°C, even a 5 M concentration of colligative antifreeze lowers the equilibrium freezing point by only ~9.3°C. Consequently, the presence of a physical barrier to inoculation is crucial. The difference in the limited supercooling abilities of earthworms after hatching compared to the potentially extended capabilities of insects, with their wax-coated, chitinous cuticle, illustrates this point. In some insects, however, it is becoming apparent that there are seasonal adaptations to inhibit inoculation. Some species that are immune to inoculative freezing in winter are readily inoculated at temperatures only slightly below the body fluid freezing point in summer (Rojas et al., 1992). This suggests seasonal variation in waxes and/or other cuticular modifications. Gehrken (1992) showed that in *Ips acuminatus* beetles, hemolymph antifreeze proteins protect against inoculative freezing to a greater extent than expected based on the measured thermal hysteresis of the hemolymph.

Freeze-avoiding species in which the body surface has limited abilities to prevent inoculative freezing may rely upon cocoons to provide some barrier to inoculation. This is the case in the northern lumbricid earthworm *Dendrobaena octaedra*. Postembryonic individuals are readily inoculated, but the cocoons (egg capsules) provide protection for this freeze-sensitive species (Holmstrup, 1992).

## Adaptations That Promote Supercooling

If an animal is able to stop inoculative freezing, then mechanisms that promote supercooling and depress the ice nucleation temperature become critical, potentially allowing the organism to survive in environments with much lower temperatures. Removal of ice nucleators from body fluids and/or production of antifreezes are common adaptations.

### Colligative Antifreezes

Production of high concentrations of compatible solutes, usually polyols, which function as colligative antifreezes has long been recognized as an important adaptation in freeze avoidance. In his classic studies of parasitic larvae of the wheat stem sawfly, *Bracon cephi*, which supercool to –45°C, Salt (1959) demonstrated that glycerol concentrations as high as 5 M are accumulated. However, one must be cautious in attributing all of an observed winter supercooling point depression to colligative antifreeze, since in studies of intact organisms, or even of their hemolymph, it cannot be ascertained whether additional factors (i.e., removal of ice nucleators, presence of antifreeze proteins) may also be responsible for a portion of the observed seasonal extension of the capacity to supercool.

In freeze-avoiding species, it is advantageous to maximize supercooling point depression resulting from accumulation of colligative antifreezes. Yet, while in most freeze-tolerant species high concentrations of cryoprotectants are common, supercooling must often be inhibited by extracellular ice nucleators (Zachariassen and Hammel, 1976; Zachariassen 1982, 1992). In contrast to the situation in freeze-avoiding organisms, extension of supercooling by cryoprotectants is counterproductive in freeze-tolerant animals. Therefore, questions arise as to whether freeze-avoiding species have evolved mechanisms to increase the effects of colligative antifreezes on supercooling beyond those in freeze-tolerant species, or whether freeze-tolerant species have evolved mechanisms to limit the effects of colligative antifreezes on supercooling.

The apparent answers to the above questions are yes, but the effect of the colligative antifreezes appears to depend upon whether potent ice nucleator activity is present. Studies on well-defined systems, in which components other than water and added antifreezes are identified, are limited. MacKenzie (1977) demonstrated that colligative antifreezes depress the homogeneous nucleation temperature of water (ΔNT) by approximately twice the magnitude of the colligative melting point depression (ΔMP), ΔNT/ΔMP = 2:1. The effect was independent of the nature of the added solute (glucose, glycerol, sucrose, NaCl, $NH_4F$, urea, ethylene glycol), except for polymers (PEG, PVP), which had a significantly greater effect (ΔNT/ΔMP = ~5:1). Franks (1981) presented a theoretical argument that indicated that the depression of homogeneous nucleation temperature produced by a given solute is directly proportional to melting point depression. These data on effects of solutes on homogeneous nucleation are potentially applicable to species that lack ice nucleators in winter and that produce significant concentrations of colligative antifreezes (Miller, 1982; Ring, 1982; Miller and Werner, 1987). Likewise, Block and Young (1979) showed that addition of glycerol to droplets of water resulted in ΔNT/ΔMP = ~2:1. Since the nucleation temperature of the water droplets without glycerol was approximately –21°C (range of –15 to –30°C), the authors correctly considered this result to illustrate the effect of glycerol on heterogeneous nucleation, although the ice nucleators present in the water were not potent ones. Reid et al. (1985) obtained similar results investigating effects of solutes (sucrose, glycerol, ethylene glycol, and dimethylsulfoxide) on heterogeneous nucleation in "polished" water droplets, ΔNT/ΔMP = ~5:3 to 2:1, except that the ratio for glycerol was ~2:3. In the presence of more active ice nucleators, colligative antifreezes appear to lower nucleation temperature by an amount equal to the colligative depression of the melting point. Lusena (1955) showed that glycerol or NaCl depresses the heterogeneous nucleation temperature initiated by silver iodide crystals by an amount equal to the melting point depression. This is at odds with Reid et al. (1985), who showed that in the presence of glycerol, ethylene glycol, or DMSO, the ΔNT/ΔMP obtained using silver iodide to induce heterogeneous nucleation was ~5:3, whereas sucrose resulted in a ratio of ~1:2. The reason for this discrepancy between the two studies is not obvious. To our knowledge, the effect of colligative antifreeze on heterogeneous nucleation in an aqueous solution of a purified biological (protein) ice nucleator has not been determined. However, studies on insect hemolymph containing potent ice nucleators demonstrate that nucleation temperature is depressed by addition of colligative antifreeze by an amount equal to melting point depression. Lee et al. (1981) extended an earlier study by Zachariassen and Hammel (1976) by demonstrating that addition of any of 11 solutes (glucose, glycerol, sucrose, lactate, fructose, NaCl, trehalose, proline, ethylene glycol, mannitol, and PVP) to hemolymph of the freeze-tolerant beetle *Eleodes blanchardi* (which contains potent ice nucleators) resulted in ΔNT/ΔMP = ~1:1.

The above suggest that addition of solutes in the absence of ice nucleators or in the presence of ice nucleators that are not especially potent will produce ΔNT/ΔMP ratios of ~2:1. In contrast, with the exception of the study by Reid et al. (1985), it appears that addition of solutes to aqueous solutions containing potent ice nucleators results in ΔNT/ΔMP ratios of ~1:1.

Numerous studies have correlated melting point depression with supercooling point depression in intact insects. Notable early studies are those of Salt (1959) on *Bracon cephi* (ΔNT/ΔMP ratio of ~1:1) and Sömme (1967) on several species (ΔNT/ΔMP ratio of ~3.7:1). Block and Young (1979) found a ratio similar to those of Sömme (1967) in the freeze-avoiding mite *Alaskozetes antarcticus*. A pattern has emerged that indicates that colligative antifreezes in freeze-avoiding insects usually depress nucleation temperature more than melting point (ratios usually 2:1 to 3:1). In freeze-tolerant species, depression of nucleation temperature is generally equal to melting point depression (Sömme 1982; Zachariassen 1985, 1992; Block 1990). Thus, freeze-avoiding species appear to maximize effects of colligative antifreezes on depression of nucleation temperatures, while freeze-tolerant species seem to minimize the effects. Based on the simplified in vitro systems described earlier, the key to whether supercooling effects are maximized (ΔNT/ΔMP of 2:1 to 3:1) or minimized (ΔNT/ΔMP of ~1:1) appears to depend upon whether potent ice nucleators are present (most freeze-tolerant insects) or not (most freeze-avoiding species in winter). In addition to their function in freeze-tolerant species of raising the absolute nucleation temperature, potent ice nucleators also appear to limit the supercooling effect of cryoprotectants so that ΔNT/ΔMP ratios are ~1:1. In freeze-avoiding species, removal of potent ice nucleators from gut and hemolymph may be required to achieve the maximum supercooling effect from colligative antifreezes.

### Antifreeze Proteins

Antifreeze proteins lower the nonequilibrium freezing point of water, but not the melting point, thereby producing a difference between the freezing and melting points, termed thermal hysteresis. Although antifreeze proteins were first discovered in polar marine fish (DeVries, 1968, 1971), they have also been identified in certain arthropods including insects (Duman 1977a; Duman et al., 1991b, 1993), spiders (Duman, 1979), mites (Block and Duman, 1989), and centipedes (Tursman et al., 1994). Theede et al. (1976) reported thermal hysteresis activity in the intertidal mussel *Mytilus edulis* from Europe; however, *M. edulis* collected in January on Cape Cod did not exhibit thermal hysteresis activity and lacked the recrystallization inhibition activity associated with antifreeze proteins (Duman and Knight, *unpublished*). Most invertebrates that produce thermal hysteresis proteins are freeze avoiding, and production of the proteins is seasonal. Correlations between increased thermal hysteresis activity and lowered supercooling points are generally obvious in most studies of antifreeze proteins in terrestrial arthropods, which led to speculation that antifreeze proteins were responsible for extended supercooling. However, correlations between antifreeze accumulation and extended supercooling do not prove a cause and effect, and the magnitude of the effect attributable to antifreeze may be difficult to ascertain. In spite of this caution, such correlations between antifreeze proteins and extended supercooling were particularly enticing in those antifreeze-protein-producing species that do not accumulate polyols in winter (Duman 1977a; Patterson and Duman, 1978).

The first experimental evidence that antifreeze proteins might extend supercooling was provided by Zachariassen and Husby (1982), who demonstrated a relationship between amount of thermal hysteresis measured and size of the seed crystal used in the measurement. These data were extrapolated to embryo crystal dimensions to suggest that antifreeze proteins could explain the level of supercooling seen in hemolymph samples (approximately −20°C). The first direct evidence that antifreeze proteins could inhibit ice nucleators was provided by Parody-Morreale et al. (1988), who showed that addition of glycoprotein antifreeze from Antarctic fish shifted the ice nucleation spectrum of water containing ice nucleating bacteria to lower temperatures. Similarly, purified antifreeze proteins from larvae of the beetle *Dendroides canadensis* inhibit the activity of certain hemolymph ice nucleating proteins of this same species (Duman et al., 1991b, 1992). These same antifreeze proteins, however, did not inhibit the activity of hemolymph lipoprotein ice nucleators from freeze-tolerant larvae of the cranefly *Tipula trivittata* (Wu and Duman, 1991). Baust and Zachariassen (1983) found that thermal hysteresis proteins from *Rhagium inquisitor* beetles did not inhibit ice nucleators from the same species. Similarly, antifreeze proteins from the beetle *Iphthimus laevissimus* did not inhibit hemolymph ice nucleators from freeze-tolerant *Eleodes blanchardi* beetles. Thus, antifreeze proteins can inhibit some ice nucleator proteins but not others.

**Removal of Ice Nucleators**

Theoretically, if an organism removed all ice nucleators and prevented inoculative freezing, freezing could be avoided down to homogeneous nucleation temperatures. In addition, colligative antifreezes might further extend supercooling beyond the homogeneous nucleation temperature. If potent ice nucleators are removed, colligative antifreezes can lower the heterogeneous nucleation temperature by approximately twice the melting point depression. Also, antifreeze proteins appear to be capable of inhibiting some ice nucleators, but not others, and therefore the removal of unaffected ice nucleators may be critical. Consequently, removal of ice nucleators, either on a seasonal or evolutionary time scale, appears to be an important adaptation in freeze-avoiding species.

It has long been recognized (Wallace and Beard, 1943; Salt, 1953) that insects may ingest ice nucleators. Therefore, cessation of feeding and clearing the gut are important adaptations in many freeze-avoiding species. Krunic and Radovic (1974) suggested that a 20°C depression in nucleation temperature of the solitary bee *Megachile rotundata* was primarily due to gut evacuation. While there may be exceptions to this (Baust and Rojas, 1985), and while the percentage of the observed supercooling enhancement attributable to any given adaptation is difficult to determine in an intact organism, the evidence is overwhelming that, in many freeze-avoiding species, removal of ice nucleators from the gut prior to the onset of cold temperatures is critical. (See the reviews by Sömme, 1982; Zachariassen, 1985; Cannon and Block, 1988). Perhaps the most convincing evidence comes from experiments in which the organism is fed potent ice nucleators which induce a significant increase in nucleation temperature. For example, starvation of field collected Collembola, *Cryptopygus antarcticus*, during the Antarctic summer lowered their supercooling point. Subsequent feeding of an homogenate of moss turf containing nucleators raised supercooling points, while feeding on purified green algae lacking ice nucleators did not (Sömme and Block, 1982). Recent studies, which have implicated ice nucleating bacteria in gut fluid nucleation, have reinforced this point (see Chapter 14). Supercooling points of lady beetles, *Hippodamia convergens*, fed water containing ice nucleating bacteria (*Pseudomonas syringae* or *Erwinia herbicola*) increased from −16 to–4°C (Strong-Gunderson et al., 1990). Ice nucleating bacteria (*Enterobacter agglomerans* and *E. taylorae*) were isolated from guts of summer-collected beetles, *H. convergens* and *Ceratoma trifurcata* (Lee et al., 1991). Feeding these to *H. convergens* caused supercooling points of the insects to increase from −16 to −3°C. These results suggest that ice nucleating bacteria must be removed or masked in freeze-avoiding insects prior to winter. Larvae of the beetle *Dendroides canadensis* have ice nucleating bacteria in the gut in summer and through part of the autumn. By winter, the gut has been evacuated and ice nucleating bacteria can no longer be isolated from the gut (Olsen and Duman, 1992). Also, recent studies have shown that purified *D. canadensis* antifreeze protein inhibits the activity of an ice nucleating active isolate of *Pseudomonas fluorescens* from *D. canadensis* gut (Olsen and Duman, *unpublished*). Since antifreeze proteins are accumulated in the gut of *D. canadensis* by late September (Duman, 1984), it appears that a combination of antifreeze protein plus gut evacuation to remove ice nucleators protects gut fluid. Application of freeze-dried *P. syringae* to stored wheat or corn containing insect pests resulted in increases in supercooling points of 4.7 to 11.9°C in various insect species, suggesting that bacteria might be used as biological insecticides to kill overwintering stored grain insect pests (Lee et al., 1993a).

In contrast to the seasonal removal of gut ice nucleators, which has been investigated for many years, removal of ice nucleators from compartments other than gut has only recently been recognized. Zachariassen (1982) demonstrated that the beetle *Bolitophagus reticulatus* removes ice nucleators, thought to be intracellular, in winter. Winter beetles that were warm acclimated, without being fed, rapidly (3–4 days) increased nucleation temperature from −30 to −10°C, indicating induction of endogenous ice nucleators. Also, Bakken (1985) showed that nucleation temperatures decreased significantly in two alpine beetles, without antifreeze production, via removal of unidentified ice nucleators. A similar situation occurs in freeze-avoiding larvae of the stag beetle *Ceruchus piceus*, which lowered supercooling points from −7°C in summer to below −20°C in winter without antifreeze production. These larvae both clear the gut and remove hemolymph lipoproteins with ice nucleator activity (Neven et al., 1986). Since metabolism is greatly reduced in winter, the normal lipid shuttle function of hemolymph lipoproteins is apparently not essential at this time, and the lipoprotein can be removed. The hemolymph levels of this lipoprotein ice nucleator are controlled by hormones, including adipokinetic hormone and probably juvenile hormone (Xu et al., 1990). It is likely that the endogenous ice nucleators removed seasonally in other freeze-avoiding species may also be proteins. However, there are other possibilities. Lee et al. (1992) identified crystalloid spheres of tribasic calcium phosphate with ice nucleating activity in Malpighian tubules of overwintering larvae of *Eurosta solidaginis*.

## Examples of Freeze-Avoiding Strategies

Whatever the location of ice nucleators, it seems critical that they be removed or their activity masked in winter. The inability to do so requires that a freeze-avoiding organism accumulate very high concentrations of antifreeze if it is to survive even moderately low winter temperatures. Table 1 illustrates this with three freeze-avoiding species of beetles that we have studied. All are common in northern Indiana, overwinter in similar microhabitats, and are exposed to similar temperatures. Yet their adaptations are quite diverse and apparently dependent upon the ability, or lack thereof, to remove ice nucleators. All clear the gut in winter, thereby removing ice nucleators. As noted earlier, in winter, *Ceruchus piceus* larvae remove a hemolymph lipoprotein ice nucleator (Neven et al., 1986). Antifreeze proteins are not produced, and only small amounts of polyols (sorbitol) are accumulated. Yet the level of supercooling (below the freezing point) is ~25°C, attributable primarily to removal of ice nucleators. *Dendroides canadensis* larvae produce both antifreeze protein and polyols (glycerol and sorbitol) that promote supercooling. Recall that the *D. canadensis* antifreeze proteins inhibit hemolymph ice nucleators (Duman et al., 1991b, 1992). In addition, the concentration of hemolymph ice nucleating proteins also decreases in winter (Olsen and Duman, 1992). In contrast, *Uloma impressa* accumulates both antifreeze proteins and very high concentrations of colligative antifreezes (glycerol and sorbitol) but retains active ice nucleators. Note

**Table 1.** Examples of freeze-avoiding insects with different strategies for depressing the supercooling point in winter

| Species<br>Season | Melting point (°C) | Freezing point (°C) | Thermal hysteresis[a] | Supercooling point (°C) | Supercooling capacity below the freezing point (°C) |
|---|---|---|---|---|---|
| *Ceruchus piceus*[b] | | | | | |
| Summer | −0.5 | −0.5 | 0 | −7 | 6 |
| Winter | −1.1 | −1.1 | 0 | −26 | 25 |
| *Dendroides canadensis*[c] | | | | | |
| Summer | −0.6 | −1.7 | 1.1 | −9 | 7 |
| Winter | −2.5 | −5.7 | 3.2 | −27 | 21 |
| *Uloma impressa*[d] | | | | | |
| Summer | −0.9 | −2.0 | 1.1 | −6 | 4 |
| Winter | −9.9 | −14.7 | 4.8 | −21 | 6 |
| *Rhabdophaga strobiloides*[e] | | | | | |
| Summer | −1.2 | ... | ... | −26.5 | 25 |
| Winter | −19.3 | ... | ... | −56 | 37 |

[a]Thermal hysteresis indicates the presence of antifreeze protein.
[b]Data from Neven et al. (1986).
[c]Data from Duman (1984).
[d]Data from Duman (1979).
[e]Data from Miller (1982).

that the hemolymph melting point is approximately −10°C and the hysteretic freezing point −15°C in winter. Yet the beetles supercool just 6°C below the hysteretic freezing point. Evidently these beetles contain very active ice nucleators, which are not removed in winter and which antifreeze protein cannot inhibit. Therefore, accumulation of very high levels of antifreezes is required to lower the freezing point to −15°C.

An example of the efficacy of combining removal of ice nucleators with extensive antifreeze accumulation is provided by willow cone gall larvae, *Rhabdophaga strobiloides* from Alaska (Miller, 1982). These undercool by ~25°C in summer, indicating a lack of potent ice nucleators (Table 1). In winter, the hemolymph melting point is decreased to −19.3°C, at least partly due to extensive polyol accumulation (glycerol ~4.8 M). The larvae undercool by ~37°C, and the supercooling point of these exceptionally freeze-avoiding insects is reduced to –56°C.

**An Evolutionary Perspective**

Since endogenous ice nucleators present problems for freeze-avoiding species, the question arises as to why such negative selection pressure did not result in removal of these nucleators, or at least of the activity, over evolutionary time. Apparently the ice nucleator activity evolved when this negative selection pressure did not exist. The proteins (or perhaps other types of nucleators) assumed certain structures, required by their particular functions, which by chance happened to result in surface water being structured in an icelike fashion. This should not be surprising since, as Darwin recognized in the *Origin of Species,* "An organ built under the influence of selection for a specific role may be able as a consequence of it's structure, to perform many other, unselected functions, as well" (Darwin, 1859). Substitute protein for organ, and the statement may apply to ice nucleator proteins. In some cases, the protein function may require this embryo crystal promoting structure and thus it cannot now be eliminated. While this is a plausible scenario, there do seem to be examples of invertebrates that lack endogenous ice nucleating activity, thus showing that ice nucleating structures, at least not highly active ones, are not required for normal activity. Aphids as a group generally are capable of fairly extensive supercooling even in summer and apparently lack active ice nucleators (O'Doherty and Bale, 1985; Knight and Bale, 1986; O'Doherty and Ring, 1987). Miller and Werner (1987) described the overwintering adaptation of three species of freeze-avoiding willow gall insects from Alaska, including *R. strobiloides* discussed above. Mean supercooling points in summer were –25 to –30°C, indicating the absence of potent ice nucleators. However, in winter, supercooling points decreased to –58°C. Glycerol accumulated to 4–6 M and can account for a major portion of the increased supercooling (based on ΔNT/ΔMP = 2:1 to 3:1 in the absence of active ice nucleators), but it is likely that removal of even the minimally active ice nucleators present in summer may also occur. Other Alaskan (Miller, 1982) and alpine (Ring, 1982) species that supercool to temperatures of –50 to −60°C have been reported. Interestingly, some of these can survive freezing at these low nucleation temperatures, although it is likely that such temperatures would only rarely be encountered. It seems that if selection pressure is sufficient (i.e., the extreme cold of the northern Rockies or central Alaska), endogenous ice nucleating activity can be eliminated over evolutionary time.

## Freeze-Tolerant Species

There are a number of adaptations recognized as contributing to the ability of animals to survive freezing of their body fluids (see Ring, 1980; Zachariassen, 1985; Baust and Rojas, 1985; Storey and Storey, 1988; Cannon and Block, 1988; Duman et al., 1991a; Lee and Denlinger, 1991 for various perspectives). However, one of the basic tenets of freezing tolerance is that intracellular freezing is lethal. With a few notable exceptions, such as fat body cells of the goldenrod gall fly *Eurosta solidaginis* (Salt, 1959; Lee et al., 1993b), this concept appears to hold. Extensive supercooling, followed by rapid freezing, can result in intracellular ice formation (Mazur, 1977, 1984) and in damage from osmotic stress (Zachariassen, 1992). Consequently, inhibition of extensive supercooling is an important adaptation in many freeze-tolerant invertebrates.

### Advantage of Freezing at a High Temperature

One of the most interesting discoveries concerning natural freeze tolerance adaptations was that by Zachariassen and Hammel (1976) describing hemolymph ice nucleators in freeze-tolerant beetles that initiate freezing at relatively high temperatures. With supercooling controlled at just a few degrees below the hemolymph freezing point, the plasma membrane is generally able to prevent inoculative freezing of cytoplasm. As solutes are excluded from extracellular ice, the osmotic concentration of unfrozen extracellular water increases, thus generating an osmotic outflux of water from cells, which lowers freezing and supercooling points of intracellular water and thereby decreasing the chance for intracellular freezing. Also, freezing at higher temperatures lessens osmotic imbalances between intracellular and extracellular compartments, minimizing the potential for osmotic shock (Zachariassen, 1992).

Another possible advantage of freezing at higher temperatures is energy savings. Scholander et al. (1953) demonstrated that frozen insects had much reduced metabolic rates compared with unfrozen supercooled individuals at the same temperatures. More time spent frozen over the winter results in less energy stores being utilized; therefore, more energy is available for growth and/or reproduction in spring.

An additional advantage of freezing at higher temperatures may be water conservation. Since overwintering terrestrial invertebrates may not eat or drink, negative water balance could result in serious desiccation. Respiratory water loss varies directly with metabolic rate. Since freezing results in reduced metabolism, freezing at higher temperatures reduces respiratory water loss. Also, the body fluids of a supercooled organism in a hibernaculum containing ice have a higher vapor pressure than ice. Consequently, the organism will lose water. In contrast, if the organism is frozen (resulting in vapor pressure equilibrium), water will be conserved (Zachariassen, 1992).

Some or all of these potential advantages of extracellular ice nucleators are undoubtedly important to the overwintering success of freeze-tolerant species. In addition, early in the evolution of cold tolerance in a particular species, endogenous ice nucleators may have been present, and it may have been impossible, or energetically costly, to remove or mask these. Consequently, evolution of freeze tolerance and perhaps further development of increased ice nucleator activity, rather than freeze avoidance, may have resulted. In some freeze-tolerant insects, ice nucleator activity in hemolymph increases in winter, indicating that ice nucleation is their primary function (Zachariassen, 1980, 1992). However, this is not always the case. For example, the hemolymph lipoprotein ice nucleator in freeze-tolerant larvae of the cranefly *Tipula trivittata* is the major hemolymph lipoprotein during all seasons, and therefore almost certainly functions to shuttle lipid (Neven et al., 1989).

## Inoculative Freezing in Freeze-Tolerant Organisms

The preceding arguments for hemolymph ice nucleators in freeze-tolerant organisms could be applied to inoculative freezing of extracellular fluids by external ice, the ultimate ice nucleator. There could even be additional advantages such as not having to maintain endogenous ice nucleators and potentially to eliminate any supercooling whatsoever. Recently, a few insects (Tanno, 1977; Fields and McNeil, 1986; Shimada and Riihimaa, 1988; Gehrken et al., 1991) and a centipede (Tursman et al., 1994) were described that survive freezing only if inoculative freezing occurs. Overwintering larvae of the moth *Cisseps fulvicollis* supercool to −12 to −17°C under dry conditions; however, they do not survive this freezing. In contrast, if cooled while in contact with external ice, they freeze above −4°C but survive (Fields and McNeil, 1986). A similar situation pertains in overwintering adult Bolitophagus reticulatus beetles, which can supercool to −30°C but then die if frozen. However, if in contact with ice, the beetles freeze inoculatively above −6°C and survive freezing (Gehrken et al., 1991). Overwintering *Lithobius forficatus* centipedes have potent hemolymph ice nucleators, which initiate nucleation at approximately −3°C. However, the centipedes do not survive even this minimal supercooling. If frozen in contact with exterior ice, freezing occurs at the hemolymph freezing point (approximately −1°C), and the centipedes survive (Tursman et al., 1994). Centipedes lack the wax coating of the cuticle of insects, and therefore the cuticle apparently does not present a barrier to inoculative freezing. The overwintering microhabitats of these insects and centipedes are generally damp, and consequently winter ice formation in the hibernaculum is common. Therefore, it is likely that they are frozen for periods of time each winter.

## Other Sites of Nucleation

In addition to inoculative freezing, other nonhemolymph sites of nucleation initiate freezing in some freeze-tolerant species. Winter acclimated, freeze-tolerant larvae of the arctic moth *Gynaephora groenlandica* have a hemolymph supercooling point of −20°C. However, intact larvae nucleate at −7°C; therefore, it is obvious that hemolymph nucleators do not initiate nucleation (Kukal et al., 1988). Ice nucleating bacteria in the gut of freeze-tolerant organisms are possible sites of nucleation.

Some freeze-tolerant populations of the gall fly *Eurosta solidaginis* apparently have hemolymph ice nucleators (Sömme, 1978; Lee et al., 1981; Zachariassen et al., 1982). However, Bale et al. (1989) reported that a New York population lacked hemolymph ice nucleators and presented evidence that frass in the gall has potent ice nucleator activity and initiates freezing. Also, as mentioned earlier, Lee et al. (1992) showed that *E. solidaginis* hemolymph supercooled to −18°C, but crystals of tribasic calcium phosphate in the Malpighian tubules initiate nucleation at temperatures as high as −7.8°C.

**Table 2.** Examples of variability in the amount of supercooling below the hemolymph freezing point seen in various freeze-tolerant terrestrial arthropods[a]

| Species | Organismal freezing temperature (°C) | Supercooling |
|---|---|---|
| *Lithobius forficatus*[b] | −1 | <1 |
| *Eleodes blanchardi*[c] | −6.3 | 5 |
| *Tipula trivitatta*[d] | −6.5 | 5 |
| *Antherea polyphemus*[e] | −21.1 | 20 |
| *Bracon cephi*[f] | −47 | 32 |
| *Pytho deplanatas*[g] | −54 | 30–40? |
| *P. americanus*[g] | −6 | 4–5 |

[a] All species, except for the centipede *L. forficatus,* supercool to some extent. The amount of supercooling in the *Pytho* species are estimates, as hemolymph freezing point was not determined.
[b] Data from Tursman et al. (1994).
[c] Data from Zachariassen and Hammel (1976).
[d] Data from Duman et al. (1985).
[e] Data from Duman et al. (1991b).
[f] Data from Salt (1959).
[g] Data from Ring (1982).

## Variations in Amount of Supercooling

While many freeze-tolerant terrestrial arthropods gain the advantages of freezing extracellularly at high temperatures, there are some that freeze after considerably greater supercooling. Table 2 presents examples of this variability. The centipede *Lithobius forficatus*, owing to inoculative freezing in the typically frozen hibernaculum, probably does not supercool at all (Tursman et al., 1994). The beetle *Eleodes blanchardi* is a freeze-tolerant species with strong hemolymph ice nucleators (Zachariassen and Hammel, 1976). The 5°C of supercooling is typical of many freeze-tolerant organisms, including other insects such as the larvae of the cranefly *Tipula trivittata* (Duman et al., 1985). However, others supercool much more. Pupae of the silk moth *Antheraea polyphemus* represent a number of silk moths (Duman et al., 1991b) and other Lepidoptera that supercool by ~20°C. Miller (1982) described several freeze-tolerant insects from central Alaska with similar capacities for supercooling. *Bracon cephi* (Salt, 1959) undercools as much as 32°C below it's −15°C hemolymph freezing point, yet survives freezing. Ring (1982) described the freeze-tolerant beetle *Pytho deplanatus* from the Canadian Rockies with a nucleation temperature of −54°C. Since the hemolymph freezing point was not reported, the amount of undercooling cannot be calculated but was likely to be 30–40°C and therefore close to the homogeneous nucleation temperature, indicating that all ice nucleators had been eliminated. This is especially interesting because a related species, *P. americanus*, found in the Rockies and the arctic had a nucleation temperature of −6°C, underscoring the point that even closely related species from similar habitats can exhibit widely different overwintering strategies.

## An Evolutionary Perspective

In view of the advantages of limited supercooling to freeze-tolerant animals, the large range of undercooling is difficult to understand. One potential explanation comes from the cryopreservation literature (Farrant, 1980). When attempting to cryopreserve a particular cell type, it is commonly observed that a certain freezing rate is optimal, with survival decreasing rapidly at slower and/or faster rates. This optimal rate often varies dramatically with different organisms and cell types. The rate of ice growth following nucleation at a certain degree of supercooling is perhaps comparable to freezing at various rates in a controlled cryopreservation experiment. While organisms in the field have little control over temperature changes, except by their choice of hibernaculum, the presence or absence of ice nucleators and their activities may afford some control over initial rates of crystal growth. The evolution of freeze tolerance in a particular species may be fine tuned, with combinations of ice nucleators of particular potencies (or perhaps their complete absence) and antifreezes to provide the optimal undercooling for that particular species.

In spite of these examples of variability in undercooling, recal! that many freeze-tolerant animals have fairly potent ice nucleators in their extracellular fluids. Thus, it appears that there has been selection for the evolution of these ice nucleators, perhaps because of the theoretical advantages discussed earlier.

## Biochemistry of Invertebrate Ice Nucleators

As discussed earlier, the ice nucleators identified from invertebrates (insects and intertidal invertebrates) are either bacterial (gut), tribasic calcium phosphate crystals from Malpighian tubules of *Eurosta solidaginis* (Lee et al., 1993b), or hemolymph proteins. Since bacterial ice nucleators are treated elsewhere in this book, and little information is available on the Malpighian tubule crystals, this section will discuss hemolymph proteins.

### Hemolymph Proteins

Zachariassen and Hammel (1976) suggested that hemolymph ice nucleators in freeze-tolerant beetles were proteins since they could be inactivated by heat. Sömme (1978) presented a similar argument for proteinaceous hemolymph ice nucleators in *Eurosta solidaginis*. Likewise, the intertidal mussel *Mytilus edulis* produces hemolymph ice nucleators that are inactivated by heat (Aunaas, 1982). Also, ice nucleating activity in the body fluid of the freeze-tolerant tardigrade *Adorybiotus coronifer* is destroyed by heating to 68°C (Westh et al., 1991). The first hemolymph ice nucleator purified was a 74-kDa protein from freeze-tolerant queens of the hornet *Vespula maculata*. It was suggested that side chains of this very hydrophilic protein (20 mol% glutamate/glutamine) could be important in hydrogen bonding water into an embryo crystal (Duman et al., 1984).

The best characterized hemolymph ice nucleator is a lipoprotein from freeze-tolerant larvae of the cranefly *Tipula trivittata* (Duman et al., 1985; Neven et al., 1989). The globular 800-kDa lipoprotein ice nucleator (LPIN) has a diameter of ~135 Å (Neven et al., 1989). The LPIN contains 45% protein, 51% lipid, and 4% carbohydrate. Two apolipoproteins are present, Apo-I (265 kDa) and Apo-II (81 kDa). The lipid components are characteristic of insect lipophorins, except that the LPIN contains phosphatidylinositol (PI). PI had not been previously identified in an insect hemolymph lipoprotein, and so its presence suggested that PI was important for ice nucleating activity. In addition, Warner (1962) had suggested that inositol might order water in an icelike fashion via hydrogen bonding by its hydroxyl groups. This was intriguing since phospholipids generally form a monolayer on the surface of insect lipophorins (Katagiri, 1985), and therefore the inositol residues would be positioned to interact with water. Also, early studies on bacterial ice nucleators indicated that PI was required (Kozloff et al., 1984); however, it now appears that PI serves to anchor the protein in the bacterial membrane rather than to order water (Turner et al., 1990). However, PI in the *Tipula* LPIN, along with the apoproteins, appears to be essential for ice nucleator activity and is likely to be directly involved in ordering the embryo crystal. To investigate the structure-function relationships of the LPIN, the effects of various treatments (enzymes such as PI-specific phospholipase C, etc.) on ice nucleator activity of the LPIN were investigated. Also, when the LPIN is delipidated and then reconstituted into proteoliposomes, ice nucleator activity is regained; thus, it is possible to selectively reconstitute proteoliposomes with certain components present or absent. These manipulations showed that both apoproteins and PI are required for activity. In fact, proteoliposomes composed only of Apo-I, Apo-II, and PI have activity (Neven et al., 1989). Substitution of either PI-4 monophosphate or PI-4,5 diphosphate for PI in the liposomes, or other manipulations of the hydroxyl residues of PI, result in greatly reduced activity. This indicates that the integrity of the hydroxyl groups is required, perhaps suggesting they are involved in organizing the embryo crystal (Duman et al., 1991c). Surprisingly, the nature of the fatty acids in PI seems not to matter. Even PI treated with phospholipase A2, which cleaves the C-2 fatty acid of PI, is effective in reconstituted proteoliposomes (Duman et al., 1991c).

In addition to PI, both apoproteins are required for activity. Western blots showed immunological similarities between LPIN apoproteins and bacterial (*Pseudomonas fluorescens*) ice nucleator protein. Polyclonal antibodies to the LPIN cross-reacted with *P. fluorescens* protein, and antibodies to bacterial ice nucleator protein cross-reacted with both Apo-I and Apo-II of the LPIN (Duman et al., 1991c). Because over 80% of the bacterial protein consists of a repeat structure (Green and Warren, 1985), these immunological similarities suggest that LPIN proteins also contain some amount of the repeat structure. This was confirmed when antibodies prepared to a synthetic peptide of the consensus octapeptide repeat of the bacterial protein cross-reacted on Western blots with *Tipula* LPIN apoproteins (Duman et al., 1991c). These data indicate that the apoproteins may be directly involved in ordering water into the embryo crystal.

Another lipoprotein with ice nucleator activity was isolated from hemolymph of the freeze-avoiding larvae of the beetle *Ceruchus piceus* (Neven et al., 1989). Recall that this LPIN is removed in winter to allow supercooling. The *Ceruchus* LPIN is very fragile and therefore has not been studied in detail. Antibodies to the *Tipula* LPIN or to the bacterial ice nucleator protein do not cross-react with the *Ceruchus* LPIN, indicating that the *Ceruchus* LPIN is immunologically distinct and consequently lacks the bacterial repeat sequence. PI is not present in the *Ceruchus* LPIN.

In addition to the LPIN, two other hemolymph proteins with ice nucleator activity were purified from overwintering *T. trivittata* larvae (Duman and Jerva, *unpublished*). Both of these 77-kDa proteins have unusual amino acid compositions. One has 40 mol% tyrosine, suggesting that it may be a storage protein. The other is 39 mol% glycine. On Western blots, neither cross-reacts with antibodies to *Tipula* LPIN or to bacterial ice nucleator protein. It appears that in *T. trivittata* different hemolymph proteins with various functions have evolved to also function as ice nucleators.

A protein ice nucleator was isolated from hemolymph of the freeze-tolerant intertidal gastropod *Melampus bidentatus* (Loomis, 1991). This 600-kDa protein is composed of 18–20 kDa subunits.

## Chain-Forming Behavior of the LPINs

While one might assume that a single LPIN molecule is sufficient to induce nucleation in a given volume of water, this does not seem to be the case. The mean nucleation temperature of 1-µl droplets of aqueous buffer containing LPIN at a concentration of $10^{-11}$ M was −19°C, the same as that of the buffer. An LPIN concentration of $10^{-9}$ M was required to produce a significant increase in nucleation temperature (to approximately −13°C). At this concentration, each 1-µl droplet contains ~$10^8$ molecules of LPIN. Additional increases in LPIN concentration produce progressively higher nucleation temperatures, up to a maximum of −6°C at concentrations at or above $1.7 \times 10^{-7}$ M (Duman et al., 1992). This indicates that either only a small percentage of the population of LPIN molecules has ice nucleator activity, or some number of LPIN molecules must cooperate (perhaps in a stochastic fashion) to produce an increase in nucleation temperature. It is likely that there is a limited area on the surface of the globular LPIN molecule that organizes an embryo crystal. The appropriate arrangement of juxtaposed LPINs might allow for the organization of a larger embryo crystal, which could then nucleate at a

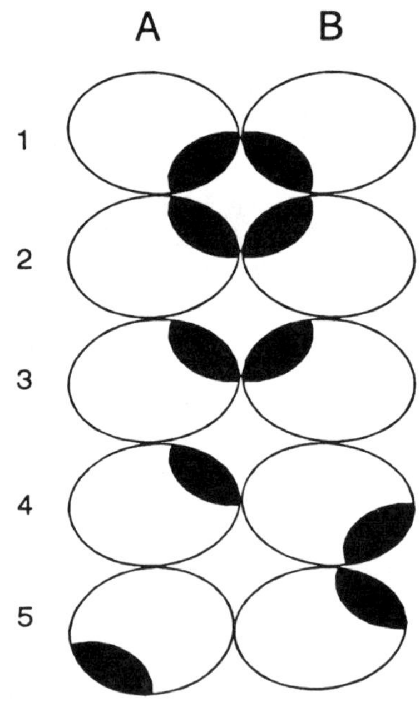


**Figure 1.** Diagrammatic representation of chain-forming behavior, based on scanning tunneling microscopy studies (Yeung et al., 1991), of the *Tipula* lipoprotein ice nucleator (LPIN) molecules . The possible cooperative arrangement between adjacent LPINs could result in larger embryo crystals and consequently increased nucleation temperatures. Each sphere indicates a globular LPIN molecule, and the darkened region on each represents the region of the surface that organizes water into an embryo crystal. Note how LPINs 1-A, B, and 2-A, B are arranged such that their water-organizing regions cooperate to form a larger embryo crystal than the others.

higher temperature than the smaller embryo crystal of a single LPIN molecule. Cooperation between nearby proteins has also been suggested for bacterial ice nucleating proteins (Wolber and Warren, 1989; Southworth et al., 1988; Mueller et al., 1990). A scanning tunneling microscopy study of the *Tipula* LPIN showed that individual LPINs are organized into chain structures (Yeung et al., 1991). Usually two chains are arranged side-by-side as shown in Figure 1. Chain-forming aggregation behavior was also observed in scanning tunneling microscope studies of the *Ceruchus piceus* LPIN (Yeung and Duman, *unpublished*). In contrast, hemolymph lipoproteins from larvae of *Manduca sexta*, which lack ice nucleator activity, did not demonstrate any organizational tendency. The chain-forming behavior of the LPINs lends some credence to the possible cooperative action of LPINs. As illustrated in Figure 1, four LPINs might, in a stochastic fashion, reach the optimal cooperative arrangement of their active sites and thereby induce nucleation at a higher temperature.

If nucleation at higher temperatures requires cooperative action of LPIN molecules provided by chain-forming behavior, it is pertinent to ask what components of the LPIN are required to form chains. (Recall that only Apo-I, Apo-II, and PI are necessary for ice nucleation activity.) Treatment of the LPIN with periodate does not disturb chain-forming behavior (Yeung and Duman, *unpublished*), but it does destroy ice nucleator activity (periodate cleaves the C-5–C-6 carbon bond of the inositol head group and oxidizes the C-5 and C-6 hydroxyls to aldehydes.) Proteoliposomes consisting only of Apo-I, Apo-II, and PI have both ice nucleating activity and chain-forming behavior. Also, reconstitutions of proteoliposomes consisting of Apo-I, Apo-II, and PIP2, which lack nucleating activity, still formed chains. This indicates that the integrity of the inositol lead group, while required for nucleator activity, is not necessary for chain-forming behavior. Proteoliposomes containing PI must have both Apo-I and Apo-II for chain forming. Thus, although Apo-I, Apo-II, and PI are all required for nucleator activity, PI (at least the inositol) is not needed for chain formation. This suggests that the hydroxyls of the inositol are involved directly in embryo crystal formation and that Apo-I and Apo-II are involved in chain formation. However, these data do not rule out the possibility that the apoproteins may also be involved directly in embryo crystal organization as suggested earlier.

## Future Studies

This chapter should underscore the fact that much remains to be learned about invertebrate ice nucleators. Essentially every section of the chapter stopped well short of providing satisfactory answers to the topic of discussion.

With regard to freeze-avoiding species, attention should be given to quantification of the portion of the seasonal supercooling point depression attributable to particular adaptations operating in each species and the cues that control the timing of the adaptations. Also, further studies into the details of the relationships between antifreezes and ice nucleators should be conducted to answer questions such as, Why do antifreeze proteins inhibit certain ice nucleators, but not others? Why do colligative antifreezes cause a $\Delta NT/\Delta MP$ ratio of 2:1 to 3:1 in some systems and of 1 in others? An important component of these answers will rely upon understanding of the structure-function relationships of ice nucleators. The hemolymph LPIN from *T. trivittata* is the only animal ice nucleator for which there is even a partial answer to the structure–ice nucleator function question. Yet other, obviously different protein ice nucleators are known. In freeze-tolerant species, an important question that could provide considerable insight into freeze tolerance is why there is so much species variation in amount of supercooling prior to freezing. Investigations of these and related questions will greatly increase our understanding of cold tolerance.

## Literature Cited

Aunaas, T. 1982. Nucleating agents in the haemolymph of an intertidal mollusc tolerant to freezing. Experientia 38:1456-1457.

Bakken, H. 1985. Cold hardiness in the alpine beetles *Patrobus septentrionis* and *Calathus melanocephalus*. J. Insect Physiol. 31:447-453.

Bale, J. S. 1987. Insect cold hardiness: Freezing and supercooling—An ecological perspective. J. Insect. Physiol. 33:899-908.

Bale, J. S., Hansen, T. N., and Baust, J. G. 1989. Nucleators and sites of nucleation in the freeze tolerant larvae of the gallfly *Eurosta solidaginis* (Fitch). J. Insect Physiol. 35:291-298.

Baust, J. G., and Rojas, R. R. 1985. Review—Insect cold hardiness: Facts and fancy. J. Insect Physiol. 31:755-759.

Baust, J. G., and Zachariassen, K. E. 1983. Seasonally active cell matrix associated ice nucleators in an insect. Cryo Lett. 4:65-71.

Block, W. 1990. Cold tolerance of insects and other arthropods. Phil. Trans. R. Soc. Lond. B 326:613-633.

Block, W., and Duman, J. G. 1989. Presence of thermal hysteresis producing antifreeze proteins in the Antarctic mite, *Alaskozetes antarcticus*. J. Exp. Zool. 250:229-231.

Block, W., and Young, S. R. 1979. Measurements of supercooling points in small arthropods and water droplets. Cryo Lett. 1:85-91.

Cannon, R. J. C., and Block, W. 1988. Cold tolerance of microarthropods. Biol. Rev. Camb. Philos. Soc. 63:23-77.

Darwin, C. 1859. On the Origin of Species by Means of Natural Selection. John Murray, London.

DeVries, A. L. 1968. Freezing resistance in some antarctic fishes. Ph.D. thesis. Stanford University, Stanford, California.

DeVries, A. L. 1971. Glycoproteins as biological antifreeze agents in antarctic fishes. Science 172:1152-1155.

Duman, J. G. 1977a. The role of macromolecular antifreeze in the darkling beetle *Meracantha contract*. J. Comp. Physiol. 115B:279-286.

Duman, J. G. 1979. Subzero temperature tolerance in spiders: The role of thermal hysteresis factors. J. Comp. Physiol. 131:347-352.

Duman, J. G. 1984. Thermal hysteresis antifreeze proteins in the midgut fluid of overwintering larvae of the beetle *Dendroides canadensis*. J. Exp. Zool. 230:355-361.

Duman, J. G., Morris, J. P., and Castellino, F. J. 1984. Purification and composition of an ice nucleating protein from queens of the hornet, *Vespula maculata*. J. Comp. Physiol. 154B:79-83.

Duman, J. G., Neven, L. G., Beals, J. M., Olson, K. O., and Castellino, F. J. 1985. Freeze tolerance adaptations, including haemolymph protein and lipoprotein ice nucleators, in larvae of the cranefly *Tipula trivittata*. J. Insect Physiol. 31:1-9.

Duman, J. G., Wu, D. W., Xu, L., Tursman, D., and Olsen, T. M. 1991a. Adaptations of insects to subzero temperatures. Q. Rev. Biol. 66:387-410.

Duman, J. G., Xu, L., Neven, L. G., Tursman, D., and Wu, D. W. 1991b. Hemolymph proteins involved in insect subzero temperature tolerance: Ice nucleators and antifreeze proteins. Pages 94-127 in: Insects at Low Temperature. R. E. Lee and D. L. Denlinger, eds. Chapman and Hall, New York, London.

Duman, J. G., Wu, D. W., Wolber, P. I., Mueller, G. M., and Neven, L. G. 1991c. Further characterization of the lipoprotein ice nucleator from freeze tolerant larvae of the cranefly *Tipula trivittata*. Comp. Biochem. Physiol. 99B:599-607.

Duman, J. G., Wu, D. W., and Yeung, K. L. 1992. Hemolymph proteins involved in the cold tolerance of terrestrial arthropods: Antifreeze and ice nucleator proteins. Pages 282-300 in: Water and Life: Comparative Analysis of Water Relationships at the Organismic, Cellular and Molecular Level. G. N. Somero, C. B. Osmond, and C. L. Bolis, eds. Springer-Verlag, Berlin.

Duman, J. G., Wu, D. W., Olsen, T. M., Urrutia, M., and Tursman, D. 1993. Thermal-hysteresis proteins. Adv. Low-Temp. Biol. 2:131-182.

Farrant, J. 1980. General observations on cell preservation. Pages 1-8 in: Low Temperature Preservation in Biology and Medicine. M. J. Ashwood-Smith and J. Farrant, eds. University Park Press, Baltimore.

Fields, P. G., and McNeil, J. N. 1986 Possible dual cold-hardiness strategies in *Cisseps fulvicollis* (Lepidoptera: Arctiidae). Can. Entomol. 118:1309-1311.

Franks, F. 1981. The nucleation of ice in undercooled aqueous solutions. Cryo Lett. 2:27-31.

Gehrken, U. 1992. Inoculative freezing and thermal hysteresis in the adult beetles *Ips acuminatus* and *Rhaquim inquisition*. J. Insect Physiol. 38:519-524.

Gehrken, U., Strømme, A., Lundheim, R., and Zachariassen, K. E. 1991. Inoculative freezing in overwintering Tenebrionid beetle, *Bolitophagus reticulatus* Panz. J. Insect Physiol. 37:683-687.

Green, R. L., and Warren, G. J. 1985. Physical and functional repetition in a bacterial ice nucleation gene. Nature 317:645-648.

Holmstrup, M. 1992. Cold hardiness strategy in cocoons of the Lumbricid earthworm *Dendrobaena octaedra*. Comp. Biochem. Physiol. 102A:49-54.

Katagiri, C. 1985. Structure of lipophorin in insect blood: Location of phospholipid. Biochim. Biophys. Acta 834:139-143.

Knight, J. D., and Bale, J. S. 1986. Cold hardiness and overwintering of the grain aphid *Sitobion avenae*. Ecol. Entomol. 11:189-197.

Kozloff, L. M., Lute, M., and Westaway, D. 1984. Phosphatidylinositol as a component of the ice nucleating site of *Pseudomonas syringae* and *Erwinia herbicola*. Science 226:845-846.

Krunic, M., and Radovic, L. 1974. The effect of gut content evacuation on the increase in cold-hardiness in *Megachile rotundata*. Entomol. Exp. Appl. 17:526-528.

Kukal, O., Serianni, A. S., and Duman, J. G. 1988. Glycerol production in a freeze tolerant arctic insect, *Gynaephora groenlandica*: An in vivo $^{13}$C NMR study. J. Comp. Physiol. 158:175-183.

Lee, R. E., and Denlinger, D. L., eds. 1991. Insects at Low Temperature. Chapman and Hall, New York.

Lee, R. E., Zachariassen, K. E., and Baust, J. G. 1981. Effect of cryoprotectants on the activity of hemolymph nucleating agents in physical solutions. Cryobiology 18:511-514.

Lee, R. E., Strong-Gunderson, J. M., Lee, M. R., Grove, K. S., and Riga, T. J. 1991. Isolation of ice nucleating active bacteria from insects. J. Exp. Zool. 257:124-127.

Lee, R. E., Mugnano, J. A., and Taylor, R. T. 1992. Endogenous crystalloid spheres regulate the supercooling point of the gall fly, *Eurosta solidaginis*. Cryobiology 29:750-751.

Lee, R. E., Lee, M. R., and Strong-Gunderson, J. M. 1993a. Insect cold-hardiness and ice nucleating active microorganisms including their potential use for biological control. J. Insect Physiol. 39:1-12.

Lee, R. E., McGrath, J. J., Morason, R. T., and Taddeo, R. M. 1993b. Survival of intracellular freezing, lipid coalescence and osmotic fragility in fat body cells of the freeze-tolerant gall fly *Eurosta solidaginis*. J. Insect Physiol. 39:445-450.

Loomis, S. H. 1991. Comparative invertebrate cold hardiness. Pages 301-317 in: Insects at Low Temperature. R. E. Lee and D. L. Denlinger, eds. Chapman and Hall, New York.

Lusena, C. V. 1955. Ice propagation in systems of biological interest. III. Effects of solutes on nucleation and growth of ice crystals. Arch. Biochem. Biophys. 57:277-284.

MacKenzie, A. P. 1977. Non-equilibrium freezing behaviour of aqueous systems. Philos. Trans. R. Soc. Lond. B 278:167-189.

Mazur, P. 1977. The role of intracellular freezing in the death of cells cooled at supraoptimal rates. Cryobiology 14:251-274.

Mazur, P. 1984. Freezing of living cells: Mechanisms and implications. Am. J. Physiol. 247:C125-C142.

Miller, L. K. 1982. Cold hardiness strategies of some adult and immature insects overwintering in interior Alaska. Comp. Biochem. Physiol. 73A:595-604.

Miller, L. K., and Werner, R. 1987. Extreme supercooling as an overwintering strategy in three species of willow gall insects from interior Alaska. Oikos 49:253-260.

Mueller, G. M., Wolber, P. K., and Warren, G. J. 1990. Clustering of ice nucleation protein correlates with ice nucleation activity. Cryobiology 27:416-422.

Neven, L. G., Duman, J. G., Beals, J. M., and Castellino, F. J. 1986. Overwintering adaptations of the stag beetle, *Ceruchus piceus*: Removal of ice nucleators in winter to promote supercooling. J. Comp. Physiol. 156:707-716.

Neven, L. G., Duman, J. G., Low, M. G., Sehl, L. C., and Castellino, F. J. 1989. Purification and characterization of an insect hemolymph lipoprotein ice nucleator: Evidence for the importance of phosphatidylinositol and apolipoprotein in the ice nucleator activity. J. Comp. Physiol. 159:71-82.

O'Doherty, R., and Bale, J. S. 1985. Factors affecting cold hardiness of the peach-potato aphid *Myzus persicae*. Ann. Appl. Biol. 106:219-228.

O'Doherty R., and Ring, J. S. 1987. Supercooling ability of aphid populations from British Columbia and the Canadian Arctic. Can. J. Zool. 65:763-765.

Olsen, T. M., and Duman, J. G. 1992. Factors affecting supercooling in overwintering larvae of the Pyrochroid beetle, *Dendroides canadensis*. Cryobiology 29:750.

Parody-Morreale, A., Murphy, K. P., DiCerca, E., Fall, R., DeVries, A. L., and Gill, S. J. 1988. Inhibition of bacterial ice nucleators by fish antifreeze glycoproteins. Nature 333:782-783.

Patterson, J. L., and Duman, J. G. 1978. The role of thermal hysteresis producing proteins in the low temperature tolerance and water balance of larvae of the mealworm, *Tenebrio molitor*. J. Exp. Biol. 74:37-45.

Ramsay, R. A. 1964. The rectal complex of the mealworm, *Tenebrio molitor* L. (Coleoptera, Tenebrionidae). Philos. Trans. R. Soc. Lond. B 248:279-314.

Reid, D. S., Folin, A. T., and Lem, C. A. 1985. The effect of solutes on the temperature of heterogeneous nucleation of ice from aqueous solution. Cryo Lett. 6:189-198.

Ring, R. A. 1980. Insects and their cells. Pages 187-217 in: Low Temperature Biology in Medicine and Biology. M. J. Ashwood-Smith and J. Farrant, eds. University Park Press, Baltimore.

Ring, R. A. 1982. Freezing tolerant insects with low supercooling points. Comp. Biochem. Physiol. 73A:605-612.

Rojas, R. R., Charlet, L. D., and Leopold, R. A. 1992. A differential scanning calorimetric analysis of inoculative freezing in an insect. Cryo Lett. 13:355-362.

Salt, R. W. 1953. The influence of food on cold-hardiness in insects. Can. Entomol. 85:261-269.

Salt, R. W. 1959. Role of glycerol in the cold hardening of *Bracon cephi* (Gahan). Can. J. Zool. 37:59-69.

Scholander, P. F., Flagg, W., Hock, R. J., and Irving, L. 1953. Studies on the physiology of frozen plants and animals in the Arctic. J. Cell. Comp. Physiol. 42:1-56.

Shimada, K., and Rhiihimaa, A. 1988. Cold acclimation, inoculative freezing and slow cooling: Essential factors contributing to the freeze tolerance in diapause larvae of *Chymomyza costata* (Diptera: Drosophilidae). Cryo Lett. 9:5-10.

Sömme, L. 1967. The effect of temperature and anoxia on the haemolymph composition and supercooling in three overwintering insects. J. Insect Physiol. 13:805-814.

Sömme, L. 1978. Nucleating agents in the haemolymph of the third instar larvae of *Eurosta solidaginis* (Fitch) (Diptera:Tephritidae). Norw. J. Entomol. 25:187-188.

Sömme, L. 1982. Supercooling and winter survival in terrestrial arthropods. Comp. Biochem. Physiol. 73A:519-543.

Sömme, L., and Block, W. 1982. Cold-hardiness of Collembola at Signy Island, Maritime Antarctic. Oikos 38:168-176.

Southworth. M. W., Wolber, P. K., and Warren, G. J. 1988. Nonlinear relationship between concentration and activity of a bacterial ice nucleation protein. J. Biol. Chem. 263:15211-15216.

Storey, K. B., and Storey, J. M. 1988. Freeze tolerance in animals. Physiol. Rev. 68:27-84.

Strong-Gunderson, J. M., Lee, R. E., Lee, M. R., and Riga, T. J. 1990. Ingestion of ice-nucleating active bacteria increases the supercooling point of the lady beetle *Hippodamia convergens*. J. Insect Physiol. 36:153-157.

Tanno, K. 1977. Ecological observation and frost-resistance in overwintering pre-pupae, *Sciara* sp. (Sciaridae). Low Temp. Sci., Ser. B 35:63-74.

Theede, H. R., Schneppenheim, R., and Bevess, L. 1976. Frostschutz-glycoproteine bei *Mytilus edulis*. Mar. Biol. 36:183-189.

Turner, M. A., Arellano, F., and Kozloff, L. M. 1990. Three separate classes of bacterial ice nucleation structures. J. Bacteriol. 172:2521-2526.

Tursman, D., Duman, J. G., and Knight, C. A. 1994. Freeze tolerance adaptations in the centipede *Lithobius forficatus*. J. Exp. Zool. 268:347-353.

Wallace, P. P., and Beard, R. L. 1943. The effect of low temperature upon mortality of larvae of *Scolytus multistriatus*. Marsham. Bull. Conn. Agric. Exp. Stn. 472:291-304.

Warner, D. T. 1962. Some possible relationships of carbohydrates and other biological components with the water structure at 37°C. Nature 196:1055-1058.

Westh, P., Kristiansen, J., and Hvidt, A. 1991. Ice nucleating activity in the freeze tolerant tardigrade *Adorybiotus coronifer*. Comp. Biochem. Physiol. 99A:401-404.

Wolber, P. K., and Warren, G. J. 1989. Bacterial ice nucleation proteins. Trends Biochem. Sci. 14:179-182.

Wu, D. W., and Duman, J. G. 1991. Activation of antifreeze proteins from the beetle *Dendroides ca-*

*nadensis*. J. Comp. Physiol. 161:279-283.

Xu, L., Neven, L. G., and Duman, J. G. 1990. Hormonal control of hemolymph lipoprotein ice nucleators in overwintering freeze susceptible larvae of the stag beetle, *Ceruchus piceus*: Adipokinetic hormone and juvenile hormone. J. Comp. Physiol. 160:51-59.

Yeung, K. L., Wolf, E. E., and Duman, J. G. 1991. A scanning tunneling microscopy study of an insect lipoprotein ice nucleator. J. Vac. Sci. Technol. B 9:1197-1201.

Zachariassen, Z. E. 1980. The roles of polyols and nucleating agents in cold-hardy beetles. J. Comp. Physiol. 140A:227-234.

Zachariassen, K. E. 1982. Nucleating agents in cold-hardy insects. Comp. Biochem. Physiol. 73A:557-562.

Zachariassen, K. E. 1985. Physiology of cold tolerance in insects. Physiol. Rev. 65:799-832.

Zachariassen, K. E. 1992. Ice nucleating agents in cold-hardy insects. Pages 262-281 in: Water and Life: Comparative Analysis of Water Relationships at the Organismic, Cellular and Molecular Level. G. N. Somero, C. B. Osmond, and C. L. Bolis, eds. Springer-Verlag, Berlin, Heidelberg.

Zachariassen, K. E., and Hammel, H. T. 1976. Nucleating agents in the haemolymph of insects tolerant to freezing. Nature 262:285-287.

Zachariassen, K. E., and Husby, J. A. 1982. Antifreeze effect of thermal hysteresis agents protects highly supercooled insects. Nature 298:865-867.

Zachariassen, K. E., Baust, J. B., and Lee, R. E. 1982. A method for quantitative determination of ice nucleating agents in insect hemolymph. Cryobiology 19:180-184.

CHAPTER 12

# Supercooling and Ice Nucleation in Vertebrate Ectotherms

**Jon P. Costanzo and Richard E. Lee, Jr.**

The tissues of nearly all vertebrates are irreparably damaged when ice forms and propagates within them. However, ice nucleation is a problem of ecological relevance primarily in ectotherms which, unlike endotherms, lack calorigenic resistance to body cooling. Our discussion addresses the processes and consequences of ice nucleation in ectothermic species, particularly those occupying terrestrial habitats.

Freezing risk is largely governed by climatological factors and generally increases with altitude and latitude. Consequently, relatively few vertebrate ectotherms occur in regions where frost long persists and deeply penetrates the soil, although the exceptions, which are represented by most major orders of the Amphibia and Reptilia, are well adapted to such environments. Freezing risk may vary substantially among populations of species having broad geographic ranges. For example, painted turtles (*Chrysemys picta*) in eastern North America hibernate in permanent ponds and sloughs, where they usually are protected from freezing, even during severe winters (Crawford, 1991). In contrast, conspecifics inhabiting the southwestern plains overwinter in shallow, ephemeral ponds that are susceptible to freezing (Christiansen and Bickham, 1989).

Freezing risk is also influenced by local topographical relief and prevailing microenvironmental conditions. The amount of snow cover is a particularly important influence on the thermal regimen within terrestrial hibernacula (e.g., Breitenbach et al., 1984). The physical medium is also critical because thermal conductance and inertia differ considerably among air, water, and soil. Organisms hibernating in deep or lotic waters probably are protected from freezing, whereas those occupying terrestrial refuges, particularly at shallow soil depths, are more vulnerable to frost (Ultsch, 1989).

Behavioral avoidance of subfreezing body temperatures is a primary defense against frost exposure. Terrestrial species minimize freezing risk during transient chill episodes through shelter-seeking, burrowing, huddling, and other protective behaviors (Spellerberg, 1972; Gregory, 1982; Lillywhite, 1987; Ultsch, 1989; Pinder et al., 1992). Protection against sustained, seasonal cold is effected either by migration to permanent aquatic sites or caves, or burrowing or tunneling to depths below the frost. Suitable hibernacula become progressively scarce with increasing altitude and latitude, a trend that restricts the geographic distribution of some species (Gregory, 1982; Ultsch, 1989; Pinder et al., 1992). Complex behavioral adaptations, such as the homing mechanism involved in annual migration to communal hibernacula (e.g., Spellerberg, 1976; Gregory, 1982; Lillywhite, 1987), permit some species to thrive in such climates.

Despite these protective mechanisms, vertebrate ectotherms at least occasionally encounter subfreezing body temperatures in nature. Field studies show that extraordinary winter conditions, such as severe cold in the absence of snow cover, may cause freezing damage or death in isolated individuals, as well as entire aggregations (Gregory, 1982; Lillywhite, 1987). Such events probably are most deleterious for terrestrial hibernators, although they also may affect species overwintering in more thermally buffered aquatic sites (Christiansen and Bickham, 1989). In some locales, freezing risk is not restricted to winter dormancy, but exists during the normal activity season, particularly as a consequence of rapidly declining environmental temperatures. Instances of extensive freezing mortality have been attributed to unseasonable frosts occurring in autumn and spring (Spellerberg, 1976; Gregory, 1982).

## Cold-Sensitive Versus Cold-Hardy Species

The outcome of a given chilling episode hinges on both the severity of the exposure (e.g., cooling rate, minimum temperature achieved, duration of exposure) and the physiological cold-hardiness of the individual. "Cold-sensitive" species do not survive prolonged exposure to subfreezing body temperatures either because they simply cannot tolerate such low temperatures (e.g., Lotshaw, 1977), or because they cannot supercool extensively under field conditions and thus succumb when ice propagates within their tissues (Weigmann, 1929; Kalabukhov, 1958; Lowe et al., 1971; Spellerberg, 1972). On the contrary, some amphibians and reptiles are quite "cold-hardy" and readily endure even lengthy, multiple exposures to subfreezing body temperatures. For these species, survival is attributable to a capacity for supercooling or a tolerance of ice forming within body tissues.

Surprisingly little is known about the hibernaculum microenvironment of cold-hardy vertebrates. Schmid (1986) recorded temperatures within the terrestrial microenvironment in which the freeze-tolerant tree frog (*Hyla versicolor*) overwinters. His data suggest that frogs would be exposed to temperatures as low as −7°C during a series of eight discrete freezing episodes, each lasting from perhaps several days to nearly 12 weeks. On the other hand, MacArthur and Dandy (1982) suggested that the freeze-tolerant chorus frog (*Pseudacris triseriata*) hibernates at the soil-snow interface where, even in Canada, minimum winter temperatures remain above −5°C. Claussen et al. (1991) periodically measured body temperatures of adult eastern box turtles (*Terrapene carolina*) in terrestrial hibernation burrows, reporting the occurrence of at least two instances of subzero body temperatures during each of the three winters of record. These turtles, which demonstrably tolerate tissue freezing to temperatures ≤−4°C (Costanzo and Claussen, 1990), survived these exposures to temperatures ranging from −0.3 to −1.4°C. Subzero temperatures, some as low as −11°C (Woolverton, 1963), have been recorded inside natal nests in which hatchling painted turtles successfully overwintered. Our field study of this species in western Nebraska indicated that during the 1990–1991 winter of record, hatchlings survived 14 discrete freeze-thaw episodes, which lasted from 6 to 320 h and exposed turtles to minimum temperatures ranging from −1 to −12°C (J. Costanzo, J. Iverson, M. Wright, and R. Lee, *unpublished*).

The evolution of physiological cold-hardiness, particularly freeze tolerance, in certain temperate species has not only promoted their overwintering survival, but has also provided a mechanism for extending seasonal activity into the cooler months. For example, the eastern box turtle may remain abroad until early November, long after the first autumnal frost, and resumes activity well before the last vernal frost (Claussen et al., 1991). Similarly, in Canada the eastern garter snake (*Thamnophis sirtalis*) commonly encounters subfreezing body temperatures while active in spring and fall (Vincent and Secoy, 1978). Its tolerance of the freezing of ≥35% of its body water for periods lasting ≥48 h likely is of primary importance in surviving outside the hibernaculum, rather than during winter per se, since this species typically hibernates in frost-protected sites (Costanzo, 1988; Costanzo et al., 1988).

## Early Studies of Supercooling in Vertebrates

The investigation of supercooling phenomena in living organisms was initiated by primarily German and Russian scientists early in the 20th century (see reviews by Weigmann, 1929; Kalabukhov, 1958). These workers achieved limited success in laboratory efforts to supercool various amphibians, reptiles, and mammals (e.g., bats, ground squirrels). Such initiatives collectively were aimed at developing protocols for preserving (particularly mammalian) specimens in the supercooled state but had little relevance to the natural history of cold-hardy animals.

Subsequently, some workers, extrapolating from the established role of supercooling in the overwintering biology of certain invertebrates (see Chapter 11), speculated that supercooling might also promote cold-hardiness in vertebrate ectotherms. Laboratory results initially supported this contention; both Lowe et al. (1971) and Spellerberg (1972) reported that many specimens remained unfrozen at temperatures as low as −8°C. Indeed, Lowe et al. (1971), who gathered supercooling data for 45 species of reptiles representing 12 families in all four major orders, categorically stated "...it is no longer surprising that reptiles voluntarily and successfully occupy shallow subsurface soil and rock crevice retreats where they may naturally be subjected to low temperatures that commonly fall well below freezing."

Unfortunately the results of these seminal studies have been misinterpreted to mean that an organism is perpetually protected from freezing so long as its body temperature remains even slightly above the so-called "supercooling point" reported for the taxon. This assertion, implicit in the statement by Lowe et al. (1971), "...the supercooling limit is measurable in a repeatable fashion and appears to represent a temperature that is characteristic for given species," has fostered untenable conclusions about the efficacy of supercooling under field conditions. The results of supercooling trials assessed under idealized laboratory conditions are not necessarily indicative of the precise role of supercooling in a species' cold-hardiness strategy. However, because supercooling phenomena are practically studied only in the laboratory, it is instructive to consider the pertinent investigative techniques.

## Laboratory Methods for Investigating Supercooling Phenomena

In common practice, the specimen is gently restrained, instrumented with a temperature probe, and subsequently cooled to a temperature below the anticipated ice crystallization temperature ($T_c$). Moist-skinned animals, such as amphibians, must be dried of excess surface moisture before testing. The temperature probe (typically a thermocouple or thermistor) is used to measure $T_c$ by detecting the heat of fusion liberated during crystallization. This "exotherm" may continue for hours or days, so long as a substantial amount of ice continues to form. The principal biophysical factors influencing the dynamics of body heat, water, and solutes in vertebrates during freezing were discussed by Claussen and Costanzo (1990).

**Table 1.** Crystallization temperatures ($T_c$) of various freeze-tolerant vertebrates (means are shown ±1 SE)

| Species | *n* | $T_c$ (°C) | Experimental cooling rate (°C/h) | Body mass (g) | Age class |
|---|---|---|---|---|---|
| Amphibians | | | | | |
| *Rana sylvatica* (wood frog) | | | | | |
| Minnesota[a] | 3 | −1.9 ± 0.1 | 0.6 | ... | ... |
| Ohio[b] | 43 | −2.2 ± 0.1 | ≈3 | 14.4 ± 2.1 | adult |
| Ontario[c] | 3 | −2.0 ± 0.1 | ... | ... | adult |
| Ontario[d] | 9 | −3.0 ± 0.04 | ≈20 | 1.6 ± 0.3 | juvenile |
| *Pseudacris triseriata* (chorus frog) | | | | | |
| Manitoba[e] | 7 | −3.0 | 4.0 | 1.1 | ... |
| Ontario[c] | 2 | −2.1 | ... | ... | adult |
| *Pseudacris crucifer* (spring peeper) | | | | | |
| Minnesota[a] | 4 | −2.0 ± 0.1 | 0.6 | ... | ... |
| Ontario[c] | 2 | −1.9 | ... | ... | adult |
| *Hyla versicolor* (gray treefrog)[a] | 6 | −2.2 ± 0.1 | 0.6 | ... | ... |
| *Hyla chrysoscelis* (Cope's treefrog)[f] | 3 | −2.0 ± 0.2 | ... | ... | ... |
| Reptiles | | | | | |
| *Thamnophis sirtalis* (garter snake)[g] | 11 | −1.4 ± 0.2 | 3.7 | 52.1 ± 1.7 | adult |
| | 3 | −1.7 ± 0.6 | 7.2 | 14.4 ± 3.0 | juvenile |
| *Chrysemys picta* (painted turtle) | | | | | |
| Ohio[h] | 13 | −0.4 ± 0.02 | 1.8 | 394 ± 16 | adult |
| Ohio[i] | 8 | −1.2 | 1.8 | 328 | adult |
| Ohio[i] | 6 | −1.1 | 21.9 | 5.3 | hatchling |
| Ontario[j] | 4 | −3.3 ± 0.2 | 21.9 | 3.8 ± 0.2 | hatchling |
| Ontario[k] | 8 | −2.0 ± 0.5 | ... | 4.8 ± 0.1 | hatchling |
| Manitoba[k] | 6 | −3.7 ± 0.3 | ... | 4.3 ± 0.1 | hatchling |
| New Jersey[l] | 9 | −4.8 ± 0.7 | ≈28 | 3.8 ± 0.04 | hatchling |
| *Terrapene carolina* (box turtle) | | | | | |
| Ohio[m] | 19 | −0.8 ± 0.2 | 1.8 | 341 ± 23 | adult |
| Ohio[n] | 3 | −1.6 ± 0.1 | ... | ... | adult |

[a] Data from Schmid (1982).
[b] Data from Layne and Lee (1987).
[c] Data from Storey and Storey (1987).
[d] Data from Storey and Storey (1985).
[e] Data from MacArthur and Dandy (1982).
[f] Data from Schmid (1986).
[g] Data from Costanzo et al. (1988).
[h] Data from Johnson (1990).
[i] Data from Claussen and Zani (1991).
[j] Data from Storey et al. (1988).
[k] Data from Churchill and Storey (1992a).
[l] Data from DePari (1988).
[m] Data from Costanzo and Claussen (1990).
[n] Data from Storey et al. (1993).

It is critical that temperature data so measured accurately represent the overall thermal status of the animal. In practice, the probe sensor is placed in the cloaca (Lowe et al., 1971; Spellerberg, 1972; Lotshaw, 1977; Costanzo and Claussen, 1990; Claussen and Zani, 1991), esophagus (Weigmann, 1929; MacArthur and Dandy, 1982), or under the skin (Spellerberg, 1972; DePari, 1988). Schmid (1982) cautioned against using invasive probes because they may induce nucleation, although this problem has not been reported. An alternative practice, particularly convenient with smaller specimens, is to position the probe on the body surface. This configuration is somewhat less satisfactory because the probe senses both ambient and body temperature, although this problem may be ameliorated simply by adding overlying insulation (e.g., Costanzo and Claussen, 1990; Costanzo et al., 1991).

Supercooling protocols involving vertebrates often are problematic because the relatively large size of these animals promotes the development of thermal gradients within the body during cooling. Adult turtles, for example, may have core-to-surface temperature differentials of several degrees or more, even when they are cooled at moderate rates (Costanzo and Claussen, 1990; Claussen and Zani, 1991). Accordingly, surface-mounted and invasive temperature probes may register very different temperatures, which, depending on the actual anatomical site of nucleation, may or may not accurately represent $T_c$. It is critical to the interpretation of supercooling data that in reporting cooling protocols authors distinguish among core body temperature, body surface temperature, and ambient temperature.

Large body mass may also confound the results of supercooling studies if the experimental cooling rate is not carefully controlled. Typically, moderately high rates are used to expedite the collection of data (e.g., Table 1). However, because nucleation is a time-dependent phenomenon, investigators should consider that cooling rate may markedly influence the outcome of supercooling experiments. For example, whereas rapid cooling commonly depresses $T_c$, some experimental results show that exceptionally high cooling rates may actually stimulate nucleation, perhaps by promoting the development of sharp thermal gradients within the body (Kalabukhov, 1958). Lotshaw (1977) reported that cooling at −30°C/h virtually eliminated supercooling in the wood frog. Some authors (e.g., Packard and Packard, 1993a) have justifiably maintained that cooling protocols should be based on ecologically relevant rates, although such parameters have not been determined for many species.

## Freeze Avoidance as a Cold-Hardiness Strategy

Supercooling may be an important element of the cold-hardiness strategies of some terrestrial ectotherms. Packard and Packard (1993a,b) argued that supercooling, rather than freeze tolerance, is critical to the winter survival of hatchling painted turtles. This contention is based on marked differences in the lower limit of thermal tolerance of supercooled (approximately −12°C) and frozen (approximately −4°C) turtles, of which only the former concurs with the lowest temperatures survived in nature. Supercooling apparently promotes survival of the mountain spiny lizard (*Sceloporus jarrovi*), a small, winter-active species that aggregates in shallow rock crevices at high elevations in Arizona and Mexico. In its native habitat, environmental temperatures during winter may fall to −2°C, but these extremes are well within the supercooling capacity reported for this species (Lowe et al., 1971; Halpern, 1979).

One fundamental benefit of freeze avoidance as a cold-hardiness strategy is that supercooling, unlike freeze tolerance, invokes relatively little physiological stress. Aside from the occasional death (e.g., Weigmann, 1929; Claussen et al., 1990), supercooling to moderate subzero temperatures is compatible with life processes (e.g., Halpern and Lowe, 1968) and thus is readily tolerated by many temperate species. Spellerberg (1972) supercooled 11 species of lizards and snakes, reporting that all fully recovered if they were warmed before nucleation occurred; however, crystallization invariably caused permanent damage, the severity of which depended on the duration of the freezing episode. Indeed, supercooled turtles can recover from many days of chilling at temperatures of –10°C and lower with no ill effects (Packard and Packard, 1993a). Another benefit is that reanimation and physiological recovery from supercooling are relatively rapid. Lowe et al. (1971) reported a "spontaneous revival" of various reptiles, including several semitropical species, supercooled to body temperatures as low as –8°C. In contrast, complete recovery from freezing may require days or weeks (Storey, 1989; Costanzo et al., 1993). Perhaps the most significant advantage of supercooling over freeze tolerance as a cold-hardiness strategy is that it offers a potentially greater range of thermal protection. Such protection is not guaranteed, however, owing to the metastability of supercooled fluid. Additionally, extensive supercooling is not an option for many species owing to inherent and extrinsic constraints, such as the presence of endogenous nucleators and seeding of freezing by environmental ice crystals.

## Biophysical Influences on Supercooling Capacity

Freeze avoidance is an important strategy permitting some organisms to survive otherwise lethal exposure to low temperature. Unlike the case with insects, some of which can remain unfrozen at temperatures of –60°C and below (see Chapters 11 and 14), vertebrate ectotherms lack well-developed supercooling capacities (e.g., Table 1) and generally crystallize at body temperatures only several degrees below their equilibrium freezing point ($FP_{eq}$), the highest temperature at which freezing is possible. With the exception of certain cold-water fishes (which produce specific antifreezes or cryoprotectants), these species apparently do not regulate $T_c$ significantly through physiological mechanisms. Nevertheless, the efficacy of supercooling as a survival mechanism is strongly influenced by several biophysical factors.

### Equilibrium Freezing Point

A solution's $FP_{eq}$ is reduced by –1.86°C for every 1,000 mosM increase in osmolality. According to nucleation theory, a reduction of a solution's $FP_{eq}$ will effectively depress $T_c$ (Vali, 1991; see Vali, Chapter 1). This principle is exploited by a variety of invertebrates that enhance supercooling capacity by accumulating osmolytes in high concentrations (see Chapter 11). In contrast, there is little evidence that cold conditioning depresses $FP_{eq}$ in amphibians and reptiles. Halpern (1979) reported a slight, albeit significant, reduction in $FP_{eq}$ from summer to winter lizards, although this shift probably reflected changes in water balance rather than the production of specific osmolytes and was not accompanied by a corresponding decrease in $T_c$. Cold-hardiness strategies of vertebrates apparently do not involve a depression of $FP_{eq}$ via the accumulation of solutes within tissues. DePari (1988) investigated hatchling painted turtles, reporting that overwintering specimens accumulated glucose, which in some tissues attained levels fourfold higher than normal, but found no correlation between animal $T_c$ and tissue levels of this osmolyte. On the contrary, many species, particularly those hibernating in moist microenvironments, become edamatous and thus have a relatively higher $FP_{eq}$ during winter.

Colligative regulation of $FP_{eq}$ is probably not an option for most vertebrates, since (unlike some insects which attain solute concentrations of up to 5 molal) they generally do not tolerate the requisite osmotic adjustment. For example, the equation describing the relationship between $T_c$ and $FP_{eq}$, empirically determined for intact reptiles (Lowe et al., 1971),

$$T_c\,(°C) = -4.7 + 1.9\,FP_{eq}\,(°C)$$

predicts that the $T_c$ of a typical reptile (blood osmolality = 310 mosM; $FP_{eq} = -0.58°C$) is −5.8°C. A depression of $T_c$ by only one degree would require a doubling of blood osmolality, a result that likely would be harmful. This rationale contradicts the conclusion of MacArthur and Dandy (1982) that the chorus frog adaptively enhances its supercooling capacity during winter by seasonally increasing plasma osmolality. Indeed, the observed increment, approximately 50 mosM, would depress blood $FP_{eq}$ only <0.1°C and thus by itself would have very little impact on supercooling capacity.

## Thermal Acclimation and Acclimatization

Thermal history and acclimatization strongly influence many physiological aspects of amphibian (e.g., Ultsch, 1989; Pinder et al., 1992) and reptilian (e.g., Spellerberg, 1972; Vincent and Secoy, 1978; Lillywhite, 1987; Gregory, 1982) cold-hardiness, prompting some workers to question whether these factors also regulate supercooling capacity. It seems reasonable that supercooling capacity would be enhanced by cold acclimation, especially during the cooler seasons. However, the limited available data do not support this hypothesis. Conversely, some data suggest that supercooling capacity may actually *decrease* during cold acclimation. For example, Lotshaw (1977) noted reduced supercooling in 5°C-acclimated, relative to 25°C-acclimated, wood frogs. In the chorus frog, such response may reflect a larger fluid volume, which in turn results from diminished kidney function at low temperature (MacArthur and Dandy, 1982). Halpern's (1979) extensive study of the mountain spiny lizard, in which specimens were acclimated to various temperatures between 7 and 30°C, clearly indicated that supercooling is enhanced by chronic exposure to warm, rather than cold, environments.

The available data also suggest that seasonal conditioning has little influence on supercooling phenomena. For example, $T_c$ is the same in autumn and spring for several species of freeze-tolerant frogs (Storey, 1985; Storey and Storey, 1987). Similarly, supercooling capacity in the garter snake is comparable in all seasons (Table 2). Moreover, no differences in $T_c$ occurred between winter and summer samples of mountain spiny lizards, even though this species depends on supercooling for winter survival (Halpern, 1979).

Although most studies to date indicate that acclimation temperature and season have little effect on supercooling capacity, some contradictory results have been reported. The mean $T_c$ for hatchling painted turtles recently excavated from subterranean hibernacula was −1.1°C ($n = 8$), but this value decreased by 3.7°C in an additional sample ($n = 4$) held 3 days at 5°C in the laboratory (Churchill and Sto-

**Table 2.** Crystallization temperatures ($T_c$) of the freeze-tolerant garter snake, *Thamnophis sirtalis,* in different seasons[a]

| | Spring | Summer | Autumn | Winter | P |
|---|---|---|---|---|---|
| $T_c$ (°C) | | | | | |
| Mean ± SE | −1.7 ± 0.2 | −2.1 ± 0.3 | −2.2 ± 0.3 | −1.4 ± 0.2 | 0.180 |
| Range | −0.7 to −2.8 | −0.7 to −3.7 | −0.7 to −4.0 | −0.7 to −2.9 | . . . |
| Body mass (g) | 32.3 ± 3.4 | 45.7 ± 5.0 | 43.5 ± 3.9 | 40.6 ± 8.0 | 0.287 |

[a]Specimens ($n$ = 11–14) were acclimated to 5°C for ≈10 days prior to testing. Means were statistically compared using analyses of variance. Data extracted from Costanzo (1988).

rey, 1992a). These authors (1992b) also studied garter snakes, reporting a substantial increase in $T_c$, from −5.5°C in a freshly collected autumn sample to −0.9°C during the several weeks specimens were kept in the laboratory under simulated winter conditions. Unfortunately, it cannot be discerned from the report whether the apparent loss of supercooling capacity was an actual effect of acclimation or merely an artifact of the large sample variability in body mass.

## Ecological Correlates

In their seminal paper on supercooling in vertebrates, Lowe et al. (1971) suggested that $T_c$ represents a temperature that is characteristic for a given species. This statement is consistent with a popular view that supercooling capacity is a phenotypic attribute that has been enhanced through natural selection for its current role in promoting cold hardiness. The available data do not support this contention. Lotshaw (1971), for example, found little difference in $T_c$ between the cold-adapted wood frog and the warm-climate bullfrog (*Rana catesbeiana*), and species with diverse cold-hardiness strategies (e.g., freeze-tolerant vs. freeze-intolerant) reportedly have comparable supercooling capacities (Schmid, 1982; Storey and Storey, 1986). Additionally, Halpern (1979) compared the supercooling capacities of mountain spiny lizards in Arizona, finding no differences between lizards living at 2,377 m in the Chiricahua Mountains (which rely on supercooling for winter survival) and 1,829 m in the relatively temperate Pinaleño Mountains. Perhaps most conclusive are the data presented by Spellerberg (1972), which suggest no ecological correlates for supercooling capacity among 11 species of southeast Australian reptiles representing several zoogeographic subregions and thermal zones.

## Body Size

On the basis of the proportional relationship between nucleator quantity and fluid volume (see Chapter 1), the probability of heterogeneous nucleation in animals must increase with body mass. This effect has been clearly demonstrated among various invertebrates but is often obscured in studies of vertebrates, perhaps owing to small sample sizes (e.g., Claussen and Zani, 1991) or insufficient variation in body mass (e.g., Costanzo, 1988). Nevertheless, the expected trend of decreasing range and average $T_c$ with increasing body mass is revealed in data pooled for 17 species of reptiles representing all major orders (Fig. 1). Notably, reptiles weighing approximately 40 g or more have little, if any, supercooling capacity (see also Table 1).

Interestingly, reptiles inhabiting cold climatic regions are generally smaller than reptiles inhabiting warmer ones (Spellerberg, 1976), a trend that may enhance the efficacy of supercooling. Additionally, the relationship between $T_c$ and body mass

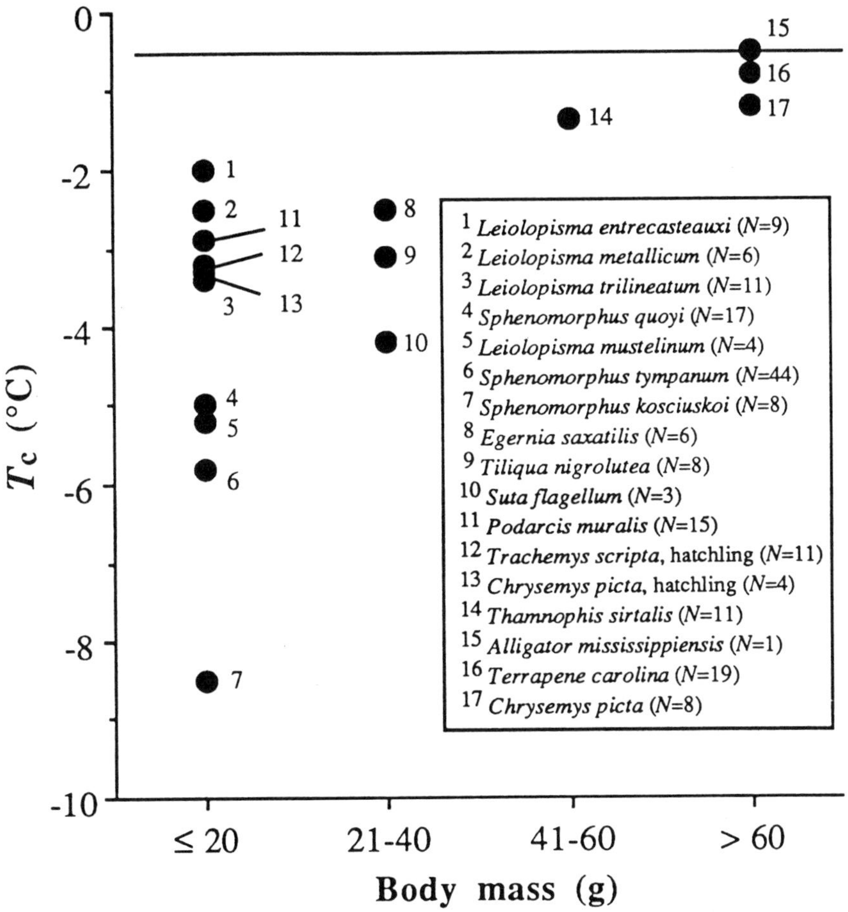


**Figure 1.** Relationship between body mass and temperature of ice crystallization ($T_c$) for 17 species of reptiles representing all major orders. Data points represent sample means collectively based on 181 individuals weighing from 1 to 600 g. The horizontal line at the top of the graph represents the approximate $FP_{eq}$. 1–10, Data are from Spellerberg (1972); 11, Claussen et al. (1990); 12, Churchill and Storey (1992); 13, Storey et al. (1988); 14, Costanzo et al. (1988); 15, Lowe et al. (1971); 16, Costanzo and Claussen (1990); and 17, Claussen and Zani (1991).

indicates a marked potential for ontogenetic differences in cold-hardiness strategy. For example, whereas turtle hatchlings weighing approximately 4 g can supercool for long periods at low temperatures (e.g., Packard and Packard, 1993a), the adults, which are nearly 10-fold heavier, usually crystallize at body temperatures at, or only slightly below, $FP_{eq}$ (Johnson, 1990; Claussen and Zani, 1991). The latter hibernate underwater where temperatures are more moderate than those encountered by the terrestrially hibernating hatchlings.

## Hydration State

Experimental desiccation demonstrably enhances supercooling in some vertebrates (e.g., Kalabukhov, 1958), perhaps by depressing organismal $T_c$ through a reduction of both $FP_{eq}$ (owing to the concentration of solutes) and body fluid vol-

**Table 3.** Crystallization temperatures ($T_c$) and supercooling durations of hatchling painted turtles (*Chrysemys picta*) experimentally desiccated by acclimation to low water potential at 4°C (mean values are reported ±1 SE)[a]

| $n$ | Body water content (%) | $T_c$ (°C) | Supercooling duration (h) |
|---|---|---|---|
| 6 | 77.2 ± 0.6 | −2.8 ± 0.1 | 4.3 ± 0.1 |
| 6 | 74.9 ± 0.2 | −3.4 ± 0.1 | 5.6 ± 0.1 |
| 9[b] | 70.6 ± 0.7 | ... | >32.0 ± 0.2 |

[a] Data from J. Costanzo, J. Iverson, M. Wright, and R. Lee (*unpublished*).

[b] Eight specimens, ultimately cooled to an equilibrium body temperature of −6.9 ± 0.4°C, remained supercooled for the duration of the trial.

ume. It thus seems plausible that adjustments in water balance may provide some control over supercooling capacity under natural conditions. Halpern (1979) found a high correlation between $T_c$ (range, −5.7 to −4.9°C) and body water content (range, 67 to 79% of total mass) of winter-collected lizards. Because even a slight experimental dehydration (i.e., 5% loss of body mass) significantly improved supercooling, she concluded that high variability in $T_c$ among field samples of reptiles may ultimately reflect natural variation in hydration state.

Hydration state appears especially important in regulating $T_c$ in hatchlings of the painted turtle, a species whose cold-hardiness strategy includes both supercooling (Packard and Packard, 1993a,b) and freeze tolerance (Storey et al., 1988; Churchill and Storey, 1992a) capacities. In our recent study, fully hydrated specimens remained supercooled only at high body temperatures and for short periods, whereas dehydrated turtles could supercool much more extensively (Table 3). Apparently desiccation, normally considered a potentially serious threat to successful overwintering, may under certain circumstances enhance cold hardiness. Future research should elucidate the relationship between hydration state (particularly in response to environmental water) and the efficacy of supercooling as a cold-hardiness strategy.

## Induced Nucleation

Contact of the skin by environmental ice abruptly terminates supercooling in some amphibians by seeding the freezing of body fluids. Laboratory tests (Layne et al., 1990; Layne, 1991) show that such "inoculative freezing" invariably commences within seconds after ice crystals contact the body. Thus amphibians, which typically select moist hibernacula to reduce desiccation (Ultsch, 1989; Pinder et al., 1992), probably do not supercool extensively under field conditions. Inoculative freezing also may occur in reptiles, including turtles (Packard and Packard, 1993a,b; Packard et al., 1993), snakes (Costanzo, 1988; Churchill and Storey, 1992b), and lizards (J. Costanzo, C. Grenot, and R. Lee, *unpublished*). Although their heavily keratinized skin poses a more formidable barrier to transcutaneous inoculation, body orifices (e.g., nostrils, cloaca, eyes) or wounds nevertheless may allow seeding by environmental ice (Kalabukhov, 1958; Lowe et al., 1971; Spellerberg, 1972). Reptilian skin apparently is not as effective a barrier as is fish skin (Valerio et al., 1992).

The significance of inoculative freezing in vertebrates under field conditions is an interesting subject requiring additional study. Halpern (1979) suggested that contact with crystallized environmental water (e.g., rainfall, melted snow) during periods of extreme cold may inhibit supercooling in the mountain spiny lizard. Inoculative freezing may also limit supercooling in hatchling painted turtles (Storey et al., 1988; Packard and Packard, 1993a,b), which hibernate in intimate contact with the soil matrix within natal nests (Breitenbach et al., 1984; DePari, 1988). Susceptibility to inoculative freezing apparently bears significantly on this species' cold-hardiness; hatchlings can tolerate temperatures of −13°C and below while supercooled (Packard and Packard, 1993a) but only as low as −4°C in the frozen state (Storey et al., 1988; Churchill and Storey, 1992a).

The environmental factors influencing the susceptibility of vertebrates to inoculative freezing are largely unknown. Studies of lizards and snakes (Spellerberg, 1972) and turtles (Packard and Packard, 1993a,b) clearly illustrate that the probability of nucleation increases with decreasing environmental temperature. Substrate moisture tension is another important factor. Hatchling painted turtles enveloped in a frozen clay soil may resist inoculative freezing for many hours (Packard and Packard, 1993a,b), whereas specimens similarly tested in a damp silica sand nucleated virtually immediately after ice formed around them (J. Costanzo, J. Iverson, M. Wright, and R. Lee, *unpublished*). This discrepancy may reflect large differences in the water potential between clay and sand substrates. Thus, in addition to ambient temperature, the susceptibility to inoculative freezing under field conditions may depend on factors such as substrate composition, texture, and water content.

In addition to contact with environmental ice crystals, supercooling capacity also is limited by the presence of potent nucleators within the gut. Kalabukhov (1958) noted that excess fluid in the stomach and intestines hindered laboratory efforts to supercool various vertebrates. Reasoning that food and dust within the gut function as ice nucleators, Halpern (1979) concluded that reduced feeding of the mountain spiny lizard during winter is beneficial. Storey and Storey (1987) speculated that the springtime loss of freeze tolerance in the chorus frog coincides with its resumption of feeding, perhaps because crystallization, which reportedly commences in the extremities, is induced by ingested ice nucleators.

## Stochastic and Temporal Factors

The $T_c$ of any fluid sample is subject to stochastic variation (Vali, 1991). This principle also applies to intact animals and is manifest in the high intraspecific variability in $T_c$ evident in some data sets (e.g., Spellerberg, 1972; Costanzo, 1988). In one case, $T_c$ ranged from −0.6 to −6.4°C (mean ± SD = −2.9 ± 2.0°C) in a sample of 15 specimens of the wall lizard, *Podarcis muralis* (Claussen et al., 1990). The stochasticity of heterogeneous nucleation is also evident in the high variability of supercooling phenomena within individuals. Making duplicate measurements for each of 10 garter snakes, Costanzo (1988) found that $T_c$ values determined at intervals ranging from 24 h to 5 weeks differed by as much as 1.9°C. This is a substantial margin given that the full range of supercooling for the group was only 3.3°C. Although the common experimental practice is to represent each specimen by the result of a single cooling trial, workers should consider that $T_c$ data thus obtained are not necessarily definitive.

Spontaneous nucleation may occur in an animal held at thermoequilibrium, so long as body temperature is less than or equal to tissue $FP_{eq}$, although the likelihood of crystallization increases over time. Unfortunately, $T_c$ is typically measured by acute cooling trials which, because they ignore the temporal aspects of nucleation, may exaggerate supercooling capacity.

Several reports suggest that some vertebrates, particularly reptiles, have the potential to remain supercooled for extended periods. Mountain spiny lizards exposed to –3°C can remain unfrozen for over 30 h (Halpern, 1979). This capacity seems ecologically significant because, in their native habitat, subfreezing ambient temperatures rarely prevail continuously for more than 10–15 h. Spellerberg (1972) kept various lizards supercooled for as long as 8 days (Spellerberg, 1972), and hatchling painted turtles can remain supercooled for 10 days or more (Packard and Packard, 1993a). Remarkably, European common lizards (*Lacerta vivipara*) indigenous to the Cevennes, France, can remain supercooled for at least several weeks (J. Costanzo, C. Grenot, and R. Lee, *unpublished*).

Declining body temperature should in theory sharply reduce supercooling duration (Vali, 1991), although the nature of this relationship in vertebrates is poorly understood. Spellerberg (1972) investigated supercooling duration in several reptile species exposed to various subfreezing body temperatures. Notably, for the scincid lizards, *Sphenomorphus tympanum*, approximately 8 days elapsed before nucleation occurred in 50% of a sample kept at –1.7°C, whereas the half-time for a sample exposed to –3°C was only 6.3 h. Clearly, judicious evaluations of supercooling as a cold-hardiness strategy must include criteria based on both thermal and temporal capacities for freeze avoidance.

## Freeze Tolerance as a Cold-Hardiness Strategy

Only 10 vertebrate species are known to survive the freezing of their body tissues under thermal and temporal conditions of ecological relevance. The adaptive significance of freeze tolerance in vertebrates was only recently described (Schmid, 1982). Subsequent research has focused on both the biochemical (review by Storey, 1989) and physiological (reviews by Lee and Costanzo, 1993; Costanzo and Lee, 1994) aspects of this phenomenon. Freeze tolerance in the wood frog is promoted by an accumulation of the cryoprotectant glucose (Storey, 1989; Costanzo et al., 1991; 1993) and the evacuation and redistribution of water from organs and tissues (Costanzo et al., 1992; R. Lee et al., 1992). Wood frogs can survive freezing episodes lasting >2 weeks and tolerate body temperatures as low as –5 to –7°C (Schmid, 1982; Storey, 1989) with corresponding body ice contents of up to 65-70% (Layne and Lee, 1987; Costanzo et al., 1993). Presumably this species tolerates ice forming exclusively in extracellular spaces, since intracellular freezing reportedly is lethal (Storey, 1989).

Although freeze tolerance may be critical to the winter survival of some species, this protective mechanism is not without substantial risks and costs. Freezing survival is sharply constrained by tolerable limits of body temperature, ice content, and time. Freezing and thawing induce severe osmotic and metabolic perturbations which may persist a week or more after thawing and necessitate cell and tissue repair (Costanzo et al., 1993; Lee and Costanzo, 1993). Also, because glycosuria develops rapidly during thawing, a substantial energy loss may occur coincident with the excretion of potentially large quantities of cryoprotectant (R. Lee and Costanzo, 1993).

## Control of Nucleation in Freeze-Tolerant Vertebrates

Control of $T_c$ is an important problem for freeze-tolerant vertebrates, because spontaneous nucleation following extensive supercooling initiates a rapid accumulation of ice which may be damaging or lethal. This principle derives empirical support from studies of wall lizards (Claussen et al., 1990) and garter snakes (Churchill and Storey, 1992b) in which cryoinjury was greatest in the specimens that supercooled most extensively (approximately −5°C) before crystallizing. Rapid freezing may also compromise survival by interfering with normal cryoprotective responses (Costanzo et al., 1991, 1992). Because the quantity of "burst ice" forming is proportional to the degree of supercooling (Claussen and Costanzo, 1990), ice nucleation ideally should occur at $FP_{eq}$, the highest body temperature at which crystallization is possible.

It is believed that extensive supercooling is not a problem for most freeze-tolerant vertebrates, simply because they have relatively large bodies (Costanzo et al., 1988; Costanzo and Claussen, 1990; Claussen and Zani, 1991). However, small species (e.g., tree frog, chorus frog) and the young of larger species (e.g., painted and box turtles) may be at risk. Fortunately, freeze-tolerant vertebrates have mechanisms that effectively limit supercooling under various environmental conditions.

### Inoculation by Ice in the Environment

Perhaps the most important mechanism limiting supercooling in most freeze-tolerant amphibians is their susceptibility to inoculative freezing. The induction of crystallization by contact with environmental ice is both rapid and reliable (Layne et al., 1990; Layne, 1991), particularly in amphibians, which generally select moist overwintering sites (Ultsch, 1989; Pinder et al., 1992). Inoculative freezing also seems important to certain freeze-tolerant reptiles, such as turtle hatchlings. Supercooling in these animals is promoted by their small size (e.g., 3–6 g) as well as the desiccating effect of the sandy soil in which they overwinter. However, a sudden increase in environmental moisture (e.g., rainfall or snowmelt) may rehydrate these hatchlings, promoting a concomitant loss of supercooling capacity and shift to freeze tolerance. Under these conditions it is fortuitous, and perhaps adaptive, that contact with environmental ice can seed freezing at high subfreezing body temperatures.

### Ice Nucleation–Active Bacteria

As discussed in several chapters of this book, ice nucleation–active bacteria are potent initiators of nucleation at high subzero temperatures. Although bacteria with the ice nucleation–active phenotype ($Ina^+$) have been known since the early 1970s, these microorganisms were associated primarily with the epiphytic flora on plants. Only recently were $Ina^+$ bacteria first reported as normal gut flora in animals (M. Lee et al., 1992; Lee et al., 1993).

Several varieties of $Ina^+$ bacteria recently were isolated from the gut of winter-collected wood frogs (M. Lee et al., 1992). Ice nucleating activity was identified phenotypically in 13 strains of *Pseudomonas fluorescens*, four strains of *Pseudomonas putida*, and two strains of *Enterobacter agglomerans*. The highest nucleation temperatures were found in aqueous suspensions of the four *P. putida* strains (range, −1.6°C to −3.0°C). The ice-nucleating activity of *P. putida* was tested in vivo by feeding these bacteria to the freeze-tolerant spring peeper. All 10 frogs that ingested bacteria froze during a 24-h exposure to –2°C, compared with only 20% of the control sample that ingested the water vehicle. Furthermore, five of the frogs ingesting *P. putida* froze within 10 min after cooling to –2°C. Consequently, *P. putida* has the potential to markedly limit supercooling in freeze-tolerant frogs and thus may promote ice formation within their tissues. Even though feeding ceases in autumn, $Ina^+$ bacteria can be retained in the gut of wood frogs throughout hibernation. Thus, if ice nucleation is not initiated via inoculation across the skin, $Ina^+$ bacteria may ensure that protective freezing occurs at relatively high body temperatures.

## Blood Nucleators

Several workers have reported finding ice-nucleating agents in the blood of two freeze-tolerant vertebrates, the wood frog (Wolanczyk et al., 1990) and the painted turtle (Storey et al., 1991). These agents are active at approximately –8°C and retain nearly full activity at dilutions of up to 1 in 500. Their sensitivity to heat and acid suggests that they are proteinaceous, although they are not degraded by various nonspecific proteases. There are intriguing similarities in properties between these agents and those produced by certain tardigrades (Westh et al., 1991), mollusks (Hayes and Loomis, 1985), and insects (see Duman et al., Chapter 11) whose role is to initiate ice formation at high subzero temperatures.

Regarding the presence of nucleating agents in wood frogs and painted turtles, Storey et al. (1991) asserted that "...one of the critical factors in the evolution of freeze tolerance as a strategy for long-term hibernation by ectothermic vertebrates has been the selective expression of specific ice-nucleating agents in the blood." This contention raises several concerns. First, because nucleating proteins do not occur universally among freeze-tolerant vertebrates, they clearly are not requisite for freezing survival. Secondly, the activity temperature of these agents (approximately –8°C) is substantially lower than the minimum temperature that

**Table 4.** Crystallization temperatures ($T_c$) of plasma samples, measured before and after heat denaturation, from various freeze-tolerant and freeze-intolerant vertebrates[a,b]

| Species | *n* | $T_c$ (°C) Unheated | $T_c$ (°C) Heated | *P* |
|---|---|---|---|---|
| Freeze-tolerant | | | | |
| *Rana sylvatica* (wood frog) | 6 | −7.3 ± 0.6 | −14.7 ± 0.8 | <0.001 |
| *Pseudacris crucifer* (spring peeper) | 4 | −7.7 ± 0.4 | −12.9 ± 2.1 | 0.063 |
| *Thamnophis sirtalis* (garter snake) | 5 | −8.1 ± 1.0 | −14.7 ± 1.0 | <0.001 |
| *Terrapene carolina* (box turtle) | 4 | −9.1 ± 0.6 | −13.2 ± 0.9 | 0.064 |
| Freeze-intolerant | | | | |
| *Carassius auratus* (goldfish) | 4 | −13.4 ± 1.1 | −16.8 ± 0.7 | 0.071 |
| *Notophthalamus viridescens* (newt) | 6 | −8.4 ± 0.3 | −14.7 ± 0.8 | <0.001 |
| *Rana pipiens* (leopard frog) | 5 | −6.8 ± 0.2 | −13.7 ± 0.5 | <0.001 |
| *Bufo americanus* (American toad) | 5 | −6.8 ± 0.8 | −10.5 ± 1.0 | 0.005 |
| *Gallus domesticus* (leghorn chicken) | 6 | −12.3 ± 1.2 | −14.7 ± 1.1 | 0.009 |
| *Mus musculus* (white mouse) | 4 | −8.1 ± 0.4 | −13.9 ± 0.4 | <0.001 |
| Sterile saline (115 mM NaCl) | 5 | −17.0 ± 1.3 | −18.5 ± 0.9 | 0.193 |

[a] Data from J. Costanzo, K. Powell, and R. Lee (*unpublished*).
[b] Samples (7 μl) from each specimen were tested in triplicate. Means (shown ± SE) within each row were compared using Student's *t* test for dependent samples.

can be survived in the frozen state by either species. Thirdly, other studies suggest that blood nucleators are not unique to freeze-tolerant vertebrates, but rather are ubiquitous among some vertebrate taxa. Lowe et al. (1971) measured $T_c$ of whole blood samples from nine vertebrates (including endotherms) and reported activity as high as –7.7°C. Additionally, Spellerberg (1972) measured $T_c$ in samples of both blood and plasma from 11 species of freeze-intolerant Australian reptiles, observing nucleating activity at temperatures as high as –8°C. Our own data further suggest that heat-labile (proteinaceous?) blood nucleators occur in freeze-intolerant, as well as freeze-tolerant, vertebrates (Table 4). The latter group includes, notably, the garter snake and box turtle, two freeze-tolerant species that Storey et al. (1991) determined lack such agents. Finally, there is no empirical evidence that the activity of these agents is expressed in vivo. To the contrary, rather than crystallizing at temperatures indicative of nucleator activity, hatchling painted turtles can supercool to body temperatures lower than –13°C (Packard and Packard, 1993a).

Considering the available evidence, it seems unlikely that blood nucleators play a biologically significant role in vertebrate freeze tolerance. Indeed, most freeze-tolerant species supercool only minimally (Table 1) and are susceptible to inoculative freezing. We concur with Storey's earlier (1985) conclusion that "...the very limited supercooling capacity of [freeze-tolerant] frogs is probably the strongest argument against a need for nucleating agents." Nevertheless, the structure and behavior of biological nucleators are of academic interest. Layne (1994) recently found nucleation activity at temperatures ranging from –3 to –7°C not only in the body fluids (plasma and urine) but also in the isolated organs (liver, skeletal muscle, skin, intestine) of both freeze-tolerant and freeze-intolerant frogs. Interestingly, tissue fluids and homogenates of the winter flounder (*Pseudopleuronectes americanus*), a species that survives by avoiding freezing, can resist crystallization at –4°C (Valerio et al., 1992). Future research should seek to elucidate the taxonomic associations of nucleating proteins among vertebrates and determine their mechanism(s) of action.

## Acknowledgments

M. F. Wright and J. R. Layne, Jr., kindly commented on the manuscript. This research was supported in part by NIH grant #NIDDKD 1R15DK43958-01 and NSF IBN-93005809.

## Literature Cited

Breitenbach, G. L., Congdon, J. D., and van Loben Sels, R. C. 1984. Winter temperatures of *Chrysemys picta* nests in Michigan: Effects on hatchling survival. Herpetologica 40:76-81.

Christiansen, J. L., and Bickham, J. W. 1989. Possible historic effects of pond drying and winterkill on the behavior of *Kinosternon flavescens* and *Chrysemys picta*. J. Herpetol. 23:91-94.

Churchill, T. A., and Storey, K. B. 1992a. Natural freezing survival by painted turtles *Chrysemys picta marginata* and *C. picta bellii*. Am. J. Physiol. 262:R530-R537.

Churchill, T. A., and Storey, K. B. 1992b. Freezing survival of the garter snake *Thamnophis sirtalis parietalis*. Can. J. Zool. 70:99-105.

Claussen, D. L., and Costanzo, J. P. 1990. A simple model for estimating the ice content of freezing ectotherms. J. Therm. Biol. 15:223-231.

Claussen, D. L., and Zani, P. A. 1991. Allometry of cooling, supercooling, and freezing in the freeze-tolerant turtle *Chrysemys picta*. Am. J. Physiol. 261:R626-R632.

Claussen, D. L., Daniel, P. A., Jiang, S., and Adams, N. A. 1991. Hibernation in the eastern box turtle, *Terrapene c. carolina*. J. Herpetol. 25:334-341.

Claussen, D. L., Townsley, M. D., and Bausch, R. G. 1990. Supercooling and freeze-tolerance in the European wall lizard, *Podarcis muralis*, with a revisional history of the discovery of freeze-tolerance in vertebrates. J. Comp. Physiol. B 160:137-143.

Costanzo, J. P. 1988. Eco-physiological adaptations to overwintering in the eastern garter snake, *Thamnophis sirtalis sirtalis*. Ph.D. thesis. Miami University, Oxford, OH. 353 pp.

Costanzo, J. P., and Claussen, D. L. 1990. Natural freeze tolerance in the terrestrial turtle, *Terrapene c. carolina*. J. Exp. Zool. 254:228-232.

Costanzo, J. P., and Lee, R. E. 1994. Biophysical and physiological responses promoting freeze tolerance in vertebrates. New. Physiol. Sci. 9:252-256.

Costanzo, J. P., and Claussen, D. L., and Lee, R. E. 1988. Natural freeze tolerance in a reptile. Cryo Lett. 9:380-385.

Costanzo, J. P., Lee, R. E., and Lortz, P. H. 1993. Glucose concentration regulates freeze tolerance in the wood frog *Rana sylvatica*. J. Exp. Biol. 181:245-255.

Costanzo, J. P., Lee, R. E., and Wright, M. F. 1991. Glucose loading prevents freezing injury in rapidly cooled wood frogs. Am. J. Physiol. 261:R1549-R1553.

Costanzo, J. P., Lee, R. E., and Wright, M. F. 1992. Cooling rate influences cryoprotectant distribution and organ dehydration in freezing wood frogs. J. Exp. Zool. 261:373-378.

Crawford, K. M. 1991. The winter environment of painted turtles, *Chrysemys picta*: Temperature, dissolved oxygen, and potential cues for emergence. Can. J. Zool. 69:2493-2498.

DePari, J. A. 1988. Overwintering in the nest chamber by hatchling painted turtles, *Chrysemys picta*, in northern New Jersey. Ph.D. thesis, Rutgers University, Newark, NJ. 125 pp.

Gregory, P. T. 1982. Reptilian hibernation. Pages 53-154 in: Biology of the Reptilia, Vol. 13. C. Gans and F. H. Pough, eds. Academic Press, New York.

Halpern, E. A., and Lowe, C. H. 1968. Metabolism of the iguanid lizard *Uta stansburiana* in the supercooled state. Physiol. Zool. 41:113-125.

Halpern, E. A. 1979. Supercooling in the iguanid lizard *Sceloporus jarrovi*. Physiol. Zool. 52:190-204.

Hayes, D. R., and Loomis, S. H. 1985. Evidence for a proteinaceous ice nucleator in the hemolymph of the pulmonate gastropod *Melampus bidentarus*. Cryo Lett. 6:418-421.

Johnson, B. B. 1990. Freeze tolerance in the adult painted turtle, *Chrysemys picta*, and the effects of freezing on the cardiac cycle. M. S. thesis, Miami University, Oxford, OH. 57 pp.

Kalabukhov, N. I. 1958. The problem of freezing, undercooling and vitrifying of animal organism. Pages 101-118 in: Inst. Biol. Int. Symp. on Freezing and Drying, 2nd ed. Blackwell Scientific Publications, Oxford.

Layne, J. R. 1991. External ice triggers freezing in freeze-tolerant frogs at temperatures above their supercooling point. J. Herpetol. 25:129-130.

Layne, J. R. 1994. Crystallization temperatures of frogs and their indvidual organs. J. Herpetol. (In press.)

Layne, J. R., and R. E. Lee. 1987. Freeze tolerance and the dynamics of ice formation in wood frogs (*Rana sylvatica*) from southern Ohio. Can. J. Zool. 65:2062-2065.

Layne, J. R., Lee, R. E., and Huang, J. L. 1990. Inoculation triggers freezing at high subzero temperatures in a freeze tolerant frog (*Rana sylvatica*) and insect (*Eurosta solidaginis*). Can. J. Zool. 68:506-510.

Lee, M. R., Lee, R. E., and Strong-Gunderson, J. M. 1992. Isolation of ice nucleating active bacteria from a freeze-tolerant frog: Identification of *Pseudomonas putida* strains active in ice nucleation. Cryobiology 29:759.

Lee, R. E., and Costanzo, J. P. 1993. Integrated physiological responses promoting anuran freeze tolerance. Pages 501-510 in: Life in the Cold: Ecological Physiological and Molecular Mechanisms, C. Carey, G. L. Florant, B. A. Wunder, and B. Horwitz, eds. Westview Press, San Francisco.

Lee, R. E., Costanzo, J. P., Davidson, E. C., and Layne, J. R. 1992. Organ dehydration and the redistribution of water during freezing and thawing in a freeze-tolerant frog (*Rana sylvatica*). J. Therm. Biol. 17:263-266.

Lee, R. E., Lee, M. R., and Strong-Gunderson, J. M. 1993. Insect cold-hardiness and ice nucleating active microorganisms including their potential use for biological control: A review. J. Insect Physiol. 39:1-12.

Lillywhite, H. B. 1987. Temperature, energetics, and physiological ecology. Pages 422-477 in: Snakes: Ecology and Evolutionary Biology, R. A. Seigel, J. T. Collins, and S. S. Novak, eds. MacMillan Publishing Company, New York.

Lotshaw, D. P. 1977. Temperature adaptation and effects of thermal acclimation in *Rana sylvatica* and *Rana catesbeiana*. Comp. Biochem. Physiol. 56A:287-294.

Lowe, C. H., Lardner, P. J., and Halpern, E. A. 1971. Supercooling in reptiles and other vertebrates. Comp. Biochem. Physiol. 39A:125-135.

MacArthur, D. L., and Dandy, J. W. T. 1982. Physiological aspects of overwintering in the boreal chorus frog (*Pseudacris triseriata maculata*). Comp. Biochem. Physiol. 72A:137-141.

Packard, G. C., and Packard, M. J. 1993a. Hatchling painted turtles (*Chrysemys picta*) survive exposure to subzero temperatures during hibernation by avoiding freezing. J. Comp. Physiol. B 163:147-152.

Packard, G. C., and Packard, M. J. 1993b. Delayed inoculative freezing is fatal to hatchling painted turtles (*Chrysemys picta*). Cryo Lett. 14:273-284.

Packard, G. C., Ruble, K. A., and Packard, M. J. 1993. Hatchling snapping turtles overwintering in natural nests are inoculated by ice in frozen soil. J. Therm. Biol. 185-188.

Pinder, A. W., Storey, K. B., and Ultsch, G. R. 1992. Estivation and hibernation. Pages 250-274 in: Environmental Physiology of the Amphibians. M. E. Feder and W. W. Burggren, eds. University of Chicago Press, Chicago.

Schmid, W. D. 1982. Survival of frogs in low temperature. Science 215:697-698.

Schmid, W. D. 1986. Winter ecology. Ekologiya Sofia 6:29-35.

Spellerberg, I. F. 1972. Temperature tolerances of southeast Australian reptiles examined in relation to reptile thermoregulatory behavior and distribution. Oecologia (Berl.) 9:23-46.

Spellerberg, I. F. 1976. Adaptations of reptiles to cold. Pages 261-285 in: Morphology and Biology of Reptiles, A. d'A Bellairs and C. B. Cox, eds. Linn. Soc. Symp. Ser. 3.

Storey, K. B. 1985. Freeze tolerance in terrestrial frogs. Cryo Lett. 6:115-134.

Storey, K. B. 1989. Biochemistry of natural freeze tolerance in animals: Molecular adaptations and applications to cryopreservation. Biochem. Cell Biol. 68:687-698.

Storey, K. B., and Storey, J. M. 1986. Freeze tolerance and intolerance as strategies of winter survival in terrestrially-hibernating amphibians. Comp. Biochem. Physiol. 83A:613-617.

Storey, K. B., and Storey, J. M. 1987. Persistence of freeze tolerance in terrestrially hibernating frogs after spring emergence. Copeia 1987:720-726.

Storey, K. B., Layne, J. R., Cutwa, M. M., Churchill, T. A., and Storey, J. M. 1993. Freezing survival and metabolism of box turtles, *Terrapene carolina*. Copeia 1993:628-634.

Storey, K. B., McDonald, D. G., Duman, J. G., and Storey, J. M. 1991. Blood chemistry and ice nucleating activity in hatchling painted turtles. Cryo Lett. 12:351-358.

Storey, K. B., Storey, J. M., Brooks, S. P. J., Churchill, T. A., and Brooks, R. J. 1988. Hatchling turtles survive freezing during winter hibernation. Proc. Natl. Acad. Sci. USA 85:8350-8354.

Ultsch, G. R. 1989. Ecology and physiology of hibernation and overwintering among freshwater fishes, turtles, and snakes. Biol. Rev. 64:435-516.

Valerio, P. F., Kao, M. H., and Fletcher, G. L. 1992. Fish skin: An effective barrier to ice crystal propagation. J. Exp. Biol. 164:135-151.

Vali, G. 1991. Nucleation of ice. Pages 131-192 in: Atmospheric Particles and Nuclei, G. Götz, E. Mészáros, and G. Vali, eds. Akadémiai Kiadó, Budapest.

Vincent, T. K., and Secoy, D. M. 1978. The effects of annual variation in temperature on cold resistance in a northern population of the red-sided garter snake, *Thamnophis sirtalis parietalis* (Reptilia, Serpentes, Colubridae). J. Herpetol. 12:291-294.

Weigmann, R. 1929. Die wirkung starker Abkühlung auf Amphibien und Reptilien. Z. Wiss. Zool. 134:641-692.

Westh, P., Kristiansen, J., and Hvidt, A. 1991. Ice-nucleating activity in the freeze-tolerant tardigrade *Adorybiotus coronifer*. Comp. Biochem. Physiol. 99A:401-404.

Wolanczyk, J. P., Storey, K. B., and Baust, J. G. 1990. Ice nucleating activity in the blood of the freeze-tolerant frog, *Rana sylvatica*. Cryobiol. 27:328-335.

Wolverton, E. 1963. Winter survival of hatchling painted turtles in northern Minnesota. Copeia 1963:569-570.

CHAPTER 13

# Control of Epiphytic Ice Nucleation–Active Bacteria for Management of Plant Frost Injury

**Steven E. Lindow**

## Ecological Basis for Plant Frost Control

This chapter considers the biological factors that increase the likelihood of frost damage to frost-sensitive plants and discusses measures by which such damage can be reduced. Frost-sensitive plant species, such as many herbaceous vegetable crops, the flowers of deciduous fruit crops, and subtropical plant species must avoid ice formation to escape frost damage (Burke et al., 1976; Cary and Mayland, 1970; Chandler, 1958). The role of ice formation in initiating frost damage to sensitive plant species is reviewed in Chapter 7. It is noteworthy that nonwoody plant species seldom contain significant numbers of intrinsic ice-nucleating agents active at temperatures above –5°C (Cary and Mayland, 1970; Lindow et al., 1978a, 1982a; Marcellos and Single, 1976; see also Chapter 8), so that these species are capable of supercooling to this temperature and avoiding frost damage (Lindow, 1982, 1983a; Lindow et al., 1978a, 1982a; Marcellos and Single, 1976, 1979; Proebsting et al., 1982). The presence of ice nucleation-active ($Ice^+$) bacteria on these plants is therefore the primary determinant of frost damage at temperatures in the range of 0 to –5°C (Lindow, 1982, 1983a; Lindow et al., 1978a, 1982a). This chapter will address the mechanisms and procedures by which $Ice^+$ bacterial populations on sensitive plants can be managed to increase the probability of supercooling.

The population sizes of $Ice^+$ bacteria on frost-sensitive plants often vary dramatically with time (see Chapter 3). While physical and environmental factors greatly influence the population sizes (Hirano and Upper, 1986, 1990), the populations generally increase with the age of the leaf. The population size of $Ice^+$ bacteria on the leaves of herbaceous species shortly after emergence from the soil is often very low (Hirano and Upper, 1986; Kaneda, 1986; Lindow, 1982; Lindow et al., 1978b,c, 1983b). Similarly, $Ice^+$ bacterial populations, principally *Pseudomonas syringae*, increase rapidly on flowers of deciduous tree fruit crops such as pear after they emerge (Cody et al., 1987; Gross et al., 1983; Hirano and Upper,

1983; Lindow, 1982, 1983b,c, 1985a,b, 1987; Montesinos and Vilardell, 1991). Newly emerged pear flowers are often nearly devoid of bacteria. The population size of Ice$^+$ bacteria, however, can increase very rapidly in such flowers; populations have increased more than 1,000-fold within 10 days (Cody et al., 1987; Gross et al., 1983; Lindow, 1982, 1983b,c, 1985a,b, 1987). Similarly, within 20 days of emergence, Ice$^+$ bacterial populations increased by over 100-fold on corn and snapbean leaves (Hirano and Upper, 1986; Goto et al., 1989; Lindow et al., 1978b,c, 1983b). Large seasonal variations in populations of Ice$^+$ bacteria are also observed on perennial plants in Mediterranean climates. For example, population sizes of epiphytic bacteria, including Ice$^+$ *P. syringae* and *Erwinia herbicola* strains, on navel orange leaves are less than about 100 cells per gram during the summer months, when maximum air temperatures often exceed 40°C and rainfall is absent. These populations increase by 100 to 1,000-fold during winter and spring months in California and Greece; maximum populations coincide with the period of maximum likelihood of frost damage (Constantinidou et al., 1991; Lindow, 1982).

The incidence of frost damage to sensitive plant species is inversely proportional to the logarithm of the population size of Ice$^+$ bacteria on the plants at the time of freezing (Lindow et al., 1978a, 1982a,b; Lindow, 1982). Similarly, the supercooling potential of a plant part is inversely proportional to the logarithm of the population size of Ice$^+$ *P. syringae* strains (Gross et al., 1984; Hirano et al., 1985; Lindow, 1985b, 1993). The mean supercooling point for bean leaves harboring $10^3$ *P. syringae* cells per gram is about –3.5°C, whereas leaves harboring $10^6$ cells per gram supercool only to about –2°C (Lindow, 1993). The population size of Ice$^+$ bacteria varies greatly between frost-sensitive species. For example, the population size of Ice$^+$ bacterial epiphytes on almond and pear (often greater than $10^6$ cells/g fresh weight) is higher than that on potato and tomato ($10^3$ to $10^4$ cells/g), and far higher than that on many waxy leaved plants such as navel orange (often less than 100 cells/g) (Constantinidou et al., 1991; Lindow, 1982; Lindow et al., 1978b). It is thus clear why some frost-sensitive plants species such as navel orange supercool substantially more than species such as almond. Navel orange leaves and fruit will often supercool to –4°C (Hendershott, 1962; Lucas, 1954; Yelenosky, 1983, 1991; Young, 1966; Lindow, *unpublished*), whereas bean leaves and almond spurs harboring high numbers of Ice$^+$ bacteria frequently supercool only to about –2°C (Gross et al., 1984; Lindow, *unpublished*). Thus the control of frost damage to plant species conferred by a reduction in Ice$^+$ bacterial populations should be expected to occur within a discrete "temperature window" that is distinctive for a given plant species.

With progressive reductions in the population size of Ice$^+$ bacterial species on frost-sensitive plants, the degree of supercooling will increase and hence the incidence of frost damage at a given temperature will decrease. However, because of the log-linear relationship between the population size of Ice$^+$ bacterial cells and frost damage, Ice$^+$ bacterial populations must be decreased considerably to achieve substantial frost protection. For example, a 10-fold reduction in the population size of Ice$^+$ bacterial species will result in only about a 0.3 to 0.5°C reduction in the supercooling point of a leaf (Lindow, 1993). The enhancement of the supercooling ability of a given plant species by frost control measures thus operates within the "temperature window" dictated by normal epiphytic Ice$^+$ bacterial populations. If we assume that intrinsic (nonmicrobial) nucleators begin to limit the supercooling ability of plant species between –5 and –6°C, then reducing the normal epiphytic Ice$^+$ bacterial populations on citrus will increase its supercooling ability only from –4 to –6°C. In contrast, complete eradication of Ice$^+$ bacteria from snapbeans (which frequently harbor high Ice$^+$ bacterial populations) would enhance the supercooling ability of these leaves from –2 to –6°C. In practice, however, if a 100-fold reduction in Ice$^+$ bacterial populations is conferred by biological or chemical agents, then the supercooling ability of bean leaves might only improve from –2 to –3°C, while a similar reduction in population size of Ice$^+$ bacteria on navel orange would reduce the supercooling point from –4 to –5°C. A decrease in population size of Ice$^+$ bacteria on plants should result in a proportional decrease in the supercooling point. However, the supercooling temperature differs greatly between species and is dependent on the normal indigenous Ice$^+$ bacterial population sizes on that species.

Because of the rapid rates at which Ice$^+$ bacterial populations can increase on leaves, it is preferable to prevent populations from increasing, rather than trying to eradicate populations after they have developed. Hirano and Upper (1986) demonstrated that *P. syringae* populations can increase with a doubling time as short as 2 h under field conditions. Similarly rapid increases in *P. syringae* population sizes on pear and almond flowers have also been reported (Cody and Gross, 1987; Gross et al., 1984; Lindow, 1982, 1983c, 1985b; Lindow and Connell, 1984), and can occur immediately before the occurrence of subfreezing temperatures (Lindow, 1982). Many eradicative chemicals, while reducing the viability of Ice$^+$ bacterial cells, do not immediately eliminate their ice nucleation activity (Anderson and Ashworth, 1986; Lindow et al., 1982b; Maki et al., 1974). For both of these reasons, procedures that prevent the development of large population sizes of epiphytic Ice$^+$ bacterial strains are preferable to methods that attempt to mitigate the populations after they develop.

## Control of Plant Frost Injury Using Bactericides

Some chemicals that kill or prevent the growth of Ice$^+$ bacterial species can be used to reduce the likelihood of frost damage. Any chemical intended to mitigate a plant pest must be registered as a pesticide by the U.S. Environmental Protection Agency (EPA) before it can be used in the United States; similar restrictions apply elsewhere. While many different metal ions, organic compounds, and antibiotics are toxic to Ice$^+$ bacteria, only copper compounds and streptomycin are currently registered for control of these species on crop plants. Several copper salts that have very low solubility in water (including copper hydroxide and basic copper sulfate) may be used (Menkissoglu and Lindow, 1991a,b). Although insoluble copper compounds can be applied to plants at high concentrations (1 to 4 kg/ha), only the free copper ions, which are released slowly, are bactericidal (Menkissoglu and Lindow, 1991a,b). Copper ions have a very broad toxicity, killing nearly all bacterial and fungal species by binding to and inactivating proteins involved in respiration and other processes (Gadd and Griffiths, 1978). In addition to copper, the antibiotic streptomycin is registered for the control of bacterial diseases on pears, apples, and a few ornamental plants. In a few states, the antibiotic oxytetracycline (Terramycin) has also been registered for the control of fire-blight disease caused by the bacterium *Erwinia amylovora*. Both of these antibiotics have inhibitory thresholds for sensitive Ice$^+$ bacteria at concentrations below 100 ppm. The incidence of frost damage to beans, strawberry, wheat, almond, pear, and other plants has been reduced following the application of copper-containing compounds (Hirano and Upper, 1985; Lindow, 1983b,c, 1985a; Lindow et al., 1978a, 1983b; Marshall, 1988). Similarly, application of antibiotics also reduced the incidence of frost damage to various crop species (Hirano and Upper, 1985; Lindow, 1983b,c, 1985a; Lindow et al., 1978a, 1983b; Marshall, 1988; Proebsting and Gross, 1988).

## Application Frequency

The application frequency of bactericides is an important determinant for their efficacy in the control of plant frost damage. After application to plant surfaces (as a foliar spray or dust), a bactericide's concentration decreases over time (Menkissoglu and Lindow, 1991b; Olson and Jones, 1983). Copper compounds are chemically stable and do not degrade upon exposure to light or other environmental factors. However, they are removed from leaves by both rain and wind. The half-life of copper deposits on leaves varied from aproximately 10 days in the absence of rain to less than 5 days when there was significant rainfall (Menkissoglu and Lindow, 1991b; Olson and Jones, 1983). Streptomycin is photolabile, and its concentration decreases rapidly in sunlight. The concentrations of bactericides on plants are usually more than sufficient shortly after treatment, but their concentrations drop to sublethal concentrations within one or a few days, depending on the material and meteorological conditions. Frost-sensitive plants are frequently undergoing rapid growth when they are exposed to subfreezing temperatures. Since most bactericides are not systemic, movement of these compounds to newly developed plant tissue (via rain splash or wind) is inefficient, so that they will not be present on all tissues unless they are frequently reapplied. To maintain adequate concentrations of bactericides on leaves and flowers prior to freezing conditions, bactericides must be applied at a frequency that is inversely proportional to the plant growth rate at the time. Copper compounds or streptomycin need to be applied to rapidly growing bean plants or pear flowers at intervals of less than 2 to 3 days for control of Ice$^+$ bacteria. In contrast, navel orange trees exhibit little growth during winter months in Mediterranean climates when they are subjected to freezing temperatures. Application of copper compounds at from 1- to 2-month intervals is sufficient to achieve substantial frost control to orange fruit (Lindow, *unpublished*). Both the efficacy and the cost of bactericidal frost control on rapidly growing plants will increase with the frequency of application. Application frequencies of 2 to 3 days represent a compromise between efficacy and cost. The high frequency with which bactericides must be applied to many rapidly growing plant species limits their usage to only relatively high-value crops such as fruits and vegetables. The relatively low per-hectare value of field crops such as corn, soybeans, and cereals does not usually justify the costs of bactericide applications for frost control.

## Preventive and Eradicative Use of Bactericides

Bactericides are maximally effective in reducing frost damage to sensitive plant species when used in a preventive rather than a eradicative manner. If the occurrence of natural freezing events could be accurately forecast at least 1 day in advance, if it were possible to apply eradicative bactericides to all susceptible plant species within 24 h prior to freezing, and if cells that were rendered nonviable by bactericides immediately lost ice nucleation activity, then eradicative bactericides would make excellent frost control agents. Unfortunately, all these assumptions are

dubious. Freezing temperatures are frequently unexpected, and even when they can be adequately anticipated, growers often have such large acreages of susceptible crops that bactericides could not be applied in the short time between frost prediction and frost event. Most importantly, however, is that treatment of cells with antibiotics or metal ions does not cause rapid abolition of warm-temperature ice nucleation activity (Anderson and Ashworth, 1986; Lindow et al., 1982b, 1983d, 1989; Maki et al., 1974; Rogers et al., 1987; Yankofsky et al., 1981). While spectinomycin was reported to cause rapid reductions in warm-temperature ice nucleation activity, its effect may be dependent on the environmental conditions during or after such treatment (Anderson and Ashworth, 1986).

Insufficient data have been obtained from field and greenhouse experiments to test whether applications of bactericides to plants colonized with Ice$^+$ bacteria can lead to rapid reductions in frost damage. Field studies are problematic because, as noted above, the occurrence of freezing temperatures cannot be predicted accurately more than 1 day in advance: it is difficult to conduct a field experiment in which bactericides are applied at different times prior to an unforeseeable freezing event. However, one such field study was conducted in Wisconsin. Different batches of corn plants were sprayed with streptomycin at intervals starting 20 days before freezing and ending 1 day before a mild radiative frost with a minimum air temperature of about –3°C. Interestingly, the incidence of freezing damage to leaves was hardly reduced on plants that received streptomycin sprays more than about 10 days, or less than about 5 days, prior to exposure to subfreezing temperatures (Lindow, Upper, and Arny, *unpublished*). The incidence of freezing damage was most reduced on plants that were sprayed 6 days prior to the freezing event (*unpublished*). These data were interpreted to indicate that ice nucleation sites in cells exposed to streptomycin less than 5 days earlier had insufficient time to be degraded before freezing temperatures were encountered. Plants that received eradicative applications of streptomycin more than 10 days prior to freezing presumably had been recolonized by Ice$^+$ bacteria after streptomycin residues diminished. More experiments of this kind are needed in order to formulate strategies for bactericide applications. A conservative approach, however, would be to assume that bactericides cannot cause a sufficiently rapid abolition of ice nucleation activity to be used as a "day before" frost control strategy. Repeated applications of bactericides at frequencies of 2 to 7 days, however, have been demonstrated to result in epiphytic Ice$^+$ bacterial populations that range from 10- to over 1,000-fold lower than on untreated plants (Hirano and Upper, 1985; Lindow, 1983b,c, 1985a; Lindow et al., 1978a, 1983b; Marshall, 1988), thereby reducing the supercooling point of the plants between 0.5 and almost 2°C.

## Significance of Bactericide Resistance

For bactericides to be effective in reducing the incidence of frost damage, the Ice$^+$ bacteria must be sensitive to them. Streptomycin-resistant strains of many phytopathogenic bacteria, including Ice$^+$ bacterial species, have been widely reported (Burr et al., 1993; Cooksey, 1990). A survey of Ice$^+$ *P. syringae* strains on pear trees in California revealed that from 5% to nearly 90% are resistant to streptomycin at 100 ppm (Lindow, *unpublished*). The pear orchards that were surveyed had been sprayed repeatedly with antibiotics for 10 to 20 years for the control of *E. amylovora*—thus, there were substantial opportunities for streptomycin resistance to develop in *P. syringae* populations. However, streptomycin resistance can develop in *P. syringae* populations within a much shorter period. In trials in Wisconsin in 1976 and 1977, all *P. syringae* strains recovered from unsprayed corn plants were sensitive to streptomycin. However, the population sizes of *P. syringae* on plants sprayed at 7-day intervals with streptomycin were as high as those on unsprayed plants at the end of each growing season, and nearly all of the *P. syringae* strains recovered were resistant to streptomycin at 100 ppm (Lindow, Upper, and Arny, *unpublished*). Since the field sprayed with streptomycin was not in the vicinity of other crops to which streptomycin had been applied, we may conclude that *P. syringae* populations have a relatively high frequency of acquiring resistance to streptomycin. Copper-resistant strains of *P. syringae* and other phytopathogenic bacteria have also been reported recently (Adaskaveg and Hine, 1985; Andersen et al., 1991; Bender and Cooksey, 1986; Cooksey, 1990; Sundin et al., 1989). As many as 30% of the Ice$^+$ *P. syringae* strains obtained as epiphytes from asymptomatic almond and citrus leaves in California were found to be resistant to high concentrations of copper ions (Andersen et al., 1991). These strains multiplied as rapidly on copper-sprayed bean plants in the greenhouse as on untreated bean leaves. Since copper compounds had been used in the sampled orchards for the control of bacterial and fungal diseases for many years, it is not known how rapidly copper-resistant strains of Ice$^+$ species develop on plants. However, the conclusion is inescapable that (in addition to being relatively expensive) copper compounds and antibiotics probably cannot yield sustained control of frost damage to plants because of the development of resistant strains.

## Control of Frost Injury Using Competitive Bacteria

Plants can be readily colonized by inoculated bacteria under the right circumstances. The maximum bacterial population size on the leaves of different plant species, the "carrying capacity," varies greatly by plant species but is characteristic for a given species. For example, bean leaves typically harbor approximately $10^7$ cells/g, while only about $10^5$ cells/g are supported on navel orange leaves (Constantinidou et al., 1991; Hirano and Upper, 1983, 1986, 1990; Lindow, 1982). If epiphytic bacterial populations are lower than the carrying capacity for a given plant species, inoculated bacteria will usually increase after application. When applied to greenhouse- or growth chamber–grown plants, which are nearly devoid of other epiphytic bacteria, non-ice nucleation–active (non-INA) bacterial species can rapidly multiply to population sizes of nearly $10^7$ cells per gram (Kim et al., 1989; Lindemann and Suslow, 1987; Lindow et al., 1982b, 1983a). The observation that single non-INA strains can represent a large percentage of the epiphytic bacterial community suggested that biological control of frost damage could be conferred by colonization of plants by certain non-INA strains. Conceptually, the presence of large population sizes of non-INA bacterial strains should prevent large population sizes of Ice$^+$ bacterial strains from occurring on the same plant. However, more information is needed on interactions within mixed bacterial populations on plants in order to understand how such biological control can be implemented consistently.

### Role of Antibiosis

Antibiosis does not appear to be important in the interaction of bacteria on leaf surfaces. Since bacterial cells of different species or strains frequently occur in close proximity on leaf surfaces, it has often been assumed that antibiotics produced by one strain would affect the growth of another strain in this habitat (Andrews, 1985). While antibiotics, such as the bacteriocins produced by *E. herbicola*, have been demonstrated to inhibit the growth of sensitive epiphytic strains on plants (Beer et al., 1984), until recently there was little evidence that these compounds were a major determinant of the interactions that were observed on plants (Vanneste et al., 1992). The lack of a role for antibiotic production in the antagonism of *P. syringae* strains on leaf surfaces was demonstrated by Lindow (1988b). A large collection of bacterial strains that prevent the growth of *P. syringae* strains on leaf surfaces was examined for antibiotic production. Only approximately 50% of the randomly selected strains that were effective in antagonism of *P. syringae* on leaf surfaces produced detectable antibiotics active against *P. syringae* in culture (Lindow, 1988b). Of course, it cannot be ruled out that a capacity for antibiotic production was missed in some of the strains because such capacity was not expressed in culture. Antibiotic-producing strains were mutagenized by exposure to ethyl methane sulfonate, and mutants deficient in antibiotic production were identified. These mutants were compared with the parental strains for their ability to prevent the growth of *P. syringae* strains on plant surfaces (Lindow, 1988b); in all cases, mutants unable to produce antibiotics were similar to the parental strain in their ability to prevent the growth of *P. syringae* on leaves. This study suggests that other factors, such as competition for limiting resources, account for the observed antagonistic interaction between bacterial strains on leaves.

### Importance of Pre-emptive Competitive Exclusion

Pre-emptive competitive exclusion apparently mediates the antagonism observed between bacterial strains on leaf surfaces. While non-INA bacteria, such as *E. herbicola* strain M232A or *Pseudomonas fluorescens* strain A506, do not reduce the population size of *E. amylovora* or *P. syringae* substantially when co-inoculated in small numbers with these target strains, they greatly reduce the population sizes the target strains can achieve when their inoculation precedes that of the target strain (Lindow et al., 1983a; Wilson and Lindow, 1993). For example, the population size of Ice$^+$ *E. herbicola* strain #26 was reduced only about two-fold when an equal number of cells of strain M232A were coinoculated on plants (Lindow et al., 1983a). In contrast, the population size of strain #26 was reduced over 100-fold when M232A was inoculated 12 hours or more prior to strain #26. In fact, an inverse relationship is observed between the logarithm of the population size of a non-INA bacterial strain on plants and the population size of an Ice$^+$ strain on the same plant. This would suggest that both strains are competing for the same limiting resource on leaves and that the presence of large numbers of a non-INA bacterial strain usurps resources that might otherwise be utilized by Ice$^+$ bacterial strains. In fact, non-INA bacteria that had lower growth rates than an Ice$^+$ strain could still be effective in antagonizing the growth of the Ice$^+$ strain when applied preemptively.

Mutants of *P. syringae* deficient in ice nucleation have provided direct evidence of competition for limiting environmental resources as a major mechanism determining the coexistence of bacteria on leaves (Lindow, 1985a, 1987). It was hypothesized that bacterial strains of a similar genotype (and therefore of similar ecological habitat requirements) would compete more directly for limiting environmental resources than dissimilar strains (Lindow, 1985c, 1988a). Mutants of *P.*

*syringae* that were deficient in ice nucleation were selected following mutagenesis with ethyl methane sulfonate. Such near-isogenic bacterial strains were utilized in pairwise competition studies to determine the magnitude of any competition exhibited between isogenic strains. All chemically induced Ice$^-$ mutants of *P. syringae* reduced the population size of Ice$^+$ parental strains that were challenge-inoculated onto plants pretreated with mutant strains (Lindow, 1987). Thus, competition for limiting resources appears sufficient to account for the major interactions among these near-isogenic bacterial strains on leaves. Ice$^-$ mutants of *P. syringae* also decreased the population size of indigenous Ice$^+$ *P. syringae* strains when applied to pear trees in the field at flowering (Lindow, 1987). Competition thus appeared to be the major interaction occurring among presumably heterologous strains of this species under field conditions. However, the incomplete exclusion of *P. syringae* by Ice$^-$ mutants of this species raises the question of the specificity with which different strains compete for limiting environmental resources. For example, do different strains of *P. syringae* have different ecological niche requirements? It would be expected that bacterial strains with dissimilar ecological niche requirements could coexist on leaves. Any one Ice$^-$ mutant of *P. syringae* might therefore not exclude all other ice nucleation-active strains of this species. This issue will be revisited.

## Nature of Limiting Factors on Leaves

Nutrients that leak from plant tissues onto the phylloplane are apparently the limiting resource for which competition occurs. Preliminary data suggest that leaf surface colonization sites are carbon-limited. Addition of carbon compounds, but not ammonium sulfate, to leaf surfaces at the time of inoculation elevated the carrying capacity of *P. syringae* strains on bean leaves (Wilson and Lindow, *unpublished*). Addition of ammonium sulfate together with carbon sources further increased the final population size of bacteria on leaves, suggesting that *P. syringae* strains are limited by carbon sources such as sugars and organic acids, and only secondarily by nitrogen sources. Populations of epiphytic yeasts on wheat have also been increased by addition of a mixture of 2% sucrose and 1% yeast extract (Dik et al., 1992). Considerable quantities of nutrients are present on the surface of leaves (Fossard, 1981; Tukey, 1966, 1970, 1971). These results strongly suggest that the "carrying capacity" of a leaf is dictated by the amount and type of carbon compounds that leak from the leaf onto the plant surface. Presumably those plant species that support high populations of epiphytic bacteria exude more nutrients onto the leaf surface.

## Selection of Antagonistic Bacteria

Many bacteria evaluated for antagonism against other species on plants have been preselected because they produce antibiotics toxic to the target species in culture (Andrews, 1985). It has been assumed that antibiotics are also produced by antagonistic bacteria on plant surfaces, but the validity of this assumption has not been adequately addressed. Likewise, the significance of antibiotic production to antagonism observed between bacterial strains on plants is often unclear. There is a poor correlation between the extent of antibiosis exhibited by a given strain in culture and the efficacy of its antagonism of an antibiotic-sensitive target species on plants (Andrews, 1985). That is, producers of antibiotics in culture do not always achieve biological control, whereas nonantibiotic producers frequently do. Thus, preconceived notions of the most important mechanisms involved in antagonism should *not* be used to determine which antagonists are tested for biological control of frost injury.

It is possible to directly identify non-INA bacterial strains that can prevent the growth of Ice$^+$ bacterial strains on plants. By definition, those bacteria most antagonistic towards Ice$^+$ bacterial strains on leaves will result in the lowest populations of the Ice$^+$ strains when the antagonist is pre-inoculated. Non-INA bacteria isolated from the predominant microflora of healthy plants under appropriate field conditions were selected (Lindow, 1985a). These non-INA strains were applied singly onto the leaves of greenhouse-grown corn plants and allowed to multiply for 24 hours. The plants were then inoculated with a relatively low concentration ($10^3$ cells/ml) of an Ice$^+$ *P. syringae* strain and incubated under moist conditions on a greenhouse bench for 48 hours. Plants were allowed to dry and cooled to approximately −5°C, and the incidence of frost damage in the population of corn leaves was measured. Corn plants are an ideal assay plant for this procedure because they grow rapidly in the greenhouse, and, if ice forms in a leaf, it does not spread to adjacent leaves of the same plant. A single pot of corn plants provides 30 to 50 individual leaves wherein the incidence of ice formation can be estimated. Various non-INA bacteria differed greatly in their ability to prevent frost damage to corn plants in this assay (Lindow, 1985b). While the non-INA bacteria all multiply to relatively high population sizes on corn leaves, the incidence of frost damage to plants pretreated with many potential antagonists was not significantly lower than to control plants (Lindow, 1985b). In contrast, certain strains, such as *P. fluorescens* A506, greatly reduced the incidence of freezing injury compared with control plants. Thus, non-INA bacteria that were effective antagonists of *P. syringae* on leaf surfaces could be identified directly by this procedure. As noted below, several of the antagonists identified by this screening procedure have proven to be consistently effective in controlling frost injury and certain bacterial diseases under field conditions.

## Specificity of Competition

Studies of the specificity of competition of bacteria have been greatly facilitated by molecular genetic approaches that have permitted the construction of truly isogenic Ice$^-$ strains of *P. syringae*. If strains of *P. syringae* differ in ecological habitat requirements, then different strains of this bacterium should coexist to different extents when coinoculated onto plants (Lindow, 1988a). This phenomenon is most easily studied by examining the ability of different Ice$^+$ parental *P. syringae* strains to coexist with a collection of isogenic Ice$^-$ *P. syringae* strains and heterogeneous Ice$^-$ *P. syringae* strains constructed in vitro. Genes conferring ice nucleation activity have been cloned from several strains of *P. syringae*. In all cases, a single *ina* gene is sufficient to confer ice nucleation activity. These genes confer production of Ina proteins from 120 kDa to over 180 kDa in size (Lindow et al., 1989; Warren, 1987; Warren and Wolber, 1991; Wolber, 1992; see also Chapter 5). The ice nucleation protein appears to function exclusively in orienting water molecules into an ice-like lattice. The low relative abundance of Ina protein molecules in bacterial membranes suggests that they do not play a major role in maintaining membrane structure. Thus the alteration of Ina proteins by mutation should not lead to substantial changes in behavior of mutant strains. Cloned ice nucleation genes were used to produce isogenic Ice$^-$ derivatives of the Ice$^+$ bacterial strains from which they were isolated. Large deletions internal to cloned ice nucleation genes from *P. syringae* and *P. fluorescens* completely abolished ice nucleation activity (Green et al., 1988; Orser et al., 1985). These deletion-containing genes were used as site-specific mutagens. They were introduced into their parental *P. syringae* and *P. fluorescens* strains on unstable plasmids (Lindow, 1985c; Warren et al., 1987). Homologous recombination occurred between chromosomal and plasmid-borne genes, so that the deletion-containing gene and the plasmid vector were integrated into the chromosome, producing a *cis*-merodiploid strain which was still phenotypically Ice$^+$ (Lindow, 1985c, 1988a; Warren et al., 1987). After many generations of growth without selection for the plasmid vector, resolution of the *cis*-merodiploid occurred when a second recombinational event took place between homologous regions of the *ina* gene. This process of "marker-exchange mutagenesis" is commonly used to create site-directed mutations in bacteria when cloned genes are available (Ruvkun and Ausubel, 1981). However, marker-exchange mutagenesis is usually performed when selection can be made for an insertional mutation conferring a selectable trait. At the time Ice$^-$ derivatives of *P. syringae* and *P. fluorescens* were produced, it was deemed undesirable to introduce novel genes, such as antibiotic resistance determinants (Lindemann and Suslow, 1985; Lindow, 1985c). Marker-exchange mutagenesis with a deletion-containing gene rather than with an insertionally inactivated gene posed problems. Second recombinational events leading to the resolution of the plasmid vector occur at a frequency of only about $10^{-4}$ to $10^{-5}$ after prolonged incubation. Ice$^+$ cells representing *cis*-merodiploids had to be distinguished from Ice$^-$ double-recombinational derivatives by a replica-freezing procedure in which the ice nucleation activity of individual bacterial colonies on a cold surface is scored (Lindow et al., 1978b). It was necessary to screen over 20,000 colonies to identify one Ice$^-$ derivative. Since the rate of homologous recombination decreases rapidly with increasing difference in primary nucleotide sequence, marker-exchange mutagenesis using nonselectable deletions was restricted to producing site-directed mutants only in strains containing completely homologous *ina* genes.

Ice$^-$ deletion mutants of *P. syringae* did not exhibit discernible differences in any biochemical or behavioral trait evaluated in culture. Like the Ice$^+$ parental strains from which they were derived, Ice$^-$ mutants did not produce disease symptoms on plants (Lindemann and Suslow, 1985; Lindow, 1985c). Ice$^-$ deletion mutants of *P. syringae* did not differ from their Ice$^+$ parental strains in their growth characteristics and competitiveness on leaves (Lindemann and Suslow, 1985, 1987; Lindow, 1985c, 1988a). No change in the relative frequency of Ice$^+$ and Ice$^-$ strains, initially applied to plants in mixtures, was observed after growth for 10 generations on leaf surfaces (Lindemann and Suslow, 1985). Ice$^-$ strains also did not differ from parental strains in their survival in soil, their survival of freeze-thaw cycles in aqueous suspensions, or their ability to grow and survive when inoculated individually onto any of 65 plant species (Lindemann and Suslow, 1985; Lindow, 1985c). The Ice$^+$ phenotype apparently is not pivotal to the adaptations that *P. syringae* has evolved for the colonization of leaf surfaces.

Statistical analysis of competitive exclusion of a number of Ice$^+$ *P. syringae* strains with isogenic recombinant Ice$^-$ and heterologous non-INA strains revealed some interaction between the competitor and target organism (Kinkel and Lindow, 1992; Lindow, 1986). Thus, some specificity appears to exist in the competitive interaction exhibited among different strains of *P. syringae*. Most noteworthy, isogenic Ice$^-$ strains were not always superior to heterologous non-INA *P. syringae* strains in preventing population increase of a given Ice$^+$ strain (Kinkel and Lindow, 1989, 1992; Lindow, 1986). This raises the possibility that some Ice$^+$ *P. syringae* strains may escape competition against certain Ice$^-$ *P. syringae* strains. Field studies were useful in evaluating the potential for some Ice$^+$ *P. syringae* strains to coexist with Ice$^-$ recombinant *P. syringae* strains on the same plants. Plants inoculated in the field with Ice$^-$ *P. syringae* strains were used as "enrichment trap plants" to select for indigenous Ice$^+$ *P. syringae* strains that were able to coexist with applied Ice$^-$ strains (Lindow, 1988). Identification of Ice$^+$ strains that escaped competition has been useful in further studies of the nature of competitive interactions of bacteria on leaves.

In a more discriminating experimental approach to test for differential competition among Ice$^+$ bacteria, isogenic and nonisogenic *P. syringae* strain-pairs were inoculated onto leaves in different relative proportions but at a constant total concentration in order to apply de Wit replacement series models (Wilson and Lindow, 1991) to the interpretation of the experimental data. Plants were incubated at high relative humidity in a growth chamber until the carrying capacity was reached and equilibrium between the strains achieved. All isogenic *P. syringae* strain-pairs were equally competitive, and the combined epiphytic population in mixtures was equal to the sum of the expected contributions of each strain (Wilson and Lindow, 1991). In contrast, nonisogenic *P. syringae* strain-pairs often exhibited directional antagonism, where the contribution of Ice$^+$ strains to the final population was less than expected and the contribution of Ice$^-$ *P. syringae* strains was greater than expected. These results suggest that while *P. syringae* strains may compete for the same limiting resource(s) on leaves, they often compete unequally.

## Field Efficacy of Biological Control

Naturally occurring non-INA bacteria have been used successfully for the control of frost injury to plants under field conditions. In most cases, the optimal time of application of the antagonistic bacteria was determined after an analysis of the normal seasonal dynamics of total bacteria and Ice$^+$ bacteria on plants. *E. herbicola* strain M232A reduced the incidence of frost damage to corn plants when inoculated onto corn seedlings shortly after emergence (Lindow et al., 1983b). Several non-INA bacterial strains, including *P. fluorescens* A506, reduced the incidence of frost injury to pear trees when applied at initiation of flowering (Cody et al., 1987; Lindow, 1982, 1983c, 1985a,b). These and other naturally occurring bacterial strains also reduced the incidence of frost injury to potato, tomato, and navel orange in trials in California (Lindow, *unpublished*).

The extent of reduction of the incidence of frost injury by antagonistic bacteria and bactericides varies greatly from one study to another. In many trials, frost injury to treated plants was reduced by as much as fourfold in comparison to untreated plants (Lindow, 1982, 1983c, 1985a,b), while in other studies little reduction of frost injury was seen (Cody et al., 1987; Proebsting and Gross, 1988; Lindow, *unpublished*). It is likely that these differences in efficacy are due to two major factors that contribute to ice formation and propagation in plants: 1) minimum air temperature, and 2) duration of subfreezing temperatures and/or presence of free moisture on leaves at the time of freezing temperatures. As noted earlier, intrinsic ice nucleators that are active at temperatures of –5°C or colder are of significant abundance in or on plants (Lindow et al., 1982a). As the temperature of plant parts approaches –5°C, ice nucleation is likely to occur irrespective of the presence of Ice$^+$ bacteria. Only above this temperature is the incidence of frost injury proportional to the logarithm of the population size of Ice$^+$ bacteria. While the minimum air temperature in field plots can be accurately measured, the temperature of plant parts cannot be assumed to be equal to the air temperature. When the sky is clear and there is little air movement (a typical condition during radiative frost events), the temperature of plant parts can be as much as 3°C colder than the air nearby (Hendershott, 1962; Marcellos and Single, 1976; Modlibowska, 1962; Renquist, 1985). The actual temperatures of plant parts during field studies of antagonistic bacteria have not been reported and are impossible to ascertain in retrospect because of insufficient information on air movement, cloud cover, and other factors. It is possible that the minimum leaf temperature was lower than –5 to –6°C in those studies where the application of antagonistic bacteria did not result in significant frost control. It would be expected that frost control by reduction of Ice$^+$ bacteria would be maximized when subfreezing temperatures were substantially above –5 to –6°C and of short duration. Under these conditions, ice nucleation should be solely of microbial origin, and secondary propagation of ice between plant parts would be minimized. The presence of free moisture on plants, particularly when adjacent plants or plant parts are in physical contact, also facilitates propagation of ice from sites of nucleation to adjacent supercooled tissues. For example, it was observed that the incidence of frost damage was proportional to density of potato plants (Lindow, *unpublished*). While individual, spatially isolated potato shoots in a frost in 1990 exhibited an approximately 10% incidence of frost damage, individual potato shoots in plants having five or more shoots per plant had more than an 80% chance of freezing. This finding suggested that each potato shoot had an approximately 10% chance of containing a single ice nucleation event but that ice propagation was relatively common between adjacent potato shoots. Thus, the probability of freezing for any one potato shoot in a plant containing more than one shoot was the combination of the probabilities of nucleation events occurring on individual shoots. It could also be expected that plants (such as young herbaceous vegetables) or plant parts (such as newly developing flowers or flower clusters of deciduous fruit trees) that are physically separated from each other at the time of freezing should have maximum opportunity to supercool and avoid injurious ice formation. The interaction of the factors that influence ice propagation with procedures to reduce the number of ice nucleation events on plants must be further studied to better predict the amount of frost control that can be expected from biological control efforts for a given minimum temperature. Thermal video image analysis could allow us to address this issue by accurately determining not only the temperatures of individual leaves when ice nucleation occurs, but also the extent of ice propagation from leaf to leaf in plant canopies.

## Use of Recombinant Ice$^-$ *P. syringae* Strains in Field Studies

Field studies have been conducted in which Ice$^-$ recombinant *P. syringae* strains were used to study competition within this species and to assess biological frost control (Lindow and Panopoulos, 1988). Because in vitro techniques were used in their construction, special authorization was required to field-test these strains in the United States. Ice$^-$ deletion mutant strains of *P. syringae* had first been constructed late in 1982 (Lindow, 1990). At that time, authority to permit the release of recombinant bacteria from physical containment was vested in the Recombinant

DNA Advisory Committee of the National Institutes of Health (NIH) (Lindow, 1990). That isogenic Ice$^+$ and Ice$^-$ strains had similar behavioral characteristics and equal requirements for resources on leaves was at the heart of arguments used to justify the safety of releasing recombinant Ice$^-$ mutants in field studies. Because Ice$^-$ *P. syringae* strains were derived from indigenous Ice$^+$ *P. syringae* strains from northern California, their reintroduction into these same habitats was predicted not to lead to perceptible changes in the composition of indigenous populations. Both Ice$^+$ and non-INA *P. syringae* strains are common on many crops in northern California (Lindow, 1982). Thus, emigration of recombinant Ice$^-$ *P. syringae* strains from plot areas was predicted to have little potential to disrupt the population structure of Ice$^+$ *P. syringae* strains on cultivated and native plant species.

The NIH granted permission to conduct field studies of two different Ice$^-$ *P. syringae* strains for the biological control of frost injury to potato at the University of California Tulelake Field Station in northern California in 1983 (Lindow, 1990). Almost immediately, concerned citizens challenged the procedures used in the NIH regulatory process. Numerous legal challenges to these procedures were mounted in the next five years (Lindow, 1990). In late 1984, the U.S. Environmental Protection Agency (EPA) claimed jurisdiction to regulate recombinant bacteria that would be used as pesticides and required an Experimental Use Permit (EUP) for the conduct of such studies (Lindow, 1990). Considerable additional data on the environmental fate of Ice$^-$ *P. syringae* strains were collected prior to obtaining an EUP in mid-1986 (Lindow, 1990). Concerns over potential economic impact from threatened boycotts of local commodities prompted the regents of the University of California to delay experimentation through 1986 (Lindow, 1990). Lawsuits also required the University of California to prepare an Environmental Impact Report, further delaying testing until April 1987 (Lindow, 1990). Field tests of Ice$^-$ derivatives of *P. syringae* and *P. fluorescens* were also conducted in 1987 by T. V. Suslow and J. Lindemann of Advanced Genetic Sciences (now DNA Plant Technology Corp.).

As a condition of the EUP under which field studies of Ice$^-$ *P. syringae* strains were conducted, extensive monitoring was performed to determine the dispersal and survival of the bacteria in and around the plot. Bacteria were collected in the air around the plot, on the leaves of surrounding vegetation for as much as 100 m in all directions, in water of a nearby canal, on insects collected above the plot, and in soil at or near the plot (Lindow and Panopoulos, 1988). Although aerosolized bacteria were dispersed beyond the boundaries of the plot location for up to 30 m in all directions, emigrant bacterial populations were not established on surrounding vegetation, nor did they survive in soil (Lindow and Panopoulos, 1988). Recombinant Ice$^-$ strains were never detected on crop plants surrounding the plot area. Recombinant Ice$^-$ strains were detected in the soil within the plot area for only about 2 weeks after spraying of inoculum onto the soil surface during inoculation of leaves (Lindow and Panopoulos, 1988). Recombinant strains were not subsequently detected in soil or water. Thus, predictions of the habitat specificity of *P. syringae* determined in laboratory and greenhouse studies were confirmed in field studies.

Recombinant Ice$^-$ *P. syringae* strains achieved high population sizes on inoculated potato plants. The population size of Ice$^-$ strains was as high as $10^6$ cells/g on plants during the first 4 weeks after inoculation (Lindow and Panopoulos, 1988). Recombinant Ice$^-$ strains represented as much as 50% of the total bacteria on inoculated plants for the first 3 weeks after spray inoculation. Because potato plants grow rapidly during the first several weeks after emergence from soil, the population size of recombinant Ice$^-$ strains was substantially lower on leaves that emerged from plants after inoculation. The average population size of recombinant strains on inoculated plants dropped to less than $10^3$ cells/g fresh weight by about 4 weeks after inoculation (Lindow and Panopoulos, 1988). The population size of indigenous Ice$^+$ *P. syringae* strains was reduced about 50-fold on inoculated plants compared with controls during the first month after inoculation (Lindow and Panopoulos, 1988). In contrast, the population size of isogenic Ice$^+$ parental strains co-inoculated onto potato was reduced over 300-fold by the respective isogenic Ice$^-$ mutant strain (Lindow, *unpublished*). Thus, some specificity in competition for leaf surface habitats among *P. syringae* strains is evident.

Because the population size of Ice$^+$ bacteria was reduced by recombinant Ice$^-$ strains, the incidence of frost injury was reduced in field frosts. Several natural radiative frosts with minimum air temperatures in the range from –2 to –5°C occurred in the plot area between 1987 and 1991. The incidence of frost injury to potato plants treated with Ice$^-$ mutants was reduced by an average of approximately 70 to 80% compared with control plants in nine frost events (Lindow, *unpublished*). In some trials, the efficacy of naturally occurring non-INA strains of *P. syringae* and *P. fluorescens* were compared with recombinant Ice$^-$ *P. syringae* strains. Usually the control of frost injury to potato by these strains was as good as that afforded by recombinant Ice$^-$ *P. syringae* strains.

## Commercial Use of Competitive Bacteria

While several strains of epiphytic bacteria have shown potential for the biological control of frost injury to plants, their use for other than experimental purposes was restricted until 1992. The EPA has determined that any bacterium used for biological control of plant frost injury is a "pesticide," since the purpose is to mitigate Ice$^+$ bacteria, which can be considered a plant pest. Therefore, such bacteria need to be registered as "pesticides" by the EPA. The registration process requires that information be presented to document that such strains are efficacious and pose no human health or environmental risk. The data requirements for registration of a pesticide are rather large and thus costly to obtain. Even though biological pesticides are recognized as having less likelihood of being hazardous and thus must fulfill only an abbreviated set of safety tests, the costs of registration of a biological pesticide exceed $150,000, not including environmental studies and efficacy trials (T. Suslow, *personal communication*). Therefore, very few biological agents have been registered as pesticides in the United States. In 1992, a mixture of three bacterial strains (*P. fluorescens* strain A506, one naturally occurring strain of *P. syringae,* and one of *P. fluorescens*) was registered for use in frost control on several crops and for control of fire blight disease of pear and apple (caused by the bacterium *E. amylovora*) by Frost Technologies Corporation (a subsidiary of DNA Plant Technology Corp. of Oakland, California). Frost Technologies had developed, in collaboration with a company specializing in fermentation, an effective process for the large-scale production of the inoculum and its long-term storage as a freeze-dried powder. Large-scale trials of strain A506 and the bacterial mixture were conducted by this company and their cooperators beginning in 1990 when an EUP for this mixture was obtained from the EPA. The results of the field trials are promising. This item will be marketed under the tradename Frostban B by Plant Health Technologies.

## Conclusions

The capacity of plants to supercool in the range of 0 to –5°C is inversely related to the sizes of their epiphytic Ice$^+$ bacterial populations. This relationship raises the possibility of controlling plant frost injury by reducing the colonization by Ice$^+$ bacteria. Bactericides can be used to demonstrate this principle, but their cost and the emergence of resistance in bacterial populations limit their usefulness for this application

Antagonistic interactions between epiphytic bacteria hold the promise of allowing *biological* control of plant frost injury. It is well established that Ice$^+$ populations can be reduced by the introduction of particular strains of non-INA bacterial epiphytes; this phenomenon is most pronounced if the non-INA strains are able to colonize plant surfaces preemptively (before the arrival of substantial numbers of Ice$^+$ bacteria). Nutrient utilization is a mechanism by which one epiphytic population can limit another, as is well demonstrated by competition experiments involving isogenic pairs of Ice$^+$ and Ice$^-$ bacterial strains. However, an isogenic strain does not always make the best antagonist. Antibiosis has commonly been assumed to be an effective mode of antagonism between epiphytic strains, but this is doubtful in view of experiments in which antibiotic-producing strains failed to demonstrate greater antagonism than nonproducers. The efficacy of antagonism varies with weather, location, and plant host. The frost protection afforded to the host plant is still more variable, being affected also by the severity of frost events and the frequency of ice propagation between plants and plant parts.

## Literature Cited

Adaskaveg, J. E., and Hine, R. B. 1985. Copper tolerance and zinc sensitivity of Mexican strains of *Xanthomonas vesicatoria*, causal agent of bacterial spot of pepper. Plant Dis. 69:993-996.

Andersen, G. L., Menkissoglu, O., and Lindow, S. E. 1991. Occurrence and properties of copper-tolerant strains of *Pseudomonas syringae* isolated from fruit tree in California. Phytopathology 81:648-656.

Anderson, J. A., and Ashworth, E. N. 1986. The effects of streptomycin, desiccation, and UV radiation on ice nucleation by *Pseudomonas viridiflava*. Plant Physiol. 80:956-960.

Andrews, J. H. 1985. Strategies for selecting antagonistic microorganisms from the phylloplane. Pages 31-44 in: Biological Control on the Phylloplane, C. E. Windels and S. E. Lindow, eds. American Phytopathological Society, St. Paul, MN.

Bender, C. L., and Cooksey, D. A. 1986. Indigenous plasmids in *Pseudomonas syringae* pv. *tomato*: Conjugative transfer and role in copper resistance. J. Bacteriol. 165:534-541.

Beer, S. V., Rundle, J. R., and Norelli, J. L. 1984. Recent progress in the development of biological control of fire blight—A review. Acta. Hortic. 151:195-201.

Burr, T. J., Norelli, J. L., Reid, C. L., and Capron L. K. 1993. Streptomycin-resistant bacteria associated with fire blight infections. Plant Dis. 77:63-66.

Burke, M. J., Gusta, L. A., Quamme, H. A., Weiser, C. J., and Li, P. H. 1976. Freezing and injury to plants. Annu. Rev. Plant Physiol. 27:507-528.

Cary, J. W., and Mayland, H. F. 1970. Factors influencing freezing of supercooled water in tender plants. Agron. J. 62:715-719.

Chandler, W. H. 1958. Cold resistance in horticultural plants: A review. Proc. Am. Soc. Hort. Sci. 64:552-572.

Cody, Y. S., Gross, D. C., Proebsting, E. L., Jr., and Spotts, R. A. 1987. Suppression of ice nucleation active *Pseudomonas syringae* by antagonistic bacteria in fruit tree orchards and evaluations of frost control. Phytopathology 77:1036-1044.

Constantinidou, H. A., Menkissoglu, O., and Stergiadou, H. C. 1991. The role of icenucleation active bacteria in supercooling of citrus tissues. Physiol. Plant. 81:548-554.

Cooksey, D. A. 1990. Genetics of bactericide resistance in plant pathogenic bacteria. Annu. Rev. Phytopathol. 28:201-219.

Dik, A. J., Fokkema, N. J., and VanPelt, J. A. 1992. Influence of climatic and nutritional factors on yeast population dynamics in the phyllosphere of wheat. Microb. Ecol. 23:41-52.

Fossard, R. 1981. Effect of guttation fluids on growth of microorganisms on leaves. Pages 213-216 in: Microbial Ecology of the Phylloplane. J. P. Blakeman, ed. Academic Press, London.

Gadd, G. M., and Griffiths, A. J. 1978. Microorganisms and heavy metal toxicity. Microb. Ecol. 34:303-317.

Goto, M., Goto, T., and Inaba, T. 1989. Identification of ice nucleation active bacteria isolated from frost damaged vegetable leaves. Ann. Phytopathol. Soc. Jpn. 55:330-335.

Green, R., Corotto, L. V., and Warren, G. J. 1988. Deletion mutagenesis of the icenucleation gene from *Pseudomonas syringae* S203. Mol. Gen. Genet. 215:165-172.

Gross, D. C., Cody, Y. S., Proebsting, E. L., Jr., Radamaker, G. K., and Spotts, R. A. 1983. Distribution, population dynamics, and characteristics of ice nucleation active bacteria in deciduous fruit tree orchards. Appl. Environ. Microbiol. 46:1370-1379.

Gross, D. C., Proebsting, E. L., Jr., and Andrews, P. K. 1984. The effects of ice nucleation-active bacteria on the temperatures of ice nucleation and low temperature susceptibilities of *Prunus* flower buds at various stages of development. J. Am. Soc. Hort. Sci. 109:375-380.

Hendershott, C. H. 1962. The response of orange trees and fruit to freezing temperatures. Proc. Am. Soc. Hortic. Sci. 80:247-254.

Hirano, S. S., and Upper, C. D. 1983. Ecology and epidemiology of foliar plant pathogens. Annu. Rev. Phytopathol. 21:243-269.

Hirano, S. S., and Upper, C. D. 1986. Temporal, spatial and genetic variability of leaf-associated bacterial populations. Pages 235-251 in: Microbiology of the Phyllosphere. N. J. Fokkema, and J. Van Den Heuvel, eds. Cambridge University Press, London.

Hirano, S. S., and Upper, C. D. 1990. Population biology and epidemiology of*Pseudomonas syringae*. Annu. Rev. Phytopathol. 28:155-177.

Hirano, S. S., Baker, L. S., and Upper, C. D. 1985. Ice nucleation temperature of individual leaves in relation to population sizes of ice nucleation active bacteria and frost injury. Plant Physiol. 77:259-265.

Kaneda, T. 1986. Seasonal population changes and characterization of ice-nucleatingbacteria on farm fields of central Alberta. Appl. Environ. Microbiol. 52:173-178.

Kim, K. C., Kim, Y. C., and Cho, B. H. 1989. Antagonistic activity of an isolate of *Candida* species to ice nucleation active *Pseudomonas syringae*. Phytopathology 79:275-277.

Kinkel, L. L., and Lindow, S. E. 1989. The role of competitive interactions in bacterial survival and establishment on the leaf surface. Pages 634-638 in: Recent Advances in Microbial Ecology, T. Hattori, Y. Ishida, Y. Maruyama, R. Y. Morika, and A. Uchida, eds. Japan Scientific Societies Press, Kyoto.

Kinkel, L. L., and Lindow, S. E. 1992. Invasion and exclusion among intraspecificbacterial epiphytes. Phytopathology 82:1078.

Lindemann, J., and Suslow, T. V. 1985. Characteristics relevant to the question of environmental fate of genetically engineered $INA^-$ deletion mutant strains of *Pseudomonas*. Pages 1005-1012 in: Proc. Int. Conf. Plant Pathogenic Bacteria, 6th. E. Civerolo, ed. U.S. Department of Agriculture, College Park, MD.

Lindemann, J., and Suslow, T. V. 1987. Competition between ice nucleation active wild-type and ice nucleation deficient deletion mutant strains of *Pseudomonas syringae* and *P. fluorescens* biovar I and biological control of frost injury on strawberry blossoms. Phytopathology 77:882-886.

Lindow, S. E. 1982. Population dynamics of epiphytic ice nucleation active bacteria on frost sensitive plants and frost control by means of antagonistic bacteria. Pages 395-416 in: Plant Cold Hardiness. P. H. Li and A. Sakai, eds. Academic Press, New York.

Lindow, S. E. 1983a. The role of bacterial ice nucleation in frost injury to plants. Annu. Rev. Phytopathol. 21:363-384.

Lindow, S. E. 1983b. The importance of bacterial ice nuclei in plant frost injury. Curr. Top. Plant Biochem. Physiol. 2:119-128.

Lindow, S. E. 1983c. Methods of preventing frost injury caused by epiphytic ice nucleationactive bacteria. Plant Dis. 67:327-333.

Lindow, S. E. 1983d. Kinetics of changes in ice nucleation activity of *Pseudomonassyringae* following temperature shifts. Phytopathology 73:809.

Lindow, S. E. 1985a. Integrated control and role of antibiosis in biological control of fireblight and frost injury. Pages 83-115 in: Biological Control on the Phylloplane. C. Windels and S. E. Lindow, eds. American Phytopathological Society, St. Paul, MN.

Lindow, S. E. 1985b. Strategies and practice of biological control of ice nucleation active bacteria on plants. Pages 293-311 in: Microbiology of the Phyllosphere. N. Fokkema, ed. Cambridge University Press, London.

Lindow, S. E. 1985c. Ecology of *Pseudomonas syringae* relevant to the field use of Ice⁻ deletion mutants constructed in vitro for plant frost control. Pages 23-25 in: Engineering Organisms in the Environment: Scientific Issues. H. O. Halvorson, D. Pramer, and M. Rogul, eds. Am. Soc. Microbiology, Washington, DC.

Lindow, S. E. 1986. Specificity of epiphytic interactions of *Pseudomonas syringae* strains on leaves. Phytopathology 76:1136.

Lindow, S. E. 1987. Competitive exclusion of epiphytic bacteria by Ice⁻ mutants of *Pseudomonas syringae*. Appl. Environ. Microbiol 53:2520-2527.

Lindow, S. E. 1988a. Construction of isogenic Ice⁻ strains of *Pseudomonas syringae* for evaluation of specificity of competition on leaf surfaces. Pages 509-515 in: Microbial Ecology. F. Megusar and M. Gantar, eds. Slovene Society for Microbiology, Ljuvljana.

Lindow, S. E. 1988b. Lack of correlation of antibiosis in antagonism of ice nucleationactive bacteria on leaf surfaces by non-ice nucleation active bacteria. Phytopathology 78:444-450.

Lindow, S. E. 1990. Design and results of field trials of Ice⁻ recombinant *Pseudomonas syringae* strains. Pages 61-69 in: Risk Assessment in Agricultural Biotechnology: Proceedings of the International Conference. J. Marois and J. Bruhning, eds. University of California, Oakland, CA.

Lindow, S. E. 1993. Novel method for identifying bacterial mutants with reduced epiphyticfitness. Appl. Environ. Microbiol. 59:1586-1592.

Lindow, S. E., Arny, D. C., Barchet, W. R., and Upper, C. D. 1978a. The role of bacterialice nuclei in frost injury to sensitive plants. Pages 249-263 in: Plant Cold Hardiness and Freezing Stress. P. Li, ed. Academic Press, New York.

Lindow, S. E., Arny, D. C., and Upper, C. D. 1978b. Distribution of ice nucleation active bacteria on plants in nature. Appl. Environ. Microbiol. 36:831-838.

Lindow, S. E., Arny, D. C., and Upper, C. D. 1978c. *Erwinia herbicola*: A bacterial ice nucleus active in increasing frost injury to corn. Phytopathology 68:523-527.

Lindow, S. E., Arny, D. C., and Upper, C. D. 1982a. Bacterial ice nucleation: A factor infrost injury to plants. Plant Physiol. 70:1084-1089.

Lindow, S. E., Arny, D. C., and Upper, C. D. 1982b. The relationship between icenucleation frequency of bacteria and frost injury. Plant Physiol. 70:1090-1093.

Lindow, S. E., Arny, D. C., and Upper, C. D. 1983a. Biological control of frost injury I: An isolate of *Erwinia herbicola* antagonistic to ice nucleation-active bacteria. Phytopathology 73:1097-1102.

Lindow, S. E., Arny, D. C., and Upper, C. D. 1983b. Biological control of frost injury II: Establishment and effects of an antagonistic *Erwinia herbicola* isolate on corn in the field. Phytopathology 73:1102-1106.

Lindow, S. E., and Connell, J. H. 1984. Reduction of frost injury to almond by control of ice nucleation active bacteria. J. Amer. Soc. Hort. Sci. 109:48-53.

Lindow, S. E., Lahue, E., Govindarajan, A. G., Panopoulos, N. J., and Gies, D. 1989. Localization of ice nucleation activity and the *iceC* gene product in *Pseudomonas syringae* and *Escherichia coli*. Mol. Plant-Microbe Interact. 2:262-272.

Lindow, S. E., and Panopoulos, N. J. 1988. Field tests of recombinant Ice⁻ *Pseudomonas syringae* for biological frost control in potato. Pages 121-138 in: Proc. Int. Conf. on Release of Genetically Engineered Microorganisms, 1st. M. Sussman, C. H. Collins, and F. A. Skinner, eds. Academic Press, London.

Lucas, J. W. 1954. Subcooling and ice nucleation in lemon. Plant Physiol. 29:245.

Maki, L. R., Galyon, E. L., Chang-Chien, M., and Caldwell, D. R. 1974. Ice nucleation induced by *Pseudomonas syringae*. Appl. Microbiol. 28:456-460.

Marcellos, H. W., and Single, W. V. 1976. Ice nucleation on wheat. Agric. Meteorol. 16:125-119.

Marcellos, H. W., and Single, W. V. 1979. Supercooling and heterogeneous nucleation of freezing in tissues of tender plants. Cryobiol. 16:74-77.

Marshall, D. 1988. A relationship between ice nucleation active bacteria, freeze damage, and genotype in oats. Phytopathology 78:952-957.

Menkissoglu, O., and Lindow, S. E. 1991a. Relationship of free ionic copper and toxicity in solutions of organic compounds. Phytopathology 81:1258-1263.

Menkissoglu, O., and Lindow, S. E. 1991b. Chemical form of copper on leaves in relation to the bactericidal activity of cupric hydroxide deposits on leaves. Phytopathology 81:1263-1270.

Modlibowska, I. 1962. Some factors affecting supercooling of fruit blossoms. J. Hortic. Sci. 37:249-261.

Montesinos, E., and Vilardell, P. 1991. Relationships among population levels of *Pseudomonas syrin-*

*gae*: Amount of ice nuclei and incidence of blast of dormant flower buds in commercial pear orchards in Catalonia, Spain. Phytopathology 81:113-119.

Olson, B. D., and Jones, A. L. 1983. Reduction of *Pseudomonas syringae* pv. *morsprunorum* on Montmorency sour cherry with copper and dynamics of the copper residues. Phytopathology 73:1520-1525.

Orser, C. S., Staskawicz, B. J., Panopoulos, N. J., Dahlbeck, D., and Lindow, S. E. 1985. Cloning and expression of bacterial ice nucleation genes in *Escherichia coli*. J. Bacteriol. 164:359-366.

Proebsting, E. L., Jr., Andrews, P. K., and Gross, D. 1982. Supercooling young developing fruit and floral buds in deciduous orchards. Hortic. Sci. 17:67-68.

Proebsting, E. L., Jr., and Gross, D. C. 1988. Field evaluations of frost injury to deciduous fruit trees as influenced by ice-nucleation-active *Pseudomonas syringae*. J. Am. Soc. Hortic. Sci. 113:498-506.

Renquist, A. R., 1985. The extent of fruit bud radiant cooling in relation to freeze protection with fans. Agric. For. Meteorol. 36: 1-6.

Rogers, J. S., Stall, R. E., and Burke, M. J. 1987. Low temperature conditioning of the ice nucleation active bacterium, *Erwinia herbicola*. Cryobiology 24:270-279.

Ruvkun, G. B., and Ausubel, F. M. 1981. A general method for site-directed mutagenesis in prokaryotes. Nature (London) 289:85-88.

Sundin, G. W., Jones, A. L., and Fulbright, D. W. 1989. Copper resistance in *Pseudomonas syringae* pv. syringae from cherry orchards and its associated transfer *in vitro* and *in planta* with a plasmid. Phytopathology 79:861-865.

Tukey, H. B. 1966. The leaching of metabolites from above-ground plant parts and its implications. Bull. Torrey Bot. Club 93:385-401.

Tukey, H. B. 1970. The leaching of substances from plants. Annu. Rev. Plant Physiol. 21:305-324.

Tukey, H. B. 1971. Leaching of substances from plants. Pages 67-80 in: Ecology of Leaf Surface Microorganisms. T. F. Preece and C. H. Dickinson, eds. Academic Press, London.

Vanneste, J. L., Yu, J., and Beer, S. V. 1992. Role of antibiotic production by *Erwinia herbicola* Eh252 in biological control of *Erwinia amylovora*. J. Bacteriol. 174:2785-2796.

Warren, G. J. 1987. Bacterial ice nucleation: Molecular biology and applications. Biotechnol. Gen. Eng. Rev. 5:107-135.

Warren, G., and Wolber, P. 1991. Molecular aspects of microbial ice-nucleation. Mol. Microbiol. 5:239-243.

Warren, G. J., Lindemann, J., Suslow, T. V., and Green, R. L. 1987. Ice Nucleation-Deficient Bacteria as Frost Protection Agents. Pages 215-227 in: Biotechnology in Agricultural Chemistry. H. M. LeBaron, R. O. Mumma, R. C. Honeycutt, and J. H. Duesing, eds. American Chemical Society, Washington, DC.

Wilson, M., and Lindow, S. E. 1993. Interactions between the biological control agent *Pseudomonas fluorescens* A506 and *Erwinia amylovora* in pear blossoms. Phytopathology 83:117-123.

Wilson, M., and Lindow, S. E. 1991. Resource partitioning among bacterial epiphytes in the phyllosphere. Phytopathology 81:1170.

Wolber, P. K. 1992. Bacterial ice nucleation. Adv. Microbiol. Physiol. 34:205-237.

Yankofsky, S. A., Levin, Z., Bertold, T., and Sandlerman, N. 1981. Some basic characteristics of bacterial freezing nuclei. J. Appl. Meteorol. 20: 1013-1019.

Yelenosky, G. 1983. Ice nucleation active (INA) agents in freezing of young citrus trees. J. Am. Soc. Hortic. Sci. 108:1030-1034.

Yelenosky, G. 1991. Apparent nucleation and freezing in different parts of young citrus trees during controlled freezes. Hortscience 26:576-579.

Young, R. H. 1966. Freezing points and lethal temperatures of citrus leaves. Proc. Am. Soc. Hortic. Sci. 88:272-279.

CHAPTER 14

# Biological Control of Insect Pests Using Ice-Nucleating Microorganisms

**Richard E. Lee, Jr., Marcia R. Lee, and Janet M. Strong-Gunderson**

## Supercooling and Ice Nucleation in Insects

One of the most fascinating aspects of overwintering insects is their capacity for extensive supercooling of body fluids. Many species remain unfrozen to –20°C or lower. Several Alaskan gall-forming species have exceptional supercooling capacities, extending their limits to –60°C (Miller, 1982). Due to their small size, and therefore small water volume, insects have the innate capacity to supercool extensively as long as efficient ice nucleators are absent (Lee, 1991; Chapter 1). The accumulation of low molecular mass polyols and sugars such as glycerol, sorbitol, and trehalose, in sometimes multimolar concentrations, also enhances supercooling in insects (Lee, 1991).

The limit of supercooling, termed the supercooling point or the temperature of crystallization, is a significant temperature for both freeze-tolerant and freeze-intolerant insects. For those few insects able to survive extensive internal ice formation, this temperature represents a major transition to a physiological state of anaerobiosis, cellular dehydration due to freeze concentration of solute, and depressed metabolism (Storey and Storey, 1988). However, the vast majority of insects are freeze-intolerant, unable to survive the freezing of their body water. For these species, including most pest insects, the supercooling point represents the absolute, lowest temperature for survival (Lee, 1989, 1991). Nonetheless, it should be recognized that without empirical testing, this value cannot be assumed to be the lethal low temperature because some species die from chilling or cold shock injury at temperatures above the supercooling point (Lee and Denlinger, 1985; Bale, 1987).

Insects' supercooling capacity may be limited by inoculative freezing and the action of several classes of heterogeneous ice nucleators (Lee et al., 1993; Chapter 11). Inoculative freezing may occur when an insect comes in contact with external ice (Layne et al., 1990). Interestingly, in diapausing drosophilid larvae of *Chymomyza costata*, freezing is tolerated only if it occurs by inoculation at high subzero temperatures, but not if larvae supercool extensively before freezing (Shimada and Riihimaa, 1988). Fields and McNeil (1986) reported similar results with overwintering larvae of an arctiid moth. Endogenous, heterogeneous ice nucleators include ice-nucleating proteins and lipoproteins (Zachariassen and Hammel, 1976; Duman et al., 1991). Internal inorganic nucleators may also play a role in regulating the supercooling point of insects; calcium phosphate crystallike compounds located in the Malpighian tubule have ice-nucleating activity sufficient to explain the supercooling point of the freeze-tolerant gall fly, *Eurosta solidaginis* (R.E. Lee et al., 1992a). Lastly, ice nucleation-active microorganisms have been reported recently as normal flora in the gut of insects (Lee et al., 1993). This chapter will focus on the natural relationships between ice nucleation–active microorganisms and insects and their potential use for the biological control of insect pests.

## Discovery of Ice-Nucleating Microorganisms Naturally Associated with Insects

Although a number of other anatomic locations have been reported, the insect gut is the one most commonly associated with the onset of ice nucleation in insects (Lee et al., 1993). As early as 1936, Salt linked an elevation of the insect supercooling point with feeding. Various investigators have reported enhanced supercooling capacity in insects that had emptied their gut (see review by Cannon and Block, 1988). Johnston and Lee (1990) reported that the gut was the most frequent site of efficient nucleators in mealworm larvae, *Tenebrio molitor*. In prepupae of *Trichiocampus populi*, Shimada (1989) directly observed ice nucleation in the gut and the growth of the ice lattice across the gut wall using cryomicroscopy. Although Salt and others suggested that dust, accidentally ingested with food, might serve as ice nucleators in the gut (Somme, 1982; Baust and Rojas, 1985; Cannon and Block, 1988), this notion has been difficult to confirm directly. Furthermore, questions remain as to whether the ice-nucleating activity of dust particles and food materials could explain the observed supercooling points at high subzero temperatures. Nonetheless, these observations strongly suggest the presence of efficient ice nucleators in the insect gut.

Until recently, ice nucleation-active microorganisms had only been associated with plants. In the early 1970s, $Ina^+$ bacteria were first discovered in decaying leaves (see Chapter 2 and other chapters in this volume). These potent biological nucleators are restricted to gram-negative, rod-shaped asporogenous bacteria displaying ice-nucleating activity as high as −1°C. Several lichen species also have ice-nucleating activity associated with their fungal component as well as several *Fusarium* spp. (Kieft, 1988; Chapter 9).

Consequently, although no previous reports had directly linked any animal with $Ina^+$ bacteria, in the mid-1980s, we began studies to determine whether $Ina^+$ bacteria were present naturally in the insect gut (Strong-Gunderson et al., 1990a; Lee et al., 1991). During the summers of 1988 and 1989, we isolated *Enterobacter taylorae* and *E. agglomerans* from the gut of a field-collected bean leaf beetle, *Cerotoma trifurcata*, and the convergent lady beetle, *Hippodamia convergens* (Table 1). Both *Enterobacter* species exhibited maximal ice nucleation thresholds of approximately −2°C that were only slightly less than that of *Pseudomonas syringae*.

Independent investigations by Kaneko et al. (1991a,b) identified an $Ina^+$ bacterium in the gut of the diamondback moth, *Plutella xylostella*. Previously, they ob-

**Table 1.** Ice nucleating active microorganisms isolated from the gut of insects

| Microorganism | Insect | Reference |
|---|---|---|
| Bacteria | | |
| *Enterobacter agglomerans* | *Hippodamia convergens* | Strong-Gunderson et al. (1990a) |
| | *Cerotoma trifurcata* | Lee et al. (1991) |
| *E. taylorae* | *H. convergens* | Strong-Gunderson et al. (1990a) |
| | *C. trifurcata* | Lee et al. (1991) |
| *Erwinia herbicola* | *Plutella xylostella* | Kaneko et al. (1991a,b) |
| Fungi | | |
| *Fusarium* sp. | *Chilo suppressalis* | Tsumuki et al. (1992) |

served that pupae that were fed germinating radish seeds had higher supercooling points than ones fed aseptic seeds (Kaneko et al., 1989). Furthermore, supercooling points increased when pupae were held at a lower temperature. This response is consistent with the effect of low-temperature conditioning, which results in increased ice-nucleating activity in Ina$^+$ bacteria (Rogers et al., 1987). In turn, this observation led to the isolation and identification of *Erwinia herbicola* from the gut of diamondback moth pupae (Kaneko et al., 1991a,b).

In 1992, a particularly interesting report described an Ina$^+$ fungus (*Fusarium* sp.), isolated from the gut of the rice stem borer larvae, *Chilo suppressalis* (Tsumuki et al., 1992). Ingestion of a mycelial suspension caused supercooling points to increase from –20 to –5°C. This report is significant for two reasons. First, at the time this work was being done, ice-nucleating activity in a free-living fungus was unknown. (However, an independent study by Pouleur and co-workers (1991) reported ice-nucleating activity in two species of free-living fungi, *Fusarium avenaceum* and *F. acuminatum.*) Secondly, the presence of an ice-nucleating microorganism in freeze-tolerant larvae suggests a mutualistic relationship between the fungus and its host (Lee et al., 1993). Here, as is believed for ice-nucleating proteins and lipoproteins (see Chapter 11), the ice-nucleating microorganisms may function to ensure that ice nucleation occurs at high subzero temperatures and to promote slow growth of the ice lattice in the extracellular space. In fact, ice-nucleating microorganisms might function even better in this regard, since they have greater ice-nucleating activity than ice-nucleating proteins and lipoproteins.

In summary, only since 1990 have Ina$^+$ bacteria and a fungus been reported as normal flora in the gut of insects (Table 1). These investigations also identified novel ice-nucleating microorganisms with previously unreported ice-nucleating activity. Further, we have identified three species of Ina$^+$ bacteria, *Pseudomonas fluorescens, E. agglomerans* and *P. putida*, from the gut of a freeze tolerant frog (M.R. Lee et al., 1992a). Considering the few research laboratories that have searched for Ina$^+$ microorganisms associated with animals, the association of these organisms may be much more common than previously recognized.

## Biological Control of Insect Pests Using Ice-Nucleating Microorganisms

A number of investigators with primarily basic approaches to the study of insect diapause and cold-hardiness have wondered, at least as a secondary objective, whether it might be possible to interfere with these natural adaptations for overwintering survival and, thereby, diminish populations of insect pests. In the mid-1980s, we began to wonder whether the natural capacity of overwintering insects to escape lethal freezing by supercooling might be artificially reduced by introducing potent nucleators into their bodies or onto their cuticles. Although at this time there was no evidence that Ina+ microorganisms were associated with insects, or any animal for that matter, we wondered whether these biological ice nucleators, whose role in promoting frost damage in plants had been so clearly established, might be used to manipulate insect supercooling points (Lee, 1990).

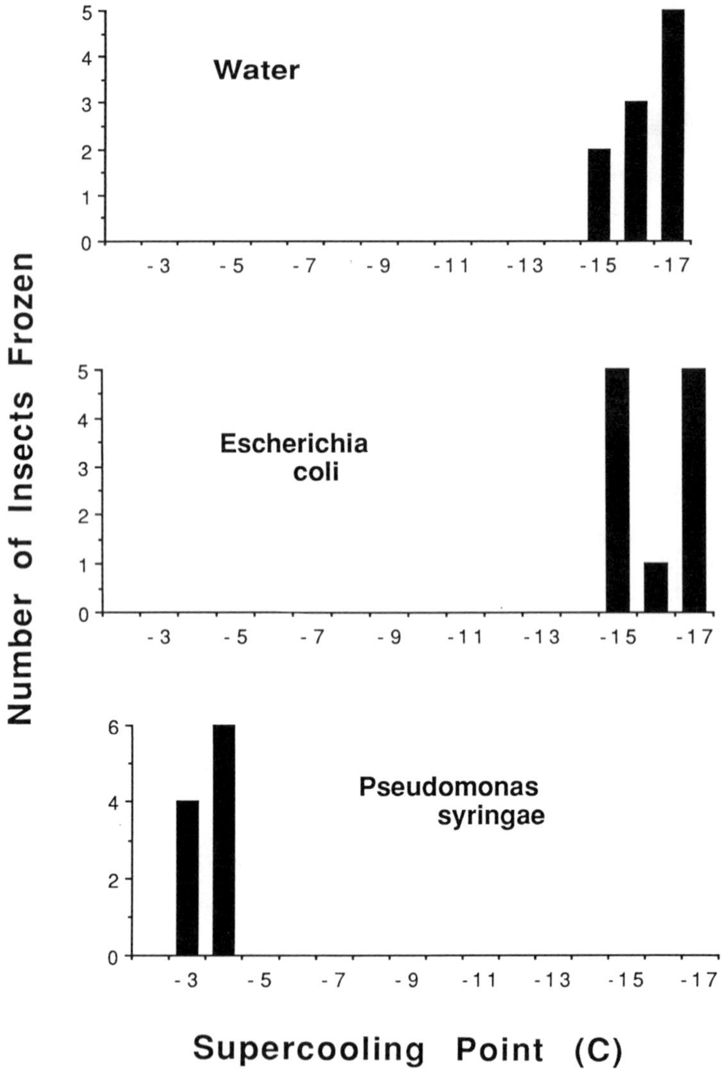


**Figure 1.** Supercooling points of adult lady beetles (*Hippodamia convergens*) determined immediately after ingesting water, or suspensions of a non–ice nucleation-active bacterium, *Escherichia coli*, or the epiphytic ice nucleating active bacterium, *Pseudomonas syringae* (both at $10^9$ bacteria/ml of water).

In our initial study we decided to use the lady beetle, *Hippodamia convergens*, as an insect model because we were familiar with its cold tolerance (Lee, 1980). This freeze-intolerant species enhances its supercooling capacity to approximately −16°C in winter. Ingestion of water or a solution of a non-Ina⁺ bacterium had no effect on the supercooling point; however, within seconds of ingesting a solution of *P. syringae*, supercooling points increased by as much as 14°C to −2°C (Fig. 1) (Strong-Gunderson et al., 1990b). Supercooling points remained elevated for at least 7 days after ingestion. In additional experiments, we were surprised that the supercooling point of beetles, whose mouths had been sealed, could be readily increased when sprayed with solutions of Ina⁺ bacteria (Strong-Gunderson et al., 1989, 1992). A possible explanation for this unexpected result comes from studies in our laboratory demonstrating that topical application of *P. syringae* to the thoracic spiracles was more effective for increasing supercooling points values than application to the abdomen and other external anatomic sites (Steigerwald et al., 1993). The elevation of insect supercooling points using Ina⁺ microorganisms has now been documented in more than 15 species from four orders of insects using five species of Ina⁺ bacteria and the fungi *Fusarium acuminatum* and *F. avenaceum* (Lee et al., 1993; Fields et al., 1993).

## Biological Control of Stored-Product Pests

Although control of stored product insects currently relies on chemical means, concerns for the toxic effects on nontarget species, the development of resistance, the removal of widely used insecticides from the market, and the problem of potential residues in food have raised interest in alternative forms of control and the use of integrated pest management strategies (Fields, 1992; Hagstrum and Flinn, 1992). Exposure to low temperature has long been used as a means for control of stored product pests because of its obvious advantages over chemical means of control (see review by Fields, 1992). Some stored product pests are very susceptible to low temperature and die even when exposed to temperatures above 0°C or only a few degrees below zero. Other stored product pests, such as the rusty grain beetle, *Cryptolestes ferrugineus*, are very cold tolerant, but not freeze tolerant, surviving weeks or months of exposure to subzero temperatures (Fields, 1990).

As a consequence, Fields (1990) proposed using Ina⁺ bacteria as a cold synergist to increase the mortality of stored product insects exposed to cold. Obviously, this approach requires that the environmental temperature of the insect pest must be lowered to that of the supercooling point either by aerating storage bins with ambient air or by refrigeration. This strategy is supported by the fact that all of the major stored product pests are freeze-intolerant. In a laboratory study, Fields (1990) used a freeze-dried, killed preparation of *P. syringae* to reduce the supercooling capacity and cold tolerance of non-cold-acclimated rusty grain beetles. Untreated control beetles have a mean supercooling point of −17°C. However, for individuals treated with 100 or 1,000 ppm of *P. syringae*, the supercooling point increased by approximately 6 and 8°C, respectively. Similar treatments caused significantly greater mortality of beetles treated with *P. syringae* than of controls when exposed to −10°C for 24 hours. In another study, Fields (1993) found that although cold-acclimated individuals of *C. ferrugineus, Sitophilus granarius* and *Oryzaephilus surinamensis* were more cold tolerant than non-acclimated ones, treatment with Ina⁺ bacteria also decreased their tolerance of subzero exposure.

**Table 2.** The effect of the Ina+ bacteria *Pseudomonas syringae* on the supercooling point (mean ± SE) of stored grain insects[a,b]

| Species | Supercooling point (°C) | |
|---|---|---|
| | Untreated control | *P. syringae* (100 ppm) |
| *Plodia interpunctella* | | |
| Indianmeal moth larvae | −10.3 ± 0.4 | −5.4 ± 0.5 |
| *Sitophilus granarius* | | |
| Granary weevil adults | −15.7 ± 1.0 | −8.0 ± 0.6 |
| *Rhyzopertha dominica* | | |
| Lesser grain beetle adults | −15.2 ± 0.6 | −3.3 ± 0.1 |
| *Tribolium castaneum* | | |
| Red flour beetle adults | −13.9 ± 0.8 | −4.7 ± 0.4 |
| *Gibbium psylloides* | | |
| Shiny spider beetle adults | −10.7 ± 0.9 | −6.0 ± 0.5 |

[a]Adapted from R. E. Lee et al. (1992b).
[b]Insects were treated with 100 ppm of *P. syringae* in wheat for 24 h at 23°C. For each species, treatment with *P. syringae* caused a statistically significant increase in the supercooling point (Student's $t$ test, $P < 0.01$).

**Table 3.** Survival of larval and adult stored grain pests exposed to various concentrations of dry, powdered *Pseudomonas syringae* in wheat for 24 h at 23°C before 24 h of exposure to −5°C[a,b]

| Species | Survival (%) | | |
|---|---|---|---|
| | | *Pseudomonas syringae* | |
| | Control | 100 ppm | 1,000 ppm |
| *Tenebrio molitor* | | | |
| Yellow mealworm larvae | 97 | 77 | 73 |
| *Sitophilus granarius* | | | |
| Grainery weevil adults | 96 | 62 | 48 |
| *Cryptolestes ferrugineus* | | | |
| Rusty grain beetle adults | 97 | 0 | 0 |
| *Cryptolestes pusillus* | | | |
| Flat grain beetle adults | 43 | 0 | 0 |
| *Plodia interpunctella* | | | |
| Indianmeal moth larvae | 96 | 36 | 3.8 |
| *Tribolium castaneum* | | | |
| Red flour beetle adults | 19 | 0 | 0 |
| *Rhyzopertha dominica* | | | |
| Lesser grain borer adults | 90 | 3.6 | 0 |
| *Gibbium psylloides* | | | |
| Shiny spider beetle adults | 93 | 25 | 0 |

[a]Adapted from R. E. Lee et al. (1992b).
[b]All treatments ($n = 30$) were significantly different ($P < 0.05$) from controls.

We conducted additional experiments using eight species of stored grain insects (R. E. Lee et al., 1992b). In five species, supercooling points of control insects ranged from –10.3 to –15.7°C; however, following treatment with 100 ppm of *P. syringae* preparation in wheat, mean supercooling points increased for all species, with *Rhyzopertha dominica* having the highest value of –3.3°C (Table 2). Complementary studies examined insect survival at subzero temperatures following treatment with Ina+ bacteria. Although some species were intolerant of –5°C exposure

**Table 4.** Cumulative mortality (%) of cold-acclimated rusty grain beetle (*Cryptolestes ferrugineus*) treated with 1,000 ppm of dry, powdered *Pseudomonas syringae* in aerated Manitoba granaries[a]

| Time after bacterial application (days) | Granary temperature (°C) | Mortality (%) | |
|---|---|---|---|
| | | Untreated | Treated |
| 7 | >0 | 3–5 | 3–7 |
| 15 | −9 to −15 | 32–53 | 98–99 |
| 22 | −9 to −15 | 62–69 | 98–100 |
| 30 | −9 to −15 | 98–99 | 100 |

[a]Data from Fields (1993).

for 24 hours even when not treated with bacteria, treatment with either 100 or 1,000 ppm of *P. syringae* in wheat caused a significant, and a dose-dependent reduction in survival for all species tested (Table 3). When the temperature of exposure was decreased to −8°C, this trend continued and survival rates decreased still further (R.E. Lee et al., 1992b).

Fields (1993) also examined the efficacy of *P. syringae* in reducing the cold tolerance of *C. ferrugineus* under field conditions in Manitoba granaries. Groups of cold-acclimated rusty grain beetle adults were treated with 1,000 ppm of *P. syringae* and placed in wheat granaries in early December. On days 15 and 22 treated beetles had significantly lower survival rates than untreated groups (Table 4). However, by day 30 nearly all individuals in both treated and untreated groups had died.

From these initial studies with stored product pests, it is clear that $Ina^+$ bacteria may be used to decrease the supercooling capacity and cold-hardiness even of cold-acclimated insects. In an attempt to standardize protocols for future studies of survival of stored product insects at extreme temperatures, Fields (1992) recommended the following: 1) tests should be conducted with insect strains that have been in the laboratory no more than 2 years, 2) the most temperature-resistant developmental stage of the pest should be used, 3) temperature-acclimated insects should be used, 4) a range of extreme temperatures should be tested so that data may be analyzed using probit analysis and that fiducial limits may be reported, and 5) the results of laboratory studies should be confirmed with field tests. Future investigations using $Ina^+$ microorganisms for the control of stored product pests should follow these criteria.

## Prospects for the Biological Control of the Colorado Potato Beetle

The Colorado potato beetle, *Leptinotarsa decemlineata* (Say), is the most serious defoliating pest of potatoes, *Solanum tuberosum* L., in North America. This species overwinters by burrowing into the soil in late summer or early autumn. When overwintering adults emerge from dormancy, they can significantly reduce yields by defoliating the early growth stages of the potato plants (Shields and Wyman, 1984). This pest is known for the wide range of pesticides, including synthetic pyrethroids, to which it has rapidly developed resistance (Casagrande, 1987). The current agricultural practice of planting extensive monocultures of potatoes further promotes the cumulative buildup of Colorado potato beetle populations from year to year. Because of these factors, current research efforts have increasingly begun to focus on alternative forms of control.

One novel approach to management of the pest uses cultural methods designed to expose Colorado potato beetles to lethal low temperatures during the winter (Milner et al., 1992; Kung et al., 1992). Trap crops of potatoes are planted in midsummer, and because these younger plants are more attractive to the adults, the beetles concentrate in these areas. In late summer, trap plants are mulched to encourage the beetles to remain at these sites over winter rather than dispersing from the fields. Insect mortality is induced by removing the mulch in midwinter immediately prior to a cold front, which causes the soil temperatures to drop rapidly. Mulching may also reduce the depth to which the beetles will burrow during the winter, increasing further their susceptibility to the lowering of soil temperature. The feasibility of this approach was recently supported by the work of Milner et al. (1992). They found that adult beetle survival was significantly lower at sites where mulch was removed in midwinter.

In collaboration with Jeffrey Wyman and Phil Kaufman, University of Wisconsin, we have begun investigating the use of Ina$^+$ microorganisms to increase the susceptibility of the overwintering beetles to low temperatures (Lee et al., 1994). Our rationale is to decrease the cold-hardiness of the beetles using these biological nucleators when applied in conjunction with the cultural control approach (Milner et al., 1992).

In our initial study, we determined that the Colorado potato beetle is a freeze-intolerant species that dies when cooled to its supercooling point (Lee et al., 1994). However, the overwintering beetles survive to temperatures immediately above their supercooling point, indicating that death is due to the onset of internal ice formation, and not low temperature per se. This result also indicates that the supercooling point may be used as a measure of the lethal low temperature, at least during short-term exposure to cold.

Considering the relatively high supercooling point of approximately –7°C for overwintering adults, it is obvious that this species lacks exceptional cold tolerance. Their limited capacity for supercooling is not surprising, however, considering their thermally protected overwintering site in the soil (Lee, 1991). Nonetheless, an elevation in the supercooling point of as little as 2 to 4 degrees in the lethal low temperature would be of major significance in decreasing the proportion of beetles surviving the winter. Colorado potato beetles exposed to –4°C had a survival rate of 54.8%, whereas only 6.2% of those exposed to –6°C survived (Kung et al., 1992).

To simulate overwintering conditions in the laboratory, we tested whether supercooling points increased in beetles that were exposed to a concentrated, freeze-dried and killed preparation of *P. syringae* mixed with soil (Lee et al., 1994). Mean supercooling points of beetles treated with *P. syringae* in concentrations ranging from 0 to 1,000 ppm were determined (Fig. 2A). In both 1991 and 1992, the supercooling point means increased significantly when beetles were exposed to increasing concentrations of *P. syringae*, ranging from –7.6 ± 0.2°C (untreated) to –3.7 ± 0.1°C (1,000 ppm). In the 1992 tests, as little as 1 ppm resulted in a supercooling point that was statistically higher than that of the untreated control. These results indicate that the effect of Ina$^+$ microorganisms on the supercooling point is dose-dependent as has been reported previously in other insects (Fields, 1990; R.E. Lee et al., 1992b).

The cumulative freezing distributions, comparable to the ice nucleation spectra that are typically used to describe the ice-nucleating activity of Ina$^+$ bacteria, were also determined for these beetles (Fig. 1B). These curves are a useful form of data presentation because they show a profile of the theoretical lethal low temperature for a population of beetles treated with various concentrations of *P. syringae* (Lee et al., 1994). For example, if beetles were exposed to −5°C, 80% of those treated with 100 ppm of *P. syringae* would be expected to freeze and die; in contrast, none or very few of the untreated control beetles would be expected to freeze at this temperature. The similarity of the 100 and 1,000 ppm curves further suggests that, under these conditions, the elevation of the supercooling point reaches a maximum near 100 ppm.

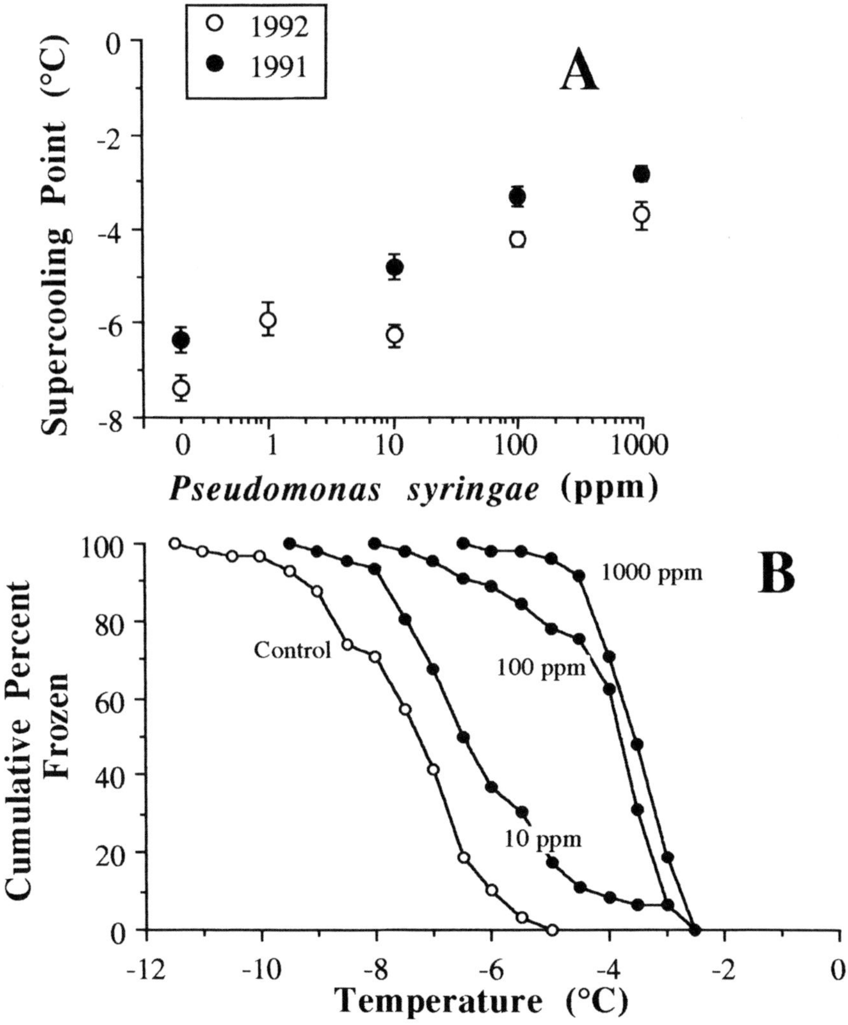


**Figure 2. A,** Effect of *Pseudomonas syringae* on the mean (± standard error) supercooling point of diapausing adults of the Colorado potato beetle. Beetles were exposed to various concentrations (0–1,000 ppm) of *P. syringae* in soil for 48 hours at 4°C. In 1991, sample sizes were $n$ = 10–11, and in 1992, $n$ = 44–58. **B,** Cumulative freezing profile for beetles exposed to various concentrations of *P. syringae* in 1992 . (Adapted from Lee et al., 1994.)

Another critical factor in the development of a biological control strategy for this species is the duration of the supercooling point elevation after application of microorganisms. The supercooling point of beetles treated with *P. syringae* was elevated significantly for 7 days after application at 4°C but was similar to that of untreated beetles by day 14 (Lee et al., 1994). Previous studies have reported a loss of bacterial ice-nucleating activity with time at temperatures above 0°C (Goodnow et al., 1990). Treatment temperature also affected supercooling point elevation of beetles following treatment with *P. syringae*. Although after 7 days the mean supercooling points for untreated beetles incubated at 4 or 10°C were similar, the values for those treated with *P. syringae* remained significantly higher at 4 than at 10°C. These data suggest that ice-nucleating activity was better retained during incubation at the lower temperature. In contrast, Fields et al. (1993) reported long-term stability of a *P. syringae* preparation held at 30°C for 8 weeks. Obviously, additional study is needed regarding the effects of temperature and duration of exposure on supercooling point elevation by *P. syringae*.

At this time it is envisioned that northern populations of the Colorado potato beetle would be treated with Ina$^+$ bacteria in late August or early September, when adults have been attracted to feed on trap crops on the edges of fields, but before they have begun to burrow into the soil to overwinter. Since ambient temperatures are still relatively high at this time of the year, the ice-nucleating activity of the bacteria may be lost before environmental temperatures drop low enough (even after applying the cultural manipulations of Milner et al. [1992]) to kill the beetles. One alternative approach to this problem would be to find ways to maintain the ice-nucleating activity of microorganisms in the gut or on the surface of the insect or to apply other Ina$^+$ microorganisms that would be retained by the beetles until environmental temperatures in their microhabitat decrease to lethal levels. Consequently, we have begun to test the efficacy of several different species of living Ina$^+$ bacteria and fungi for their effect on the beetle's supercooling point. We have also tested the effect of suspensions of living Ina$^+$ bacteria on the supercooling point. Suspensions of living *P. fluorescens, P. syringae,* and *P. putida* sprayed onto the adults all caused a significant increase in the supercooling point, indicating that living bacterial cells may also be used for supercooling point manipulation.

Another modification that may be of value in the development of biological control methods is the use of surfactants in combination with Ina$^+$ microorganisms (Lee et al., 1993). The addition of Tween 80 to *F. acuminatum* suspensions significantly increased the amount of supercooling point elevation in the beetle, *H. convergens,* compared with use of this Ina$^+$ fungus alone (M.R. Lee et al., 1992b).

In the case of the Colorado potato beetle, the application of the Ina$^+$ bacteria would be facilitated by the use of trap cropping, which concentrates the beetles in a narrow portion of the field and, thereby, would reduce the amount and cost of biological nucleators that must be applied. The successful integration of these biological ice nucleators with the cultural control strategy of Milner et al. (1992) also would allow use of this cultural control approach in areas of the country where it otherwise could not be used because winter soil temperatures would be too mild.

## Concluding Remarks

Use of Ina$^+$ microorganisms for biological control has both advantages and disadvantages. First, since it appears to be effective against a diverse range of insects, consideration must be given to avoiding detrimental effects on beneficial insects. This may be accomplished if applications can be targeted to specific areas where few nontarget insects would be exposed, such as in the trap crops for Colorado potato beetles or storage product sites. If field isolates of $Ina^+$ microorganisms can be used for control, it would circumvent problems associated with the release of genetically engineered organisms into the environment. Considering the apparent ease and rapidity with which $Ina^+$ bacteria cause an increase in the supercooling point, it may be relatively difficult for insects to develop resistance to this control measure, since it would require blocking any and all avenues of contact between the $Ina^+$ organisms and internal water. The development of resistance to transcuticular nucleation is undoubtedly more complex and is expected to be an unlikely or at least slower process than common mechanisms of resistance, such as alteration in the structural target for a toxin or the production of toxin-destroying enzymes. On the other hand, if insects move to areas (e.g., burrow more deeply into the soil) sufficiently warm to remain above the supercooling point, this approach would be ineffective. Another advantage of this treatment is the biodegradability of these preparations, which are unlikely to leave behind contaminating residues. Lastly, this approach is fully compatible with other control measures that might be used concomitantly for integrated pest management of a given species.

Although the initial studies related to the potential use of ice-nucleating microorganisms are encouraging, considerably more work is needed to determine whether this approach will prove useful for biological control.

## Acknowledgments

We thank Paul Fields for generously providing us with an advance copy of a manuscript in press and Valerie Bennett, Chuck Burks, and Paul Fields for their comments on the manuscript. This research was supported by the Cooperative State Research Service (USDA) grant #93-37302-9003, NSF grant IBN-9305809, and Genencor Intl., San Francisco.

## Literature Cited

Bale, J. S. 1987. Review. Insect cold hardiness: Freezing and supercooling-an ecophysiological perspective. J. Insect Physiol. 12:899-908.

Baust, J. G., and Rojas, R. R. 1985. Review—Insect cold hardiness: Facts and fancy. J. Insect Physiol. 31:755-759.

Cannon, R. J. C., and Block, W. 1988. Cold tolerance of microarthropods. Biol. Rev. 63:23-77.

Casagrande, R. A. 1987. The Colorado potato beetle: 125 years of mismanagement. Bull. Entomol. Soc. Am. 33:142-150.

Duman, J. G., Wu, D. W., Xu, L., Tursman, D., and Olsen, T. M. 1991. Adaptations of insects to subzero temperatures. Q. Rev. Biol. 66:387-410.

Fields, P. G. 1990. The cold-hardiness of *Cryptolestes ferrugineus* and the use of ice nucleation-active bacteria as a cold-synergist. Pages 1183-1191 in: Proc. Int. Working Conf. on Stored-Product Protection, 5th. Vol. 2. F. Fleurat-Lessard and P. Ducon, eds. Bordeaux, France.

Fields, P. G. 1992. The control of stored-product insects and mites with extreme temperatures. J. Stored Prod. Res. 28:89-118.

Fields, P. G. 1993. Reduction of cold tolerance of stored-product insects by ice-nucleating-active bacteria. Environ. Entomol. 22:470-476.

Fields, P. G., and McNeil, J. N. 1986. Possible dual cold-hardiness strategies in *Cisseps fulvicollis* (Lepidoptera: Arctiidae). Can. Entomol. 118:1309-1311.

Fields, P.G., Pouleur, S., and Richard, C. 1993. Stability of ice-nucleating bacteria and fungus as a means of reducing the cold-hardiness of insect pests. Cryobiology 30:623.

Goodnow, R. A., Harrison, M. D., Morris, J. D., Sweeting, K. B., and LaDuca, R. J. 1990. Fate of ice nucleation-active *Pseudomonas syringae* strains in alpine soils and waters in synthetic snow samples. Appl. Environ. Microbiol. 56:2223-2227.

Hagstrum, D. W., and Flinn, P. W. 1992. Integrated pest mangement of stored-grain insects. Pages 535-562 in: Storage of Cereal Grains and Their Products, 4th ed. D. B. Sauer, ed. American Association of Cereal Chemists, St. Paul, MN.

Johnston, S. L., and Lee, R. E. 1990. Regulation of supercooling and nucleation in a freeze intolerant beetle (*Tenebrio molitor*). Cryobiology 27:562-568.

Kaneko, J., Kita, K., and Tanno, K. 1989. Bacteria in the gut determines the supercooling point of the diamondback moth, *Plutella xylostella*, pupae reared on germinating radish seeds (*Raphanus satiuus* L. var. *acanthiformis* Makino). Jpn. J. Appl. Entomol. Zool. 33:82-91.

Kaneko, J., Kita, K., and Tanno, K. 1991a. Ice nucleating active bacteria isolated from the diamondback moth, *Plutella xylostella* L. pupae (Lepidoptera: Ypononeutidae). Jpn. J. Appl. Entomol. Zool. 35:7-11.

Kaneko, J., Yoshida, T., Owada, T., Kita, K., and Tanno, K. 1991b. *Erwinia herbicola*: Ice nucleation active bacteria isolated from diamondback moth, *Plutella xylostella* L. pupae. Jpn. J. Appl. Entomol. Zool. 35:247-251.

Kieft, T. L. 1988. Ice nucleation activity in lichens. Appl. Environ. Microbiol. 54:1678-1681.

Kung, K-J. S., Milner, M., Wyman, J. A., Feldman, J., and Nordheim, E. 1992. Survival of Colorado potato beetle (Coleoptera: Chrysomelidae) after exposure to subzero thermal shocks during diapause. J. Econ. Entomol. 85:1695-1700.

Layne, J. R., Lee, R. E., and Huang, J. L. 1990. Inoculation triggers freezing at high subzero temperatures in a freeze-tolerant frog (*Rana sylvatica*) and insect (*Eurosta solidaginis*). Can. J. Zool. 68:506-510.

Lee, M. R., Lee, R. E., and Strong-Gunderson, J. M. 1992a. Isolation of ice nucleating active bacteria from a freeze-tolerant frog: Identification of *Pseudomonas putida* strains active in ice nucleation. Cryobiology 29:759.

Lee, M. R., Lee, R. E., Strong-Gunderson, J. M., and Minges, S. R. 1992b. Treatment with ice nucleating active fungi and surfactants decrease insect supercooling capacity. Cryobiology 29:743.

Lee, R. E. 1980. Physiological adaptations of Coccinellidae to supranivean and subnivean hibernacula. J. Insect Physiol. 26:135-138.

Lee, R. E. 1989. Insect cold-hardiness: To freeze or not to freeze. BioScience 39:308-313.

Lee, R. E. 1990. Method and composition for increasing the supercooling point in invertebrates. U.S. patent pending.

Lee, R. E. 1991. Principles of insect low temperature tolerance. Pages 17-46 in: Insects at Low Temperature. R. E. Lee and D. L. Denlinger, eds. Chapman and Hall, New York.

Lee, R. E., Costanzo, J. P. Kaufman, P. E., Lee, M. R., and Wyman, J. A. 1994. Ice nucleating active bacteria reduce the cold-hardiness of the freeze-intolerant Colorado potato beetle (Coleoptera: Chrysomelidae). J. Econ. Entomol. 87:377-381.

Lee, R. E., and Denlinger, D. L. 1985. Cold tolerance in diapausing and nondiapausing stages of the flesh fly, *Sarcophaga crassipalpis*. Physiol. Entomol. 10:309-315.

Lee, R. E., Lee, M. R., and Strong-Gunderson, J. M. 1993. Review: Insect cold-hardiness and ice nucleating active microorganisms including their potential use for biological control. J. Insect Physiol. 39:1-12.

Lee, R. E., Mugnano, J. A., and Taylor, R. T. 1992a. Endogeneous crystalloid spheres regulate the supercooling point of the gall fly, *Eurosta solidaginis*. Cryobiology 29:750-751.

Lee, R. E., Strong-Gunderson, J. M., Lee, M. R., and Davidson, E. C. 1992b. Ice-nucleating active bacteria decrease the cold-hardiness of stored grain insects. J. Econ. Entomol. 85:371-374.

Lee, R. E., Strong-Gunderson, J. M., Lee, M. R., Grove, K. S., and Riga, T. J. 1991. Isolation of ice nucleating active bacteria from insects. J. Exp. Zool. 257:124-127.

Miller, K. 1982. Cold-hardiness strategies of some adult and immature insects overwintering in interior Alaska. Comp. Biochem. Physiol. 73A:595-604.

Milner, M., Kung, K.-J. S., Wyman, J. A., Feldman, J., and Nordheim, E. 1992. Enhancing overwintering mortality of Colorado potato beetle (Coleoptera: Chrysomelidae) by manipulating the temperature of its habitat. J. Econ. Entomol. 85:1701-1708.

Pouleur, S., Richard, C., Martin, J.-G., and Antonn, H. 1991. Ice nucleation activity in two *Fusarium* species. (Abstr.) Int. Conf. Biological Ice Nucleation, 5th. Madison, WI.

Rogers, J. S., Stall, R. E., and Burke, M. J. 1987. Low-temperature conditioning of the ice nucleation active bacterium, *Erwinia herbicola*. Cryobiology 24:270-279.

Shields, E. J., and Wyman, J. A. 1984. Effect of defoliation at specific growth stages of potato yields. J. Econ. Entomol. 7:1194-199.

Shimada, K. 1989. Ice-nucleating activity in the alimentary canal of the freezing-tolerant prepupae of *Trichiocampus populi* (Hymenoptera:Tenthredinidae). J. Insect Physiol. 35:113-120.

Shimada, K., and Riihimaa, A. 1988. Cold acclimation, inoculative freezing and slow cooling: Essential factors contributing to the freeze-tolerance in diapausing larvae of *Chymomyza costata* (Diptera: Drosophilidae). Cryo Lett. 9:5-10.

Somme, L. 1982. Supercooling and winter survival in terrestrial arthropods. Comp. Biochem. Physiol. 73A:519-543.

Steigerwald, K. A., Lee, M. R., Lee, R. E., and Marshall, J. C. 1993. Effect of ice nucleating active bacteria on insect supercooling capacity varies with the anatomic site of application. (Abstr.) Int. Conf. on Biological Ice Nucleation, 6th. Laramie, WY.

Storey, K. B., and Storey, J. M. 1988. Freeze tolerance in animals. Physiol. Rev. 68:27-84.

Strong-Gunderson, J. M., Lee, R. E., and Lee, M. R. 1989. Ice regulating bacteria promote transcuticular nucleation in insects. Cryobiology 26:551.

Strong-Gunderson, J. M., Lee, R. E., and Lee, M. R. 1990a. New species of ice nucleating active bacteria isolated from insects. Cryobiology 27:691.

Strong-Gunderson, J. M., Lee, R. E., Lee, M. R., and Riga, T. J. 1990b. Ingestion of ice nucleating active bacteria increases the supercooling point of the lady beetle *Hippodamia convergens*. J. Insect Physiol. 36:153-157.

Strong-Gunderson, J. M., Lee, R. E., and Lee, M. R. 1992. Topical application of ice nucleating bacteria decreases insect cold tolerance. Appl. Environ. Microbiol. 58:2711-2716.

Tsumuki, H., Konno, H., Maeda, T., and Okamoto, Y. 1992. An ice-nucleating active fungus isolated from the gut of the rice stem borer, *Chilo suppressalis* Walker (Lepidoptera: Pyralidae). J. Insect Physiol. 38:119-125.

Zachariassen, K. E., and Hammel, H. T. 1976. Nucleating agents in the haemolymph of insects tolerant to freezing. Nature 262:285-287.

CHAPTER 15

# Ice Nucleation Genes as Reporters

**Nickolas J. Panopoulos**

## Uses of Reporter Genes in Biological Research

One avenue for the investigation of biological processes is to determine how and when relevant genes are turned on or off, and how their activity is modulated by environmental and cellular signals. Since most genes in living organisms do not have easily assayable functions, investigation of their temporal and spatial patterns of expression presents difficulties. The concept of using a gene with an easily assayable product to "report" the expression of another gene was a breakthrough and has served as a basis for the development of analytical tools in molecular biology for nearly 25 years. At present, a wide variety of genes and proteins are used as reporters of transcription, as probes for the dissection of protein topology, localization, and secretion, and as indicators of other cellular, genetic, and environmental processes (Broome-Smith et al., 1990; Gallagher, 1992; Koncz et al., 1990; Manoil and Beckwith, 1986; Silhavy and Beckwith, 1985; Shaw et al., 1987; Stewart and Williams, 1992, 1993). Ice nucleation genes are a recent addition to the ranks of genetic reporters. This chapter will discuss the features of ice nucleation genes that are germane to their use as "ice nucleation activity (INA) reporters."

## Conventional Versus INA Reporters

The most frequently used reporters in biological research encode enzymes whose activity can be easily detected and quantified, usually by employing a chromogenic, fluorogenic, radioactive, or immunodetectable substrate. Such genes, here referred to as conventional reporters, include *lacZ*, *phoA*, *cat*, *uidA* (*gusA*), *xylE*, and *lux*, which respectively encode β-galactosidase, alkaline phosphatase, chloramphenicol aceyltransferase, β-glucuronidase, catechol 2,3-oxygenase, and luciferase. The luciferase reporters differ from the others only in that the measured reaction products are the photons emitted during the conversion of substrate, rather than a chemical substance. INA reporters differ from conventional reporters in a fundamental sense: the signal detected is not due to enzymatic catalysis but is, instead, a physical phenomenon (the liquid-to-solid phase transition of water). Another important difference between INA and conventional reporters is the nature of the dose-response curve obtained when the amount of the reporter protein is plotted against the signal it generates; conventional reporters typically produce linear responses, whereas INA reporters give a nonlinear response (Lindgren et al., 1989; Southworth et al., 1988; see also Chapter 5).

In comparison to conventional reporters, INA reporters have been used in relatively few instances, and all published examples involve gram-negative prokaryotes. However, INA reporters can be employed in a broad spectrum of organisms: all gram-negative bacteria studied to date, some gram-positive bacteria, yeasts, and plants can express functional ice nuclei upon introduction of bacterial *ina* genes (Bartlein et al., 1992; D. A.Bartlein, S. E. Lindow, N. J. Panopoulos, S. P. Lee, and T. H. H. Chen, *unpublished*; S.E. Lindow, *unpublished*; and D. Pridmore, *personal communication*). The use of an ice nucleation gene as a transcriptional reporter was first explored in 1989 (Lindgren et al., 1989). The objective of this and subsequent studies in this laboratory was to establish the transcriptional organization of genes in the 22-kb *hrp* cluster of the plant pathogen *Pseudomonas syringae* pv. *phaseolicola* and to analyze their temporal pattern of expression during pathogenesis and hypersensitive necrosis. Fusions between most of these genes and the *lac* operon produced by insertional mutagenesis with pHoHo1 (Tn*3-lacZYA*; Stachel et al., 1985) showed no detectable expression in vitro (Lindgren and Panopoulos, *unpublished*). The *hrp* genes of *P. syringae* pv. *phaseolicola* have particular pH, medium, and plant signal requirements for expression (Rahme et al., 1992), which were unknown at that time. Furthermore, *lacZ* would not be a useful reporter in green leaves, especially at early stages of infection when the bacterial population is small (usually less than $10^6$ CFU/$cm^2$). The *inaZ* reporter employed in this early study has been used in several other laboratories in research on plant-bacterium interactions (Table 1; Andersen, 1993; Fellay et al., 1991; Frederick, 1989; Georgakopoulos,1993; Grimm and Panopoulos, 1989; Huyhn et al., 1989; Lee, 1993; Loper and Lindow, 1994; Ma et al., 1991; Rahme et al., 1991, 1992).

## Properties of *ina* Genes Relevant to Their Use as Reporters

Various reporters satisfy the needs of particular applications to different degrees (Meighen, 1991; Shaw et al., 1987; Silhavy and Beckwith, 1985; Stewart and Williams, 1993). Most have limitations or disadvantages, and none satisfies all theoretical and practical requirements of different experimental needs to the same degree. Accordingly, there is a continuing need for new reporters that may be appropriate in particular situations. One essential prerequisite of any reporter application is the absence of endogenous activities that would interfere with the assay. INA reporters satisfy this requirement in a large number of cases, since endogenous ice nucleators active at warm assay temperatures are confined to a few organisms that carry *ina* genes (see Chapter 3). Also, warm-temperature ice nuclei (active above –5°C) are rare or absent from environmental samples, such as soil, as well as from plants, when these are not colonized by Ice$^+$ bacteria (as is true for plants grown in the greenhouse). This is especially important for the use of INA reporters in the study of gene activity in plant pathogens or epiphytes, in soil, or in other environmental samples. A second important requirement for reporters is the ability to express a phenotype or signal that can be sensitively detected and easily quantified with minimal sample processing in individual cells, in cell populations, or in various natural settings under a wide variety of conditions. Besides these requirements, reporters ideally should not give false positives, require expensive equipment, em-

**Table 1.** Examples of ice nucleation genes used as reporters

| Organism | Target genes | Experimental setting | References |
|---|---|---|---|
| *Pseudomonas syringae* | | | |
| pv. *phaseolicola* | *hrp, avr* | bean, tobacco leaves; pH, osmotic conditions, nutrients | Lindgren et al. (1989), Rahme et al. (1991, 1992) |
| pv. *glycinea* | *avr* | soybean leaves, nutrient effects | Huyhn et al. (1989) |
| pv. *tomato* | *cor* | soybean leaves, nutrient effects | Ma et al. (1991) |
| *P. syringae* | *pvd, est* | bean phyllosphere | Loper and Lindow (1994), Andersen (1993), Lindow (1993) |
| *P. fluorescens* | *pvd* | bean rhizosphere | Loper and Lindow (1994) |
| *P. aureofaciens* | *pnz* | germinating seeds, plant rhizosphere | Georgakopoulos (1993) |
| *Agrobacterium tumefaciens* | *vir* | acetosyringone induction | Lindgren et al. (1989) |
| *Erwinia herbicola* | *iaa* | phytohormone production | Brandl and Lindow (*unpublished*) |
| *Zymomonas mobilis* | *pdc, bla, hrp* | extremophilic culture conditions | Drainas et al. (1995) |

ploy hazardous chemicals, or involve time-consuming procedures for the assays or for sample preparation. INA reporters clearly surpass most conventional reporters in regard to the speed of the assay commonly used (the assay itself takes about 1–5 min). Furthermore, they are much more sensitive than conventional reporters ($\leq 10^5$-fold more sensitive than *lacZ*), and at least in some applications they have comparable sensitivity to luciferase reporters (Lindgren et al., 1989).

## Tools and Methods for Constructing *ina* Gene and Ina Protein Fusions

### General-Purpose Transposons and Vectors

The range of tools currently available for the construction of *ina* gene fusions is somewhat limited. Most research applications of INA reporters utilize the Tn*3*-Spice transposon originally constructed by Lindgren et al. (1989). A few other gene fusion vehicles are available or are currently under construction. Their salient features are briefly reviewed here. When more flexibility or a particular structural design is needed for different applications, new vehicles and gene fusion tools can be easily made, taking advantage of the absence of cleavage sites for many restriction endonucleases in *inaZ* and/or other *ina* genes. The transposons described below are used for the construction of *inaZ*::target gene fusions by in vivo insertional mutagenesis of cloned DNA fragments. These fusions can serve the needs of both mutational and expression analysis of the target gene. A few plasmid vectors exist that carry the *inaZ* gene in various replicons and allow for the construction of such fusions by cloning a DNA fragment upstream of an ice nucleation gene. In most cases, the ice nucleation gene lacks its native promoter.

#### Tn*3*-Spice1, 2, and 3

The three Tn*3*-Spice transposon elements are derived from pHoHo1, a Tn*3*-*lacZYA* transposon (Stachel et al., 1985), by replacement of the *lacZYA* segment with a promoterless *inaZ* gene cloned from plasmid pMWS10 (Wolber et al., 1986). They also contain the 2.0-kb $Sp^r/Sm^r$-encoding "Omega" fragment (Prentki and Krish, 1984) to provide better selectable markers for bacteria, such as *Pseudomonas,* that are relatively resistant to β-lactam antibiotics. The Tn*3* transposase gene (*tnpA*) is inactive in all three Tn*3*-Spice elements due to the truncation of its C-terminus in the pHoHo1 progenitor and the insertion of the Omega fragment in the *ClaI* site of pTn*3*-Ice. Tn*3*-Spice elements are used as part of a two-plasmid system, along with the pSShe plasmid ($Cm^r$), which provides the Tn*3* transposase function in trans, as in pHoHo1 mutagenesis (Stachel et al., 1985).

The three Tn*3*-Spice transposons differ from each other in regards to whether or not the *inaZ* gene can be fused to upstream sequences in frame. Tn*3*-Spice1 is the transposon originally described by Lindgren et al. (1989) as Tn*3*-Spice. In this element, the segment preceding the *inaZ* gene has translational stop codons in all three reading frames; as a result, only *transcriptional* fusions can be made. This is a desirable feature in routine genetic analysis, since translational fusions could produce InaZ proteins with diverse N-terminal extensions and, therefore, unpredictable activity. Tn*3*-Spice2 and Tn*3*-Spice3 (M.N. Mindrinos, L.G. Rahme and N.J. Panopoulos, *unpublished*) differ from Tn*3*-Spice1 in that one of the three reading frames in the segment upstream of *inaZ* does not carry stop codons; therefore, the fusions generated by these elements can be either transcriptional (the majority) or translational. The specialized uses of Tn*3*-Spice2 and Tn*3*-Spice3 remain to be explored.

**Tn*3*-Nice**

Tn*3*-Nice is equivalent to Tn*3*-Spice1, with a neomycin-resistance gene from Tn*903* replacing the $Sp^r/Sm^r$ fragment (J. Kraus and J. Loper, *unpublished*).

**Tn*5*-*ina* Derivatives**

Tn*5*-*ina* derivatives derive from the Tn*5*-*lac* transposons described by Simon et al. (1989) by insertion of the *inaZ* gene near the terminus of the left inverted repeat ($IR_L$). They carry the cognate transposase gene in the same element as *inaZ* and will permit the generalized insertion of the reporter in the bacterial chromosome at random by suicide mutagenesis, which is not currently possible with the Tn*3*-Spice transposons.

**pLAFR6-*inaZ***

The pLAFR6-*inaZ* plasmid is derived from the broad host range vector pRK290 (Ditta et al., 1980) and thus can replicate in a broad range of gram-negative bacterial hosts. It contains a promoterless *inaZ* gene plus a multilinker sequence and synthetic transcriptional terminators on both sides (D. Dahlbeck and B.J. Staskawicz, *unpublished*). The terminators insulate the *inaZ* gene from the transcriptional activity of vector sequences; as a result, the basal level of *inaZ* expression is very low (although this may vary with the host in which the plasmid is propagated). When a promoter-containing DNA fragment is cloned in front of *inaZ* in the proper orientation, the level of *inaZ* expression increases accordingly.

**pVSP61-*inaZ***

pVSP61 is a hybrid plasmid consisting of the pVS1 and pACYC184 replicons (W. Tucker, *personal communication*) and is very stable in *Escherichia coli* and *Pseudomonas* species grown in culture, the rhizosphere, or the phylloplane. The promoterless *inaZ* was inserted into the pUC8-derived multilinker of pVSP61 in a direction opposite to *lacZ* and a fragment carrying the *pvd* (pyoverdine) promoter from *P. syringae* was placed upstream. This reporter expresses iron-regulated INA and exemplifies the possibility of biosensing using INA reporters (Loper and Lindow, 1994).

## Procedures Used in Tn*3*-Spice Mutagenesis

The Tn*3*-Spice1 transposon is the most frequently used *ina* fusion vehicle. The procedures for its use are described in Lindgren et al. (1989) and are based on those used for Tn*3*-*lac* mutagenesis using pHoHo1 (Stachel et al., 1985). The pMB8-derived replicon of the transposon donor plasmid can be maintained in most *E. coli* strains but displays thermosensitive replication in a $polA^{ts}$ mutant C2110 (also designated SF800) and is unable to replicate in nonenteric bacterial hosts. Accordingly, if the target DNA fragments are carried on *polA*-independent vectors, the transposon insertion derivatives of the target plasmid are specifically selected in the final step (see below). A general outline of the strain constructions and steps followed in the isolation of Tn*3*-Spice1 insertions in target DNA is given below:

1) The target DNA is cloned in a vector whose replication in *E. coli* is *polA*-independent. Cosmid clones in the commonly used broad host range vectors may be used directly for pTn*3*-Spice1 mutagenesis.

2) The target DNA–containing plasmid is introduced into an *E. coli* strain that carries pTn*3*-Spice1 (transposon donor plasmid) and pSShe (encoding transposase). This transfer may be accomplished by triparental mating, for example using the helper plasmid pRK2013 (Ditta et al., 1980).
3) The strain carrying all three plasmids (the target, pTn3-Spice1, and pSShe) is allowed to conjugate with the *polA*$^{ts}$ recipient, again mobilizing the target plasmid with pRK2013, in a triparental mating.
4) The cells in the conjugation mixture are plated on a medium that contains nalidixic acid, to which only the *polA*$^{ts}$ (recipient) strain is resistant, plus an antibiotic to which the *polA*$^{ts}$ strain is sensitive and against which the transposon confers resistance (e.g., Sm, Sp, or Ap). The agar plates are then incubated at 42°C. Almost all colonies obtained by this procedure are progeny of the *polA*$^{ts}$ strain carrying derivatives of the target plasmid into which the Tn*3*-Spice1 element has transposed.

Although triparental conjugation works quite well in practice, the plasmid transfers in steps 2 and 3 can be accomplished by other means: i) DNA may be extracted from the donor strain and introduced into the recipient by transformation or electroporation; or ii) when the target plasmid carries the *cos* site of phage lambda and has the appropriate size for packaging, it may be transduced at step 2; or iii) when the target plasmid is 10–12 kb less than the packaging capacity of lambda phage (48 kb ± 5%), the product of Tn*3*-Spice transposition will have the appropriate size for packaging in vivo; thus, it may be transduced at step 4.

Once the final selection step is accomplished, the mutagenized target plasmids are analyzed to map the position and orientation of the Tn*3*-Spice insertion. The physical separation of the Tn*3* transposase gene and the Tn*3*-Spice transposons in different replicons in the system described above precludes further transposition after step 4.

### Expression Analysis of Tn*3*-Spice Insertions

In most cases, it is possible to obtain initial information about the transcribed regions in the target DNA by screening the colonies containing plasmids with individual insertions for ice nucleation activity. Insertions in which *inaZ* is oriented opposite to the direction of target gene transcription usually express little or no activity, while those in the correct orientation express much higher activity. The levels of activity expressed by plasmids that carry oppositely oriented insertions in a given target region may differ up to 10,000-fold. The transcriptional orientation of the target gene inferred from this preliminary screening has usually been correct in that it matched the direction in which the target genes were transcribed, as determined by other methods. When the expression of the target gene can be measured in the induced and uninduced states, only the correctly oriented fusions increase under the inducing conditions. Unfortunately, this option is often not available, either because the gene in question is not under inductive control, or the inducing treatments are unknown at the time of investigation. In such cases, it is recommended that the mutations be transferred to the bacterial chromosome by marker exchange techniques. This manipulation always causes a significant reduction in the level of ice nucleation activity expressed by a given fusion. In our experience, the activity levels of fusions having the wrong orientation in the target gene, unlike those with the correct orientation, usually become undetectable (less than $10^{-8}$ ice nuclei/cell) and do not respond to inductive treatments. This strategy generally allows the correct orientation to be determined unequivocally.

## Assay Methods

Ice nucleation activity can be readily detected and quantified in bacterial cultures, leaf or root washings, or macerated plant tissues by fairly simple methods. The most commonly used method is based on the droplet-freezing assay (Vali, 1971; see also Chapter 1) carried out at a constant temperature (e.g. –5 or –9°C). However, leaf disks or whole leaves that have been inoculated with the bacteria can be assayed for ice nucleation activity directly without any processing steps (by tube freezing assays). A detailed description of these methods is given by Lindow (1990). Some relevant points are illustrated by the investigation of *hrp* gene expression in vitro and during infection and endophytic multiplication of *P. syringae* pv. *phaseolicola* on plant leaves. Starter inoculum is prepared by growing the bacterial cultures in a medium in which the expression of most *hrp* genes is repressed (Rahme et al., 1992). Fresh inoculum is prepared by transferring a small aliquot of this culture to fresh medium and adjusting the cell density to 0.05 to 0.1 $OD_{600}$ units. Incubation of the cultures resumes under the same conditions, and aliquots are removed at different times, serially diluted, and assayed for INA at –9°C. Plant leaves are inoculated with bacterial suspensions prepared from the overnight cultures at the desired cell density—by vacuum infiltration in the case of bean leaves (this option is limited to small plants or detached plant parts for technical reasons) or by injecting the suspension into the intercellular spaces of leaves. Injection can be easily accomplished with a thin hypodermic needle or by simply appressing the mouth of a needleless plastic syringe or plastic Pasteur pipette onto a point on the leaf surface where a small hole or incision has been made with a needle or a razor, while supporting the leaf with the finger (this is routine with leaves of many plants). The intercellular leaf spaces of many dicotyledonous plants have a capacity to receive from 4 to 10 ml of fluid/$cm^2$ of leaf. After inoculation, the plants are kept in a growth chamber at 24°C with an appropriate photoperiod and relative humidity. The water-soaked areas of the leaves usually lose the intercellular water within 30 to 60 minutes, provided the stomata do not close. For analysis of the INA reporter gene's activity, the leaf disks may be cut at different times and easily disrupted by grinding or other methods to release the bacteria; a standard droplet-freezing assay is then performed.

### Calculation of INA and Quantification of Gene Induction

Ice nucleation activities are frequently expressed in terms of ice nuclei per cell, i.e., the number of ice nuclei normalized for the number of bacterial cells present. The mathematical basis for INA calculations is the formula given by Vali (1971; see also Chapter 1). Because INA values computed in this way are very small numbers, they are usually expressed as a logarithm or an exponent of 10. Sometimes ice

**Technical note:** When studying gene *repression* with INA reporters, one must keep in mind that the lifetime of the ice nuclei can be quite long, and this may confound the analysis. Detailed analysis of the kinetics of ice nucleus turnover is available only in one *E. coli* strain (Watanabe et al., 1990). In some *P. syringae* strains, ice nucleus turnover appears to be much slower than in *E. coli* (S.E. Lindow and N.J. Panopoulos, *unpublished*) and different situations may be encountered in other hosts.

nucleation is not detected: no droplets freeze in the most concentrated sample. Since Vali's equation cannot be solved in such cases, an arbitrary default value of 1 (frozen droplet) is used, and the value obtained is treated as an upper limit of the INA expression. For Ice$^-$ or non-INA bacterial cultures, the INA limit values obtained in this way are typically less than $10^{-8}$/cell.

The quantitative interpretation of INA values requires consideration of the non-linear relationship between the amount of InaZ protein (which is proportional to the level of *inaZ* transcription) and the signal it generates. Previous studies established a log-linear relationship in *E. coli* (Southworth et al., 1988; see also Chapter 5) and *P. syringae* pv. *phaseolicola* (Lindgren et al., 1989): the INA values determined at –9°C vary with the square (i.e., the second power) of intracellular InaZ protein concentration, up to about 0.4 nuclei/cell. Above this value, the activity continues to increase with InaZ protein concentration but in a more linear fashion, up to the theoretical maximum of one nucleus per cell, when the assay is carried out in non-disrupted cells. For example, when the activities do not exceed 0.4 nuclei/cell, a 100-fold increase in activity indicates a 10-fold increase in transcription.

Accordingly, in the case of an inducible gene, the degree of induction can be approximated by taking the square root of the ratio between the INA values of the induced and uninduced states:

$$\text{(induced/uninduced transcription)} = \sqrt{\text{(induced/uninduced INA)}}$$

However, when activities measured fall in the linear range of the *inaZ* reporter system (above 0.4 nuclei/cell), the degree of induction will be underestimated by this formula. If higher assay temperatures are employed, it should be borne in mind that toward higher assay temperatures, the INA increases with the cube (i.e., the third power) of InaZ protein concentration (Southworth et al., 1988).

On occasion, INA values have been obtained that exceed the theoretical maximum of one nucleus per cell (Lindgren et al., 1989; Huyhn et al., 1989; R.D. Frederick, C. Drainas, and N.J. Panopoulos, *unpublished*). In such cases, the *inaZ* reporter was carried on plasmids. Two possible scenarios may account for this: 1) a significant fraction of the cell population had lost plasmids but retained functional ice nuclei, and cell number was estimated by a plating method selective for plasmid markers; or 2) the bacteria released cell-free nuclei, as do some naturally Ice$^+$ *Erwinia* strains (see Chapter 4).

## Advantages and Disadvantages of Ice Nucleation Genes as Reporters

Compared to conventional reporters, there are several advantages for *ina* genes. First, the speed and simplicity of INA assays makes the use of INA reporters attractive; the only specialized equipment (a refrigerated bath) is not particularly expensive. Second, the sensitivity of INA reporters is very high (it is comparable in its sensitivity to luciferase assays, which are approximately 1,000-fold more sensitive than CAT assays [Berthold, 1990], and may be comparable to fluorogenic GUS assays). Third, the log-linear dose-response relationship holds over a large range of INA values (more than six orders of magnitude), at least for assays carried out at –9°C (Southworth et al., 1988; Lindgren et al., 1989).

One limitation to the use of ice nucleation reporters is that the particular organisms in which they are used must lack endogenous ice nucleation genes and any other biotic or abiotic sources of –9°C ice nuclei. Ice nucleation genes are absent from most microorganisms, and in strains that have endogenous *ina* genes, they can be inactivated by gene replacement methods (see Chapter 13).

Ice nucleation reporters have been instrumental in studies with *hrp* genes (Lindgren et al., 1989; Rahme et al., 1991, 1992) and in the monitoring of bacterial gene activity in the rhizosphere (Georgakopoulos, 1993; Loper and Lindow, 1994) and the phylloplane (Andersen, 1993; Y.-A. Lee, 1993; Loper and Lindow, 1994). In all these cases, the INA reporters permitted the detection of low levels of gene expression under conditions where conventional reporters would not have detected any significant activity above background levels.

The theoretical maximum of one detectable nucleus per cell and the log-linear response of INA reporters may discourage their broader use in laboratories unfamiliar with ice nucleation. Because of the log-linear relationship between the detected signal and the nucleation protein concentration, INA reporters are sensitive to vector promoter effects on cloned genes, vector copy number, transcriptional termination efficiency of upstream genes, degree of polarity within operons, generation of artificial promoters at fusion points, and RNA turnover in the cells. Another disadvantage for certain experiments is the long lifetime of the ice nucleation proteins once they are synthesized in certain cells. Finally, the temperature requirements for ice nucleus formation preclude the application of INA reporters to organisms growing above 30°C.

## Conclusions and Prospects

As the use of molecular tools in biological and environmental research continues to expand, reporter genes are having a major impact on the analysis of many biological processes and systems. It is likely that reporter technologies will play a role in agricultural and environmental biology, such as the study of the interactions among organisms and their environments, both in the laboratory and in the more complex situations outside. Despite some disadvantages, ice nucleation reporters can provide useful tools for many applications. The development of an iron sensor based on a *pvd* (pyoverdin) promoter *inaZ* fusion (Loper and Lindow, 1994) and the use of INA-based selections for bacterial mutants sensitive to environmental stresses during epiphytic growth (Andersen, 1993) provide examples of novel applications. I hope that this chapter will help stimulate interest among readers to explore the applications of INA reporters.

## Acknowledgments

I thank my collaborators and colleagues for permission to describe unpublished work and Amalia Afendra for assistance in the timely preparation of the manuscript.

## Literature Cited

Andersen, G. A. 1993. Molecular characterization of an epiphytic fitness locus in *Pseudomonas syringae* pathovar *syringae*. Ph.D. thesis. University of California, Berkeley.

Bartlein, D. A., Lindow, S. E., Panopoulos, N. J., Lee, S. P., Mindrinos, M. N.,and Chen, T. H. H. 1992. Expression of a bacterial ice nucleation gene in plants. Plant Physiol. 100:1730-1736.

Berthold, A. 1990. Advances in measurement of reporter genes: Chloramphenicol acetyltransferase and firefly luciferase. Biomed. Biochim. Acta 49:1243-1245.

Broome-Smith, J. K., Tadayyon, M., and Zhang, Y. 1990. β-lactamase as a probe of membrane protein assembly and protein export. Mol. Microbiol. 4:1637-1644.

Drainas, C., Vartholomatos, G., and N. J. Panopoulos. 1995. The ice nucleation gene from *Pseudomonas syringae* as a sensitive reporter for promoter analysis in *Zymomonas mobilis*. Appl. Environ.

Microbiol. (In press.)

Ditta, G., Stanfield, S., Corbin, D., and Helinski, D. 1980. A broad host range DNA cloning system for gram negative bacteria: Construction of a gene bank for *Rhizobium meliloti*. Proc. Natl. Acad. Sci. USA 77:7347-7351.

Fellay, R., Rahme, L. G., Mindrinos, M. N., Frederick, R. D., Pisi, A., and Panopoulos, N. J. 1991. Genes and signals controlling the *Pseudomonas syringae* p. *phaseolicola*-Plant interaction. Pages 45-52 in: H. Hennecke and D. P. S. Verma, eds. Advances in Molecular Genetics of Plant-Microbe Interactions. Kluwer Academic Publishers, Dordrecht, The Netherlands.

Frederick, R. D. 1989. Molecular characterization and nucleotide sequence od two *hrp* loci from *Pseudomonas syringae* pv. *phaseolicola*. Ph.D. thesis. University of California, Berkeley.

Gallagher, S. R., ed. 1992. GUS protocols: Using the GUS gene as a reporter of gene expression. Academic Press, New York.

Georgakopoulos, D. G., Henson, M. Panopoulos, N. J., and Sehroth, M. N. 1994. Cloning of a phenazine biosynthetic locus of *Pseudomonas aureofaciens* PGS12 and analysis of its expression in vitro with the ice nucleation reporter gene. Appl. Environ. Microbiol. 60:2931-2938.

Grimm, C. G., and Panopoulos, N. J. 1989. The predicted protein product of a pathogenicity locus from *Pseudomonas syringae* pv. *phaseolicola* is homologous to a highly conserved domain of several procaryotic regulatory proteins. J. Bacteriol. 171:5031-5038.

Huyhn, T. V., Dahlbeck, D., and Staskawicz, B. J. 1989. Bacterial blight of soybean: Regulation of a pathogen gene determining host cultivar specificity. Science 245:1374-1377.

Koncz, C., Langridge, W. H. R., Olsson, O. J. Shell, and Szalay, A. A. 1990. Bacterial and firefly luciferase genes in transgenic plants: Advantages and disadvandages of a reporter gene. Dev. Genet. 11:224-232.

Lee, Y.-A. 1993. Studies on copper resistance and quinate metabolism in *Xanthomonas campestris* pv. *juglandis*, the cause of walnut blight. Ph.D. thesis. University of California, Berkeley.

Lindgren, P. B., Frederick, R., Govindarajan, A. G., Panopoulos, N. J., Staskawicz, B. J., and Lindow, S. E. 1989. An ice nucleation reporter gene system: Identification of inducible pathogenicity genes in *Pseudomonas syringae* pv. *phaseolicola*. EMBO J. 8:2990-3001.

Lindow, S. E. 1993. Novel method for identifying bacterial mutants with reduced epiphytic fitness. Appl. Env. Microbiol. 59:1586-1592.

Lindow, S. E. 1990. Bacterial ice nucleation activity. Pages 428-434 in: Methods in Phytobacteriology. Z. Klement, K. Rudolph, and D. C. Sands, eds. H. Stillman Publishers, Boca Raton, FL.

Loper, J. E., and Lindow, S. E. 1994. A biological sensor for iron available to bacteria occupying microhabitats on plant surfaces. Appl. Environ. Microbiol. 60:1934-1941.

Ma, S.-W., Morris, V. L., and Cuppels, D. A. 1991. Characterization of a DNA region required for production of the phytotoxin coronatine by *Pseudomonas syringae* pv. *tomato*. Mol. Plant-Microbe Interact. 4:69-74.

Manoil, C., and Beckwith, J. 1986. A genetic approach to analyzing protein topology. Science 233:1403-1408.

Meighen, E. A. 1991. Molecular biology of bacterial bioluminescence. Microbiol. Rev. 55:123-142.

Prentki, P., and Krisch, H. M. 1984. In vitro insertional mutagenesis with a selectable DNA fragment. Gene 29:303-313.

Rahme, L. G. 1991. The *hrp* ("harp") gene of *Pseudomonas syringae* pv. *phaseolicola*: Organization and expression, signalling and role in the plant-bacterium interaction. Ph.D. thesis. University of California, Berkeley.

Rahme, L. G., Mindrinos, M. N., and Panopoulos, N. J. 1991. The genetic and transcriptional organization of the *hrp* gene cluster of *Pseudomonas syringae* pv. *phaseolicola*. J. Bacteriol. 173:575-586.

Rahme, L. G., Mindrinos, M. N., and Panopoulos, N. J. 1992. Plant and environmental sensory signals control the expression of *hrp* genes in *Pseudomonas syringae* pv. *phaseolicola*. J. Bacteriol. 174:3499-3507.

Shaw, J. J., Rogowsky, P., Close, T. J., and Kado, C. I. 1987. Working with bacterial bioluminescence. Plant Mol. Biol. Rep. 5:225-236.

Silhavy, T. J., and Beckwith, J. R. 1985. Use of *lac* fusions for the study of biological problems. Microbiol. Rev. 49:398-418.

Simon, R., Qandl, J., and Klipp, W. 1989. New derivatives of transposon Tn*5* suitable for mobilization of replicons, genration of operon fusions and induction of genes in gram-negative bacteria. Gene 80:161-169.

Southworth, M. W., Wolber, P. K., and Warren, G. J. 1988. Nonlinear relationship between concentration and activity of a bacterial ice nucleation protein. J. Biol. Chem. 263:15211-15216.

Stachel, S. E., Gynheung, A., Flores, C., and Nester, E. W. 1985. A Tn*3-lacZ* transposon for the random generation of β-galactosidase gene fusions: Application to the analysis of gene expression in *Agrobacterium*. EMBO J. 4:891-898.

Stewart, G. S. A. B., and Williams, P. 1992. *lux* genes and the applications of bacterial bioluminescence. J. Gen. Microbiol. 138:1289-1300.

Stewart, G. S. A. B., and Williams, P. 1993. Shedding new light on food microbiology. ASM News 59:241-246.

Vali, G. 1971. Quantitative evolution of experimental results on the heterogeneous freezing nucleation of supercooled liquids. J Atmos. Sci. 28:402-409.

Watanabe, N. M., Southworth, W. M., Warren, G. J., and Wolber, P. K. 1990. Rates of assembly and degradation of bacterial ice nuclei. Mol. Microbiol. 4:1871-1879.

Wolber, P. K., Deininger, C. A., Southworth, M. W., Vandekerhove, J., Van Montagu, M., and Warren, G. J. 1986. Identification and purification of a bacterial ice nucleation protein. Proc. Natl. Acad. Sci. USA 83:7256-7260.

CHAPTER 16

# Transduction of *ina* Genes for Bacterial Identification

**P. K. Wolber, R. L. Green, W. T. Tucker, N. M. Watanabe, C. A. Vance, R. A. Fallon, C. Lindhardt, and A. J. Smith**

A key component of public health–based disease control is the ability to detect pathogens. Of particular interest to this chapter is the monitoring of food samples for the presence of foodborne pathogens. Such surveillance is crucial for the prevention of diseases transmitted via foods. There is a need for assays that can rapidly detect pathogenic bacteria in diverse food samples that typically contain many varieties and substantial numbers of other, nonpathogenic bacteria. This need has been amplified by the recent move towards the "hazard analysis and critical control points" (HACCP) approach to microbiological quality control of foods (O'Donnell, 1992), since the implementation of HACCP in food production requires prompt monitoring of relevant pathogens.

Numerous new technologies have been applied to the detection of pathogenic bacteria in foods (Veld et al., 1988). Detection is usually based on bacterial biochemistry (e.g., media-based methods), surface properties (e.g., immunoassays), or genomic properties (e.g., DNA probe hybridization assays). All these approaches (as currently configured) require microbiological preenrichment and selective enrichment of the target organism before reliable detection can be achieved in "dirty" (complex) samples like foods. Also, all are limited by the signal-to-noise ratio of the recognition step employed; their recognition systems have an intrinsic tendency to produce false signals (noise), and this limits the amplification that can be employed before the tests give false positives at an unacceptable level. The prompt detection of some key types of bacteria (e.g., *Salmonella*, *Listeria monocytogenes*, *Escherichia coli* O157:H7) is made even more difficult by the requirements that only live bacteria be detected (since these pathogens do not contaminate foods with exotoxins) and that the tolerated levels of viable bacteria may be as low as a single colony-forming unit (CFU) per 25 g of food (Andrews et al., 1992).

Several groups have attempted to measure levels of viable bacteria by phage-mediated introduction of a reporter gene that imparts a detectable novel phenotype. The reporter genes employed so far have been the *luxAB* bioluminescence gene cassette from *Vibrio fischeri* (Kodikara et al., 1991; Stewart and Williams, 1992;

Turpin et al., 1993; Ulitzur and Kuhn, 1989), the *inaW* ice nucleation gene of *Pseudomonas fluorescens* (Gutterson et al., 1993; Wolber, 1993; Wolber and Green, 1990a,b), and the firefly luciferase gene (Jacobs et al., 1993). Of these, the *inaW* gene has so far proven to be the most sensitive. In this chapter, we outline the considerations for applications of *ina* genes in transduction assays and describe methods and results for the particular purpose of detecting *Salmonella* in food samples. Finally, we discuss the potential application of *ina*-transduction assays to the detection of other organisms.

## Underlying Facts and Concepts

The basis of *ina*-transduction assays is the use of a species-specific bacteriophage to introduce a highly expressed *ina* gene into an Ice$^-$ target organism (Gutterson et al., 1993; Wolber, 1993; Wolber and Green, 1990a, b). Under such circumstances, the majority of bacterial cells that receive an *ina* gene express "Type 3" ice nucleation sites with thresholds between –8°C and –10°C (Yankofsky et al., 1981) within 30 minutes of the start of gene expression (Watanabe et al., 1990). The Bacterial Ice Nucleation Diagnostic (BIND) *ina*-transduction assay is shown diagrammatically in Figure 1.

*ina*-Transduction assays are quite specific, due to the species-specificity of bacteriophage infection, and quite sensitive, due to the large amplification produced by an ice nucleation event. Under some circumstances, such assays can detect a single bacterium in the presence of millions of non–target species cells (see Detection of *Salmonella* by P22-Mediated Transduction of *inaW*). However, it is the absence of background noise (in the form of potentially interfering ice nucleation sites) in most food samples (Table 1) that is most crucial for achieving high sensitivity in real systems. The few foods bearing significant numbers of ice nucleation sites are mostly derived with minimal processing from plants that might be expected to support populations of ice nucleation-active bacteria. Even these samples can be analyzed after suitable alteration of the assay protocol (e.g., dilution and further enrichment). Other factors favorably affecting sensitivity are the small amount of Ina protein needed to produce an active Type 3 nucleation site (Lindgren et al., 1989; Southworth et al., 1988; Watanabe et al., 1990) and the stability of the resulting Ice$^+$ phenotype.

The way in which phages are used adds further sensitivity. The phage strains used in *ina*-transduction assays are engineered so that nearly every phage particle contains an *ina* gene. In addition, since the Ice$^+$ phenotype of Type 3 sites is stable to subsequent cell death or lysis, the phage can be used at very high multiplicities of infection. The killing of cells by super-infection does not matter, provided that the first few phages to infect a cell cause the assembly of at least one ice nucleation site. Thus, enormous excesses of transducing phage can be used to overwhelm interfering substances that nonspecifically bind phage. Transduction can be viewed simply as a biochemical reaction that is rate-limited by diffusion of phage particles to bacterial surfaces. For the phage concentrations used in *ina*-transduction assays, this diffusion time is on the order of 30 minutes.

The Ice$^+$ phenotype is most readily employed in qualitative assays, since the usual methods for quantitative measurement of nucleation sites require large numbers of replicate samples at several orders of dilution (Vali et al., 1976). The organisms that are the most suitable targets for qualitative assays are those that must be completely absent from a food sample of some set size ("zero tolerance" organisms). Examples include all *Salmonella* species and *Listeria monocytogenes*. These organisms are the first two targets for *ina*-transduction assays; the assay for *Salmonella* is best developed and will be used as an example throughout this article.

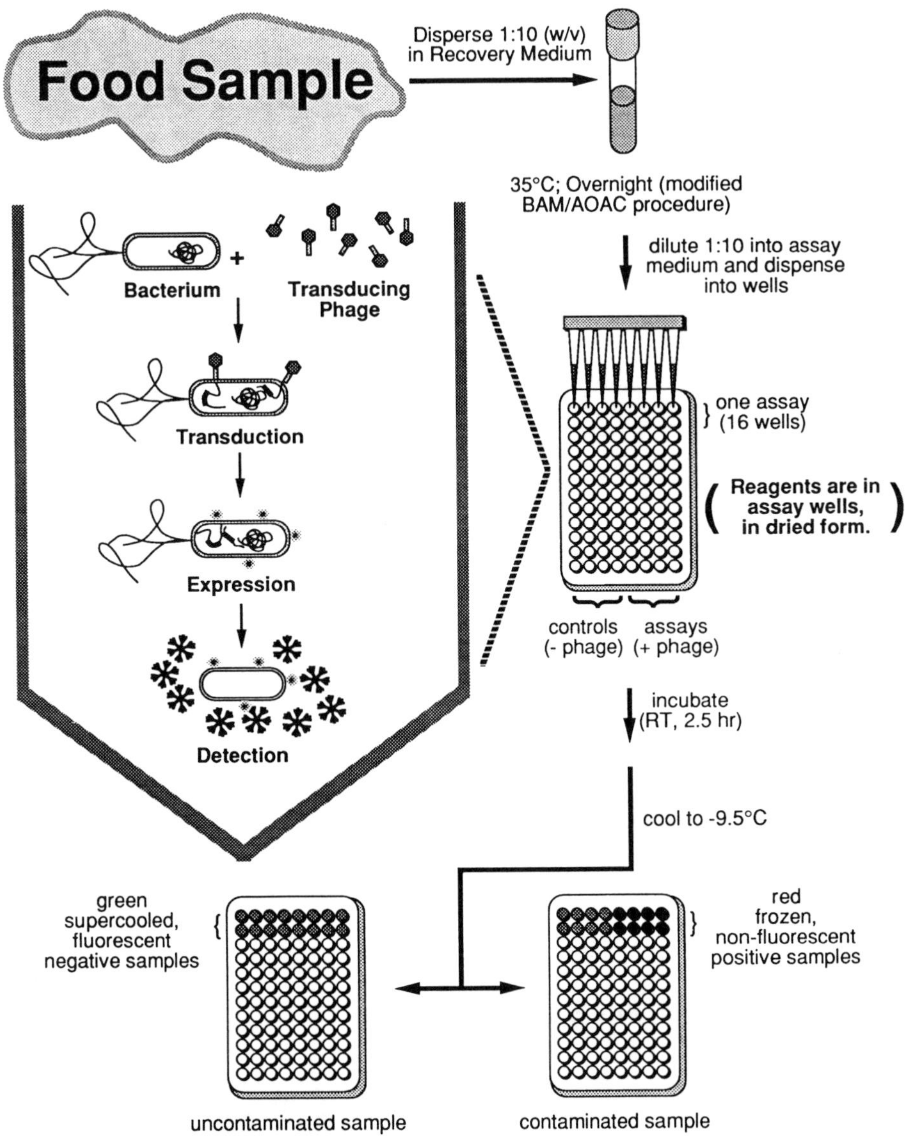


**Figure 1.** The bacterial ice nucleation diagnostic (BIND) assay system. The main figure shows a schematic diagram of the assay protocol; the inset shows a cartoon of the events occurring in a positive assay well. BAM/AOAC procedure: the procedure sanctioned by the Association of Official Analytical Chemists (Andrews et al., 1992).

**Table 1.** Intrinsic ice nucleation activity of nonselective food preenrichments

| Sample[a] | Background signal (frozen/total wells)[b] |
|---|---|
| Black pepper | 0/8 |
| Borril | 0/8 |
| Breaded fish fingers | 0/8 |
| Cadbury's dairy milk bar | 0/8 |
| Cheddar cheese | 0/8 |
| Chicken livers | 0/8 |
| Chicken nuggets | 0/8 |
| Chilled trifle (raspberry) | 0/8 |
| Cocoa powder | 0/8 |
| Cod in batter | 0/8 |
| Custard powder | 0/8 |
| Danish blue cheese (raw milk) | 0/8 |
| Dried egg white | 0/8 |
| Fresh pasteurized milk | 0/8 |
| Frozen gateaux | 0/8 |
| Frozen loose corn | 0/8 |
| Frozen peas | 0/8 |
| Frozen prawns | 0/8 |
| Garden vegetable dried soup | 0/8 |
| Hollandaise sauce powder | 0/8 |
| Mars bar | 0/8 |
| Milk chocolate bar | 0/8 |
| Minced beef with vegetable in gravy | 0/8 |
| Minced turkey | 0/8 |
| Ovaltine | 0/8 |
| Paté | 0/8 |
| Powdered whole egg | 0/8 |
| Ready-made custard | 0/8 |
| Salami | 0/8 |
| Sausage | 0/8 |
| Sesame chicken with rice | 0/8 |
| Vanilla ice cream | 0/8 |
| Chili con carne | 1/8 |
| Chilled ham and mushroom pizza | 1/8 |
| Dried skimmed milk powder | 1/8 |
| Fresh salad mix | 1/8 |
| Mayonnaise | 1/8 |
| Chicken and mushroom noodle pot | 2/8 |
| Chopped pork | 2/8 |
| Cole slaw | 2/8 |
| Dog food | 2/8 |
| Minced beef | 2/8 |
| Muesli bar | 2/8 |
| Pork pie | 2/8 |
| Cake mix (Victoria sponge) | 3/8 |
| Fresh bean sprouts | 8/8 |
| Gerbil food | 8/8 |
| Paprika | 8/8 |
| Rosemary | 8/8 |
| Scotch egg | 8/8 |
| Thyme | 8/8 |

[a]Samples were analyzed after 24 hr of nonselective preenrichment (1:10, w/v, buffered peptone and water).
[b]Fifty-microliter wells at −9.5°C.

In the current format of the BIND *ina*-transduction assay (Fig. 1), multiple microtiter plate wells are used to perform the measurement, and an equal number of control (no phage) wells are used to compensate for the possible presence of low concentrations of background ice nucleation sites. The wells also contain a fluorescent dye that is quenched and changes color when a well freezes (Warren and Wober, 1988). This provides a simple visual indication of a positive result. The BIND format requires no washing (i.e., separation) steps, unlike all other rapid assay formats for zero-tolerance organisms; in diagnostics parlance the BIND format is "homogeneous." This last feature greatly reduces the labor involved in conducting an assay.

## Detection of *Salmonella* by P22-Mediated Transduction of *inaW*

### Method of Phage Construction

The short-tailed, icosahedral, temperate phage P22 (Rutila and Jackson, 1981; Susskind and Botstein, 1978) infects *Salmonella* serogroups A, B, and $D_1$ (LeMinor, 1984); approximately 70% of the *Salmonella* cases reported to the Centers for Disease Control (CDC) during the period 1977–1987 were caused by members of these serogroups (CDC, 1988). The P22 genome has been mapped and mostly sequenced (Rutila and Jackson, 1981), and P22 is capable of transducing nonphage genes (Susskind and Botstein, 1978). P22 packages terminally redundant (~103% of the P22 genome), double-stranded DNA into phage heads by a "head-stuffing" mechanism (Susskind and Botstein, 1978). There is a facile approach for making high-frequency transducing particles (TPs) from any phage that packages its DNA in this way. If a location in the phage genome can be found that is tolerant of small (i.e., shorter than the terminal redundancy) insertions, then such an insertion can be used as a target for homologous recombination of an *ina* gene expression cassette into a "tagged" prophage.

The reason for making insertions in two steps is that a head-stuffing phage can tolerate additions shorter than the genome terminal redundancy, provided that essential coding regions are not interrupted. This is not true of insertions longer than the terminal redundancy (e.g., an *ina* expression cassette), since the resulting phage genome exceeds the maximum length imposed by the volume of the phage head. However, if the *ina* expression cassette is recombined into a previously introduced, short sequence in a tagged-phage lysogen, the prophage is not subject to size constraints, since these are only imposed by the phage head during DNA packaging; the "oversize" prophage is fully capable of transition to the lytic state, excision, and packaging, since the original tagged phage was selected for these properties. The TPs produced by such a prophage will usually be defective for further rounds of lytic or lysogenic growth, because the inserted *ina* gene is longer than the phage genome terminal redundancy. However, they will all be capable of transducing an *ina* gene, and a supply of TPs can be obtained by propagation of the recombinant prophage-containing lysogen.

The organization of the P22 genome is well enough understood that potential insertion sites have been selected on the basis of their function. However, genetic methods must be employed to find appropriate insertion sites in uncharacterized temperate phage strains. A useful approach is the insertion into the phage of a short, *cis*-activated transposon that contains a selectable gene. Two suitably small genes that have been used successfully are the *supF* suppressor tRNA gene (Phadnis et al., 1989) and the *lacZ-α* gene (Gutterson et al., 1993; Vieira and Messing, 1982; Zabin and Fowler, 1980). Insertion of the former gene is detected by employing an amber mutant phage; insertion of the *lacZ-α* gene is detected by subsequent lysogenization of a *lacZYA* ΔM15 host with selection for lactose utili-

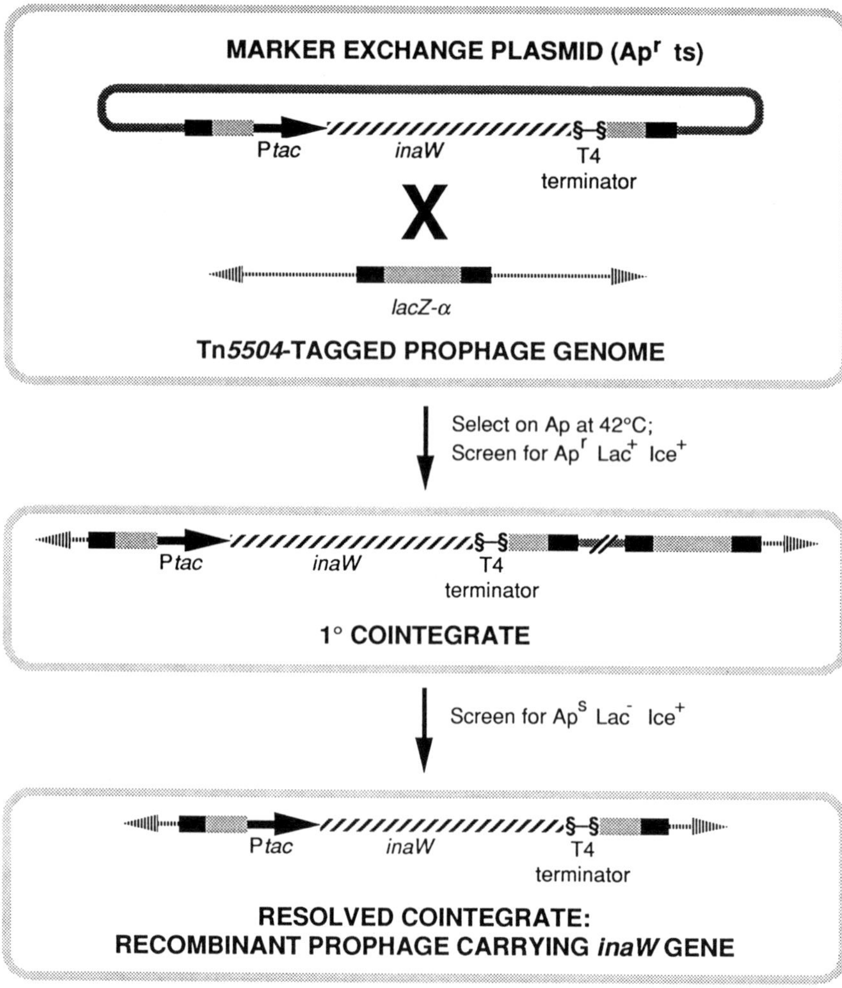


**Figure 2.** Construction of a recombinant prophage containing the *inaW* gene by marker exchange. **Top,** An ampicillin-resistant (Ap$^r$), temperature-sensitive (ts) plasmid recombines with a prophage previously tagged by a small transposon (Tn*5504*, which contains *lacZ*-α). The plasmid contains a derivative of Tn*5504* formed by insertion of an Ina expression cassette. The expression cassette consists of the *inaW* gene driven by the strong promoter P*tac*, with transcription terminated downstream of *inaW* by a sequence from phage T4. **Middle,** The primary recombinant, a prophage-plasmid cointegrate, selected by requiring lysogen growth under ampicillin selection at the nonpermissive temperature for plasmid replication. Homologous recombination between *lacZ*-α sequences is responsible for both the formation and the resolution of the cointegrate. **Bottom,** The resolved cointegrate is isolated by screening lysogens for loss of lactose utilization and antibiotic resistance but retention of the Ice$^+$ phenotype.

zation (Gutterson et al., 1993; Zabin and Fowler, 1980). Recombinant prophages have been constructed from phage P22 by this method; the chain of recombination events (with relevant phenotypes of each stage noted) is shown in Figure 2. The primary cointegrate (merodiploid) and resolved prophages are denoted as P22::pWTT1069 and P22::*inaW*, respectively; the entire class of recombinant prophages is denoted by P22::*ina*.

## Assay Performance in Model Systems

Transducing particles derived from P22::*ina* constructs have been tested in a variety of model systems. The key points examined have been sensitivity, specificity, the effect of P22 lysogeny on detection, and the degree of interference by dead target organisms. These investigations are summarized in Table 2; by comparison, other rapid methods for *Salmonella* detection require $10^4$ to $10^6$ target CFU/ml and are sensitive to the presence of live nontarget organisms and/or dead target organisms. In general, the sensitivity and specificity of detection by *ina*-transduction are good. More surprising are the assay's comparative insensitivity to P22 lysogeny of the target (which causes only a 10-fold decrease in sensitivity) or to the presence of a $10^7$-fold excess of dead target cells (which has no effect). These counter-intuitive results may be understood by reference to the large excess of P22-derived TPs in the assay mixture. Although P22 lysogens present a surface barrier to reinfection, the barrier is not perfect and is eventually overwhelmed by a high multiplicity of infection. Once the phage DNA has entered the host, the *inaW* gene is not subject to prophage-encoded repressors, because a nonphage promoter drives *inaW* transcription. In the presence of dead targets, the phage concentration is high enough to ensure that all targets, live or dead, eventually bind transducing phage.

## Assay Performance and Noise Levels in Real Systems

The performance of the *ina*-transduction assay when real samples are assayed is of considerable practical importance. The results of 20 assays on five food samples, deliberately contaminated by low levels of *S. typhimurium*, *S. enteritidis*, *S. agona*, or uncontaminated (control) are shown in Table 3; the standard microbiological assay method was used for comparison. The P22-based *ina*-transduction assay detected all P22-sensitive *Salmonella* (including a natural *S. enteritidis* contaminant present in the chicken liver sample) after overnight nonselective preenrichment. The levels of background ice nucleation observed from these samples were low

**Table 2.** Properties of P22::*ina*

| Property examined | P22::*ina* result summary |
|---|---|
| Sensitivity | Observed detection limits range between single cell detection and $10^4$ CFU/ml, depending upon *Salmonella* strain. |
| Specificity | No effect on *S. typhimurium* LT2 detection limit observed in the presence of $10^7$ CFU/ml of *E. coli, Bacillus subtilis,* or *S. havana* (which does not bear the receptor for P22). |
| Effect of P22 lysogenization | Detection limit for *S. typhimurium* LT2 $(P22)^+$ is 10-fold reduced from (single-cell) detection limit for *S. typhimurium* LT2 wild-type. |
| Effect of presence of dead bacteria | No effect on *S. typhimurium* LT2 detection limit observed in the presence of $10^7$ CFU/ml of heat-killed *S. typhimurium* LT2; heat-killed bacteria alone were not transduced to $Ice^+$ phenotype. |

**Table 3.** Comparison of culture and *ina*-transduction as methods of detecting *Salmonella* spp. inoculated into commercial products

| Product type | Strain added to product (~10 CFU/sample) | INA following P22::*ina* infection | Colonies following selective culture |
|---|---|---|---|
| Milk powder | *S. enteritidis* | + | + |
| | *S. typhimurium* LT2 | + | + |
| | *S. agona* | + | + |
| | Control | − | − |
| Chicken liver[a] | *S. enteritidis* | + | + |
| | *S. typhimurium* LT2 | + | + |
| | *S. agona* | + | + |
| | Control | + | + |
| Chocolate drink powder | *S. enteritidis* | + | + |
| | *S. typhimurium* LT2 | + | + |
| | *S. agona* | + | + |
| | Control | − | − |
| Minced beef in gravy[a] | *S. enteritidis* | + | + |
| | *S. typhimurium* LT2 | + | + |
| | *S. agona* | + | + |
| | Control | − | − |
| Pepper | *S. enteritidis* | − | − |
| | *S. typhimurium* LT2 | − | − |
| | *S. agona* | − | + |
| | Control | − | + |

[a] Bacteria subjected to freeze-thaw stress.

(often, no background was observed), in keeping with the results reported in Table 1. Finally, two of the pepper samples were naturally contaminated with *Salmonella* from P22-insensitive serogroups. Only the natural contaminants were recovered (many spices possess bactericidal properties that interfere with recovery of deliberately added organisms). The P22-insensitive contaminants were not detected by the P22-based reagent, as expected.

## Reagent Phage Strains Based on Genus-Specific *Salmonella* Phage

### Construction of Phage Reagents

Edwards and Ewing (1972) list nearly 1,900 distinct *Salmonella* serotypes, many of which have caused reported illnesses in the U.S. during the last 15 years (CDC, 1988). In order to provide *Salmonella* surveillance comparable to currently accepted methods, a new assay must detect most *Salmonella* isolates and should not exhibit serotype-specific "holes" in detection. It is acceptable to fail to detect a few isolates but not to miss an entire serogroup or serotype. Clearly, the P22-based *ina*-transduction reagent does not detect the entire genus *Salmonella*. Therefore, coverage must be improved by the addition of transducing vehicles based on other phages.

There is one phage family, represented by the phage Felix O1 (abbreviated FO) that lyses nearly all *Salmonella* isolates (Cherry et al., 1954; Felix and Callow, 1943; Lindberg, 1977; Mäkelä, 1985; Wilkinson et al., 1972). FO is a long-tailed, icosahedral, virulent phage that recognizes an α-(*N*-acetyl-D-glucosamine)-1,2-glucose linkage at the outer edge of the lipopolysaccharide core (Lindberg, 1977; Mäkelä, 1985; Rick, 1987); the recognition site is ubiquitous within the genus *Salmonella*, but extremely rare outside this genus (Cherry et al., 1954; Lindberg, 1977).

Virulent phages are difficult to modify, because they usually alter host rules for transcription (in which case a phage promoter must be used to drive the *ina* gene), sometimes utilize chemically modified DNA, and often exhibit complex life cycles that are easily disrupted (Studier, 1973). In addition, they cannot be manipulated as stable host-phage combinations; therefore, selectable markers (usually, suppressor tRNA genes) must operate on the phage, not on the host, and recombinant phage must conform to the DNA length limit imposed by phage head size. Several *ina*-transducing reagents have been constructed from a member of the FO family despite these obstacles. The phage that served as the starting material for this construction, although derived from FO, exhibited greater infectivity than FO against some *Salmonella* serogroups, such as group E (Mäkelä, 1985).

### Reagent Specificity and Sensitivity

Two of the transducing phage reagents, denoted as ØW10-1114 and Ø10.1-28-4.1, have been extensively characterized. Both contain the *inaW* gene (driven by a phage promoter) and both are "locked" phages, capable of lytic growth in particular laboratory strains of *Salmonella* but not in wild-type strains. Both are resolved cointegrates: through homologous recombination, they have lost the sequences of the plasmid originally used to introduce the *inaW* gene into the phage (Fig. 2). Finally, the two reagents yield virtually indistinguishable detection efficiencies. This family of reagents is referred to as "*ina*-sapphire" phage.

In a survey of 438 isolates from *Salmonella* serogroups B–H (representing 98.3% of reported U.S. cases), *ina*-sapphire phage detected 435 members of the set (C. Lindhardt and N. Watanabe, *unpublished*). In a parallel examination of 155 members of the rarer serogroups I–Z, *ina*-sapphire phage detected 86 of 134 isolates. Thirty of the misses (and all of the consistent misses) occurred in serogroup O, which is known to include strains resistant to the FO phage family (Cherry et al., 1954). Finally, *ina*-sapphire phage detected 18 of 21 representatives of the rarest *Salmonella*, which bear O-antigens 51–68.

Concurrent assay of 121 different *Escherichia* isolates (mostly *E. coli*) resulted in 11 positive results (all *E. coli*); this outcome was expected from the published properties of phage FO (Cherry et al., 1954). Examination of 24 *Citrobacter* isolates and representatives of 54 other common food-contaminating bacterial species yielded no positive assay results with *ina*-sapphire phage.

### Assay Performance in Real Systems

The results (C. Lindhardt and N. Watanabe, *unpublished*) of 15 duplicate *ina*-transduction assays of 13 products which had (before storage) yielded positive *Salmonella* assay results (standard microbiological method) are shown in Table 4. In all cases, a standard microbiological assay was also performed in parallel on the sample, since real *Salmonella*-positive samples are generally contaminated at low levels and often lose their contaminant(s) during storage. Agreement was good: the two methods agreed on eight positive and 19 negative results; the *ina*-transduction assay produced positive results for two samples from which *Salmonella* could not be isolated by microbiological methods and one negative result for a sample from which a *Salmonella* strain was isolated by the microbiological method. Subsequent work indicated that all of the disagreements could be explained by the strain-dependent sensitivity of the *ina*-sapphire phage. In two cases, the phage probably detected *Salmonella* that did not survive the culturing required for confirmation. In one case, the *Salmonella* strain isolated was relatively insensitive to the *ina*-sapphire phage.

## Detection Limits

For the current *ina*-transduction assay protocol (assuming binomial distribution of frozen and unfrozen wells), the 95% confidence detection limit can be approximated (C. Lindhardt, *unpublished*) as

$$DL_{95\%} = (200/f)\,(1 + 0.25\sqrt{f})\ \text{CFU/ml},$$

where $f$ is the frequency of ice nucleation per transduction target, measured at the

**Table 4.** Detection of *Salmonella* spp. in naturally contaminated products, using *ina*-sapphire phage

| Product type | Transduced wells frozen (8 total) | Control wells frozen (8 total) | *ina*-sapphire transduction result | Culture method result | *Salmonella* isolate serogroup |
|---|---|---|---|---|---|
| Buttermilk powder | 0 | 0 | – | – | |
| | 0 | 0 | – | – | |
| | 0 | 0 | – | – | |
| | 0 | 0 | – | – | |
| Cotton seed | 0 | 0 | – | – | |
| | 8 | 0 | + | + | R |
| Cotton seed meal | 8 | 0 | + | + | E1 |
| | 8 | 1 | + | + | F |
| Fish meal | 0 | 0 | – | – | |
| | 0 | 0 | – | – | |
| Hide powder | 0 | 0 | – | +[a] | H |
| | 7 | 0 | + | + | C1 |
| Thin hide sticks | 3 | 0 | – | – | E1 |
| | 8 | 0 | + | + | E1 |
| Meat and bone meal | 0 | 0 | – | – | |
| | 0 | 0 | – | – | |
| Meat meal | 2 | 0 | – | – | |
| | 8 | 0 | + | + | C1 and 2/F/G/H |
| Milk powder | 0 | 0 | – | – | |
| | 0 | 0 | – | – | |
| Pet pellets | 4 | 4 | – | – | |
| | 4 | 3 | – | – | |
| Soya | 8 | 5 | – | – | |
| | 8 | 2 | + | –[a] | |
| Sunflower | 5 | 0 | + | + | T/U |
| | 7 | 0 | + | –[a] | |
| Whey powder | 0 | 0 | – | – | |
| | 0 | 0 | – | – | |
| | 0 | 0 | – | – | |
| | 1 | 0 | – | – | |

[a] Method disagreement.

assay readout temperature (usually −9.5°C). The $DL_{95\%}$ is the minimum concentration of bacteria in the original nonselective preenrichment needed to ensure a 95% probability of detection; the formula assumes that 8 × 50 μl samples of a 1:10 dilution of the nonselective preenrichment are actually assayed (see Fig. 1). The *ina*-transduction frequency *f* is measured by standard methods (Vali, 1971). This formula has been used to calculate $DL_{95\%}$ values for a sampling of 27 different se-

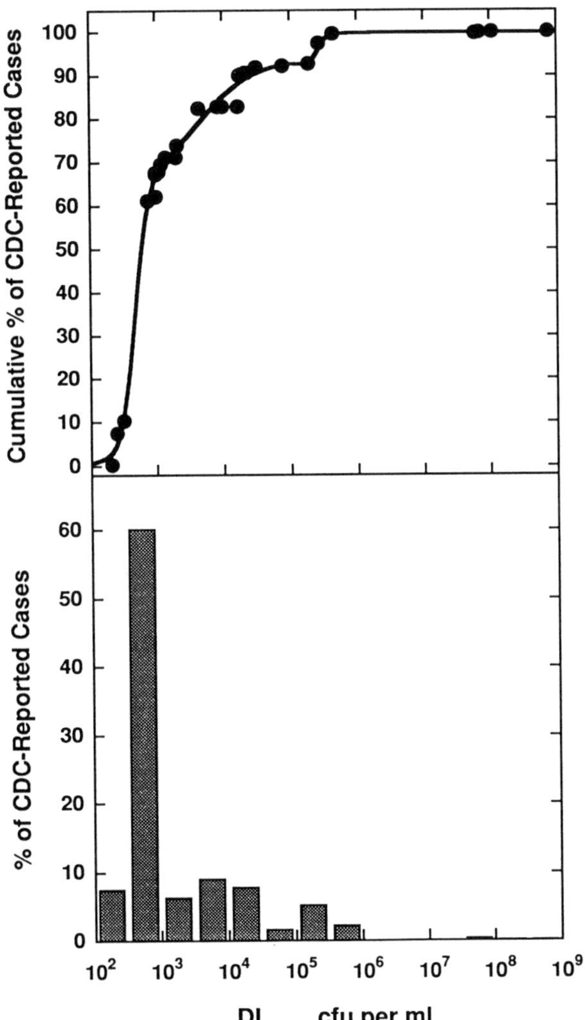


**Figure 3.** Theoretical cumulative and distributed detection of an epidemiologically weighted set of *Salmonella* serotypes. **Top,** The cumulative detection of *Salmonella* as a function of 95% confidence detection limit ($DL_{95\%}$), obtained by integrating the normalized distribution of epidemiologically-weighted $DL_{95\%}$ values. **Bottom,** Histogram of this distribution, calculated by assigning the incidence (as reported to the Centers for Disease Control) of disease caused by each serotype in the 25-member sample set to the $DL_{95\%}$ value for that serotype, collecting the values into ½-log-unit "bins" of $DL_{95\%}$ values and normalizing the resulting distribution.

rotypes of *Salmonella*; these data, in turn, have been used to produce an incidence-weighted graph of cumulative detection as a function of the concentration of *Salmonella* recovered from a nonselective preenrichment. The inferred distribution of $DL_{95\%}$ values and cumulative detection frequency are shown in Figure 3. The $DL_{95\%}$ distribution may be multimodal, although the appearance of multiple, overlapping distributions may in part be an artifact of the sample size. The four serogroup O strains examined form a distinct distribution at $DL_{95\%}$ values between $10^7$ and $10^9$ CFU/ml. This distribution is not visible in Figure 3 because of the low epidemiological incidence of these strains.

The false-negative rate of the *ina*-transduction assay (in the tested configuration) can be estimated by combining the cumulative incidence of $DL_{95\%}$ values less than $10^4$ CFU/ml (~86%) with the empirical fact that 95% of nonselective preenrichments that eventually lead to positive *Salmonella* results contain at least $10^4$ *Salmonella* CFU/ml. (M. Curiale, *personal communication*) Therefore, the probability of failure is approximately $(1 - 0.86) \times (1 - 0.95) = 0.007$, or 0.7%. This is less than the 1% false negative rate claimed by current rapid assay methods. However, this figure must be taken with caution, since it does not include the potential effects of the analytic matrix or competing microflora on the frequency of transduction. It is also clear that detection of serogroup O will have to be improved to produce an assay free of serogroup-specific weaknesses.

## Detection of Non-*Salmonella* Organisms

### Coliform Detection

Besides *Salmonella*, the most commonly assayed nonexotoxic food organisms are "coliforms" (a general grouping of enteric bacteria used to monitor general sanitation) and *Listeria monocytogenes*, a gram-positive, psychrotrophic, food-borne, opportunistic pathogen. The adaptation of *ina*-transduction assays to the detection of coliforms should prove to be straightforward, given the extensive research literature on expression of *ina* genes in gram-negative bacteria (Corotto et al., 1986; Green and Warren, 1985; Lindgren et al., 1989; Orser et al., 1985; Southworth et al., 1988) and the successful construction of prototype coliform assays based on transduction of *lux* genes (Kodikara et al., 1991; Stewart and Williams, 1992).

### Expression of *ina* Genes in Eukaryotes and Gram-Positive Bacteria

The detection of gram-positive bacteria by transduction of *ina* genes is more technically demanding. The consensus sequence requirements for high-level gene expression in gram-positive hosts are more stringent than those for gram-negative hosts (Band and Henner, 1984). Therefore, replacement of the gene promoter and Shine-Dalgarno sequences are generally required to achieve expression of gram-negative structural genes in gram-positive bacteria.

Green and Wolber (*unpublished*) have replaced the Shine-Dalgarno sequence and endogenous promoter of the *inaZ* gene (Green et al., 1988; Green and Warren, 1985) with a near-perfect, gram-positive, consensus Shine-Dalgarno sequence and a gram-positive *sp02* promoter (Schoner et al., 1983; Williams et al., 1981). The modified *ina* gene was inserted into the shuttle vector pAT18 (Trieu-Cuot et al., 1991), and the resulting plasmid was used to transform representatives of the two most pathogenic *L. monocytogenes* serogroups (1 and 4). The transformed strains were both $Ice^+$; their cumulative ice nucleation spectra are shown in Figure 4. The expression of Type 3 ice nucleation sites was quite good; no more than a 10-fold increase in ice nucleation frequency, corresponding to a two to fourfold increase in Ina protein (Lindgren et al., 1989; Southworth et al., 1988; Watanabe et al., 1990; see also the chapter by Warren, this volume), will be needed to provide sensitivity of detection comparable to that already seen in *Salmonella*.

Finally, *ina* genes could potentially be used for detecting either eukaryotic pathogens or viruses of eukaryotes. The first step needed for such an application, the expression of *ina* genes in eukaryotes, has already been taken by transforming plants with *ina* genes modified for plant expression (Baertlein et al., 1992). The initial data suggest that a sufficiently strong promoter yields plant cells bearing from 0.1 to 1 Type 3 nucleation sites per cell.

Detection of eukaryotic pathogens by *ina*-transduction requires the availability of suitably specific viral vectors; unfortunately, for many of the organisms of interest, such vectors are not readily available. However, reporter genes have already been used to detect viruses, such as HIV, that utilize *trans*-activators as molecular switches of the viral life cycle (Felber and Pavlakis, 1988; Schwartz et al., 1990). In such an assay, an "indicator" cell line is first constructed by transforming a virus-susceptible cell line with the reporter gene fused to a promoter responsive to the viral *trans*-activator. If regulation of the resulting construct is sufficiently tight, the indicator strain can be used to produce a quantitative signal indicating the presence of infective virus long before the infected cells show classical symptoms of viral infection. So far, no one has attempted construction of such an assay utilizing an *ina* gene.

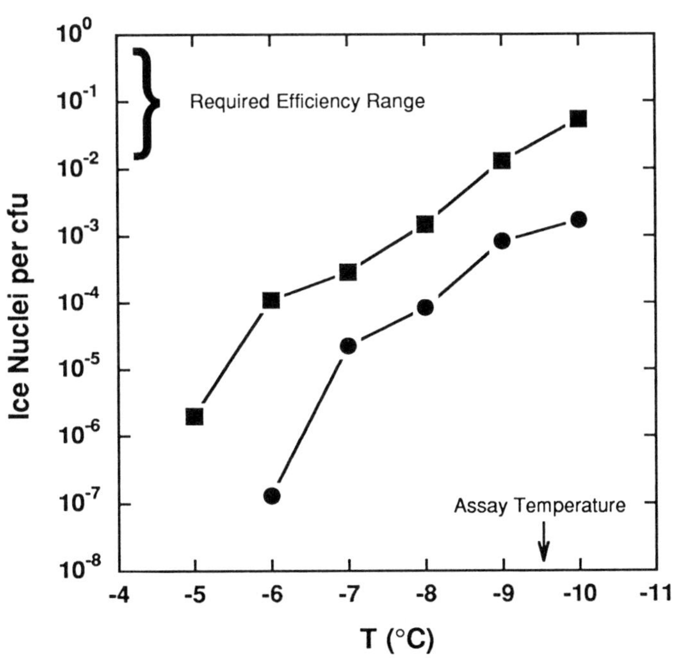


**Figure 4.** Ice nucleation activities of *Listeria monocytogenes* strains transformed with pRLG89 as a function of temperature. Data are shown for two transformed *L. monocytogenes* strains, BL174 (●) and BL210 (■).

## Concluding Remarks

To date, three coliphage (L. Corotto, G. Warren, and P. Wolber, *unpublished*; R. Green and P. Wolber, *unpublished*) and six *Salmonella*-infecting phage species (J. Dove, R. Fallon, R. L. Green, W. T. Tucker, C. A. Vance, N. M. Watanabe, and P. K. Wolber, *unpublished*) have been modified to transduce *ina* genes. Most of these phages have yielded at least one *ina*-transduction reagent capable of detecting its bacterial targets with high sensitivity. A combined family of transducing phages for *Salmonella* is now undergoing rigorous validation and should lead to commercialization of an assay for *Salmonella* during 1995. Assays for other important bacterial pathogens will follow. The authors hope that gene transduction assays will eventually become a well-established and generally accessible approach to the detection of bacterial pathogens and that we all will be safer for that fact.

## Acknowledgments

The authors gratefully acknowledge helpful conversations with J. Kuhn, M. Suissa, S. Ulitzur, B. Stocker, G. Stewart, W. Studier, N. Gutterson, J. Bedbrook, and G. Warren. We also acknowledge the work carried out in the laboratory of M. Suissa, J. Kuhn, and S. Ulitzur in the identification of marker gene insertion strategies for virulent phages and their provision of materials and information on which the work reported here with genus-specific *Salmonella* phage was based. P. Wolber, R. Green, N. Watanabe, and C. Vance were supported in part by NIH SBIR Grant 2 R44 AI29222-02A1.

## Literature Cited

Andrews, W. H., Bruce, V. R., June, G., Satchell, F., and Sherrod, P. 1992. *Salmonella*. Pages 51-69 in: FDA Bacteriological Analytical Manual, 7th ed. L. A. Tomlinson, ed. Association of Official Analytical Chemists International, Arlington, VA.

Baertlein, D. A., Lindow, S. E., Panopoulos, N. J., Lee, S. P., Mindrinos, M. N., and Chen, T. H. H. 1992. Expression of a bacterial ice nucleation gene in plants. Plant Physiol. 100:1730-1736.

Band, L., and Henner, D. J. 1984. *Bacillus subtilis* requires a "stringent" Shine-Dalgarno region for gene expression. DNA 3:17-21.

CDC. 1988. Annual summary 1987: Centers for Disease Control *Salmonella* surveillance. U.S. Department of Health and Human Services—Public Health Service, Washington, DC.

Cherry, W. B., Davis, B. R., Edwards, P. R., and Hogan, R. B. 1954. A simple procedure for the identification of the genus *Salmonella* by means of a specific bacteriophage. J. Lab. Clin. Med. 44:51-55.

Corotto, L. V., Wolber, P. K., and Warren, G. J. 1986. Ice nucleation activity of *Pseudomonas fluorescens*: Mutagenesis, complementation analysis and identification of a gene product. EMBO J. 5:231-236.

Edwards, P. R., and Ewing, W. H. 1972. Identification of Enterobacteriaceae. Burgess Publishing Company, Minneapolis.

Felber, B. K., and Pavlakis, G. N. 1988. A quantitative bioassay for HIV-1 based on trans-activation. Science 239:184-187.

Felix, A., and Callow, B. R. 1943. Typing of paratyphoid B bacilli by means of Vi bacteriophages. Brit. Med. J. 2:127-130.

Green, R. L., Corotto, L. V., and Warren, G. J. 1988. Deletion mutagenesis of the ice nucleation gene from *Pseudomonas syringae* S203. Mol. Gen. Genet. 215:165-172.

Green, R. L., and Warren, G. J. 1985. Physical and functional repetition in a bacterial ice nucleation gene. Nature (London) 317:645-648.

Gutterson, N. I., Tucker, W. T., and Wolber, P. K. 1993. Transducing particles and methods for their production. U.S. patent 5,187,061.

Jacobs, W. R., Barletta, R. G., Udani, R., Chan, J., Kalkut, G., Sosne, G., Kieser, T., Sarkis, G. J.,

Hatfull, G. F., and Bloom, B. R. 1993. Rapid assessment of drug susceptibilities of *Mycobacterium tuberculosis* by means of luciferase reporter phages. Science 260:819-822.

Kodikara, C. P., Crew, H. H., and Stewart, G. S. A. B. 1991. Near on-line detection of enteric bacteria using *lux* recombinant bacteriophage. FEMS Microbiol. Lett. 83:261-266.

LeMinor, L. 1984. Pages 427-458 in: Bergey's Manual of Systematic Microbiology. N. R. Kreig and J. G. Holt, eds. Williams and Wilkins, Baltimore, MD.

Lindberg, A. A. 1977. Bacterial surface carbohydrates and bacteriophage adsorption. Pages 289-356 in: Surface carbohydrates of the prokaryotic cell. I. W. Sutherland, ed. Academic Press, London.

Lindgren, P. B., Frederick, R., Govindarajan, A. G., Panopoulos, N. J., Staskawicz, B. J., and Lindow, S. E. 1989. An ice nucleation reporter gene system: Identification of inducible pathenogenicity genes in *Pseudomonas syringae* pv. *phaseolicola*. EMBO J. 8:1291-1301.

Mäkelä, P. H. 1985. The use of bacteriophages and bacteriocins in the study of bacterial cell surface structures. Pages 155-178 in: Enterobacterial Surface Antigens: Methods for Molecular Characterization, T. K. Korhonen, E. A. Dawes, and P. H. Mäkelä, eds. Elsevier, New York.

O'Donnell, C. D. 1992. Beyond the hype of HACCP. Prepared Foods, April, pp. 29-30.

Orser, C., Staskawicz, B. J., Panopoulos, N. J., Dahlbeck, D., and Lindow, S. E. 1985. Cloning and expression of ice nucleation genes in *Escherichia coli*. J. Bacteriol. 164:359-366.

Phadnis, S. H., Huang, H. V., and Berg, D. E. 1989. Tn*5*supF, a 264-base-pair transposon derivative of Tn*5* for insertion mutagenesis and sequencing DNAs cloned in phage l. Proc. Natl. Acad. Sci. USA 86:5908-5912.

Rick, P. D. 1987. Lipopolysaccharide biosynthesis. Pages 648-662 in: *Escherichia coli* and *Salmonella typhimurium*: Cellular and molecular biology, F. C. Neidhardt, J. L. Ingraham, B. Magasanik, K. B. Low, M. Schaechter, and H. E. Umbarger, eds. American Society for Microbiology, Washington, DC.

Rutila, J. E., and Jackson, E. N. 1981. Physical map of the bacteriophage P22 genome. Virol. 113:769-775.

Schoner, R. G., Williams, D. M., and Lovett, P. S. 1983. Enhanced expression of mouse dihydrofolate reductase in *Bacillus subtilis*. Gene 22:47-57.

Schwartz, O., Virelizier, J.-L., Montagnier, L., and Hazan, U. 1990. A microtransfection method using the luciferase-encoding reporter gene for the assay of human immunodeficiency virus LTR promoter activity. Gene 88:197-206.

Southworth, M. W., Wolber, P. K., and Warren, G. J. 1988. Nonlinear relationship between concentration and activity of a bacterial ice nucleation protein. J. Biol. Chem. 263:15211-15216.

Stewart, G. S. A. B., and Williams, P. 1992. *lux* genes and applications of bacterial bioluminescence. J. Gen Microbiol. 138:1289-1300.

Studier, F. W. 1973. Genetic analysis of non-essential bacteriophage T7 genes. J. Mol. Biol. 79:227-236.

Susskind, M. M., and Botstein, D. 1978. Molecular genetics of bacteriophage P22. Microbiol. Rev. 42:385-413.

Trieu-Cuot, P., Carlier, C., Poyart-Salmeron, C., and Courvalin, P. 1991. Shuttle vectors containing a multiple cloning site and a *lacZ-$\alpha$* gene for conjugal transfer of DNA from *Escherichia coli* to gram-positive bacteria. Gene 102:99-104.

Turpin, P. E., Maycroft, K. A., Bedford, J., Rowlands, C. L., and Wellington, E. M. H. 1993. A rapid luminescent-phage based MPN method for the enumeration of *Salmonella typhimurium* in environmental samples. Lett. Appl. Microbiol. 16:24-27.

Ulitzur, S. Y., and Kuhn, J. C. 1989. Detection and/or identification of microorganisms in a test sample using bioluminescence or other exogenous genetically-introduced marker. U.S. patent 4,861,709.

Vali, G. 1971. Quantitative evaluation of experimental results on the heterogeneous freezing nucleation of supercooled liquids. J. Atmos. Sci. 28:402-409.

Vali, G., Cristensen, M., Fresh, R. W., Galyan, E. L., Maki, L. R., and Schnell, R. C. 1976. Biogenic ice nuclei. II. Bacterial sources. J. Atmos. Sci. 33:1565-1570.

Veld, J. H. I., Hartog, B., and Hofstra, H. 1988. Changing perspectives in food microbiology: Implementation of rapid microbiological analyses in modern food processing. Food Rev. Int. 4:271-329.

Vieira, J., and Messing, J. 1982. The pUC plasmids, an M13mp7-derived system for insertion mutagenesis and sequencing with synthetic universal primers. Gene 19:259-268.

Warren, G. J., and Wolber, P. K. 1988. Ice nucleation immunoassay. U. S. patent 4,784,943.

Watanabe, N. M., Southworth, M. W., Warren, G. J., and Wolber, P. K. 1990. Rates of assembly and degradation of bacterial ice nuclei. Mol. Microbiol. 4:1871-1879.

Wilkinson, R. G., P. Gemski, J., and Stocker, B. A. D. 1972. Non-smooth mutants of *Salmonella typhimurium*: Differentiation by phage sensitivity and genetic mapping. J. Gen. Microbiol. 70:527-

554.

Williams, D. M., Duvall, E. J., and Lovett, P. S. 1981. Cloning restriction fragments that promote expression of a gene in *Bacillus subtilis*. J. Bacteriol. 146:1162-1165.

Wolber, P. K. 1993. Bacterial ice nucleation. Pages 203-235 in: Advances in Microbial Physiology. A. H. Rose, ed. Academic Press, London.

Wolber, P. K., and Green, R. L. 1990a. Detection of bacteria by transduction of ice nucleation genes. Trends Biotechnol. 8:276-279.

Wolber, P. K., and Green, R. L. 1990b. New rapid method for detection of *Salmonella* in foods. Food Sci. Technol. 1:80-82.

Yankofsky, S. A., Levin, Z., Berfold, T., and Sandlerman, N. 1981. Some basic characteristics of bacterial freezing nuclei. Appl. Meteorol. 20:1013-1019.

Zabin, I., and Fowler, A. V. 1980. β-galactosidase, the lactose permease protein and thiogalactoside transacetylase. Pages 89-122 in: The Operon. J. H. Miller and W. S. Reznikoff, eds. Cold Spring Harbor Laboratory, Cold Spring Harbor, NY.

CHAPTER 17

# Applications of Bacterial Ice Nucleation Activity in Food Processing

**Michiko Watanabe and Soichi Arai**

In several food processing applications, it is desirable to *minimize* the number of nucleation events in order to form larger ice crystals. Bacterial ice nuclei have utility in such applications because they make it possible to achieve freezing at a relatively warm temperature—approximately –5°C. A nucleation event always tends to prevent other nucleation events in its vicinity because it releases the latent heat of fusion. At warmer temperatures, this inhibition tends to affect a larger volume. Hence, toward warmer temperatures, there will be fewer nucleation *events* per unit volume, even though a large number of ice nucleators may be present.

The freeze-concentration technique is gaining ground in food technology because it avoids the loss of volatile flavor components and the chemical deterioration that may accompany methods which involve heating. Larger crystals are desirable because they drastically improve the efficiency with which the frozen and unfrozen phases can be separated. Ice nuclei have been applied to freeze-concentration with advantage (Watanabe and Arai, 1987; Watanabe et al., 1989, 1991b; Honma et al., 1993; Kumeno et al., 1993, 1994).

Freeze-texturing is a technique that relies on ice crystals to cause changes that, after thawing, impart food with anisotropic mechanical properties. The larger crystal sizes resulting from use of ice nuclei cause useful changes in the scale of the anisotropy (Arai and Watanabe, 1986).

## Bacterial Sources of Ice Nucleation Activity

In food applications, the identity of any bacteria that are used in processing is important. *Pseudomonas* strains are potentially phytopathogenic (Lindow, 1983), and *Erwinia* strains are members of the Enterobacteriaceae, which makes their use in food processing undesirable. On the other hand, bacteria of the genus *Xanthomonas* have been permitted for food use (Jeanes, 1974); a nonphytopathogenic Ice$^+$ strain of *Xanthomonas* would be a preferred source of ice nuclei. Recently, we isolated *X. campestris* INXC-1 from shoots of tea (Watanabe et al., 1993). Cells of this strain have a cell size of 0.5 × 1.6 μm, and possess one flagellum. The strain produces a yellow pigment but no brown pigment; under aerobic conditions, it produces acids from a variety of sugars and tolerates heating to 50°C but loses its ice nucleation activity (INA) above 27°C. Its growth is promoted by the addition of glycerol, succinate, lactate, and sucrose to a yeast extract–peptone medium.

Strain INXC-1 produces xanthan gum. When gum accumulates in the medium, the bacterial cells are entrapped, and their separation from the medium becomes difficult. Fortunately, the addition of lactose inhibits gum production while promoting cell growth. Figure 1 shows a time-course change in the INA measured by a drop-freezing assay (Vali, 1971). Cells in logarithmic growth phase (24 hr after the start of cultivation) generated a freezing spectrum from –3.5 to –8°C. Nucleation temperatures became higher with cultivation time, and at 72 hr of cultivation almost all nucleation occurred at or above –5°C. Prolonged incubation evidently helps to maximize INA. A possible reason might be that Ina protein molecules continue to accumulate in membrane-bound aggregates. Recently, M. Watanabe and J. Watanabe (*unpublished*) attempted to find inducers of bacterial INA that were also normal components of food; they identified 4-hydroxy-3-nitrophenylacetic acid, isolated from mustard seeds, as an inducer effective at levels of 0.1–1.0 parts per million.

## Preliminary Treatments of INA Bacteria for Applications

For practical applications of bacterial ice nuclei, containment and/or sterilization of the cells is desirable for the sake of manufacturing hygiene. Cells can be contained by cellophane and still nucleate water at –5°C; secondary nucleation occurs

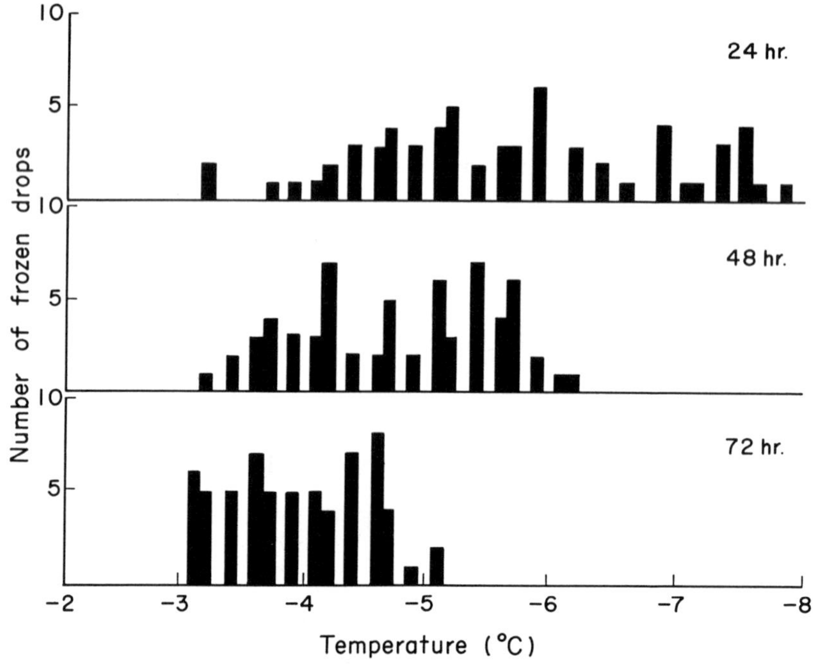


**Figure 1.** Histograms of drop freezing temperatures of the cells of *Xanthomonas campestris* INXC-1 in 24-, 48-, and 72-hr cultures. Ice nuclei with higher threshold temperatures accumulate with prolonged incubation. Each 5-µl drop contained approximately $10^4$ cells; cooling was at the rate of 1°C/min. Reprinted, with permission, from Watanabe et al. (1993).

through the cellophane membrane. Cells can also be entrapped in calcium alginate gel (Watanabe et al., 1989). In cellophane containment and gel entrapment, a desired specific gravity can be obtained by including air and/or glass powder together with the cells. This makes it possible to fix the tube or the gel at any particular position of the sample to be frozen. When entrapped cells are fixed at an appropriate position in a sample, ice nucleation is observed to start at that position.

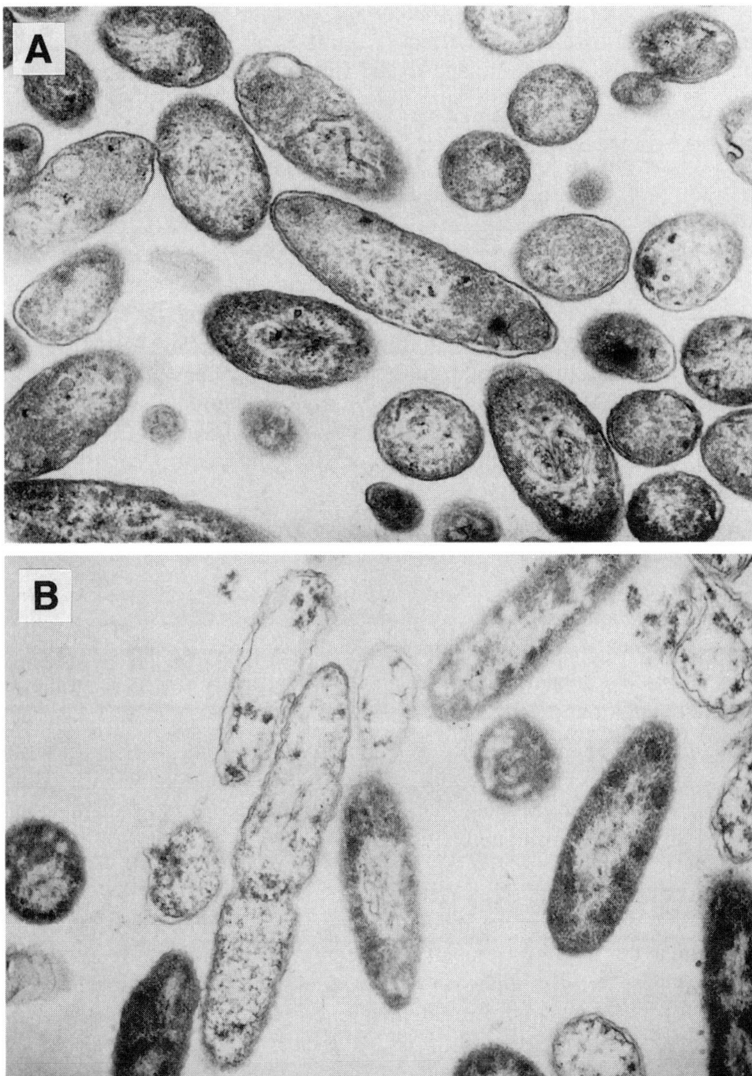


**Figure 2.** Transmission electron micrographs showing cross-sections of intact (**A**) and pressure-sterilized (**B**) cells of *Xanthomonas campestris* INXC-1. The observed differences suggest damage to the cell membrane and leakage of cytoplasm. The sterilized cells underwent pressurization at 300 MPa for 10 min at 5°C in a pressure generator (Mitsubishi Heavy Industries, MFP-7000). Reprinted, with permission, from Honma et al. (1993).

Hayashi (1989) proposed that high pressure could be used to kill microorganisms. This approach is applicable to the sterilization of $Ice^+$ bacteria. We have demonstrated high-pressure sterilization of the $Ice^+$ strains, *E. ananas* IN-10 and *X. campestris* INXC-1. Each cell suspension was packed in a plastic bottle and then compressed up to 200–500 MPa at 20°C and at 5°C. Both species were sterilized at 300 MPa, irrespective of temperature. The pressurization at 20°C resulted in a decrease of the ice nucleation activity, although the cells regained a high activity if chilled at 0°C for more than 1 day (Watanabe et al., 1991a). The pressurization at 5°C was favorable for retaining activity (Honma et al., 1993). The electron micrographs in Figure 2 show the damage to the sterilized cells in comparison to intact cells.

## Freeze Concentration of Food Materials with Bacterial Ice Nuclei

Freeze-concentration is based on the principle that substances dissolved or dispersed in water are excluded and thus concentrated during the growth of ice. This technique has been applied, particularly in the coffee and vinegar industries (Omran and King, 1974), because it causes minimal loss of flavor, slows the deterioration of heat-labile substances, and inhibits the activities of enzymes and microorganisms. In one currently used method of freeze-concentration, the material to be concentrated is first cooled to a subzero temperature and then seeded with ice chips as a primary nucleation agent. This procedure, however, has the following disadvantages:

1) The temperature of the material is not always homogeneous, so some seeded ice crystals may melt.
2) Any stirring process used to attain a homogeneous temperature distribution may cause secondary nucleation (Shirai et al., 1985).
3) The formation of smaller ice crystals due to secondary nucleation results in an increase in the ice crystal surface area and, therefore, an increase in the adsorption of the concentrated substances.

To solve these problems, it is usual to strictly control the temperature and stirring (Shirai et al., 1986); however, use of bacterial ice nuclei makes the addition of ice chips and stirring unnecessary. We applied this approach to the freeze-concentration of raw egg white, milk, and fruit juice.

### Freeze-Concentration of Egg White

We examined whether or not the bulk water in egg white freezes at a subzero temperature near the melting point of ice. Figure 3 shows freezing exotherms and melting endotherms in egg white in the presence and absence of the $Ice^+$ bacterial cells. In the freezing process, a great difference was found between the samples with and without the cells. Egg white with the cells froze at about −3°C, whereas that without the cells supercooled to about −18°C. In the melting process, however, there was no difference between the two samples. These results indicate that bacterial cells are useful as ice nuclei for freeze-concentration.

Egg white is hardly denatured, even when frozen (Erikson and Borden, 1955). Thus freeze concentration can be safely applied to it. We freeze-concentrated egg white on a kilogram scale. Entrapped cells were attached to the wall of a beaker, and an egg white sample was placed there. The solids content of the freeze-concentrated product was controlled by changing the time of standing at –5°C. It was favorable to position the entrapped cells on the wall of the beaker near the liquid surface, because heat conductivity of glass is better than that of air, and the specific gravity of ice is lower than that of egg white. After freezing, the ice produced was separated with a sieve until the solids content of the residual egg white sample reached 15 or 20%. Because of high viscosity, the use of centrifugation was necessary to obtain a higher solids content (Table 1). The solids recoveries of the freeze-concentrated egg white samples were satisfactory, and no significant change in foamability was found for the concentrated product (Table 1).

Concentrated egg white is used to make heat-induced gels and foam foods. Among the factors affecting the properties of heat-induced protein gels, solids con-

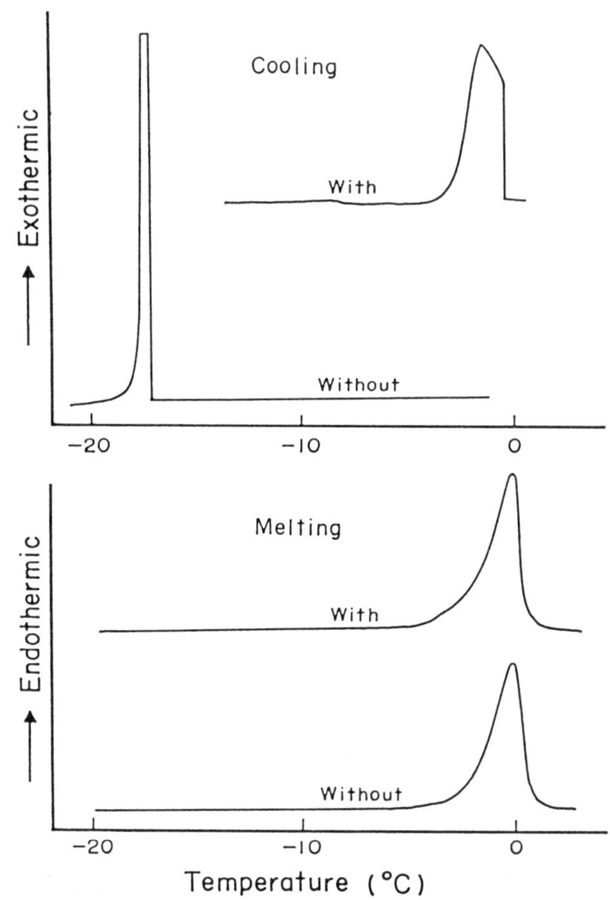


**Figure 3.** Freezing and melting thermograms for egg white with and without Ice$^+$ bacterial cells. The cells elevated the temperature of the freezing exotherm but had no effect on the melting endotherm. Cooling and warming were both at the rate of 1°C/min. Reprinted, with permission, from Watanabe et al. (1989).

tent, gelation temperature, and pH are thought to be important. We first examined a possible effect of the solids content and then the effects of temperature and pH. Donovan et al. (1975) observed an increase in viscosity of egg white at 62–65°C and the beginning of gelation at 70°C. We tentatively set the gelation temperature at 90°C.

Table 2 summarizes the solids content, gelation temperature, and pH effects on the viscoelastic and rupture parameters of the gels from egg white samples with and without freeze-concentration. Increasing solids content gave gels with increasing values of $E_H$ (elastic modulus of a Hookean body), $E_V$ (elastic modulus of a Kelvin-Voigt body), $\eta_N$ (viscosity coefficient of a Newtonian body), rupture stress, and rupture energy. Temperature and pH affected gel properties to a smaller extent than solids content. The results suggest that solids content is the most important factor influencing rheological properties of egg white gels: higher solids contents give harder gels.

To clarify why a hard and fragile gel formed from concentrated egg white, microstructures of the gel were compared with those of a soft gel. Scanning electron micrographs (Fig. 4) show that small ice crystals formed in the hard gel, probably because of its dense structure (Fig. 4, column 1). Tight aggregation of egg white components was also observed in the hard gel (Fig. 4, column 2). Nuclear magnetic resonance parameters $T_1$ and $T_2$ decreased, and the molar ratio of bound water to total water increased with increasing solids content (Watanabe et al., 1990). It should be noted that the bound water content per unit weight of solids was smaller in the hard gel than in the soft gel. This means a decrease in a molecular surface area and supports the occurrence of aggregation among egg white components.

**Table 1.** Solids recoveries, apparent viscosities, and foamabilities of freeze-concentrated egg white and lemon juice[a]

| Material | Solids content (%) | Solids recovery[b] (%) | Apparent viscosity[c] (mPa·s) | Foamability[d] |
|---|---|---|---|---|
| Egg white | 11[e] | 100 | 12 | 2.8 |
| | 15 | 96 | 13 | 2.9 |
| | 20 | 91 | 14 | 2.7 |
| | 25 | 87[f] | 26 | 2.7 |
| | 30 | 75[f] | 77 | 2.7 |
| Lemon juice | 12[e] | 100 | 3 | N.D. |
| | 24 | 95 | 7 | N.D. |
| | 36 | 92 | 9 | N.D. |

[a] Reprinted, with permission, from Watanabe et al. (1989).

[b] Aliquots (1 g each) of the materials were dried at 105°C for 3 h and then weighed to determine solids recovery.

[c] Materials were subjected to apparent viscosity measurement with a viscometer (Haake, Rhotovisco RV-3) at the shear rate of 117/s. Standard errors were less than 1%. mPa·s: millipascal-seconds.

[d] Egg white before and after freeze-concentration was adjusted to a solids content of 10% by adding water. Each sample (20 ml) was maintained at 25°C for 20 min and then stirred for another 3 min at 3,000 rpm with a blender (Nihon Seiki AM-5) with six rods (5 mm in diameter). The whipped sample was immediately transferred to a cylinder, and, 0.5 min later, the foam volume was measured. Foamability is expressed as the ratio of foam volume to initial volume (20 ml). Standard errors were less than 0.3.

[e] Starting material.

[f] Centrifuged at 5,000 *g* for 30 min at −5°C.

Next, we tried to obtain a viscoelastic gel from the hard gel by partially dissociating the aggregation and by introducing more water into the gel networks. When a surfactant was added as a hydrophobic bond dissociator, the gel gradually changed to a paste. When the gel was dipped in 2 M urea or 10% L-arginine as a hydrogen bond dissociator overnight at ambient temperature, it swelled into a viscoelastic gel. The treatment with arginine was more favorable than that with urea.

It is believed that the surface tension and viscosity of egg white are important determinants of its foamability and foam stability. The surface tension and apparent viscosity at a shear rate of 1 $s^{-1}$ of the freeze-concentrated egg white were $4.3 \times 10^{-2}$ N/m and $1.4 \times 10^{-2}$ Pa·s, respectively, and the corresponding values for the intact egg white were $4.7 \times 10^{-2}$ N/m and $1.2 \times 10^{-2}$ Pa·s. The low surface tension and high viscosity of the freeze-concentrated egg white indicate that it can form stable foams easily. We compared the foaming power and foam stability of the concentrated egg white with those of the intact egg white. The maximum foaming power of the intact sample was reached after a short period of whipping (3 min) and decreased thereafter, whereas the concentrated sample required whipping for 4 min to obtain its maximum foaming power, which was then maintained despite further whipping. The concentrated sample elicited a high foaming power, and the resultant foam was highly stable both in the presence and absence of sucrose (Kumeno et al., 1994).

In the presence of sucrose, the concentrated egg white produced a hard foam, whereas without sucrose the foam had low storage and loss moduli, which were

**Table 2.** Effects of solids content, gelation temperature, and pH on viscoelastic and rupture parameters of heat-induced egg white gels[a,b]

| Solids content (%) | Gelation temperature (°C) | pH | Viscoelastic parameter[c] $E_H$ | $E_V$ | $\tau$ | $\eta_N$ | Rupture parameter[d] Strain | Stress | Energy |
|---|---|---|---|---|---|---|---|---|---|
| 10 | 90 | 9 | 0.27 | 0.73 | 25 | 0.36 | 0.37 | 0.28 | 0.29 |
| | 105 | 9 | 0.13 | 0.49 | 19 | 0.31 | 0.50 | 0.20 | 0.32 |
| | 120 | 7 | 0.28 | 0.90 | 19 | 0.36 | 0.38 | 0.20 | 0.29 |
| | | 9 | 0.14 | 0.45 | 15 | 0.25 | 0.49 | 0.23 | 0.33 |
| 20 | 90 | 9 | 0.58 | 2.51 | 23 | 2.02 | 0.52 | 1.53 | 2.56 |
| | 105 | 9 | 1.11 | 3.85 | 18 | 1.77 | 0.47 | 0.82 | 1.33 |
| | 120 | 7 | 1.67 | 5.33 | 21 | 2.18 | 0.49 | 1.14 | 2.05 |
| | | 9 | 0.98 | 3.52 | 19 | 1.59 | 0.46 | 0.93 | 1.46 |
| 30 | 90 | 9 | 1.95 | 12.35 | 20 | 5.56 | 0.47 | 4.19 | 7.58 |
| | 105 | 9 | 3.70 | 14.96 | 20 | 6.26 | 0.59 | 3.44 | 7.01 |
| | 120 | 7 | 4.67 | 18.29 | 18 | 7.21 | 0.60 | 4.26 | 9.62 |
| | | 9 | 3.26 | 12.05 | 15 | 6.26 | 0.55 | 3.16 | 6.50 |

[a] Reprinted, with permission, from Watanabe et al. (1990).

[b] Units of the parameters: $E_H$, $\times 10^5$ N/m$^2$; $E_V$, $\times 10^5$ N/m$^2$; $\tau$, seconds; $\eta_N$, $\times 10^8$ Pa·s; rupture strain, cm/cm; rupture stress, $\times 10^5$ N/m$^2$; and rupture energy, $\times 10^4$ J/cm$^3$. Each datum is expressed as a mean of the measured values.

[c] Each sample gel was formed into a cylinder (2.2 × 2.0 cm). A parallel plate viscometer (Toyo Seiki Seisakusho) was used to obtain a creep curve of the formed gel. Viscoelastic constants were analyzed by the graphical procedure. Measuring time and temperature were set at 5 min and 25°C.

[d] A stress-strain curve of the gel (2.2 × 1.5 cm) was obtained with a Dynagraph (Iio Denki) with parameters as follows: plunger diameter, 1.3 cm; compressive speed, 6.0 cm/min; and deformation ratio, 80%. Rupture parameters were deduced from the stress-strain curve (Kobayashi and Nakahama, 1986).

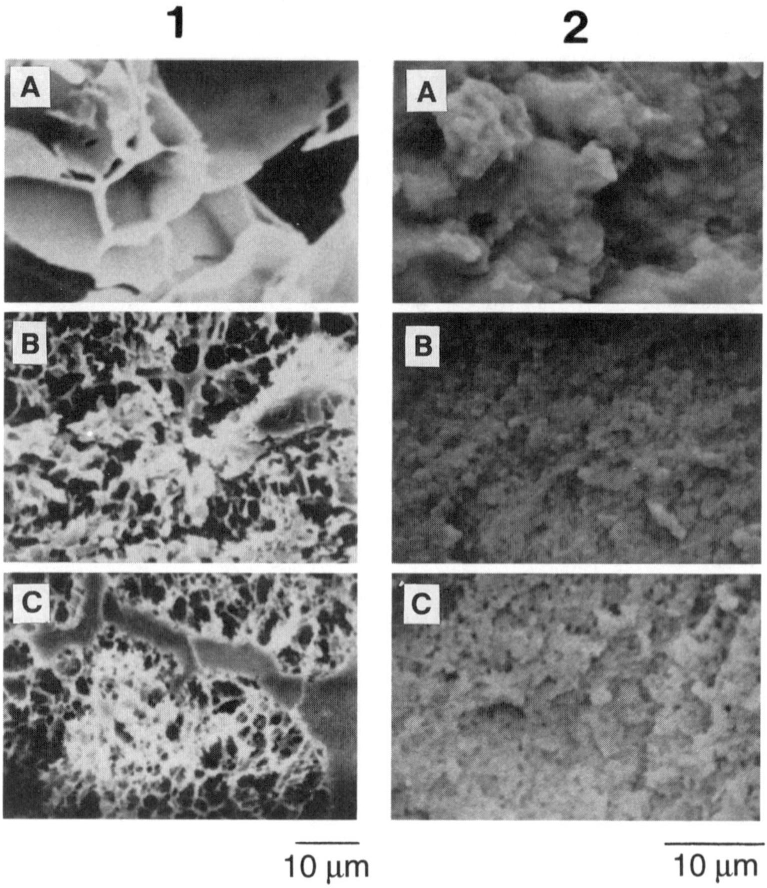


**Figure 4.** Scanning electron micrographs of freeze-dried (column 1) and wet (column 2) sections of heat-induced gels. The comparisons at left (1) indicate smaller ice crystals in the hard gel; at right (2) they show tighter aggregation of egg white components. Observations were made at 1,000× magnification. Solid contents were 10% in A, 20% in B, and 30% in C. Reprinted, with permission, from Watanabe et al. (1990).

probably caused by the high foaming power. As shown in Figure 5, the size distribution of the bubbles of the concentrated egg white foam had a narrow range, irrespective of the addition of sucrose. The concentrated sample produced a foam with thin film, and, when sucrose was added, the resultant foam had even smaller bubble diameters and thinner films. The small-sized foams with thin films are probably attributable to the flexibility of the concentrated egg white foam.

The results obtained with egg white confirm that functionalities of a material can be changed by freeze-concentration.

## Freeze-Concentration of Milk

Since Hayashi (1987) proposed pressure-processing of foods, the suggestion has been taken up extensively. High-pressure technology is useful to change properties

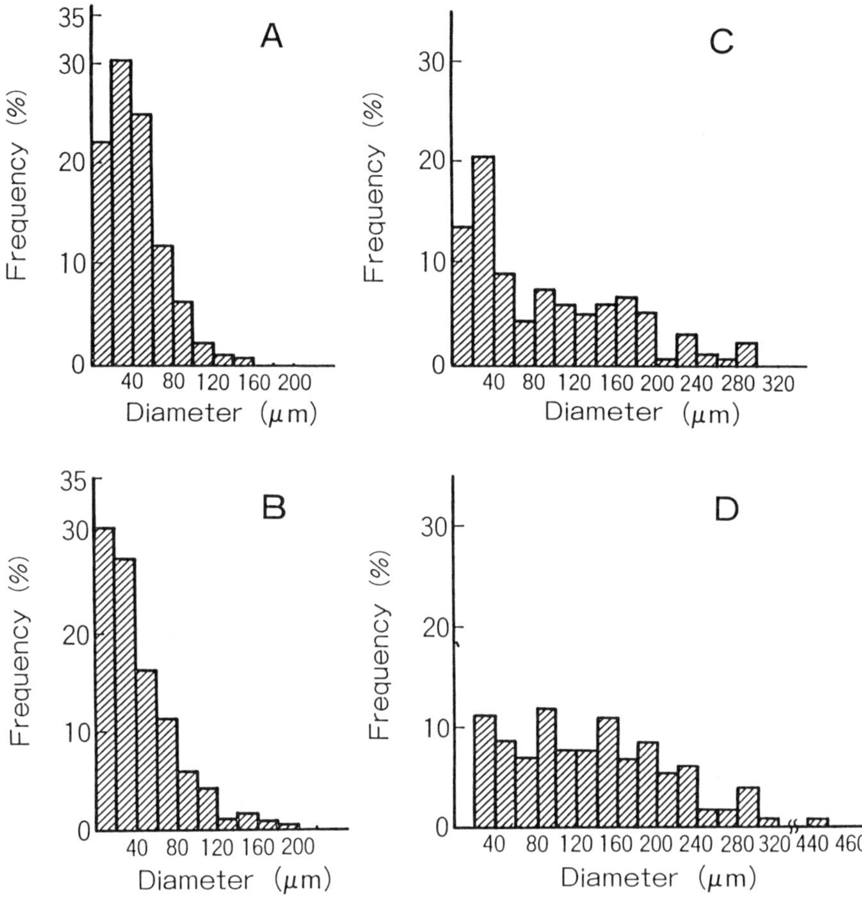


**Figure 5.** Size distributions of bubbles in foams, comparing samples derived from freeze-concentrated and intact egg white, both with and without sucrose. **A**, Freeze-concentrated, with sucrose; **B**, not concentrated, with sucrose; **C**, freeze-concentrated, without sucrose; and **D**, not concentrated, without sucrose. The histograms indicate a narrowing of the bubble size distribution in the foams made from concentrated egg white. The foams were observed at 40× magnification under a polarizing microscope, and an image analyzer (Carl Zeiss IBAS) was used to determine the size distribution of the bubble diameters. Reprinted, with permission, from Kumeno et al. (1993a).

of some food materials. It can denature proteins to obtain gels in the cases of egg (Bridgman, 1914; Hayashi et al., 1989) soy protein (Matsumoto and Hayashi, 1990), meat (Suzuki et al. 1990), and milk. Pressure-processing also has the merit that fresh flavors and colors are usually retained.

Our freeze-concentration process with bacterial ice nuclei also retains latent functions of fresh materials (Watanabe et al., 1991b). This section illustrates the synergy of the two methods in the production of milk gels produced by a combination of the freeze-concentration and pressurization techniques.

Freeze-concentrated milk (both raw and market milk) is able to gel upon pressure treatment, whereas neither unconcentrated nor freeze-dried milk will gel unless sucrose is also added. The gels produced from freeze-concentrated raw milk were stronger and more viscoelastic than those from the freeze-concentrated market milk. These results suggest that very labile components in milk are involved in gelation. In this sense, freeze-concentration is favorable for concentration because it avoids protein denaturation (Watanabe et al., 1989).

The textural properties of pressure-induced gels from freeze-concentrated raw milk depended on pressure. Pressurization at more than 300 MPa induced gelation, and increasing pressure made a gel stronger and more viscoelastic.

When gels were heated, storage modulus decreased with temperature. The temperature at which storage modulus approaches zero corresponds to the melting temperature of the gel (Shimizu et al., 1991). A remarkable decrease in the modulus was observed at about 40°C, and the modulus reached zero at 63–75°C regardless of pressure and sugar concentration. As is well known (Bridgman, 1914; Hayashi et al., 1989; Matsumoto and Hayashi, 1990; Okamoto et al., 1990; Suzuki et al., 1990; Yamamoto et al., 1990; Ohmori et al., 1991; Suzuki and Suzuki, 1992), no protein denatures at 300 MPa. We speculate that native proteins contribute to pressure-induced gelation.

The gel prepared from milk only by the combined method of freeze-concentration and pressurization is novel as a food product. It is distinguished from milk jelly (produced with the aid of gelling agents) by its whiteness, brightness, and fresh milk flavor. An optimized procedure to make a pressure-induced milk gel is to use freeze-concentrated raw milk, to add sugar at 10% to the concentrated milk, and to compress it at more than 300 MPa at a low temperature.

## Freeze-Concentration of Lemon Juice to Reduce Volume

Lemon juice was also freeze-concentrated with the aid of bacterial ice nuclei. The presence of polyols reduces the amount of water that freezes during cooling (Franks, 1985). This means that ice formation is more difficult in a sugar solution. Lemon juice, however, could be efficiently concentrated with high solids recoveries (Table 1). On freeze-concentration, constituents with low boiling points were well retained (Fig. 6), whereas they decreased on vacuum-concentration, causing a change from the original flavor. We regarded limonen (peak L) and citral (peak C)

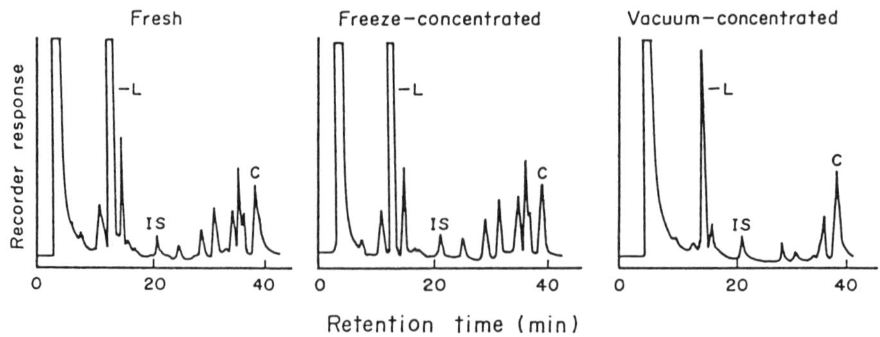


**Figure 6.** Gas chromatograms of fresh, freeze-concentrated, and vacuum-concentrated lemon juice samples demonstrating the retention of volatile limonen (peak L) in comparison to the less volatile citral (peak C) in freeze-concentrated juice. (IS denotes the peak produced by the internal standard, *n*-pentadecane). Freeze-concentrated lemon juice had a solid content of 36% (see Table 1); lemon juice vacuum-concentrated at 10 torr also had a solid content of 36%. Reprinted, with permission, from Watanabe et al. (1989).

as easily volatile and not easily volatile components, respectively, and calculated the ratios of limonen to citral (L/C) to be 8.63, 8.60, and 1.46 for the fresh, freeze-concentrated, and vacuum-concentrated materials, respectively. Panelists in a sensory test answered that the freeze-concentrated product was not significantly different in flavor from fresh material.

## Freeze-Concentration of Strawberry Juice

In the conventional production of strawberry jam, a strawberry paste is mixed with granulated sugar and then evaporated by heating at atmospheric pressure. The product loses its fresh flavor owing to evaporation of certain flavor components. Pressurization of the fruits (Horie et al., 1991) obviates the need for heat treatment, but the jam's texture, particularly its thickness, must then be provided by adding large amounts of thickening agents such as pectin, because the method lacks a concentration step.

Freeze-concentration can be used to remove excess water before pressurizing to sterilize the product and still retain a fresh flavor. We have compared jam made in the conventional way with jam formed by freeze-concentrating strawberry juice at 5°C overnight (using bacterial ice nuclei) followed by pressurization to 400 MPa.

We measured the viscoelastic properties of the jam samples. None of the measures differed significantly between samples (Table 3). Thus, a jam sample with normal viscoelastic and textural properties was produced without the use of any heating process. The nonheated jam is superior in brightness (*L* value) and red color

**Table 3.** Viscoelastic, textural, and color properties of a nonheated strawberry jam sample[a]

| Property | Nonheated (Mean ± SE) | Heated[b] (conventional) (Mean ± SE) |
|---|---|---|
| Viscoelastic parameter[c] | | |
| Apparent viscosity (Pa·s) | | |
| at 14.6/s shear rate | 1.34 ± 0.30 | 1.03 ± 0.21 |
| at 58.5/s shear rate | 0.65 ± 0.08 | 0.67 ± 0.05 |
| Yield stress (Pa) | 3.65 ± 1.23 | 2.52 ± 0.35 |
| Consistency index (Pa·s$^n$) | 0.73 ± 0.19 | 0.51 ± 0.11 |
| Flow behavior index | 0.52 ± 0.05 | 0.61 ± 0.03 |
| Texture[d] | | |
| Hardness (RU) | 0.143 ± 0.012 | 0.117 ± 0.030 |
| Cohesiveness (RU) | 0.629 ± 0.008 | 0.635 ± 0.025 |
| Adhesiveness (RU) | 0.165 ± 0.013 | 0.115 ± 0.025 |
| Springiness (RU) | 2.367 ± 0.058 | 2.781 ± 0.695 |
| Color[e] | | |
| Brightness (*L* value) | 7.8 | 7.2 |
| Red (*a* value) | 15.2 | 13.4 |
| Yellow (*b* value) | 5.8 | 5.4 |

[a] Reprinted, with permission, from Watanabe et al. (1991b).
[b] Jam sample produced by the conventional heating method.
[c] Measured with a viscometer (Haake, Rhotovisco RV-3) at 25°C. No significant difference was observed in any of the parameters between the nonheated and the conventional samples.
[d] Measured with a Rheolometer (Iio Denki). The conditions for measurement were as follows: sample temperature, 25°C; cycle speed, 12 cycles/min; clearance, 2.0 mm; sample height, 1.2 cm; plunger diameter, 4.0 cm; motion number, 2; and load range, 0.5 kg. No significant difference was observed in any of the parameters between the nonheated and the conventional samples.
[e] Measured with a color difference meter (Nihon Denshoku, ND-1001-DP).

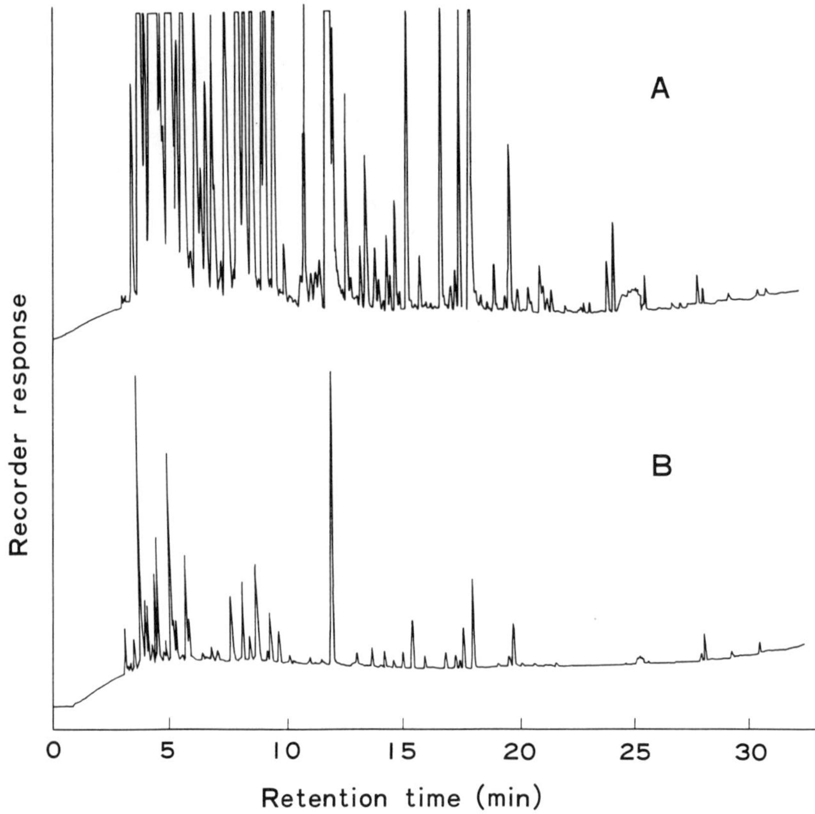


**Figure 7.** Gas chromatograms of nonheated (**A**) and conventional (**B**) strawberry jam samples. Comparison indicates that more volatiles were retained during production of the nonheated jam. Headspace volatiles were trapped in a Tenax TA trapping tube; the headspace gas was injected into a gas chromatograph using a fused silica capillary column coated with PEG 20M, with a flame ionization detector. Reprinted, with permission, from Watanabe et al. (1991b).

(*a* value) to the conventional jam, probably because the Maillard reaction does not proceed during production of the nonheated jam. Flavor was evaluated using the Tenax TA trapping technique (Tsugita et al., 1980) and gas chromatography (Tsugita et al., 1979). Figure 7 shows chromatograms obtained with the nonheated and the conventional jam samples, suggesting that the former retained the original flavor components. Evidently, freeze-concentration with the aid of bacterial ice nuclei can permit the manufacture of a nonheated jam with superior color and flavor. Similar procedures could be generally applied for the purpose of producing processed food items retaining original color and flavor qualities.

## Freeze-Drying with Bacterial Ice Nuclei

Demand is growing for a means of minimizing supercooling in the freeze-drying of foods. The higher the temperature at which freezing can be achieved, the more efficient the process. Moreover, the formation of larger ice crystals imparts properties to the product that facilitate the powdering process subsequent to drying. Watanabe and Arai (1987) experimented with the use of bacterial ice nuclei to reduce supercooling in food freeze-drying.

This application is best illustrated in the freeze-drying of two food items that are difficult to freeze: soy sauce and soybean paste. In the case of soy sauce, a supercooling of only 2°C was observed in the presence of bacterial ice nucleators; in soybean paste, a greater degree of supercooling was seen. For each of the food items, a porous structure with large cavities was observed when the sample with the $Ina^+$ cells was freeze-dried. The porous products from both the soy sauce and the soybean paste could be easily pulverized. Both types of sample supercooled approximately 10 degrees more in the absence of the $Ina^+$ cells.

Applying bacterial ice nuclei to the freeze-drying of food items that are difficult to freeze under proper conditions made it possible to shorten their freezing times and to obtain powdered products with greater efficiency. Nagashima and Suzuki (1985) stated that the dilution of a sample with water was necessary for effective freezing of high-salt food items. The addition of the $Ina^+$ bacterial cells to such samples is expected to allow similarly effective freezing without dilution. It may result in lowered processing costs.

## Anisotropic Texturing with Bacterial Ice Nuclei

Texturing of proteins has been studied actively by many researchers for the purpose of producing meat analogs from plant materials (Feeney et al., 1982). A method of freeze-texturing has long been practiced in Japan in making frozen soybean curd. Generally, the water in an aqueous dispersion or hydrogel of proteins tends to supercool to a great degree. To break down such a supercooled state for ice crystallization, the temperature has to be maintained much below the melting point of the ice in the same system. This is economically disadvantageous. Another disadvantage is the disordered formation of ice crystals and their omnidirectional growth, which produces an isotropic texture perceived as spongelike. Although the first disadvantage may be inevitable when conventional freezing is used, the second one can be removed by two devices for making ice crystals propagate in a particular direction. One is to use fibrous materials such as fish muscles (Lillford, 1985) and the other is to force the texture of a frozen slurry to turn to one direction by melting part of the ice under added pressure (Wilding et al., 1984).

Arai and Watanabe (1986) undertook experiments based on the idea that bacterial ice nuclei would prevent supercooling of the bulk water in aqueous dispersions and hydrogels of food proteins and polysaccharides, and that this might modify ice crystallization so as to improve their textures. As seen above (Fig. 3), egg white freezes at a warmer temperature in the presence of bacterial ice nuclei. When texture was examined microscopically, bacterially nucleated frozen egg white was seen to contain large, long, mutually parallel ice crystals.

The parallelism of ice crystals suggests that an isotropic liquid such as egg white can be textured into a film or flake state. We attempted to impart textures to isotropic liquids or gels of five proteinaceous and four polysaccharide compositions.[1] A hot setting step was sometimes necessary to prevent the breakdown of frozen textures when water was added. Control samples without the ice nucleation-active

[1]**Technical note:** The use of degassed water to disperse some materials is important because dissolved gases are excluded from ice, forming bubbles that perturb the resulting texture.

bacterial cells were similarly prepared with addition of water instead of the bacterial cell suspension.

The products frozen with the bacterial cells generally possessed directionally arranged textures, whereas the textures of the products without the cells were spongelike. Directional textures were obtained from raw egg white, bovine blood, 5–15% soybean protein isolate, soybean curd, milk curd, 0.5–2% agar, 5–20% corn starch paste, and 0.5–2% glucomannan. However, such a texture was not obtained from a calcium-bridged glucomannan gel at the concentrations tested (0.25–2%); this may be due to its strong gel structure, which interfered with the directional growth of ice crystals.

## Conclusions

Among various techniques for removal of water from foods, freeze-concentration is unique in employing the principle that solutes in aqueous media are excluded during the growth of ice crystals. In the presence of added ice nucleation-active bacterial cells as ice nuclei, the bulk water in foods freezes at a subzero temperature near the melting point of ice. For applications, we thus concentrated raw egg white and found that the product formed a hard gel when heated and also gave a fine foam when whipped. The freeze-concentrated product from fresh milk was characterized by forming a gel when pressurized. A similar technique was applied to fresh lemon juice to obtain a concentrated product still retaining its original flavor. As another application, strawberry paste was separated into juice and pulp fractions, and the juice fraction alone was freeze-concentrated. The pulp fraction was then put back together with sugar, pectin, and citric acid. We thus succeeded in making a jam without heating, which, compared with conventional jam, was almost equal in texture and superior in fresh flavor and color.

## Literature Cited

Arai, S., and Watanabe, M. 1986. Freeze texturing of food materials by ice-nucleation with the bacterium *Erwinia ananas*. Agric. Biol. Chem., 50:9-175.

Bridgman, P. W. 1914. The coagulation of albumin by pressure. J. Biol. Chem., 19:511-512.

Donovan, J. W., Mapes, C. J., Davis, J. G., and Garibaldi, J. A. 1975. A differential scanning calorimetric study of the stability of egg white to heat denaturation. J. Sci. Food Agric. 26:73-83.

Erikson, S. E., and Borden, R. E. 1955. Frozen eggs in relation to their use in cooked products. Kentucky Agric. Exp. Stn. Bull. 635:3-23.

Feeney, R. E., Yamasaki, R. B., and Geoghegan, K. F. 1982. Chemical modification of proteins: An overview. Adv. Chem. 198:3-55.

Franks, F. 1985. Complex aqueous system at subzero temperatures. Properties of Water in Foods. D. Simatos and J. L. Multon, eds. NATO ASI (Adv. Sci. Inst.) Ser. Ser. E Appl. Sci. 90:497-509.

Hayashi. R. 1987. Possibility of high pressure technology for cooking, sterilization, processing and storage of foods. Shokuhin to Kaihatsu, 22(7):55-62.

Hayashi, R. 1989. Use of high pressure in food processing and preservation. Pages l-30 in: Use of High Pressure in Foods. R. Hayashi, ed. San-Ei Shuppan, Kyoto.

Hayashi, R., Kawamura, Y., Nakasa, T., and Okinaka, O. 1989. Application of high pressure to food processing: Pressurization of egg white and yolk, and properties of gels formed. Agric. Biol. Chem. 53:2935-2939.

Honma, K., Makino, T., Kumeno, K., and Watanabe, M. 1993. High-pressure sterilization of ice nucleation-active *Xanthomonas campestris* and its application to egg processing. Biosci. Biotechnol. Biochem. 57:1091-1094.

Horie, U., Kimura, K., and Hori, K. 1991, Development of processed foods from fruits by pressurization. Nippon Nogeikagaku Kaishi 65:706-707 .

Inukai, S., and Matsuda, N. 1980. Factors affecting heat resistance of bacterial spores. Part 1. Effect of pH on heat resistance of genus *Bacillus* spores. Canners J. 59:219-224.

Jeanes, A. 1974. Extracellular microbial polysaccharides: New hydrocolloids of interest to food industry. Food Technol. 28:34-40.

Kobayashi, M., and Nakahama, N. 1986. Rheological properties of mixed gels. J. Texture Stud. 17:161-174.

Kumeno, K., Kurimoto, K., Nakahama, N., and Watanabe, M. 1994. Functional properties of freeze-concentrated egg white foam and its applicatios to food processing. Biosci. Biotechnol. Biochem. 58:447-450.

Kumeno, K., Nakahama, N., Honma, K., Makino, T., and Watanabe, M. 1993. Pressure-induced gelation of freeze-concentrated milk. Biosci. Biotechnol. Biochem. 57:750-752.

Lindow, S. E. 1983. The role of bacterial ice nucleation in frost injury to plants. Annu. Rev. Phytopathol. 21:363-384.

Matsumoto, T., and Hayashi, R. 1990. Properties of pressure-induced gels of various soy protein products. Nippon Nogeikagaku Kaishi 64:1455-1459.

Nagashima, N., and Suzuki, E. 1985. Freezing curve by broadline pulsed NMR and freeze-drying. Refrigeration Sci. Technol. 1985-1:65-70.

Ohmori, T., Shigehisa, T., Taji, S., and Hayashi, R. 1991. Effect of high pressure on the protease activities in meat. Agric. Biol. Chem. 55:357-361.

Okamoto, M., Kawamura, Y., and Hayashi, R. 1990. Application of high pressure to food processing: Textural comparison of pressure- and heat-induced gels of food proteins. Agric. Biol. Chem. 54:183-189.

Omran, A. M., and King. C. J. 1974. Kinetics of ice crystallization in sugar solutions and fruit juices. AIChE J. 20:795-803.

Shimada, S., and Takada, Y. 1989. Effect of high pressure on the microbial behavior and death. Pages 31-52 in: Use of High Pressure in Foods. R. Hayashi, ed. San-Ei Shuppan, Kyoto.

Shimizu, A., Kitabatake, N., Higasa, T., and Doi, E. 1991. Melting of the ovalbumin gels by heating: Reversibility between gel and sol. Nippon Shokullin Kogyo Gakkaishi 38:1050-1056.

Shirai, Y., Nakanishi, K., Matsuno, R., and Kamikubo, T. 1985. On the kinetics of ice crystallization in batch crystallizers. AIChE J. 31:676-682.

Shirai, Y. Sasaki, K. Nakanishi. K., and Matsuno, R. 1986. Analysis of ice crystallization in continuous crystallizers based on a particle size dependent growth rate model. Chem. Eng. Sci. 41:2241-2246.

Suzuki, C., and Suzuki, K. 1992. The protein denaturation by high pressure. Changes of optical rotation and susceptability to enzymic proteolysis with ovalbumin denatured by pressure. J. Biochem. 52:67-71.

Suzuki, A., Watanabe, M., Iwamura, K. Ikeuchi, Y., and Saito, M. 1990. Effect of high pressure treatment on the ultrsstructure and myofibrillar protein of beef skeletal muscle. Agric. Biol. Chem. 54:3085-3091.

Tsugita, T., Imai, T., Doi, Y., Kurata, T., and Kato, H. 1979. CC and GC-MS analysis of headspace volatiles by Tenax GC trapping techniques. Agric. Biol. Chem. 43:1351-1354.

Tsugita, T., Kurata, T., and Kato, H. 1980. Volatile components after cooking rice milled to different degrees. Agric. Biol. Chem. 44:835-840.

Vali, G. 1971. Quantitative evaluation of experimental results on the heterogeneous freezing nucleation of supercooled liquid. J. Atmos. Sci. 28:402-409.

Watanabe, M., and Arai, S. 1987. Freezing of water in the presence of the ice nucleation active bacterium *Erwinia ananas*, and its application for efficient freeze-drying of foods. Agric. Biol. Chem. 51: 557-563.

Watanabe, M., Watanabe, J., Kumeno, K., Nakahama, N., and Arai, S. 1989. Freeze concentration of some foodstuffs using ice nucleation-active bacterial cells entrapped in calcium alginate gel. Agric. Biol. Chem. 53:2731-2735.

Watanabe, M., Kumeno, K., Nakailama, N., and Arai, S. 1990. Heat-induced gel properties of freeze-concentrated egg white produced using bacterial ice nuclei. Agric. Biol. Chem. 54:2055-2059.

Watanabe, M., Makino, T., Kumeno, K., and Arai, S. 1991a. High-pressure sterilization of ice nucleation-active bacterial cells. Agric. Biol. Chem. 55:291-292.

Watanabe, M., Arai, E., Kumeno, K., and Honma, K. 1991b. A new method for producing a non-heated jam sample: The use of freeze concentration and high-pressure sterilization. Agric. Biol. Chem. 55:2175-2176.

Watanabe, M., Watanabe, J., Makino, T., Honma, K., Kumeno, K., and Arai, S. 1993. Isolation and cultivation of a novel ice nucleation-active strain of *Xanthomonas campestris*. Biosci. Biotechnol. Biochem. 57:994-995.

Wilding, P., Lillford, P. J., and Regenstein, J. M. 1984. Functional properties of proteins in foods. J. Chem. Technol. Biotechnol. B 34:182-189.

Yamamoto, K., Miura, T., and Yasui, T. 1990. Gelation of myosin filament under high hydrostatic pressure. Food Microstruct. 9:269-277.

CHAPTER 18

# The Role of Nucleation in Cryopreservation

Gregory M. Fahy

In 1985, I organized a symposium for the Society for Cryobiology entitled "Physical Events at Low Temperatures." My first speaker, Don Rasmussen, spoke on the subject of "Nucleation." In discussing the concept of the symposium before it took place, a certain senior cryobiologist offered the opinion that nucleation is really beside the point since, first of all, there is nothing to be done about it and, secondly, it is not really significant because it is what happens after nucleation that kills cells, not nucleation itself. As we now more generally appreciate, nucleation can in fact be modified and even largely avoided, and the site of nucleation, which can determine whether cells die after nucleation takes place, is also under our control. Therefore, nucleation is indeed highly pertinent to cryopreservation, and it is the purpose of this chapter to provide a brief introduction to this subject.

## Nucleation Site Determines Survival of Frozen Cells and Defines Slow and Rapid Cooling

Figure 1 depicts the three possible fates that can befall a cell during cooling to cryogenic temperatures. Each fate is defined by the nature of the nucleation process involved. If nucleation is exclusively extracellular, the cell shrinks and tends to survive. If nucleation is intracellular, ice crystals form inside the cell and damage its internal structure, usually leading to cell death. Either situation may arise during continuous freezing or during discontinuous freezing in which abrupt temperature steps are imposed (see below). The third fate is that nucleation fails to occur either intracellularly or extracellularly—instead of freezing, the cell and its surrounding medium pass into the glassy state, i.e., they vitrify (Fahy et al., 1984). Very often, vitrified cells survive on warming (Fahy, 1988).

As initially explained and quantitatively analyzed by Mazur (Mazur, 1963), and as subsequently validated in detail mostly by Mazur, Cravalho, and their colleagues, the factor that determines whether cells freeze externally or internally is the cooling rate or, more precisely, the thermal history. Cooling rate affects nucleation site primarily for two reasons.

First, the extracellular space is uniformly found to contain heterogeneous nucleators of thus far undefined nature that are more effective than any putative intracellular nucleators may be (Franks et al., 1983; Rall et al., 1983; Rasmussen et al., 1975). Consequently, in almost any biologically relevant situation short of ultrarapid cooling, crystallization will always begin extracellularly.

Second, the ability of the cell to respond osmotically to extracellular freezing is not instantaneous. The ice that forms in aqueous solutions is a pure phase that rejects any solutes that may be present. Extracellular nucleation therefore results in a reduction in extracellular solvent water but not in a loss of dissolved solutes. Thus, freezing produces an increase in extracellular solute concentration by a decrease in extracellular water concentration. This reduction of extracellular water concentration creates a water concentration (more properly, water activity) gradient across the cell membrane that results in net diffusion of water from the intracellular to the extracellular space. This loss of cell water in turn concentrates the intracellular solutes and therefore lowers the freezing point of the cytoplasm until it is equal to the freezing point of the extracellular space, thus eliminating the driving force for any intracellular freezing. But when cooling proceeds too quickly, the cell cannot lose water as rapidly as it is being lost from the extracellular space, and therefore the freezing point of the cytoplasm fails to fall as rapidly as the temperature of the cytoplasm, leading to the potential for intracellular nucleation.

It should be realized that cells that can live in distilled water because they are surrounded by a cell wall dehydrate normally when frozen in distilled water. This is the case, for example, with yeast cells (Mazur, 1965b). This forces us to recognize that cellular dehydration is driven by the lower vapor pressure of ice in comparison to the vapor pressure of intracellular water: the intermediary of "osmotic pressure"

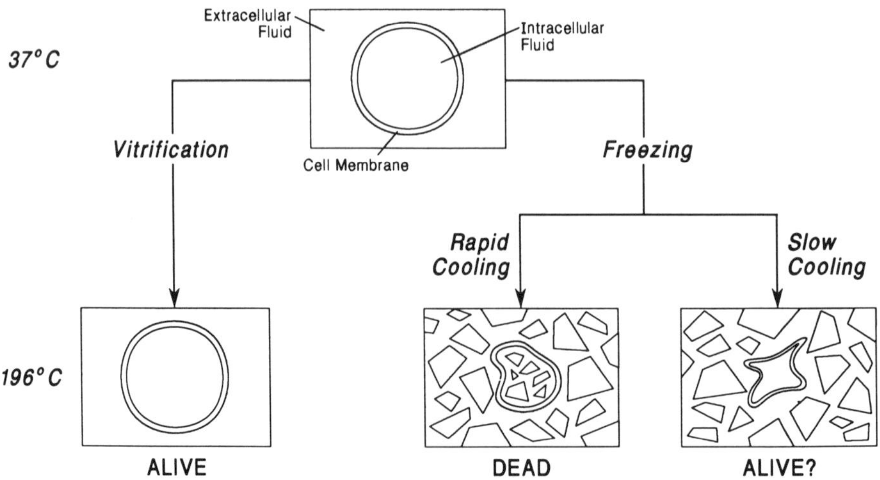


**Figure 1.** Physiochemical processes during cryoperservation of cells—possible fates of cells cooled to low temperatures. Freezing can occur exclusively in the extracellular space (right), it can occur in both the extracellular and the intracellular space (middle), or it may occur not at all (left). Freezing may be accomplished either by continuous cooling protocols or by interrupted (two-step or three-step) cooling protocols (see text). The avoidance of ice altogether requires either ultrarapid cooling or, more practically, the use of very high concentrations of cryoprotective agents (see text). Adapted from Coger and Toner (1995).

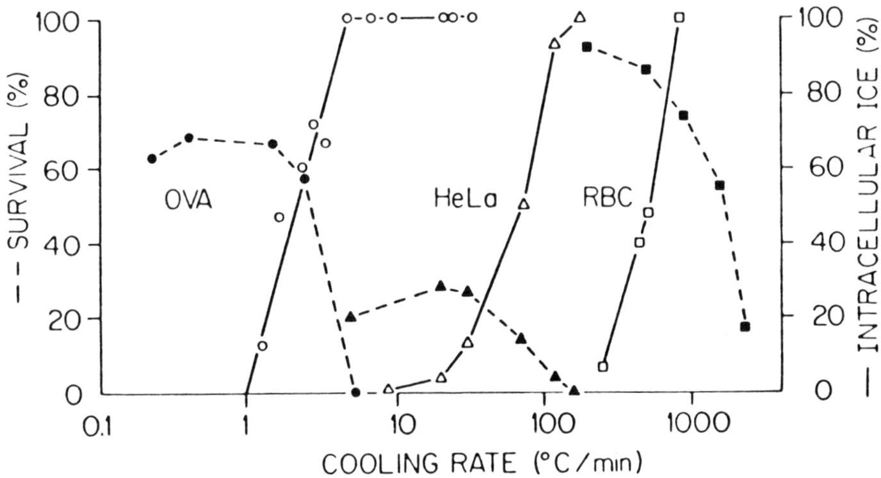


**Figure 2.** Relationship between freeze-thaw survival (closed symbols) and the incidence of intracellular nucleation (open symbols). Circles: data for ova. Triangles: data for HeLa cells. Squares: data for human red blood cells (RBC). Reprinted, with permission, from Mazur (1988).

is not necessary, though it is usually present for most cells. Furthermore, the extracellular solution, too, dehydrates entirely because its vapor pressure is greater than that of ice; once ice forms, both the extracellular and the intracellular compartments adjust themselves to it by the same mechanism. Given sufficient dilution of the water in solution by solutes as they concentrate, the vapor pressure of water in solution can be lowered to equal that of ice at any given temperature, and this is when the liquid solution comes into equilibrium with ice, and the formation of ice stops.

The biological consequences of intracellular nucleation are shown in Figure 2, which plots the survival rate of a variety of cell types against the rate at which they are cooled and relates cell death to the incidence of intracellular ice formation. Although different cooling rates are required to kill different cell types, cell death is closely associated in all cases with a rising likelihood of intracellular freezing. Because cell death is associated with intracellular nucleation rather than with any specific absolute cooling rate, Mazur has defined "fast" cooling as a cooling rate that induces intracellular nucleation and "slow" cooling as any lesser cooling rate, and it is understood that the specific cooling rate that is "fast" for one cell type may not be "fast" for another.

Based on these phenomena, it is fair to say that the field of applied cryobiology as well as the possibility of surviving freezing temperatures in nature exist largely because of the special nucleation properties of biological systems.

## Controlling Nucleation Site During Cryopreservation by Freezing

### Supercooling

As noted above, intracellular freezing is only possible when the cytoplasm is supercooled.[1] Although it is evidently possible for supercooled intracellular water to

[1]Although Franks has suggested the term "undercooling" in preference to "supercooling," the more common usage is adopted here.

freeze due to the action of weak, truly intracellular nucleating agents (Franks et al., 1983) or via homogeneous nucleation (Rall et al., 1983; Rasmussen et al., 1975), it is probably more typical for cells containing supercooled water to be nucleated by growth of ice through the cell membrane (Mazur, 1965a,b; Rall et al., 1983). This has been suggested to occur through ordinary membrane pores (Mazur, 1965b), through membrane defects resulting from a critical transmembrane hydrostatic pressure gradient (Muldrew and McGann, 1990), by direct injury to the membrane caused by contact with ice (Fujikawa, 1981), by reorganization of membrane components by extracellular ice so as to render them effective at nucleating intracellular water (Toner and Tompkins, 1992), or through other mechanisms (Steponkus, 1984; Dowgert and Steponkus, 1983; Steponkus et al., 1985). Regardless of the precise cause, the key to preventing intracellular nucleation is to minimize intracellular supercooling.

Although emphasis is usually placed on cooling rate as the factor that leads to intracellular nucleation, the importance of the initial conditions of extracellular nucleation should not be overlooked (Diller, 1975). Extensive extracellular supercooling implies extensive intracellular supercooling.

One easy method to minimize supercooling during freezing is to use commercially available ice-nucleating substances such as Snomax (freeze-dried *Pseudomonas syringae* from Snomax Technologies, Rochester, New York) in the freezing medium. If one is freezing tissue samples that require sterility or that otherwise should not be exposed to nucleating substances, manual seeding by touching the wall of the container, at a site distant from the tissue itself, with a piece of dry ice or a wand precooled in liquid nitrogen is an effective procedure and should be done as soon as the temperature falls slightly below the freezing point of the medium (for example, Fany, 1980b). Since ice propagation through the sample after manual seeding may require minutes to come to completion, it is important to allow full ice propagation before proceeding with controlled-rate freezing so that cells are able to respond fully to the lower vapor pressure of local ice crystals and thus minimize supercooling.

Certain controlled-rate freezing devices contain automated methods for inducing nucleation in samples as they are cooled to near their freezing points. This is a highly desirable feature, but different methods may not yield the same results. Merely cooling the surrounding atmosphere abruptly to a temperature well below the freezing point and then raising it after the nucleation of a dummy sample occurs or after a set period of time has elapsed should be viewed as an unreliable procedure. There is no guarantee that a dummy sample will nucleate at the same temperature as the adjacent cell sample or that, in fact, any two cell samples will nucleate at the same time. There is the further potential problem of loss of cells by intracellular freezing at the walls of their containers, since the nucleation principle employed requires extensive supercooling at the container wall for its effectiveness. Some devices are able to nucleate straws by touching all straws at the same moment with a wand that has been sufficiently precooled. In principle this accomplishes guaranteed nucleation of all samples at the time desired with minimal contact area between the nucleating wand and the container and thus with minimal local cell loss. This approach is theoretically superior and would be second in effectiveness only to direct introduction of nucleators throughout the sample.

It is important to point out that supercooling can in itself provide a means of preserving samples for significant times, particularly nonliving samples such as purified proteins in solution (Franks, 1988; Mathias, 1991). Since nucleation is difficult to prevent for lengthy periods in bulk samples, supercooling must generally be accomplished by subdividing the aqueous phase into many small droplets dispersed in a noncrystallizing organic phase (emulsification). Since each droplet has a low probability of containing a nucleator, and since homogeneous nucleation is highly improbable at temperatures above about –20°C, it is possible to maintain considerable quantities of sample in a nonfrozen state for weeks or even for years by this approach (Franks, 1988; Mathias, 1991).

## Cooling Rate

Not evident in Figure 2 is the general observation that, at lower cooling rates than indicated in Figure 2, cell survival increases as cooling rate increases, reflecting reduced damage from time-dependent, low-temperature conditions other than intracellular freezing. (Injury caused by slow freezing is usually thought to result

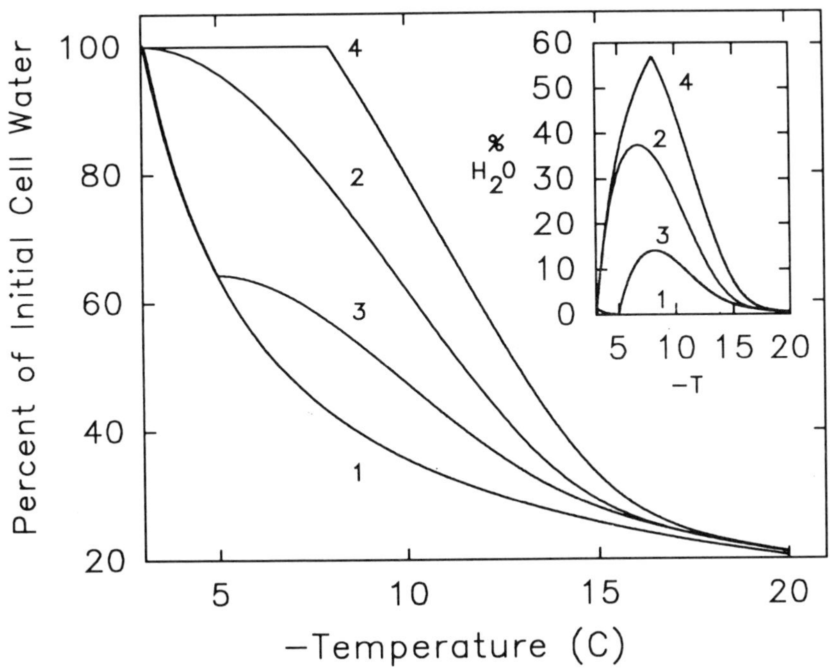


**Figure 3.** Calculated loss of freezable water from yeast cells under different freezing conditions. Curve 1 = cell water content at equilibrium; 2 = cell water content based on the typical assumption that nucleation takes place at the equilibrium freezing point (equal to $T_M$) and that cooling rate (100°C/min) then remains constant despite release of the latent heat of fusion; 3 = cell water content based on the more realistic assumption that temperature rebounds to $T_M$ shortly after nucleation (or that nucleation occurs at $T_M$) and that subsequent cooling is slow (0.5°C/min) until a constant cooling rate (again, 100°C/min) is imposed at –5°C, causing a departure from the equilibrium curve; 4 = maximal hazard assumption, i.e., that the cell supercools 5°C prior to nucleation, and cools at a constant rate (100°C/min) thereafter, without thermal rebound. Inset: Hazard of intracellular nucleation expressed as the percent of cell water that is freezable as a function of temperature and thermal history (curve labels are the same as previously described). All curves calculated as described in the legend of Figure 1 of Fahy (1981a) according to methods detailed in Fahy (1980a, 1981a,b), using 1 M DMSO as the cryoprotectant.

from alterations in the composition and properties of the extracellular solution and is therefore termed "solution effects" injury.) Therefore, the purpose of controlling cooling rate is to minimize both the total time required for freezing (thus minimizing "solution effects") and the likelihood of intracellular nucleation. However, the attainment of these goals is more complex than this simple statement may imply. The likelihood of intracellular nucleation depends on the kinetics of cellular water loss during freezing, which are decidedly nonlinear.

The basic issues are best discussed in terms of Mazur's model of cell dehydration during freezing (Mazur, 1963). As is evident from Figure 3, a cell cooling at an infinitely slow rate will generally dehydrate very rapidly within the first several degrees below the freezing point and then lose water much more slowly thereafter (curve 1). Cells cooled at constant rates, however (Fig. 3, curve 2), respond to subfreezing temperatures rather ineffectively at first, experiencing the greatest supercooling just below their nominal freezing point. On the other hand, cooling at a predetermined constant rate despite rapid ice propagation through the sample just below the freezing point is a physically unrealistic situation in most cases due to liberation of the latent heat of fusion and consequent warming of the environment back to just below the freezing point. The effect of slow dissipation of the heat of fusion by slow cooling followed by more rapid cooling is shown as curve 3 in Figure 3. Finally, the effect of 5°C of supercooling prior to freezing without thermal rebound during freezing is indicated as curve 4. The inset of Figure 3 expresses these results explicitly in terms of the amount of supercooled intracellular water, and thus the likelihood of intracellular nucleation.

This discussion makes it obvious that the occasionally expressed desire to force a preconceived constant cooling rate on samples even during the initial part of the crystallization process in order to comply with the imperative of freezing at controlled cooling rates is misplaced. The net effect of such a maneuver will be to convert curve 3 into curve 2 or even into curve 4, i.e., to bias the system in favor of intracellular nucleation. Some have suggested that "solution effects" injury must be minimized by hastening the passage through the freezing point, but this is also generally incorrect. Solution effects injury is the result of severe dehydration for most cells and is not caused by the mild dehydration that occurs in the immediate vicinity of the freezing point. For example, Fahy (1980b) reported that the "freezing point isotherm" for tissue slices frozen in 15%, w/v, dimethyl sulfoxide was over at –10°C or above (corresponding to conversion of well under half of the volume of the solution into ice), whereas injury became just detectable at –17°C (52% volume conversion), and severe only below –29°C (>62% volume conversion).

Finally, it should be understood that the commonly accepted imperative of imposing a fixed cooling rate on samples during freezing, whether in the temperature region of the freezing point or at lower temperatures, is not motivated by what is biologically optimal but by what is computationally and experimentally convenient. It was Mazur who first urged cryobiologists to use constant cooling rates, but this suggestion was motivated in part by the fact that it is computationally easier to model and compare the effects of constant cooling rates as described in Figure 3 than it is to model and compare the effects of variable cooling rates that might in fact be more biologically meaningful. The basic situation can be visualized with the aid of the schematic diagrams of Figure 4.

Figure 4 represents three possible continuous cooling protocols. For simplicity, we make the usual assumption that nucleation takes place extracellularly at the freezing point and that cooling proceeds without respect to the release of the latent heat of fusion. The situation depicted on the left is the effect of cooling at a constant rate, and the resulting cell water response resembles curve 2 of Figure 3.

The situation depicted in the middle of the figure is a commonly encountered one, in which biologists without controlled-rate freezing equipment place their cells in a low-temperature freezer surrounded by some type of insulation in the hope that this will slow the cooling rate sufficiently. What theoretically happens in this situation, of course, is that the cooling rate is initially high due to the initial large thermal gradient between the sample and the environment and then falls progressively as this thermal gradient is dissipated. The cell water response is maximally adverse: the amount of cell water at risk of nucleation is increased compared to the constant cooling rate protocol, while exposure time at lower temperatures, where "solution effects" injury proceeds at a maximum rate, is also maximized.

The situation depicted on the right in Figure 4 shows the effects of a hypothetical cooling protocol tailored to the biological properties of the cell. To minimize supercooling, the cooling rate is first kept low to permit near-equilibrium loss of cell water. As cell water content approaches the minimum permissible values, and the risk of "solution effects" injury therefore begins to rise, the cooling rate is progres-

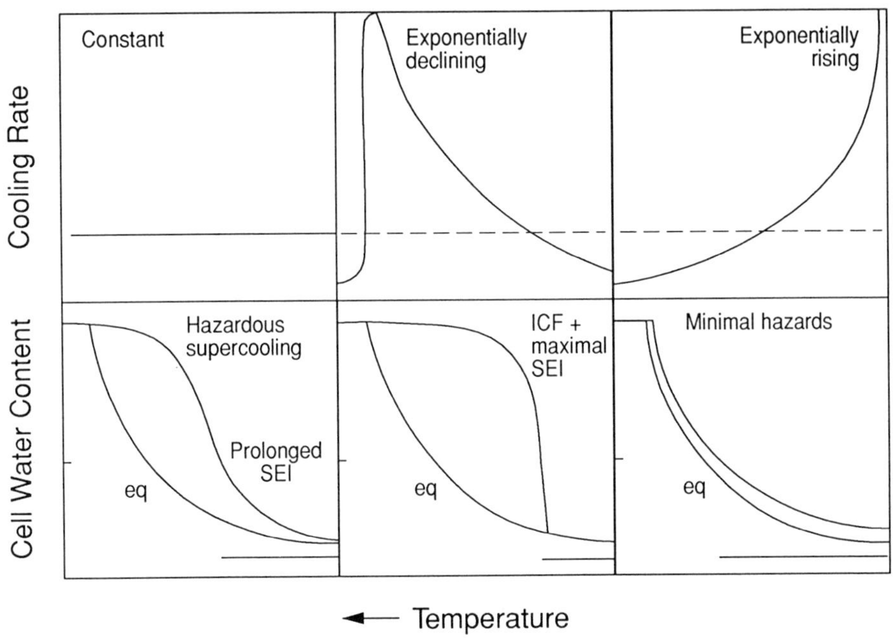


**Figure 4.** The effects of two commonly used freezing procedures (constant cooling rate, left; and exponentially declining cooling rate, middle) vs. a proposed freezing procedure (continually rising cooling rate, right) that minimizes both "solution effects" injury (SEI) and the danger of intracellular nucleation (ICF). Dashed horizontal lines in middle and right upper panels represent the same cooling rate shown at the left, for comparison. Cell water content (lower panels) is compared with the equilibrium cell water content (eq), which is theoretically attained only at very low cooling rates. Horizontal solid line begins when cell liquid volume is reduced to 50% of initial volume, marking the onset of hazard for SEI. Note that in the SEI region, cooling rates higher than indicated by the horizontal dsashed line (upper panels) signify lower exposure times and therefore less injury. For further description, see text. All curves shown are schematic.

sively increased. Slow cooling at lower temperatures can have only a marginal protective effect on the likelihood and quantity of intracellular ice formation, so there is little reason not to increase cooling rate progressively to minimize "solution effects" injury as cell dehydration permits. Figure 4 suggests that a cooling protocol similar to that proposed could give higher cell survivals than constant cooling rate protocols, and the biologically motivated protocols ought to be derivable from previously established data on the effect of constant cooling rates on cell survival. This novel idea has yet to be validated experimentally. However, it approximates what many investigators now do when they cool through the freezing point isotherm slowly, adopt a higher constant cooling rate to about –40°C, and then plunge their samples into liquid nitrogen.

## Interrupted Cooling and Warming Protocols

There are many laboratories that lack controlled-rate freezing equipment and that must preserve cells at cryogenic temperatures. Short of simple insertion into a freezer as discussed above, there are uncomplicated measures that can potentially achieve the purpose of controlled-rate freezing, that is, avoidance of intracellular nucleation with a minimum exposure time to potentially damaging conditions, without the controlled-rate freezer. Of these, possibly the most important is two-step freezing.

### Two-Step Freezing

In two-step freezing, the cryoprotected sample is first exposed directly to a subzero bath held at some temperature in the vicinity of –20 to –30°C (Walter et al., 1975). It is thought that the temperature must be chosen to be high enough to prevent intracellular nucleation upon exposure to the supercooling represented by this temperature, yet low enough so that sufficient dehydration will occur during a hold at this temperature to prevent intracellular freezing upon subsequent transfer to liquid nitrogen. Walter et al. (1975) used freeze substitution to visualize intracellular ice crystals and reported that the surviving cells were those that were massively dehydrated and lacked clear-cut signs of intracellular ice, whereas, the cells whose morphology indicated little dehydration contained intracellular ice and were killed.

Questions about this standard approach to two-step freezing arise from considering the basic physics of aqueous solutions. Cytoplasm in equilibrium with ice at a temperature of –25°C will have a homogeneous nucleation temperature of around –90°C (Fahy et al., 1984; Rall et al., 1983) (see also Figure 5), which is above the glass transition temperature of all common penetrating cryoprotectant solutions. Success of two-step protocols involving temperatures near –25°C at the first step is probably due to the slow rate of growth of nuclei formed after subsequent cooling to –90°C and below and the brief time available for growth and recrystallization during very rapid warming.

Theoretically, two-step freezing should require an initial dehydration temperature in the vicinity of about –40°C in order to avoid intracellular nucleation on subsequent plunging into liquid nitrogen (Fahy et al., 1984), but this may not be possible because direct exposure to –40°C may itself induce intracellular nucleation before dehydration has time to occur (Diller, 1975). Kuwano et al. (1994), using a modified two-step protocol that avoided this nucleation problem by using slow rather than rapid cooling to the first holding temperature, reported that several strains of *Porphyra* did show optimal survival when the initial holding temperature was −40°C as predicted from phase diagram considerations (Fahy et al., 1984). Similarly, Alink et al. (1977) found that for adult rat heart cells, two-step cooling was ineffective because the temperature required in the first step to protect against damage upon plunging into liquid nitrogen was sufficiently low to kill most cells prior to plunging. Overall, survival tended to be higher when a protocol was employed in which the first holding temperature was gradually lowered before plunging the sample into liquid nitrogen. An alternative for cells that cannot tolerate long exposure times at high subzero temperatures would be to employ two steps prior to the −196°C step, e.g., −15 to −20°C, followed by transfer to a bath in the vicinity of −40°C.

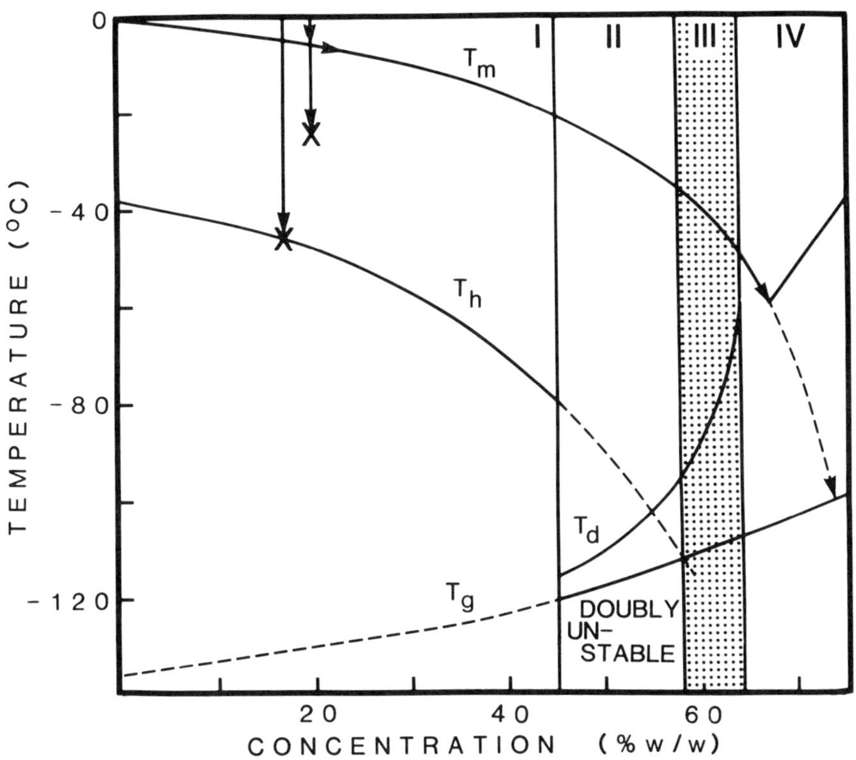


**Figure 5.** Effect of cryoprotectant concentration on the equilibrium melting/freezing point ($T_m$), the homogeneous nucleation temperature ($T_h$), the glass transition temperature ($T_g$), and the temperature of devitrification on warming ($T_d$). In region I, vitrification is essentially impossible because supercooling below $T_h$ is impossible for all practical purposes. In region II, supercooling through $T_h$ produces extensive nucleation, resulting in devitrification at relatively low temperatures during warming, but ice crystal growth rates are slow enough to permit the apparent vitrification of most of the sample and observation of $T_g$. Glass is thermodynamically metastable by its nature, but glass formed by quenching through the normal $T_h$ is even more unstable and is termed "doubly unstable" by physical chemists. At the intersection between regions II and III, $T_h$ becomes equal to $T_g$, and this is the lowest concentration that, in principle, produces a fully vitrified specimen. Despite this, devitrification is observed at concentrations above the $T_h/T_g$ intersection, in region III, perhaps because of nucleation below $T_g$ or the nucleation of cubic ice, whose nucleation behavior need not be described by the $T_h$ curve. Finally, in region IV, concentrations are so high that devitrification is not observed, and the solution is virtually stable at all temperatures. Overall, the best concentration region for vitrification of large objects is region III. The positions of $T_h$, $T_g$, and $T_d$ are influenced by cooling and warming rate. Reprinted, with permission, from Fahy et al. (1984).

## Nucleation During Cryopreservation by Vitrification

Vitrification is solidification without freezing. As illustrated in Figure 5, this is possible at attainable cooling rates, because high concentrations of cryoprotectants drive the homogeneous nucleation temperature ($T_h$) of water down and elevate the glass transition temperature ($T_g$), until at some concentration $T_h$ becomes equal to or less than $T_g$ (Fahy et al., 1984). Vitrification is particularly important for preserving complex cell assemblages such as corneas and organs (Fahy et al., 1984), which can be structurally damaged by the mechanical effects of ice, but has also proven to be a convenient way of preserving cell suspensions (Fahy, 1988). For vitrification, it is important to minimize not only homogeneous but also heterogeneous nucleation.

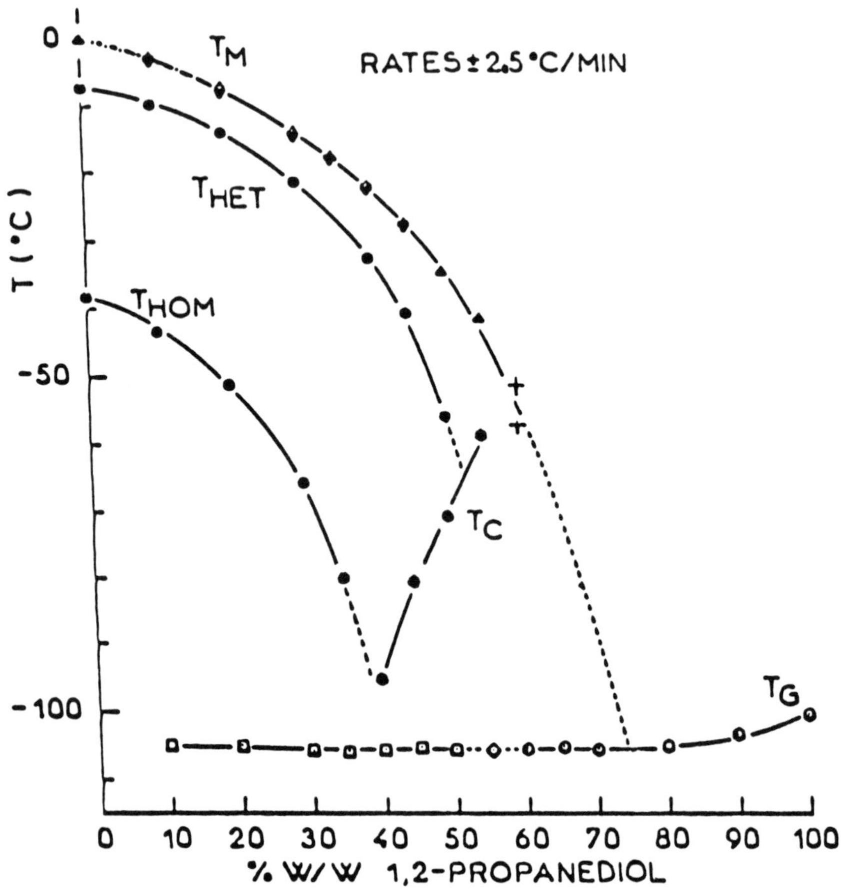


**Figure 6.** Effect of cryoprotectant (1,2-propanediol) on heterogeneous nucleation ($T_{HET}$). Aqueous samples containing or not containing freeze-dried *Pseudomonas syringae* were emulsified in a hydrocarbon phase and cooled in a differential scanning calorimeter for detection of crystallization. Traditionally, a plot of homogeneous nucleation temperature ($T_{HOM}$ in this figure) vs. $T_M$ as in Figure 9 yields a slope of about 2, i.e., $T_{HOM}$ falls 2°C for every 1°C fall in $T_M$. It is apparent here that such a plot for $T_{HET}$ vs. $T_M$ will have a slope between 1 and 2, in agreement with previous studies from other laboratories carried out in relatively dilute solutions. $T_c$ equivalent to $T_d$ of Figure 5. Reprinted, with permission, from Mehl (1992).

## Heterogeneous Nucleation of Cryoprotectant Solutions

Heterogeneous nucleation in the presence of glass-forming concentrations of cryoprotectants has not been studied extensively. Some of the most complete work is that of Mehl (1992), who obtained complete phase diagram information for 1,2-propanediol in the presence and absence of *P. syringae* nucleators (Fig. 6). In this study, heterogeneous nucleation temperature ($T_{HET}$) was found to be less sensitive to suppression by solute than was $T_h$, in agreement with previous observations. Although high pressures may well inactivate biogenic nucleators by inducing conformational changes, heterogeneous nucleation at high pressures appears not to have been studied to date.

An impression of the importance of heterogeneous nucleation for vitrification is provided in Figure 7. Crystallization in this example (a 500-ml sample of a 50%, w/w, solution of 1,2-propanediol) can be confidently assigned to heterogeneous rather than homogeneous nucleation on the basis of the fact that the expected $T_h$ is −100°C or below (compare Angell, 1982; Fahy et al., 1984). The data in Figure 7

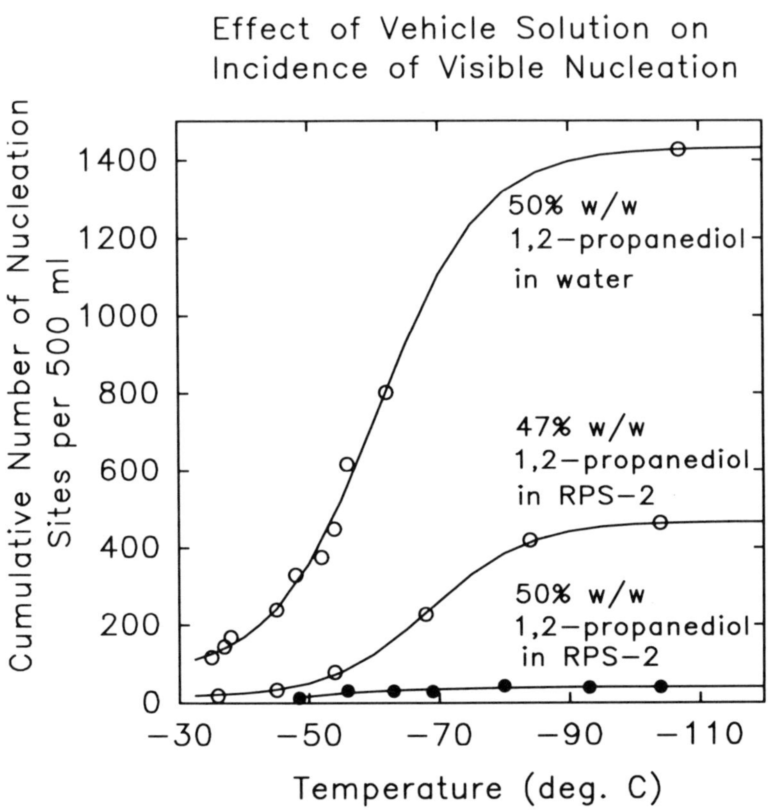


**Figure 7.** Effect of physiological solutes and small changes of cryoprotectant concentration on cumulative incidence of nucleation (CIN spectrum) of large volumes of cryoprotectant during cooling. RPS-2 is a storage solution for mammalian kidneys; its composition and other experimental details are given in Fahy et al. (1990). Data were obtained by photographing 482-ml specimens, projecting each image onto a standard 18 × 24 in. page, then tracing and numbering each crystal, and finally correcting the results to 500 ml. For discussion, see text. Photographic views of these data are given in Fahy et al. (1990).

represent the cumulative population size of visible ice crystals during vitrification of the major portion of the solution in the presence and absence of physiological vehicle solution solutes, each crystal representing the effect of a single nucleation event. As can be seen, the number of nucleation events (critical nuclei) is large in the absence of vehicle solution solutes (upper curve). Adding these physiological solutes dramatically reduces nucleation density (lowest curve), a factor that could be critical to the success of organ vitrification. However, lowering the concentration of the 1,2-propanediol again produces large numbers of nuclei (middle curve). If each crystal is capable of damaging a blood vessel in an organ during cooling (Pollock et al., 1986), the result could be massive injury upon transplantation. If, on the other hand, each crystal can produce only local injury, then a consideration of the total volume fraction occupied by the crystals (small even for the upper curve) suggests that injury could be minimal. Until it becomes possible to vitrify organs without damaging regions that are not in contact with such growing crystallization centers, it will remain difficult to determine which of these two possible outcomes is the correct one.

## Homogeneous Nucleation of Cryoprotectant Solutions

Classically, homogeneous nucleation is a rapid event that occurs at a sharply defined temperature ($T_h$) that rises as the volume of the sample increases (Angell, 1982). The distinctness of homogeneous nucleation becomes blurred, however, under conditions that impede the growth of ice, for two reasons. First, slow ice crystal growth delays and reduces the exothermic signal that serves as the means of detecting homogenous nucleation, so that the temperature at which homogeneous nucleation is detected may be less than the temperature at which it actually occurs and will be dependent on cooling rate. More fundamentally, homogeneous nucleation itself becomes distributed over a significant temperature range because the probability of nucleation per unit time and per unit volume becomes low enough to preclude all possible nucleation from taking place at the first possible opportunity. This means that the actual extent of homogeneous nucleation may depend strongly on cooling rate when concentrations are high. Under such conditions, $T_h$ per se loses its meaning, and what becomes meaningful is either the temperature at which homogeneous nucleation first begins to increase appreciably during cooling or the sum total of homogeneous nuclei that actually form under the specific circumstances at hand (the cumulative incidence of nucleation, or CIN spectrum). The former is the more conservative (and more easily determined) measure and may be extrapolated reasonably well from the more dilute portion of the $T_h$ curve (Fahy et al., 1984). The CIN spectrum is increasingly inferable from accurate theoretical models of nucleation in concentrated cryoprotectant solutions (Bronsheteyn and Steponkus, 1995; Kresin and Korber, 1991; Karlsson et al., 1994).

The complexities introduced by high concentrations (as well as by higher rates of temperature change) are extinguished at still higher concentrations, at which the probability of nucleation falls to negligible values, or growth of the resulting ice embryos becomes vanishingly small, or both, rendering homogeneous nucleation moot (Fahy et al., 1984). Experimentally, it has been found that the incidence of visible ice vanishes (i.e., vitrification becomes evidently complete) at the same concentration that brings the onset temperature for homogeneous nucleation down to $T_g$, the glass transition temperature (when both $T_g$ and the extrapolated $T_h$ are based on a cooling rate of about 10°C/min) (Fahy et al., 1984).

Despite the complexities of homogeneous nucleation, nuclei not detected on cooling can be detected by warming back to a sufficiently high temperature to permit them to grow and thus produce heat, provided concentration is low enough to permit crystal growth. An example of the application of this approach is given in Figure 8, wherein the data reflect the amount of melting observed after holding at different temperatures for 1.5 or 30 min. The 30-min peak clearly reflects nucleation more than ice crystal growth per se, since *no* signal is observed at the higher temperatures after 1.5 min, indicating no nucleation. Further, the upper limb of the peak rises as temperature falls (and crystal growth rates diminish). Combining

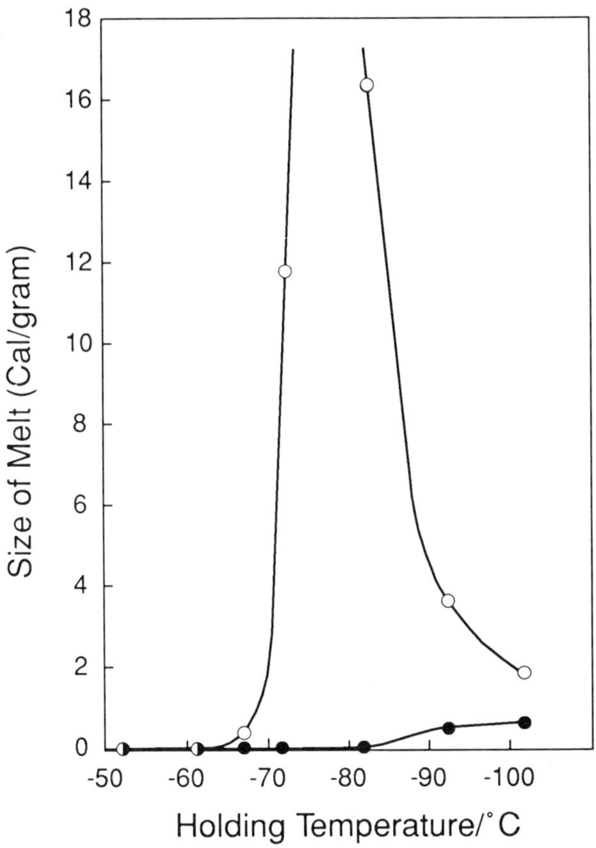


**Figure 8.** Detection of "invisible" nucleation at low temperatures by its effects during warming to temperatures permitting crystal growth. The vitrification solution used consisted of 6%, w/v, polyvinylpyrrolidone K30 (mean $M_r$, 40,000 Da) plus 49%, w/v, D(1)$FP_{10}$, which consists of 10%, w/v, 1,2-propanediol and 39%, w/v, of a mixture of dimethyl sulfoxide and formamide in a 1:1 mole ratio. Samples were cooled to the temperatures indicated at 80°C/min and held for either 1.5 min (filled circles) or 30 min (open circles) before rewarming at 80°C/min. Data shown are the number of calories detected by differential scanning calorimetry (DSC) during melting of the resulting ice. DSC traces on cooling showed no evidence of crystallization, and ice formation on warming appeared as devitrification with an onset temperature of more than –70°C; even at a warming rate of 10°C/min, devitrification became detectable only near –70°C and peaked at –53°C. Therefore, ice crystal growth is minimal below –70°C in this system, and the limiting factor for the signals shown is nucleation. Data were collected by T. Takahashi (currently at the Japanese Red Cross, Sapporo, Japan) on 17 July 1984 at the American Red Cross in Bethesda, MD.

calorimetric analysis of this kind with cryomicroscopy has allowed the contributions of nucleation and crystal growth to be separated from each other in accounting for the overall calorimetric record (Mehl, 1990, 1992, 1993).

Ice formation is accompanied by an increase in the volume occupied by water molecules undergoing crystallization. For this reason, nucleation is inhibited by high pressures in much the way it is inhibited by solutes. In fact, Figure 9 reveals that the effects of pressure and of solutes on $T_h$ are essentially identical, and this has led to extensive experimentation to attempt to exchange elevated pressure for some of the potentially toxic cryoprotectant required for vitrification.

Figure 9, however, obscures the fact that the mechanisms by which pressure and cryoprotective solutes depress $T_h$ are in opposition to each other (MacFarlane et al., 1992). Cryoprotectants form hydrogen bonds with water, in the process interfering with water-water hydrogen bonding but also increasing the effective volume occupied by each water molecule, or imposing water "structuring." The latter effect is in opposition to the structure-breaking effect of high pressure (MacFarlane et al.,

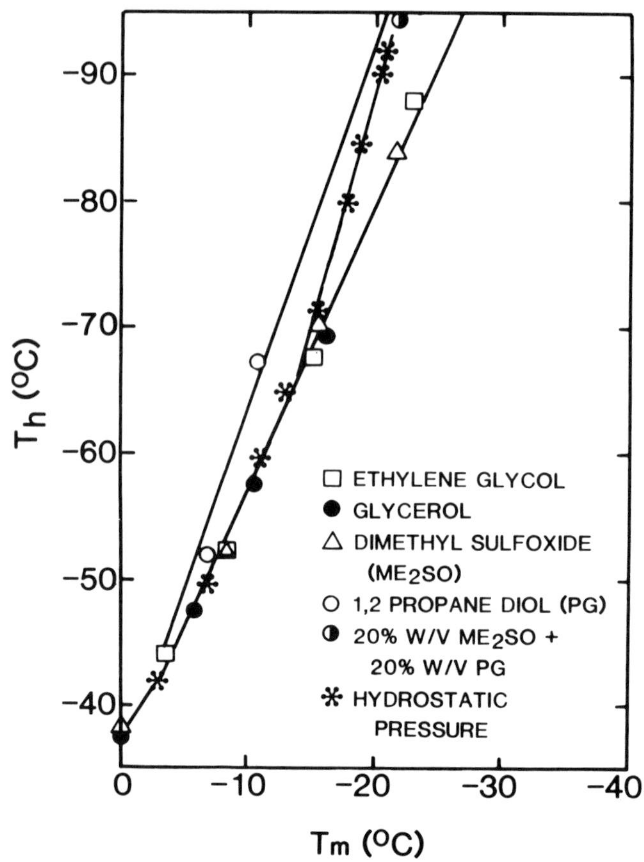


**Figure 9.** Comparison of effectiveness of pressure and cryoprotective agents in depressing the homogeneous nucleation temperature of water. The depression of $T_h$ induced by solutes or by pressure is plotted against the equilibrium freezing point depression caused by pressure or solutes. Reprinted, with permission, from Fahy et al. (1984).

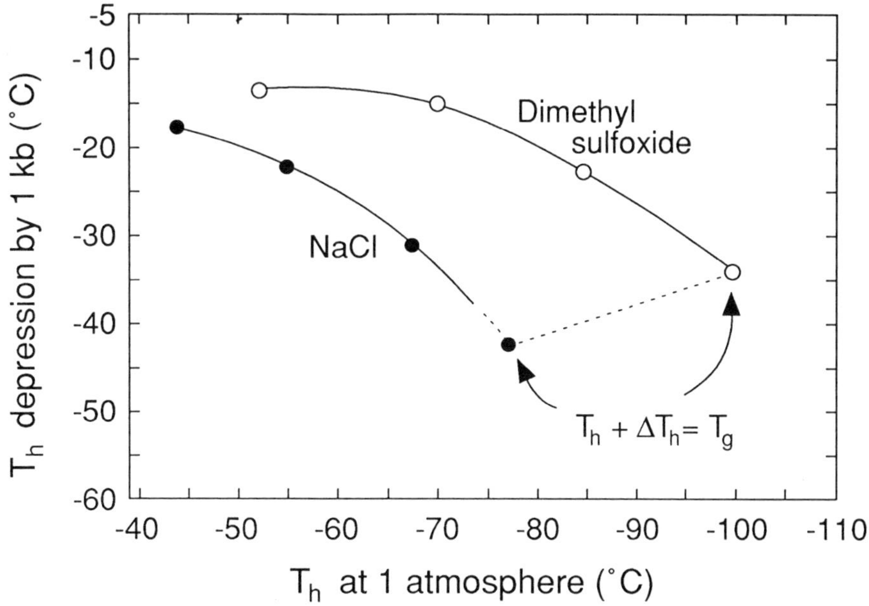


**Figure 10.** Lack of equivalence of salts and cryoprotectants in supporting depression of the homogeneous nucleation temperature of water ($T_h$) by high pressures. NaCl and dimethyl sulfoxide are compared on an equal basis by determining the change in $T_h$ produced by application of 1,000 atm (1 kb, or 100 MPa) (vertical axis) at a previously existing $T_h$ depression produced by either solute at ambient pressure (horizontal axis), without regard to the absolute concentration or specific chemical properties of either solute. The initial $T_h$ represents a universal scale for which agents can be compared based only on their net effect on water. The points and dashed line referred to as $T_h + \Delta T_h = T_g$ define the ends of the $T_h$ curves, where $T_h$ is depressed until it coincides with the glass transition temperature. NaCl data derived from Kanno and Angell (1977); DMSO data are collected from Fahy et al. (1984) and MacFarlane et al. (1981).

1992). This is strikingly reflected in the antagonism that cryoprotectants produce to the $T_h$-depressing effects of high pressure (Fahy et al., 1984). As indicated in Figure 10, high pressures depress $T_h$ much more strongly in the presence of electrolytes (which, like pressure, are chaotropic, or structure-breaking) than in the presence of "kosmotropic," or structure-making cryoprotectants such as dimethyl sulfoxide. This phenomenon limits the utility of pressure for facilitating vitrification and suggests that more ionic cryoprotectants or vehicle solutions would be more advantageous for high-pressure vitrification.

## Homogeneous Nucleation of Vitrified Solutions

The storage time of vitrified systems is expected to be limited mostly by nucleation that may continue at a slow pace even below $T_g$. Nucleation below $T_g$ is possible because $T_g$, like $T_h$ at high concentrations, is not an absolute boundary between the liquid and solid states but is dependent upon cooling rate and upon the time scale of the observation in question. Given sufficiently long time scales (which lengthen rapidly as temperature drops progressively below $T_g$), more and more nucleation can occur. "Shelf life" in the vitrified state may therefore depend on the cooling rate, the storage temperature, and the attainable warming rate.

So far, hard data on nucleation rates below the measured $T_g$ of solutions that are relevant for cryopreservation are meager (Angell and MacFarlane, 1982). Chang found devitrification temperature to fall by 5°C after only 180 min of annealing of 55%, w/w, glycerol at 8°C below the measured $T_g$ (Chang, 1992). However, this is a concentration that yields a doubly unstable glass (Fahy et al., 1984), and it is not clear that similar changes would occur at higher concentrations. Williams and colleagues (1990) have pointed out that the interfaces formed by spontaneous fracturing in vitrified aqueous solutions can serve as extremely effective nucleating surfaces for ice, and it is important to avoid this artifact in studies of solution nucleation rates below $T_g$. Mehl (1993) reported that the classically defined critical heating rate of a vitrification solution stored for 6 months at about 10°C below the measured $T_g$ was only moderately elevated in comparison to nonstored samples, due to the slow pace of additional nucleation during the 6-month storage period. On the other hand, he pointed out that if damage is associated more with the size of each growing crystal than with the number of such crystals (Mazur, 1965a, 1988), the critical warming rate for avoiding damage is essentially independent of continuing nucleation during storage, since all nuclei will be the same size and will grow at the same rate on warming (Mehl, 1990; Mehl, 1993). Clearly, much additional research on sub-$T_g$ nucleation and its biological consequences is needed.

## Avoiding Nucleation or the Effects of Nucleation

### During Vitrification for Cryopreservation

During cooling to sub-$T_g$ temperatures, nucleation may be a problem either in the temperature range of rapid ice crystal growth (about −30 to −80°C) or at temperatures closer to $T_g$, which systems such as organs may experience for prolonged times due to the low cooling rates that attend the approach to storage temperatures close to $T_g$. The former temperature range will be governed by heterogeneous nucleation and the latter by homogeneous nucleation.

#### Controlling Heterogeneous Nucleation

Several approaches to avoiding heterogeneous nucleation or its consequences can be explored. For solutions containing an air-liquid interface, nucleation at this interface will inevitably occur unless special precautions are taken, perhaps in part because water vapor in the air is capable of crystallizing and seeding the interface and in part because the surface energy of the interface favors nucleation (a factor that may explain in part why freezing-point osmometers successfully nucleate solutions by generating turbulence and bubbles in them). Whatever the mechanism, it is possible to reduce or prevent this unique type of heterogeneous nucleation by layering an immiscible hydrocarbon phase over the solution to cover the interface. Williams et al. (1990) found isopentane (2-methyl butane) to be an effective interface-protecting material; I have found methylcyclohexane and methylcyclopentane to be considerably more reliable.

For noninterfacial heterogeneous nucleation, a direct approach would be to identify, characterize, and finally to inactivate or remove all or most heterogeneous nucleators. Despite many decades of research on water and supercooling, the nucleators present in ordinary beakers of laboratory water have evidently not been identified. Once such identifications are available, technologies such as monoclonal antibody production could be applied to the inactivation of the nucleators, or the nuclei could be removed by adsorption to appropriate immobilized ligands, and so forth. Less elaborate methods may also be helpful. For example, heating and/or subjecting the solution to high pressures (e.g., 2,000 atm or more) for a few hours prior to use to inactivate organic nucleators. Another possibly feasible approach could be to emulsify a bulk aqueous solution in a reversible way, cool the emulsion to temperatures sufficient to nucleate droplets containing active heterogeneous nucleating agents, separate the frozen droplets from the unfrozen droplets by centrifugation or drop-sorting techniques, and then break the emulsion to give back the nucleator-depleted bulk solution. Even more feasibly, bulk solutions could be centrifuged while being cooled to force the nucleators trapped in the developed crystals that they have nucleated to float to the surface (Fahy et al., 1990), where they can be removed along with the ice.

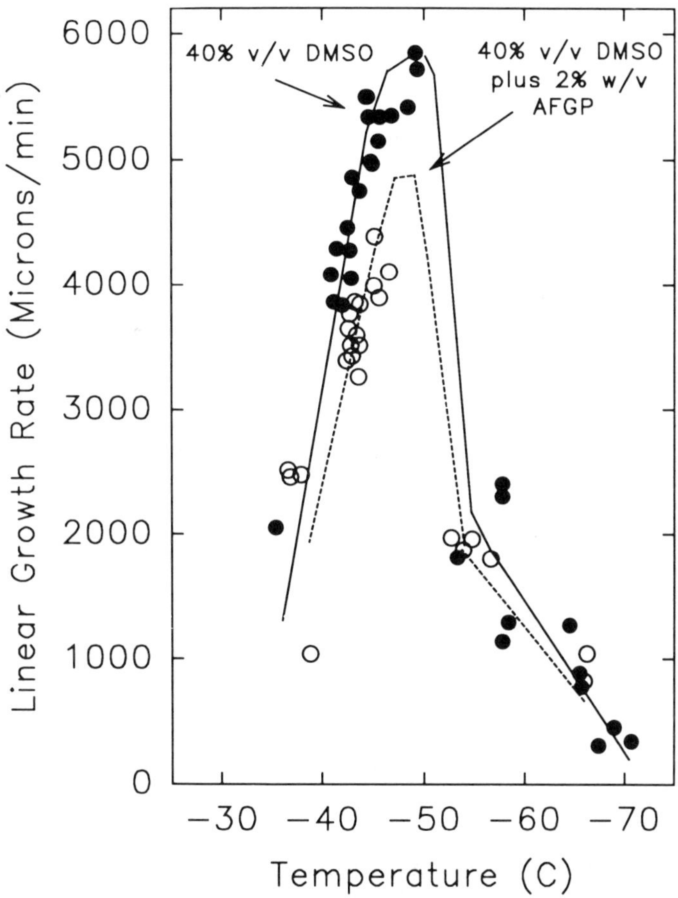


**Figure 11.** Lack of effectiveness of antifreeze glycoproteins (AFGP, consisting of a physiological mixture of AFGPs 1–8 of the Antarctic nototheniid fish *Dissostichus mawsoni*) on inhibiting ice crystallization rates in 40%, v/v, dimethyl sulfoxide in water. Filled circles, no AFGP; open circles, 2%, w/v, AFGP present. Data were obtained by timing the movement of an ice front down a standardized vertical channel in a plastic plate; the crystal orientation with respect to the axis of the channel was therefore random. AFGP was provided by A. L. DeVries.

Parody-Morreale et al. (1988) presented evidence that antifreeze glycoproteins can directly inactivate nucleators by binding to their ice-nucleating sites. Another possibility is to use such proteins to block the growth of ice that is nucleated. However, as noted in Figure 11, there may be little effect in highly concentrated aqueous solutions, whose cryoprotectant content may already produce "antifreeze" effects superior to or comparable to those of the natural proteins. Furthermore, it has been shown that using natural antifreeze proteins in combination with cryoprotectants may cause the ice that does form to become more lethal, actually increasing damage rather than decreasing it (Ishiguro and Rubinsky, 1994). This is due to crystal growth primarily in the usually unimportant c-axis direction, yielding thin needle-shaped crystals (for details, see Davies and Hew, 1990; Raymond and DeVries, 1977).

Both of these problems could be overcome by developing artificial molecules capable of inhibiting crystal growth in the direction of the c-axis. As effective as low molecular mass cryoprotectants are, their limited sizes prevent them from having the type of cooperative interactions with ice that several points of contact would provide, so it is possible to imagine specifically engineered polymers that could be more effective. Such polymers should remain of relatively low molecular weight in comparison to the proteins, however, to permit them to adsorb onto the ice crystal lattice before it grows past them, a problem that may also contribute to the failure of natural antifreezes to inhibit crystal growth at temperatures below –30°C (Fig. 11). Finally, the ability to inhibit simultaneously all crystallographic growth planes of ice, not just most such growth planes, should significantly outstrip the effectiveness of and circumvent the hazards of natural antifreeze proteins. Effectively, this would be tantamount to controlling the physics of ice and thereby changing the rules of the game of cryobiology.

#### Avoiding the Consequences of Homogeneous Nucleation

Since homogeneous nucleation will generally be of importance only during warming from the vitrified state, when homogeneous nuclei can grow to damaging sizes (devitrification), the most direct solution to this problem is rapid warming. The ability to meet the challenge of rapid warming will depend on the stability of the cryoprotectant medium, exact details of the cooling regimen, the availability of appropriate warming technology, the biological effects of different ice crystal size distributions during rapid warming, the applicability of hydrostatic pressure to slow down crystal growth and/or to prevent additional nucleation on warming, and the ability to specifically modify the growing ice crystal/solution interface so as to reduce the growth rate of the developing nuclei. With regard to the latter, Sutton and Pegg (1993) have reported that antifreeze peptides can be extraordinarily effective in blocking ice crystal growth during warming following previous homogeneous nucleation of 2,3-butanediol solutions. The control of devitrification is a complex and fertile area for future investigation.

### During Cooling for Cryoultramicrotomy

Microscopists have devised several cryo techniques to reduce the artifacts associated with frozen sections. Some of these involve the use of high concentrations of agents such as sucrose, which do not penetrate living cells and therefore must either be used on dead cells or must be allowed to kill the cells before they can be examined. However, the same vitrification techniques being pursued by cryobiologists for maintaining the viability of living systems could also allow biochemists and morphologists to obtain thin sections of living biological material without freezing artifacts and without chemical fixation or embedding. Using rabbit renal cortex and our standard vitrification solution VS41A, which vitrifies at ambient pressure (Mehl, 1993), we have been able to make sections over a very narrow temperature range (about −111.5 ± 1.0°C), tissue instability (crumbling or smearing) being observed at higher or lower temperatures. This sectioning temperature is over 10°C above $T_g$, and therefore should permit considerable nucleation during sectioning, but it is well below the crystal growth range, so visible nuclei should be absent.

### During Freezing of Complex Systems

Although I have been pursuing the cryopreservation of mammalian organs by vitrification, there may be an alternative approach to the cryopreservation of complex systems. This approach would involve the deliberate induction of very large numbers of nucleation events (at least $10^{12}/cm^3$, or about $1/\mu m^3$, or more) in the extracellular space, all at roughly the same temperature. In this way, paradoxically, the mechanical consequences of nucleation might be avoidable by inducing nucleation on an unprecedented scale.

MacFarlane and co-workers proposed one version of this approach, which in their scheme would be used to circumvent recrystallization and therefore devitrification-related injury in formerly vitrified organs (MacFarlane et al., 1981). The idea was to anneal formerly vitrified organs near $T_g$ so as to allow nucleation densities to become so great that no growth of the formed nuclei could occur during warming, since most freezable water would have already been converted into the nuclei themselves. Forsyth and MacFarlane (1986) later pointed out that very high nucleation densities can give the visual impression of vitrification, because the individual crystallites are too small to scatter light and too numerous and uniform to support substantial recrystallization (opacification) on warming.

A more general strategy could be used to allow similar benefits to be obtained in nonvitrified systems being preserved by slow-freezing procedures. This would involve introducing high quantities of nucleating agents in the simultaneous presence of large quantities of antifreeze substances such as antifreeze proteins and cryoprotectant concentrations around 3 to 6 M (the latter to limit effects such as cell shrinkage). This strategy is based on the common observation that ice per se need not be harmful to organized tissues or cells, provided its growth by normal mechanisms or through recrystallization is prevented (Mazur, 1965a, Mazur, 1988). By inducing large numbers of nuclei that cannot grow to damaging dimensions due to the presence of the antifreeze substances, extracellular ice could be rendered innocuous and inert, unable to reorganize during either cooling or warming into large grains that could damage the tissue. Even the water emerging from cells as they respond osmotically during freezing might be tied up in a series of new nucleation sites, since nucleating substances not responsible for the formation of particular ice crystals may be excluded from those crystals and therefore concentrated in the remaining unfrozen liquid.

As above in the discussion of designing specific ice growth inhibitors, this approach would essentially change the physics of ice, and thereby change many of the basic phenomena of cryobiology. Indeed, artificial, specific ice-binding substances should be more effective and more practical to use in this scheme than natural antifreeze proteins. Such a scheme would only work if the nucleators and antifreeze substances do not inactivate one another as cited above (Parody-Morreale et al.,

1988). This, however, could be arranged by sterically designing one or both of the components of this system (the nucleator or the ice-binding molecule) so as to prevent contact at the active sites.

Are there natural examples of this strategy? Nucleating proteins in freeze-tolerant vertebrates are thought not only to induce but also to at least partially control ice crystallization (Storey and Storey, 1992), and antifreeze proteins in insects are similarly thought to be important for minimizing recrystallization (Storey and Storey, 1992). The observation of both nucleating and antifreeze activity in the same organism, however, seems not to be generally recognized (Storey and Storey, 1992). Whether this is because there are no such organisms or because they have been overlooked is an open question.

## Conclusion

Nucleation plays a central role in applied cryopreservation, both for freeze preservation and for vitrification. Additional attention to the avoidance or minimization of nucleation or the control of its consequences could have major effects on the efficacy of applied cryopreservation and could help to link applied cryopreservation studies to the cryobiology of organisms in nature.

## Acknowledgments

The author thanks Joe Watson, Kris Thompson, and Melissa Consoli for photographic and graphics support, A. L. DeVries for providing antifreeze glycoprotein, M. Toner for permission to publish a modified form of his figure as Figure 1, T. Takahashi for permission to publish Figure 8, J. Matthes for counting most of the crystals of Figure 7, and P. M. Mehl for interesting discussions. Supported by grants GM-17959, BSRG 2507-RR 05737, and a grant from the G. Harold and Leila Y. Mathers Charitable Foundation.

## Literature Cited

Alink, G. M., Verheul, C. C., and Offerijns, F. G. J. 1977. Three-step cooling: A preservation method for adult rat heart cells. Cryobiology 14:409-417.

Angell, C. A. 1982. Supercooled water. Pages 1-81 in: Water, A Comprehensive Treatise. Vol. 7, Water and Aqueous Solutions at SubzeroTemperatures. F. Franks, ed. Plenum Press, New York.

Angell, C. A., and MacFarlane, D. R. 1982. Conductimetric and calorimetric methods for the study of homogeneous nucleation below both $T_h$ and $T_g$. Adv. Ceramics 4:66-79.

Bronshteyn, V. L., and Steponkus, P. L. 1995. Nucleation and growth of ice crystals in concentrated solutions of ethylene glycol. Cryobiology 32:1-22.

Chang, Z. H. 1992. The biological and physical aspects of cryopreservation by vitrification: A *Drosophila melanogaster* model. Dissertation. State University of New York at Binghamton, NY.

Coger, R., and Toner, M. 1995. Preservation techniques for biomaterials. Pages 1557-1566 in: Biomedical Engineering Handbook. CRC Press, Boca Raton.

Davies, P. L., and Hew, C. L. 1990. Biochemistry of fish antifreeze proteins. FASEB (Fed. Am. Soc. Exp. Biol.) J. 4:2460-2468.

Diller, K. R. 1975. Intracellular freezing: Effect of extracellular supercooling. Cryobiology 12:480-485.

Dowgert, M. F., and Steponkus, P. L. 1983. Effect of cold acclimation on intracellular ice formation in isolated protoplasts. Plant Physiol. 72:978-988.

Fahy, G. M. 1980a. Analysis of "solution effects" injury: Equations for calculating phase diagram information for the ternary systems NaCl-dimethylsulfoxide-water and NaCl-glycerol-water. Biophys. J. 32:837-850.

Fahy, G. M. 1980b. Analysis of "solution effects" injury: Rabbit renal cortex frozen in the presence of dimethyl sulfoxide. Cryobiology 17:371-388.

Fahy, G. M. 1981a. Simplified calculation of cell water content during freezing and thawing in nonideal solutions of cryoprotective agents and its possible application to the study of "solution effects" injury. Cryobiology 18:473-482.

Fahy, G. M. 1981b. Analysis of "solution effects" injury: Cooling rate dependence of the functional and morphological sequellae of freezing in rabbit renal cortex protected with dimethyl sulfoxide. Cryobiology 18:550-570.

Fahy, G. M. 1988. Vitrification. Pages 113-146 in: Low Temperature Biotechnology: Emerging Applications and Engineering Contributions. J. J. McGrath and K. R. Diller, eds. American Society of Mechanical Engineers, New York.

Fahy, G. M., MacFarlane, D. R., Angell, C. A., and Meryman, H. T. 1984. Vitrification as an approach to cryopreservation. Cryobiology 21:407-426.

Fahy, G. M., Saur, J., and Williams, R. J. 1990. Physical problems with the vitrification of large biological systems. Cryobiology 27:492-510.

Forsyth, M., and MacFarlane, D. R. 1986. Recrystallization revisited. Cryo Lett. 7:367-378.

Franks, F. 1988. Storage in the undercooled state. Pages 107-112 in: Low Temperature Biotechnology: Emerging Applications and Engineering Contributions. J. J. McGrath and K. R. Diller, eds. American Society of Mechanical Engineers, New York.

Franks, F., Mathias, S. F., Galfre, P., Webster, S. D., and Brown, D. 1983. Ice nucleation and freezing in undercooled cells. Cryobiology 20:298-309.

Fujikawa, S. 1981. The effect of different cooling rates on the membrane of frozen human erythrocytes. Pages 323-334 in: Effects of Low Temperatures on Biological Membranes. G. J. Morris and A. Clarke, eds. Academic Press, New York.

Ishiguro, H., and Rubinsky, B. 1994. Mechanical interactions between ice crystals and red blood cells during directional solidification. Cryobiology 31:483-500.

Kanno, H., and Angell, C. A. 1977. Homogeneous nucleation and glass formation in aqueous alkali halide solutions at high pressures. J. Phys. Chem. 81:2639-2643.

Karlsson, J. O. M., Cravalho, E. G., and Toner, M. 1994. A model of diffusion-limited ice growth inside biological cells during freezing. J. Appl. Phys. 75:4442-4455.

Kresin, M., and Korber, Ch. 1991. Influence of additives on crystallization kinetics: Comparison between theory and measurements of aqueous solutions. J. Chem. Phys. 95:5249-5255.

Kuwano, K., Aruga, Y., and Saga, N. 1994. Cryopreservation of the conchocelis of Porphyra (Rhodophyta) by applying the simple prefreezing system. J. Phycol. 30:566-570.

MacFarlane, D. R., Angell, C. A, and Fahy, G. M. 1981. Homogeneous nucleation and glass formation in cryoprotective systems at high pressures. Cryo Lett. 2:353-358.

MacFarlane, D. R., Forsyth, M., and Barton, C. A. 1992. Vitrification and devitrification in cryopreservation. Adv. Low-Temp. Biol. 1:221-278.

Mathias, S. F. 1991. Undercooling—Low temperature without freezing. Trends Biotechnol. 9:370-372.

Mazur, P. 1963. Kinetics of water loss from cells at subzero temperatures and the likelihood of intracellular freezing. J. Gen. Physiol. 47:347-369.

Mazur, P. 1965a. Physical and chemical basis of injury in single-celled micro-organisms subjected to freezing and thawing. Pages 213-315 in: Cryobiology. H. T. Meryman, ed. Academic Press, New York.

Mazur, P. 1965b. The role of cell membranes in the freezing of yeast and other single cells. Ann. New York Acad. Sci. 125:658-676.

Mazur, P. 1988. Stopping biological time: The freezing of living cells. Ann. New York Acad. Sci. 541:514-531.

Mehl, P. M. 1990. Experimental dissection of devitrification in aqueous solutions of 1,3-butanediol. Cryobiology 27:378-400.

Mehl, P. M. 1992. Cryomicroscopy as a support technique for calorimetric measurements by DSC for the study of the kinetic parameters of crystallization in aqueous solutions. Part 1. Nucleation in the water-1,2-propanediol system. Thermochim. Acta 203:475-492.

Mehl, P. M. 1993. Nucleation and crystal growth in a vitrification solution tested for organ cryopreservation by vitrification. Cryobiology 30:509-518.

Muldrew, K., and McGann, L. E. 1990. Mechanisms of intracellular ice formation. Biophys. J. 57:525-532.

Parody-Morreale, A., Murphy, K. P., Cera, E. D., Fall, R., DeVries, A. L., and Gill, S. J. 1988. Inhibition of bacterial ice nucleators by fish antifreeze glycoproteins. Nature 333:782-783.

Pollock, G. A., Pegg, D. E., and Hardie, I. R. 1986. An isolated perfused rat mesentery model for direct observation of the vasculature during cryopreservation. Cryobiology 23:500-511.

Rall, W. F., Mazur, P., and McGrath, J. J. 1983. Depression of the ice-nucleation temperature of rapidly cooled mouse embryos by glycerol and dimethyl sulfoxide. Biophys. J. 41:1-12.

Rasmussen, D. H., Macauley, M. N., and MacKenzie, A. P. 1975. Supercooling and nucleation of ice in single cells. Cryobiology 12:328-339.

Raymond, J. A., and DeVries, A. L. 1977. Adsorption inhibition as a mechanism of freezing resistance in polar fishes. Proc. Natl. Acad Sci. USA 74:2589-2593.

Steponkus, P. L. 1984. Role of the plasma membrane in freezing injury and cold acclimation. Annu. Rev. Plant Physiol. 35:543-584.

Steponkus, P. L., Stout, D. G., Wolfe, J., and Lovelace, R. V. E. 1985. Possible role of transient electric fields in freezing-induced membrane destabilization. J. Membrane Biol. 85:191-198.

Storey, K. B., and Storey, J. M. 1992. Natural freeze tolerance in ectothermic vertebrates. Annu. Rev. Physiol. 54:619-637.

Sutton, R. L., and Pegg, D. E. 1993. Devitrification in butane-2,3-diol solutions containing anti-freeze peptide. Cryo. Lett. 14:13-20.

Toner, M., and Tompkins, R. G. 1992. Transport phenomena during freezing of isolated hepatocytes. AICHE J. 38:1512-1522.

Walter, C. A., Knight, C. A., and Farrant, J. 1975. Ultrastructural appearance of freeze-substituted lymphocytes frozen by interrupting rapid cooling with a period at −26°C. Cryobiology 12:103-109.

Williams, R. J., Mehl, P., and Carnahan, D. L. 1990. Ice nucleation and growth in vitrifiable solutions. Pages 71-86 in: Cryopreservation and Low Temperature Biology in Blood Transfusion. C. Th. Smit Sibinga, P. C. Das, and H. T. Meryman, eds. Kluwer Academic, Boston.

CHAPTER 19

# Applications of Biological Ice Nucleators in Spray-Ice Technology

**Richard J. LaDuca, A. Franklin Rice, and Patrick J. Ward**

Ice nucleation-active (Ina$^+$) bacteria have been successfully utilized in a number of commercial application areas that take advantage of the ability of these microorganisms to efficiently initiate freezing events at high subzero temperatures (for reviews see Warren, 1987; Margaritis and Bassi, 1991; Gurian-Sherman and Lindow, 1992, 1993). In large part, the commercial potential and value of Ina$^+$ bacteria is directly proportional to the savings in energy costs that can be realized when freezing events are initiated at higher subzero temperatures with Ina$^+$ bacteria. Earlier chapters of this book have discussed the application of Ina$^+$ bacteria in the control of insect pests (Chapter 14), the use of ice nucleation genes as reporters (Chapter 15), and the impact of biological ice nucleators in food processing (Chapter 17). Spray-ice applications of Ina$^+$ bacteria will be the focus of this review.

The largest volume commercial use of Ina$^+$ bacteria is today found in the spray-ice application area of artificial snowmaking. In this application area, the ability of Ina$^+$ bacteria to efficiently and cost effectively generate artificial snow has been shown to offer significant advantages over non-biologically nucleated snowmaking systems. Further commercial evaluation of Ina$^+$ bacteria in spray-ice applications is the focus of efforts in both academia and industry. Issues relating to the commercial manufacture of Ina$^+$ bacteria and use of these biological ice nucleators in spray-ice-type product application areas such as artificial snowmaking, natural thermal storage in ice ponds, freeze crystallization, weather modification through cloud seeding, and arctic ice island construction will be described.

## Fermentation and Recovery of Ina$^+$ Bacteria

### Induction of the Ina$^+$ Phenotype in Ice Nucleation-Active Bacteria

Economically viable utilization of ice nucleation-active bacteria for commercial spray-ice-type product applications requires that the ice nucleation frequency of warm-temperature nuclei (defined as the ratio of cells active in ice nucleation at −5°C to the total number of bacterial cells in a population of Ina$^+$ bacteria) be as close to one as possible. Historically, the frequency of expression of the $Ina^+$ phenotype in pure cultures of ice nucleation-active bacteria has been shown to be highly variable (Maki and Willoughby, 1978; Lindow et al., 1982; Lindow, 1983; Yankofsky et al., 1983; Hirano et al., 1985). The source of this variability in the frequency and level of $Ina^+$ expression has been attributed to physiological responses of the $Ina^+$ bacteria to nutritional and environmental signals (Hirano et al., 1985; Rogers et al., 1987; Obata et al., 1990; Pooley and Brown, 1991; Lawless and LaDuca, 1992; Nemecek-Marshall et al., 1993). Lindow et al. (1982) showed that the ratio between the number of ice nuclei and the number of bacterial cells in a culture can vary with incubation temperature, growth medium composition, culture age, and genotype. Nucleation frequencies appear to be most dramatically impacted by temperature. When ice nucleation-active bacteria are allowed to incubate at low temperatures (not conducive to cell growth), a shift in the frequency of warm-temperature ice nuclei can be observed (Rogers et al., 1987). This process, perhaps the result of the maturation of the nucleation site on the cell surface, affects the number and type (Type 1, 2, or 3) of ice nucleation sites observed in the microbial population.

As $Ina^+$ bacteria progress through a growth cycle in culture, key growth-promoting nutrients are consumed at stoichiometric levels. Depletion of key nutrients signal the end of the active growth phase and progression into the stationary growth phase. Elevated nucleation frequencies in *Pseudomonas syringae* have been temporally linked to the transition of cells from the exponential to the stationary phase of growth (Lindow et al., 1982; Hirano et al., 1985; Deininger et al., 1988; Luquet et al., 1991; Pooley and Brown, 1991). Nemecek-Marshall et al. (1993) found that nutritional starvation for nitrogen, phosphorous, sulfur, or iron, but not carbon, at 32°C coupled with a shift to 14 to 18°C will lead to the rapid induction of warm-temperature ice nuclei. Expression of ice nuclei may also be regulated by other nutritional factors. Kozloff and co-workers have shown that the addition to minimal media of sugars, such as inositol, mannose, and glucosamine in combination with $Mn^{2+}$ will enhance ice nucleus formation (Turner et al., 1990; Kozloff et al., 1991a,b). They suggested that these nutrients are components of the stepwise assembly of $Ina^+$ protein aggregates in the cell membrane, a process proposed to involve covalent modification of the protein by glycosylation and anchoring to phosphatidylinositol (Kozloff et al., 1991a,b). Numerous studies have focused on physiological and genetic responses of microorganisms to changes in the availability of nutrients (Ingrahm et al., 1983; Matin et al., 1989; Blum et al., 1990). Nutrient depletion resulting from the lack of carbon, phosphorus, or nitrogen and related physiological stresses including heat shock, osmotic shock, and oxidative stress in bacteria may promote the expression of unique sets of genes that function to mitigate the effects of an environmental stress (Groat et al., 1986; Blum et al., 1990; Jenkins et al., 1991; McCann et al., 1991). Expression of these genes is usually accomplished through the use of a non-housekeeping-type transcriptional promoter. This type of regulation is similar to that observed in the expression of low-temperature (Jones et al., 1987) or starvation (Matin et al., 1989) genes. Other unique nucleotide sequences, termed gearbox-type promoters, are known in genes whose products are required at higher relative amounts when growth rates approach zero (Aldea et al., 1990; Vincente et al., 1991). Nemecek-Marshall et al. (1993) suggested that the regulatory region of the *inaZ* gene preceding the translational start codon has a high degree of homology to the –35 and –10 regions of the consensus gearbox promoter proposed by Aldea et al. (1990). Regions of homology were identified at 57 and 38 nucleotides, respectively, from the translational start codon of the *inaZ* gene. This *inaZ* putative promoter sequence, AAGCAA(13 bp)CGGCTGCT, has a high degree of homology to the consensus gearbox promoter sequence identified in *Escherichia coli*, CTGCAA(14-16 bp)CGGCAAGT (Vincente et al., 1991). The homology of these regulatory regions suggests that a gearbox-type or related promoter structure may function to temporally control expression of the $Ina^+$ protein following nutritional limitation in the culture medium (Nemecek-Marshall et al., 1993).

## Large-Scale Manufacture of $Ina^+$ Bacteria

Successful commercialization of spray-ice nucleation technology in all application areas is based on the ability to manufacture the product (i.e., $Ina^+$ bacteria) at minimal cost with the highest quality. It is essential that the development of fermentation and recovery processes include optimization with respect to total activity produced in the fermentor, recovery process yield, and minimization of total processing time. Generally, strain selection, choice of raw materials, fermentation parameter definition, recovery methodology, and ability to scale-up collectively prevail in implementing a process that is economically favorable.

The commercialization of ice nucleation bacteria started in 1985 with the development of SNOMAX® Snow Inducer (Genencor International, Rochester, New York) for the artificial snowmaking industry. Fermentation and recovery processes have been described for the large-scale production of microorganisms having high levels of Type 1 and Type 2 ice nuclei (Lynn and Noto, 1988; Hendricks et al., 1992; Lawless and LaDuca, 1992). To date, these processes are considered state-of-the-art for commercial manufacturing and will be described in some detail.

## Fermentation of $Ina^+$ Bacteria

Optimal growth conditions for microorganisms expressing ice nucleation activity have been under investigation since their discovery in the early 1970s. Traditionally, fermentation conditions have varied widely with regard to the nutritional and environmental conditions necessary to achieve cell growth and ice nucleation activity (Lindow et al., 1978; Maki and Willoughby, 1978; Kozloff et al., 1983). Extrapolation of these conditions and methods to large scale often resulted in less than optimal expression of the $Ina^+$ phenotype on a per cell basis as well as high cost for the growth medium raw materials. In the manufacture of $Ina^+$ bacteria for spray-ice applications, optimal growth conditions require the generation of high cell densities with high nucleation frequencies (i.e., the ratio of cells active in $Ina^+$ at $-5°C$ to the total cell number in the bacterial population must be as close to one as possible) in a low-cost production medium.

Initial improvements in the large-scale manufacture of $Ina^+$ bacteria focused primarily on the nutritional conditions necessary to develop high cell density fermentation processes. Hendricks et al. (1992) described a process using controlled fermentation pH, mannitol as a carbon source, and yeast extract as a nitrogen source for the generation of *P. syringae* $Ina^+$ bacteria. This process produced greater than $10^{11}$ ice nuclei per gram of dry cells in the reactor with fermentor productivities exceeding $10^{11}$ ice nuclei per liter per hour. Further medium development led to the use of a chemically defined medium for the manufacture of $Ina^+$ bacteria (Lynn and Noto, 1988). Fermentation of $Ina^+$ bacteria in a chemically defined medium, as compared to media utilizing more complex sources of carbon and nitrogen (e.g., yeast extract), has resulted in enhanced efficiencies in Ina⁺ yield and recovery in large-scale processes. Further enhancements to the fermentation process for manufacturing Ina⁺ bacteria have sought to leverage the environmental conditions that impact Ina⁺ expression in *P. syringae*.

To achieve maximal cell densities and ice nucleation frequencies in a large-scale fermentation process, programmed fermentation strategies have been designed to take into account the temperature dependence of nucleation frequency in *P. syringae* as well as to take advantage of the correlation between temporal expression of warm-temperature ice nuclei in the growth cycle and depletion of key nutrients (Lawless and LaDuca, 1992). In submerged cultures of *P. syringae* that use these strategies, maximal nucleation frequency at –5°C can approach one active nucleus out of every one to 10 cells. These levels of expression have been accomplished by supplying stoichiometrically limiting quantities of nitrogen in the fermentation medium such that at the end of the active growth phase, there is insufficient nitrogen remaining to inhibit ice nucleus expression and formation during the subsequent stationary growth phase (Lawless and LaDuca, 1992). By programming a temperature downshift during fermentation to coincide with the nutritional depletion of nitrogen, maximal nucleation frequencies and cell densities can be obtained. Figure 1 shows a schematic for this high productivity process.

To take full advantage of the economy associated with large-scale fermentation, it is essential that fermentation development work take place in order to predict the behavior of the process on a large scale. This can be accomplished by either scale-up or scale-down methodology. It involves understanding the fermentation response to changes in environmental parameters to ensure optimum performance at the increased scale. Although behavior of a population of microorganisms should be invariant to scale, the behavior of factors affecting the surrounding environment is more difficult to predict (e.g., oxygen transfer, broth mixing characteristics, and shear). Scale-up can have considerable costs in both time and equipment but incurs less risk than scale-up of a process based solely on theoretical models. Understanding of the nutritional and environmental factors that affect the induction of ice nucleation activity associated with *P. syringae* has enabled the commercialization of this particularly interesting application of industrial biotechnology.

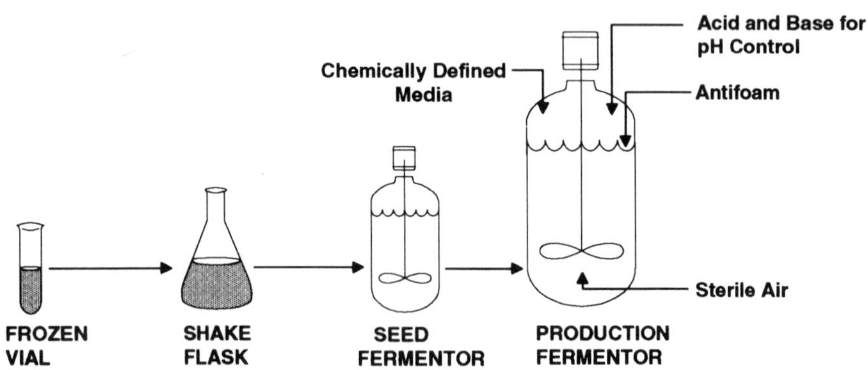


**Figure 1.** Genencor International's high-productivity process. The final production fermentor involves a two stage fermentation. First biomass is increased by fermentation at optimum growth temperature. Second, Ina⁺ induction takes place by rapid fermentor cooling when nutrient limitation brings cell mass into stationary growth phase. Fermentor harvest occurs at maximum Ina⁺ (6–10 hr post induction).

Ice nucleation-active bacteria are currently being produced by the thousands of liters per fermentation, with productivities reaching one active ice nucleation site per cell at −5°C. Associated with the large-scale manufacture of any industrial fermentation process are larger risks. One of the risks that can virtually shut down production without advanced warning is the susceptibility of bacteria to bacteriophage infection.

### *P. syringae* and Susceptibility to Bacteriophage Infections

Naturally occurring strains of *P. syringae* are susceptible to lytic infection by bacteriophage. It is clear that cell lysis due to phage infection during fermentation is detrimental to the production of ice nucleation activity and should be guarded against in the production environment. Lytic events in the large-scale manufacture of any microbial product including $Ina^+$ bacteria have generally been dealt with through two parallel approaches. In the first approach, microbial isolates that are insensitive to lytic infection by isolated species of bacteriophage are isolated as spontaneous mutants. Bacterial isolates resistant to a broad range of phage can be obtained by selecting spontaneous phage-resistant mutants following batch or sequential exposure of the phage-sensitive $Ina^+$ bacteria to numerous serotypes of bacteriophage. In this way, a family of bacteriophage-resistant $Ina^+$ producers can be generated to minimize susceptibility of the commercial process to lytic fermentation failures. A second and essential approach to minimize lytic infections has been to control, through enhanced cleaning and monitoring procedures, all avenues of bacteriophage introduction into the fermentation process. For this reason it is essential to manufacture under sterile conditions, abide by all current good manufacturing practices, and regularly monitor the air both in the fermentation area and at the intake for process air quality.

### Recovery of Ice Nucleation Activity

The design of a process for the recovery of ice nucleation activity involves the integration of many factors. Such factors can and should include recovery process yield, total processing time, and final desired product form. Of these considerations, the final product form is determined once the final customer has been identified, and the customer's needs and applications are understood. Final package size, product shelf life, environmental impact, and shipping and handling constraints must be defined before the entire downstream process can be developed. It is essential that the recovery methodology be cost effective on a large scale and meet all the customer requirements and specifications (Wheelwright, 1991).

The large-scale purification of active ice nucleation bacterial fermentation broths has presented several major purification challenges. These challenges stem from the susceptibility of the ice nucleation protein to denaturation by temperature, and susceptibility of the active site to incur damage over time in high osmotic strength, aqueous environments. An example of large-scale recovery of ice nucleating bacteria is the production of SNOMAX Snow Inducer. This product is recovered on the large scale through a series of unit operations that help to preserve the integrity of the ice nucleation protein all the way from fermentation harvest broth to delivery to the final customer. Figure 2 is a schematic of the process typical for the large-scale recovery of SNOMAX.

In the production of SNOMAX, the fermentation broth at harvest (peak $Ina^+$ activity) is sent downstream for cell separation. Since the bacterial cells are the prod-

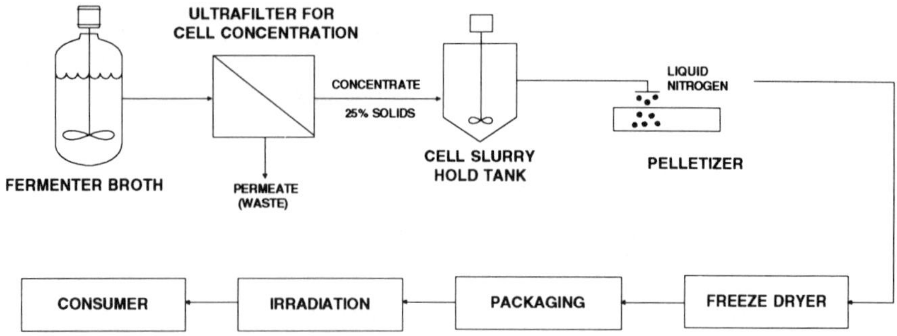


**Figure 2.** Typical commercial-scale recovery process for bacterial ice nuclei.

uct of interest, cell concentration is essential to reduce the volume needing to be processed in further downstream refinements. Typically, cell concentration can take place by centrifugation or permeable membrane separation. In the case of SNOMAX, ultrafiltration is preferred. Ultrafiltration uses microporous polymeric membranes driven by liquid pressures to separate the cells from the spent fermentation broth. Ultrafiltration takes advantage of the sieving mechanism of the membrane based on molecular weight. Thus cells and high molecular weight components are concentrated on one side of the membrane, while smaller molecules, such as water, salts, and low molecular weight proteins, pass through the porous skin of the membrane. Concentration continues until the cell slurry reaches approximately 25% dry solids. It is this high osmotic strength aqueous environment that causes Ina$^+$ denaturation to occur rapidly, damaging the ice nucleation protein. To stop the denaturation process, the cell slurry is rapidly fed under pressure in thin streams into liquid nitrogen. The interfacial surface interactions with liquid nitrogen cause the streams of cell mass concentrate to freeze as pellets. The ice nucleation activity of cells in the pellets is much more stable than in the slurry. The frozen pellets are subsequently freeze-dried. Freeze-drying at low temperature is advantageous because it minimizes loss of heat-labile products such as the Ina$^+$ protein. The freeze-dried Ina$^+$ product is relatively stable at room temperature, but shelf life can be extended through packaging and long-term storage at subzero temperatures. Packaging of the material is best conducted with an inert gas purge to prevent oxidation of the ice nucleation protein over time. It is also recommended that ultraviolet light–insensitive materials be used to store the final product.

For product applications of Ina$^+$ bacteria, some regulatory agencies require that the cell viability of *P. syringae* be reduced. This helps to ensure that the number of viable organisms in the environment does not increase beyond already existing natural Ina$^+$ bacterial populations due to the use and release of Ina$^+$ bacteria. Beta irradiation of SNOMAX powders is routinely performed to reduce cell viability and in some instances to produce a sterile product. Effective cell inactivation can be achieved through low-dose beta irradiation with minimal Ina$^+$ loss.

## Commercialization of Spray-Ice Technologies

### Snowmaking

The production of artificial snow at ski resorts worldwide has become an economic necessity to ensure adequate snow coverage under variable weather conditions throughout the ski season. To this end, it is estimated that worldwide annual snow production requires approximately 10 billion U.S. gallons (37.8 billion liters) of water to generate 60 million acre feet (70 billion cubic meters) of snow. Typical ski areas utilize an estimated 500 to 750,000 U.S. gallons of water to generate 3 to 4 acre feet (3,500 to 4,000 $m^3$) of snow each day of operation. Marginal weather conditions both early and late in the ski season typically limit the amount, as well as quality, of man-made snow that can be produced, resulting in shorter operational seasons and reduced revenues for ski operators. During the ski season, temperatures above −2°C dry bulb limit effective operation of snow-making equipment resulting in poor quality and quantity of artificial snow generated. Operators of these facilities must thus balance the costs associated with generation of artificial snow against the benefits of an extended season and the amount of skiable terrain that can be maintained throughout operational periods and extremes of weather fluctuations.

In the production of artificial snow, water is commonly sprayed through nozzles under air pressure in order to facilitate supercooling and nucleation of water droplets. The atomizing nozzles typically employed in this application of spray-ice technology promote the quick formation of fine water droplets, typically 200 to 700 μm in size, significantly enhancing air-water heat exchange rates and formation of nucleated droplets. Compressed air primarily promotes formation of atomized water droplets, however pressure changes resulting from expansion of air across the orifice of the nozzle further enhance cooling of the droplets and the crystallization process. Thus, in the absence of heterogeneous sources of ice nucleators, artificial snowmaking is primarily affected by water flow rates and temperature, air pressure, ambient temperatures, and relative humidity.

Biological sources of ice nucleators are effective in enhancing the temperature range at which artificial snow can be generated (Woerpel, 1980). The presence of biological ice nucleators in water droplets generated in standard snowmaking equipment initiates ice formation without the need to supercool the water by the amount that is typically required in their absence. Commercialization of biological ice nucleators in this application area has been accomplished by SNOMAX Technologies, a division of Genencor International, Inc., with the introduction of SNOMAX. SNOMAX is a freeze-dried powder containing $Ina^+$ *P. syringae* that facilitates efficient production of man-made snow at temperatures as high as 1.1°C dry bulb. In typical snowmaking applications, 270 g of SNOMAX is used to treat 100,000 gallons of water. Inclusion of this biological ice nucleator increases the volume and quality of man-made snow produced and extends the temperature range at which snow can be made utilizing standard snowmaking equipment (Liao and Ng, 1990). Snow densities are reduced by 10 to 15% in man-made snow generated with biological ice nucleators. Further, snow volumes generated in the presence of SNOMAX have been observed to increase by 25 to 60% (compared with snow volumes generated in the absence of the nucleator) depending on the equipment used and atmospheric conditions.

Since its introduction in 1985, SNOMAX has been used at the Calgary and Albertville Olympic Games and in 1994 was used in all machine-made snow generated for the Lillehammer Olympic games in Norway. Through the use of SNOMAX, biological ice nucleators are utilized routinely as snowmaking additives in ski areas throughout North America, South America, Australia, New Zealand, Japan, and Europe.

The application of *P. syringae* as an ice nucleator in the manufacture of artificial snow has previously raised questions concerning the release of microorganisms into alpine environments. The *P. syringae* cell, which is encapsulated in the ice crystals of man-made snow, can contact soil and natural waters from snow melt runoff. Studies designed to evaluate the fate of $Ina^+$ *P. syringae* in alpine soils, waters, and synthetic snow samples have shown that ambient environmental conditions (e.g., freeze-thaw cycles, sunlight, and soil contact) severely limit the survival of this microorganism (Goodnow et al., 1990) subsequent to its use as a snow inducer.

## Natural Thermal Storage

Natural thermal storage systems, also known as "ice ponds," are simple outdoor man-made ponds equipped with centrifugal pumping systems that atomize recirculating pond water via overhead-mounted spray system assemblies (Fig. 3). These same pumps also deliver 1.7°C ice water to an "interfacing" heat exchanger located within the facility of choice requiring cooling, replacing a conventional chlorinated fluorocarbon (CFC) chiller and associated cooling tower system. Ice ponds may be either constructed open to the environment for seasonal operation or housed completely enclosed within insulated structures for year-round operation. Ice ponds are based upon the ambient freezing, thawing, and recirculation of water, requiring a seasonal winter climate with 4,000 (or more) heating-degree days (HDD) per year typically required for economic operation.

Several attempts have been made in the past to economically capture wintertime coldness in the form of frozen water as perhaps the ultimate in environmentally compatible refrigeration. Most noteworthy is the Prudential Insurance Company's 1979 installation of a 4,000 m$^2$ ice pond for year-round comfort cooling at one of their commercial buildings (12,000 m$^2$) in Princeton, New Jersey.

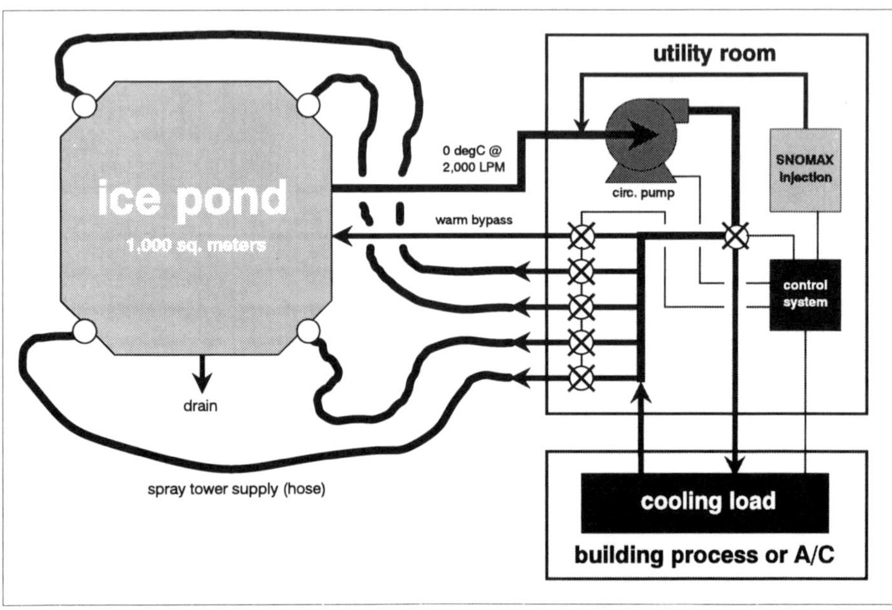


**Figure 3.** Ice pond schematic for enhanced natural thermal storage

Since the introduction of commercially available biological ice nucleation–active products, energy efficiency and economic viability of these ambient systems have been substantially improved. Unlike conventional, mechanical ice thermal storage systems, these "enhanced" ice ponds dramatically reduce both power and energy requirements while shifting electrical loads to off-peak hours of utility demand, all without the use of refrigerant gases. An additional important feature of the ice pond is its ability to reject sensible heat directly to the atmosphere without consuming valuable stored ice. It is estimated that 40% of the total heat load can thus be rejected directly to the environment via evaporative cooling.

Theoretical calculations of cost savings achieved via elimination of conventional refrigerant compressors are substantiated by two industrial installations located in upstate New York: an 80 refrigeration-ton system at Kutters Cheese (Corfu, New York) and a 500 refrigeration-ton system at Genencor International, Inc. (Rochester, New York).

At Kutters Cheese (a cheese-making company), the plant owner wanted to supplement existing CFC-based refrigeration capacity with natural thermal storage to meet seasonal peak production demands without additional investments in compressor capacity. Based on a seasonal mode of operation and a 6,900-HDD weather profile, the enhanced ice pond constructed in 1989 was operated continually for 5 months with a seasonal average energy efficiency of 0.05 kW/ton compared with existing electrical, mechanical equipment operating at 1.0 kW/ton, a 20:1 reduction in energy consumption. Similar results have been obtained at Genencor International's biochemicals production facility in Rochester, New York, where a 1,200-$m^2$ ice pond has been in seasonal operation since 1991. In both cases, energy cost savings were sufficient to offer a 3-year simple pay-back period of invested capital.

Ice ponds do have limitations. Given the 4,000-HDD economic restriction, the southern half of the United States is unsuitable for economic application of this technology. Ice ponds also have aesthetic limitations and require land—a combination not usually suited for high-density, high-rise central business district environments.

## Ambient Freeze-Crystallization and Water Purification

Natural thermal storage systems can also be effectively utilized as low-cost, freeze concentration machines for the effective separation of impurities from water. Water desalination is an excellent example of such an application.

Natural freeze concentration systems are based on the fact that many inorganic materials are preferentially excluded from ice crystals during the freezing process. In the case of sea water, for example, freezing of the brackish water will result in the formation of a low-salt ice mass and a high-salt concentrate. By linking two to four ponds in series and feeding sequential ponds with concentrate from the prior stage, water of higher potable quality can be produced in the form of ice. Removing ice from the final-stage ice pond provides both cooling and "fresh" water. In October, 1984, engineers from NOVA Inc., an engineering firm in Gaithersburg, Maryland, in conjunction with the New York State Energy Research and Development Authority (NYSERDA) constructed such a system in Greenport, New York, for the production of potable water from sea water. The study concluded that using snowmaking machines to spray and collect sea water in a series of ice ponds could produce drinking water for $1.50 to $3.00 per 1,000 gallons, or about the same unit cost as the then state-of-the-art reverse osmosis machines. The system produced 10 million gallons per year of desalinated water using a 1.5-acre reservoir and an 8,000-gallon per minute pumping system. The addition of Ina+ bacteria in these systems should provide even greater efficiencies in freeze-crystallization applications, given the wide diversity of freezing-point depressions found in aqueous systems containing various inorganic impurities.

## Weather Modification

Weather modification (cloud seeding) is the intentional treatment of individual cloud or storm systems for the purpose of altering the physical processes leading to the formation and growth of water droplets and ice crystals in clouds. Most continental precipitation is produced by ice crystal formation in supercooled clouds, known as the cold rain process. However, many regions have only a few ice nuclei in the atmosphere, which reduces the efficiency of the cold rain process. For this reason, artificial ice nuclei are introduced through conventional weather modification techniques in order to enhance the precipitation process. Once ice has formed, the ice crystals accumulate moisture and grow in size and mass until they fall as either rain or snow depending on the temperature of the atmosphere near the ground. Cloud-seeding agents have proven to be useful in fog and stratus dispersal, winter and summer precipitation augmentation, and hail suppression (Weather Modification Association, 1984).

Very much has been learned in the past 20 years regarding the ability of certain strains of naturally occurring bacteria to efficiently nucleate the formation of ice at slight supercoolings (Gregor, 1967; Schnell and Vali, 1972, 1976; Maki et al., 1974; Vali et al., 1976; Maki and Willoughby, 1978; Lindow et al., 1978, 1982; Yankofsky et al., 1981; Levin et al., 1987). Most naturally occurring ice nuclei found in the atmosphere are not active at temperatures warmer than about –10°C. For this reason, there is continued interest in searching for novel ice nuclei that can convert liquid water to ice crystals at the lowest degree of supercooling. Although dry ice is a very good source of warm-temperature ice nuclei, it has limitations in handling, storage, bulk density, and dispersal (Woodley and Henderson, 1990).

Recently, Genencor International, Inc. commercialized *P. syringae* (SNOMAX Weather Manager) for use as an atmospheric ice nucleus alternative. To date, this is the only existing commercial bacterial ice nucleation product used to induce precipitation from supercooled clouds. Figure 4 shows a series of electron micrographs of increasing magnification of an ice crystal captured after SNOMAX aerosol was introduced into a –20°C supercooled cloud. It is evident that the bacterial particle initiated ice crystal formation because the rod-shaped bacterial particle is symmetrically centered in the hexagonal ice crystal formation.

Atmospheric tests of SNOMAX Weather Manager have been performed during the past few years. Initial quantitative atmospheric tests indicate that *P. syringae* may prove to be a valuable tool for artificial nucleation in the atmosphere (Rogers et al., 1987; Ward and DeMott, 1989; Jung, 1990; Woodley and Henderson, 1990). Continued research in capturing warm-temperature Ina+ bacteria in quantity at low cost could eventually result in a displacement of silver iodide and solid carbon dioxide technologies, and become widely used in weather modification.

## Arctic Spray-Ice Construction

Low-temperature sea water is an abundant resource in arctic regions that can be used in numerous spray-ice technology applications for the construction of load-

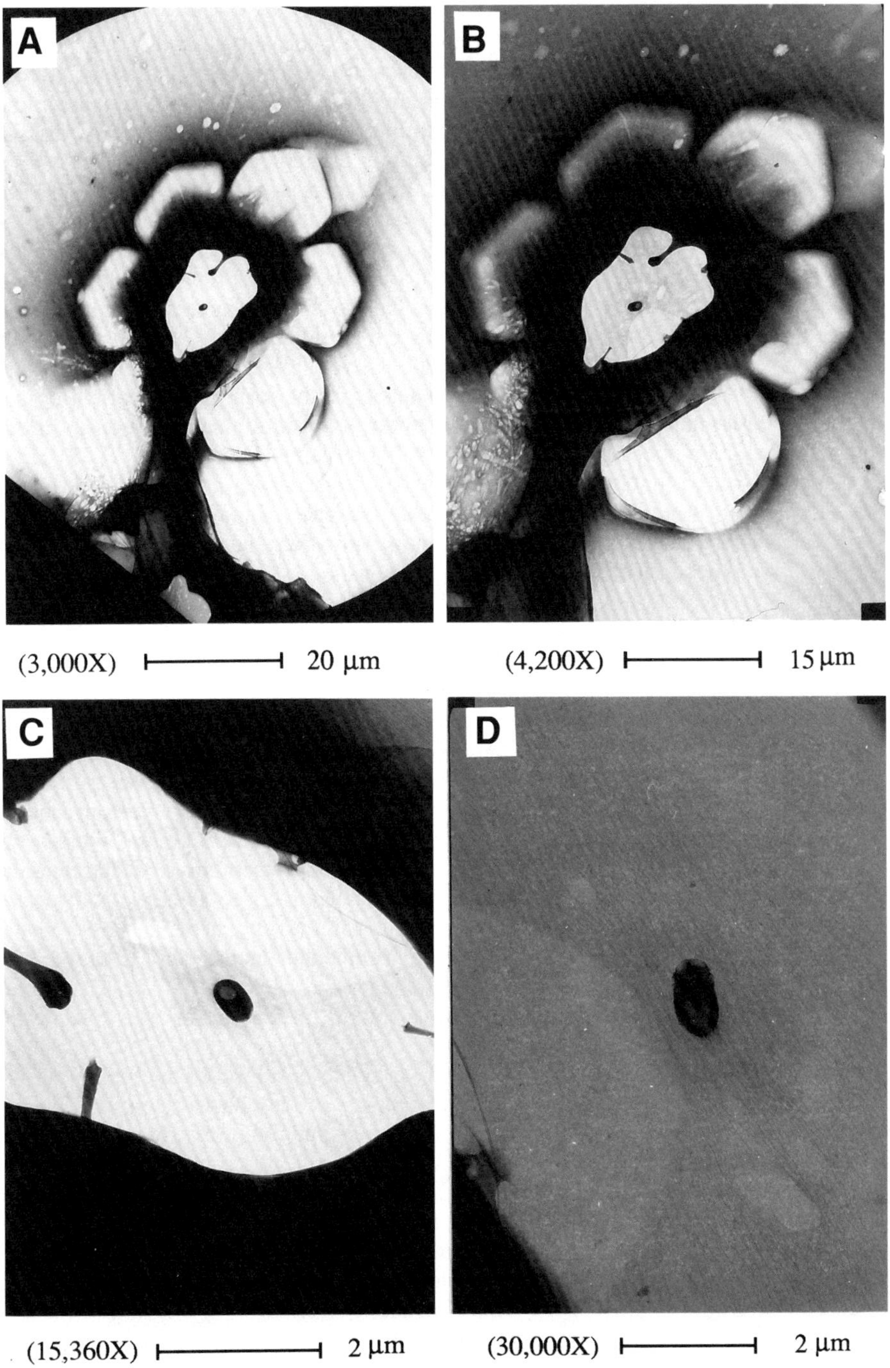


**Figure 4.** Electron micrographs of biologically seeded ice crystals. Panels A to D show increasing magnifications of a transmission electron micrograph of an ice crystal captured after SNOMAX aerosol introduction into a −20°C supercooled cloud. Reprinted, with permission, from Ward and DeMott (1989).

bearing structures including aircraft runways, roads, ice berms, and ice islands for offshore drilling. Construction of ice structures in the absence of ice nucleants can be initiated only when ambient air temperatures are sufficiently low to cause rapid freezing of sea water. The efficiency of droplet formation and freezing of the sea water in these high-volume sprays depends upon the temperature of the air, the salinity of the water, and the efficiency of heat exchange between the air and sprayed droplet. In addition, other factors, including water volume and velocity at the spray nozzle, trajectory of the stream spray, surface tension of the sea water, and wind speed, affect the efficiency of spray-ice formation. In contrast to snowmaking applications, successful ice construction requires that the frozen droplets that form in the sprayed sea water that falls to the surface have a liquid content (Owen et al., 1987). Additional freezing that occurs after impact positively influences the strength of the sprayed-ice structure. Heterologous ice nucleants in the form of ice nucleation-active *P. syringae* have been successfully applied in arctic ice construction to effectively reduce supercooling in sprayed sea water streams (Owen et al., 1987). This work demonstrated that more rapid freezing of single water droplets and of larger mass impounded waters can be obtained in the presence of the Ina$^+$ bacteria, resulting in enhanced efficiencies of ice formation as well as extended temperature ranges for use of spray ice in ice construction applications.

## Summary

The ability of biological ice nucleators to minimize supercooling of aqueous solutions and facilitate ice crystal formation at high subzero temperatures has resulted in numerous opportunities for use of these microorganisms in spray-ice applications. Efforts to identify and characterize species of Ina$^+$ bacteria in conjunction with efforts to biochemically and physiologically characterize ice-nucleating mechanisms have been critical in the design and implementation of commercial processes for the manufacture of biological ice nucleators. Fermentation and recovery processes have been established that result in high-level expression of Ina$^+$ in *P. syringae* ice nucleators. Nutritional and environmental signals appear to be key physiological effectors required for the efficient expression of Ina$^+$ in *P. syringae* and are expected to play key roles for expression of this activity in other species of Ina$^+$ bacteria.

The use of biological ice nucleators in spray-ice applications including snowmaking, weather modification, thermal storage, arctic ice construction, and freeze crystallization and effluent processing has been expanding since the commercial introduction of Ina$^+$ products in 1985. The primary role of ice nucleators has been to reduce the supercooling of aqueous solutions used in these applications, thereby elevating nucleation temperatures and reducing the amount of energy required to complete the freezing process. These energy savings have resulted in significant reductions in the costs associated with generation of spray-ice products. In addition to energy savings, the enhanced freezing efficiencies resulting from the use of biological ice nucleators in snowmaking applications also improve the quality of the final product. This combination of economic as well as quality enhancements in spray-ice applications should provide the basis for expanded use of biological ice nucleators in this application area as well as form a platform for generation of new uses of these unique microorganisms.

## Literature Cited


Aldea, M., Garrido, T., Pla, J., and Vincente, M. 1990. Division genes in *Escherichia coli* are expressed incoordinately to septum requirements by gearbox promoters. EMBO J. 9:3787-3794.

Blum, P. H., Jovanovich, S. B., McCann, M. P., Schultz, J. E., Lesley, S. A., Burgess, R. R., and Matin, A. 1990. Cloning and in vivo and in vitro regulation of cyclic AMP-dependent carbon starvation genes from *Escherichia coli*. J. Bacteriol. 172:3813-3820.

Deininger, C. A., Mueller, G. M., and Wolber, P. K. 1988. Immunological characterization of ice nucleation proteins from *Pseudomonas syringae*, *Pseudomonas fluorescens*, and *Erwinia herbicola*. J. Bacteriol. 170:669-675.

Goodnow, R. A., Harrison, M. D., Morris, J. D., Sweeting, K. B., and LaDuca, R. J.1990. Fate of ice nucleation-active *Pseudomonas syringae* strains in alpine soils and waters and in synthetic snow samples. Appl. Environ. Micro. 56:2223-2227.

Gregor, P. H. 1967. Atmospheric microbial cloud systems. Sci. Progr. Oxford 55:613-628.

Groat, R. G., Schultz, J. E., Zychlinsky, E., Buckman, A., and Matin, A. 1986. Starvation proteins in *Escherichia coli*: Kinetics of synthesis and role in starvation survival. J. Bacteriol. 168:486-493.

Gurian-Sherman, D., and Lindow, S. E. 1992. Ice nucleation and its application. Curr. Op. Biotechnol. 3:303-306.

Gurian-Sherman, D., and Lindow, S. E. 1993. Bacterial ice nucleation: Significance and molecular basis. FASEB (Fed. Am Soc. Exp. Biol.) J. 7:1338-1343.

Hendricks, D., Ward, P. J., and Orrego, S. A. Production of microorganisms having ice nucleation activity. 1992. U.S. patent 5,137,815.

Hirano, S. S., Baker, L. S., and Upper, C. D. 1985. Ice nucleation temperature of individual leaves in relation to population sizes of ice nucleation active bacteria and frost injury. Plant Physiol. 77:259-265.

Ingrahm, J. L., Maaloe, O., and Neidhardt, F. C. 1983. Regulation at the whole cell level, pages 349-370. in: Growth of the Bacterial Cell. Sinauer Associates, Inc., Sunderland, MA.

Jenkins, D. E., Auger, E. A., and Matin, A. 1991. Role of RpoH, a heat shock regulatory protein, in *Escherichia coli* carbon starvation protein synthesis and survival. J. Bacteriol. 173:1992-1996.

Jones, P. G., VanBogelen, R. A., and Niedhardt, F. C. 1987. Induction of proteins in response to low temperature in *Escherichia coli*. J. Bacteriol. 169:2092-2095.

Jung, J. A. 1990. Preliminary field experiments of SNOMAX® on cumulus mediocris clouds to artificially induce the production of ice particles. J. Weather Modification 22:153-157.

Kozloff, L. M., Schofield, M. A., and Lute, M. 1983. Ice nucleation activity of *Pseudomonas syringae* and *Erwinia herbicola*. J. Bacteriol. 153:222-231.

Kozloff, L. M., Turner, M. A., and Arellano, F. 1991a. Formation of bacterial membrane ice-nucleating lipoglycoprotein complexes. J. Bacteriol. 173:6528-6536.

Kozloff, L. M., Turner, M. A., Arellano, F., and Lute, M. 1991b. Phosphatidylinositol, a phospholipid of ice-nucleating bacteria. J. Bacteriol. 173:2053-2060.

Lawless, R. J., and LaDuca, R. J. 1992. Fermentation of microorganisms having ice nucleation activity using a temperature shift. 1992. U.S. patent 5,153,134.

Levin, Z., Yankofsky, S. A., Pardes, D., and Magal, N. 1987. Possible application of bacterial condensation freezing to artificial rainfall enhancement. J. Clim. Appl. Meteor. 26:1188-1197.

Liao, J. C., and Ng, K. C. 1990. Effect of ice nucleators on snowmaking and sprayfreezing. Ind. Eng. Chem. Res. 29:361-369.

Lindow, S. E., Arny, D. C., and Upper, C. D. 1978. *Erwinia herbicola*: A bacterial ice nucleus active in increasing frost injury to corn. Phytopathology 68:523-527.

Lindow, S. E., Hirano, S. S., Barchet, W. R., Arny, D. C., and Upper, C. D. 1982. Relationship between ice nucleation frequency of bacteria and frost injury. Plant Physiol. 70:1090-1093.

Lindow, S. E. 1983. The role of bacterial ice nucleation in frost injury to plants. Annu. Rev. Phytopathol. 21:363-384.

Luquet, M. P., Cochet, N., Bouabdillah, S., Pulvin, S., and Clausse, D. 1991. Some characteristics of a biological ice nucleating agent: *Pseudomonas syringae*. Cryo Lett. 12:191-196.

Lynn, S. Y., and Noto, G. D. 1988. Fermentation of microorganisms having ice nucleation activity using a defined medium. European patent 272669.

Maki, L. R., Galyan, E. L., Chein, M. C., and Caldwell, D. R. 1974. Ice nucleation induced by *Pseudomonas syringae*. Appl. Microbiol. 28:456-459.

Maki, L. R., and Willoughby, K. J. 1978. Bacteria as biogenic sources of freezing nuclei. J. Appl. Meteriol. 17:1049-1053.

Margaritis, A., and Bassi, A. S. 1991. Principles and biotechnological applications of bacterial ice nuclei. Crit. Rev. Biotechnol. 11:277-295.

Matin, A. E., Auger, E. A., Blum, P. H., and Schultz, J. E. 1989. Genetic basis of starvation survival in nondifferentiating bacteria. Annu. Rev. Microbiol. 43:293-316.

McCann, M. P., Kidwell, J. P., and Matin, A. 1991. The putative sigma factor KatF has a central role in development of starvation-mediated general resistance in *Escherichia coli*. J. Bacteriol. 173:4188-4194.

Nemecek-Marshall, M., LaDuca, R. J., and Fall, R. 1993. High-level expression of ice nuclei in a *Pseudomonas syringae* strain is induced by nutrient limitation and low temperature. J. Bacteriol. 175:4062-4070.

Obata, H., Tokuyama, T., Kawate, S., Hori, H., and Higashi, Y. 1990. Culture conditions of *Erwinia uredovora* in reference to its high ice-nucleating activity of the culture supernatant. Agric. Biol. Chem. 54:2171-2174.

Owen, L. B., Masterson, D. M., and Green, S. J. 1987. Rapid construction of ice structures with chemically treated sea water. U.S. patent 4,637,217.

Pooley, L., and Brown, T. A. 1991. Effects of culture conditions on expression of the ice nucleation phenotype of *Pseudomonas syringae*. FEMS (Fed. Eur. Microbiol. Soc.) Microbiol. Lett. 77:229-232.

Rogers, J. S., Stall, R. E., and Burke M. J. 1987. Low-temperature conditioning of the ice nucleation active bacterium, *Erwinia herbicola*. Cryobiology 24:270-279.

Schnell, R. C., and Vali, G. 1972. Atmospheric ice nuclei from decomposing vegetation. Nature 236:163-165.

Schnell, R. C., and Vali, G. 1976. Biogenic ice nuclei. Part I: Terrestrial and marine sources. J. Atmos. Sci. 33:1554-1564.

Turner, M. A., Arellano, F., and Kozloff, L. M. 1990. Three separate classes of bacterial ice nucleation structures. J. Bacteriol. 172:2521-2526.

Vali, G., Christensen, M., Fresh, R. W., Gayan, E. L., Maki, L. R., and Schnell, R. C.1976. Biogenic ice nuclei. Part II: Bacterial sources. J. Atmos. Sci. 33:1565-1570.

Vincente, M., Kushner, S. R., Garrido, T., and Aldea, M. 1991. The Role of the "Gearbox" in the transcription of essential genes. Mol. Microbiol. 5:2085-2091.

Ward, P. J., and DeMott, P. J. 1989. Preliminary experimental valuation of SNOMAX® snow inducer, *Pseudomonas syringae*, as an artificial ice nucleus for weather modification. J. Weather Modification 21:9-13.

Warren, G. J. 1987. Bacterial ice nucleation: Molecular biology and applications. Biotechnol. Genet. Eng. Rev. 5:107-135.

Weather Modification Association. 1984. Some facts about seeding clouds. The Association, P.O. Box 8116, Fresno, CA 93747.

Wheelwright, S. M. 1991. Protein Purification—Design and Scale-Up of Downstream Processing. Hanser Publishers, New York.

Woerpel, M. D. 1980. Snow Making. U.S. patent 4,200,228.

Woodley, W. L., and Henderson, T. J. 1990. Atmospheric tests of an organic nucleant in supercooled fog. J. Weather Modification 22:127-132.

Yankofsky, S. A., Levin, Z., Barthold, T., and Sanderman, N. 1981. Some basic characteristics of bacterial freezing nuclei. J. Appl. Meterol. 20:1013-1019.

Yankofsky, S. A., Nadler, T. N., and Levin, Z. 1983. Induction of latent freezing nucleus capability in an ice nucleation-active bacterium. Curr. Microbiol. 9:263-268.

# Glossary

**acclimation**—The ability to increase tolerance to a given environmental stress, usually triggered by an environmental cue. In reference to cold, it means an increase in tolerance to subzero temperatures.

**alpha-helix**—A term used to describe the secondary structure of a protein. The polypeptide chain turns regularly on itself every 3–6 amino acids due to hydrogen bonding between the amino acids. This results in a rigid structure and is often found in globular protein.

**amorphous layer**—An especially thickened layer of the cell wall in the vicinity of the pit membrane of a xylem ray cell. It is also referred to as the *protective layer*, although its function is poorly understood.

**anaerobic stress**—A stress imposed on an organism as a result of the absence of free oxygen or air. (*See also* ice encasement.)

**anneal**—The process of holding a substance at a temperature below the melting transition in order to change the glass transition characteristics. Annealing may either allow for complete crystallization (*see* devitrification and recrystallization), so that glasses do not form with subsequent cooling, or it may enhance the glass transition signal if the glass is particularly stable.

**antemelting**—A minor endothermic event that occurs prior to the true melting transition. This phenomenon can be detected upon warming previously frozen aqueous solutions of sugars with specific concentrations. Although antemelting events remain unexplained, it is believed that they reflect a partial melting of a highly concentrated domain. Similar pretransitions are also observed in lipids and are believed to reflect changes in crystalline structure. In this respect, antemelting events may be classified as cooperative transitions.

**antifreeze protein**—A protein that lowers the freezing point of an aqueous solution without lowering the vapor pressure (and therefore the melting temperature) of aqueous solutions. The depression of the freezing point by an antifreeze protein may exceed that of the melting point by a large factor. (*See* thermal hysteresis, thermal hysteresis protein.)

**apoplast**—The nonliving part of the plant external to the plasma membrane that comprises the extracellular spaces, the cell walls, and dead xylem elements.

**beta-sheets**—A folding pattern resulting from hydrogen-bonded lateral associations of extended polypeptide chains into sheets, formed by parts of globular proteins and primarily due to the presence of large numbers of glycine residues.

**bound water**—Water that has thermodynamic and motional properties that differ from pure water because of the association of bound water with a macromolecular surface. Also known as *vicinal water.*

**chilling injury**—Injury incurred by biological material due to low nonfreezing temperatures (e.g., 15 to 0°C or lower if supercooling occurs). In this phenomenon, ice does not occur in the tissue (contrast with Cold injury).

**chilling requirement**—Low-temperature requirement to overcome rest endodormancy (in seeds and buds).

**cold injury**—Injury incurred by biological material due to the crystallization of water at temperatures below 0°C. The term is often used interchangeably with *winter injury, freezing injury,* and *frost injury.*

**cold protection**—Methods of guarding against injury from temperatures below 0°C. The term is often used interchangeably with frost protection and commonly refers to protection of blossoms in the spring.

**cold shock**—Imposition of a brief, nonfreezing temperature resulting in a strain to the organisms; may or may not induce further acclimation.

**cooperative transition**—A phase transition in which the reorganization of molecules occurs before the actual transition temperature. This type of transition is described as an order-disorder transition in which domains of the new phase form in different parts of the old phase, and the presence of the new phase favors the formation of more of that phase. In a differential scanning calorimetry (DSC) trace of a cooperative transition, the heat capacity changes continuously before the actual phase transition temperature, rather than abruptly as is observed in true first-order transitions. Often DSC traces of water melting in biological tissues resemble cooperative transitions. In a multicomponent, metastable system, cooperative transitions can lead to phase separations. For example, in a mixed lipid system, the ordering of one lipid component can lead to the separation and ordering of a different lipid component.

**critical temperature**—The lowest ambient temperature at which the whole plant or parts of a plant can endure for 30 minutes without injury. This is not to be confused with the chemical definition of critical temperature, which is the temperature of vaporization of a liquid.

**cryopreservation**—A process in which biological materials are stored at subfreezing temperatures in order to maintain the chemical and physical integrity of the tissues. Cryopreservation implies that viable tissues are maintained in suspended animation. It is believed that the formation of aqueous glasses (vitrification) prevents the lethal formation of ice crystals and this in turn allows tissues to be cryopreserved.

**cryoprotectant**—Any compound or agent that protects cells from freezing and chilling injury. Consists of two classes: penetrating (e.g., dimethyl sulfoxide and sugars) and nonpenetrating (e.g., polyvinylpyrrolidone).

**cytorrhysis**—Collapse of the cell wall with the protoplasm attached due to a loss of water—in contrast to plasmolysis, where the protoplasm shrinks and separates from the cell wall.

**deep supercooling**—Maintenance of the liquid state at temperatures well below the freezing point of the liquid. For example, water freezes at 0°C if nucleated with ice; however, in the absence of nucleators, water can remain in a metastable liquid state to −38°C. Xylem ray parenchyma and dormant hardy flower buds deep supercool from −20 to −50°C and −10 to −40°C, respectively.

**dehardening**—Synonymous with deacclimation, primarily to warm temperatures. The loss of freezing resistance in plant tissues, organs, and cells.

**devitrification**—A crystallization process that occurs upon slow warming of a glass to a temperature above the glass transition temperature. Because the process liberates heat, it is identified by an exothermic event after the glass melting transition and before the melting transition. If the substance crystallizes (i.e., is annealed) at the devitrification temperature, a glass will not form upon subsequent cooling. Devitrification events occur because the liquid phase formed after the glass melts is unstable at the low temperature, and the solid phase is the equilibrium phase.

**differential scanning calorimetry (DSC)**—A technique that measures the change in heat capacity of a sample with temperature. In DSC, the sample temperature is kept identical to that of a reference, while the external temperature is raised or lowered. The energy that is applied to the sample or released from the sample to keep the sample isothermal with the reference is measured. Because heat capacity changes during a phase transition, DSC can be used to quantify the energy of the transition as well as the temperature of the transition.

**differential thermal analysis (DTA)**—A technique that measures a change in temperature in a sample undergoing a phase transition. In DTA, the sample temperature is compared with that of a reference while the system is cooled or warmed. During a phase transition, the temperature of the sample rises (exothermic) or falls (endothermic) compared with that of the reference. DTA can be used to determine the temperature of a phase transition.

**dynamic mechanical analysis (DMA)**—A technique that measures the viscoelastic properties of a substance. A substance is given an oscillating (dynamic) deforming stress (stretching, bending, or pressure), and the work done by deforming the object is measured (mechanical analysis) as an elastic or storage modulus. Elastic properties are dependent on the molecular properties of the material. For example, crystalline structures have larger moduli (they do not deform easily), and amorphous polymers have smaller moduli. When the substance is restored to its original shape (relaxed), energy is usually dissipated as heat. The loss modulus (also measured by DMA) describes the energy loss, and Tan δ is a function that compares the storage modulus with the loss modulus. Tan δ peaks at the glass transition temperature, at which the storage modulus is great relative to the loss modulus. A related property, strength, is the stress needed to break a material. In most cases, increased strength is due to increased crystallization.

**electrolyte conductivity test**—Loss of electrolytes from cells following an injurious stress due to a loss of the semipermeability properties of the membranes.

**electrolyte leakage**—A technique used to determine cell or tissue viability by estimating membrane integrity. Following a stress, the tissue is shaken in a given quantity of water for a predetermined period, and the electrical conductivity of the effusion is determined. The tissue is then killed, either in liquid nitrogen or by heat, shaken for a given time, and the final conductivity determined. Viability is assessed either at the stress level that results in a 50% loss in electrolytes (relative to that of killed tissue) or at the midpoint between the maximum and minimum loss of electrolytes.

**evaporative cooling**—Cooling due to the vaporization of a liquid. In the context of cold hardiness, it is the reduction in temperature of biological matter due to the loss of water to the atmosphere.

**extracellular freezing**—The crystallization of apoplastic water, i.e., water not contained within the cytoplasm of living cells.

**extraorgan or extratissue freezing**—Mechanism of freezing tolerance in which water is translocated from supercooled tissues or organs to nucleation centers in adjacent tissues (extratissue freezing) or outside the organs (extraorgan freezing).

**first-order transition**—a phase transition in which the intermolecular interactions change abruptly within a narrow temperature range ($\delta T$), where $T$ is the transition temperature. Ostensibly, this results in a discontinuity in reaction rates, molecular mobilities, and the energy content of the system. First-order transitions are defined because the first derivative of chemical potential with respect to temperature is discontinuous. An explanation follows: at the transition temperature of a first-order transition, the chemical potentials of both phases are equal (by definition). However, the change in the chemical potential with respect to temperature (holding pressure constant) $(\delta\mu/\delta T)_{p.}$, is different on each side of the transition temperature. This difference $(\delta\Delta\mu/\delta T)_p = -\Delta S_m$ (the subscript m denotes "per mole"). Thus, first-order transitions are associated with a discrete change in entropy. Equilibrium exists at the transition temperature; thus, $\Delta S_m = \Delta H_{t,m}/T \neq 0$ (the subscript t denotes "transition"). Thus, first-order transitions are also associated with a change in enthalpy (phase transition enthalpy). Since $H$ is discontinuous at the transition temperature ($\Delta H \neq 0$), its derivative $(\delta H/\delta T)_p$ or the heat capacity ($C_p$) is infinite at that temperature. Differential scanning calorimetry measures heat capacity and first-order transitions are depicted by a peak at the transition temperature. Freezing and melting transitions are first-order transitions.

**flooding**—The partial or complete covering of an organism by water, resulting in anaerobic stress.

**free water**—Water that can be removed from polymeric surfaces gravimetrically. Its properties resemble water in dilute solutions. Also known as *bulk water*.

**freeze avoidance**—Lack of water crystallization in tissues or cells at subzero temperatures due to the absence of either intrinsic or extrinsic nucleators. The tissue supercools and thereby escapes injury due to the effects of ice.

**freeze dehydration**—Loss of symplastic water due to the vapor pressure deficit created by ice in the extracellular spaces. The degree of dehydration is a function of temperature; as the temperature decreases, freeze-dehydration increases, resulting in a concentration of the cell contents.

**freeze desiccation**—*See* Freeze dehydration.

**freeze tolerance**—The ability of a cell or tissue or whole organism to tolerate the presence of ice in the apoplastic spaces.

**freezing injury**—Injury incurred by biological material due to the presence of ice.

**freezing nucleation assay**—Detection and quantitative assessment of ice nucleation active substances by dispersing them in distilled water and determining the freezing temperatures of drops or other units of the sample.

**freezing point**—The temperature at which freezing nucleation is initiated.

**frost**—A deposit of one of several forms of ice crystals as a result of sublimation of water vapor on objects colder than 0°C.

**advection frost**—Occurs from the movement of large cold air masses into an area for several days resulting in severe low temperatures and often accompanied by strong winds.

**black frost**—A dry freeze that occurs when the dew point is low, preventing air vapor crystallization on objects and resulting in the internal freezing of vegetation.

**hoarfrost/white frost**—A deposit of interlocking ice crystals formed by direct sublimation of air vapor on objects.

**radiation frost**—Occurs on calm, clear nights when there is unimpeded radiation from the earth resulting in strong temperature inversions. Usually occurs in the early morning hours and is characterized by relatively mild subfreezing temperatures.

**frost hardening**—An increase in the freeze tolerance of a plant. Depending upon the plant species and environmental stimulus, frost hardening can be divided into one, two, or three stages. In many species, frost hardening is restricted to the second stage.

**first stage**—Triggered by short day environment, which stimulates the production of a translocatable hardiness-promoting factor. Predominantly an active metabolic process. Found in woody plants, e.g., red-osier dogwood.

**second stage**—Triggered by low temperatures of less than 10°C. Both metabolic and physical changes are involved. Found in winter annuals and woody plants not sensitive to photoperiod.

**third stage**—Triggered by subzero temperatures. Found in hardy and some herbaceous plants, e.g., winter cereals, that have been exposed to prolonged freezing temperatures, resulting in physical alterations. In some cases, cryoprotective agents are excreted from the symplast to the apoplast to protect the plasma membrane or modify ice growth.

**frost heaving**—Partial or complete uplifting of a surface due to expansion of ice in underlying material or tissue.

**frost plasmolysis**—Separation of the dead protoplast from the cell wall following a lethal stress. Primarily due to the inability of the cell to reabsorb and maintain turgor following a lethal freezing stress.

**frost protection**—Methods of guarding against freezing injury. Commonly refers to protection of blossoms in the spring. The term is often used interchangeably with cold protection.

**gel phase (polar lipid)**—In the context of a polar lipid, the gel phase is a molecular crystal with up to three two-dimensional (2-D) lattices (rectangular, oblique and hexagonal), each with specific chain-packing subcells. (The 2-D hexagonal lattice should not be confused with the hexagonal phase, which is a 1-D lamellar liquid crystal.) In the rectangular and oblique lattice, the hydrocarbon chains are tightly packed in layered arrays that are held together by van der Waals forces. In the hexagonal lattice, the aliphatic chains lose some of the specific chain-chain interactions, are more loosely packed, and have greater rotational mobility. The crystalline structure with nonspecific chain packing may be stable at higher temperatures than the more tightly packed crystals with specific chain-chain interactions. Because the van der Waals interactions that make the crystal are relatively weak, melting of the crystal is associated with a relatively small enthalpy change. The melting temperature is also rather low compared with other crystals. Polar lipids with shorter chain lengths and a greater proportion of unsaturated fatty acids have fewer intermolecular interactions, less stable crystalline structure, and lower melting temperatures than do polar lipids with longer chain lengths and a smaller proportion of unsaturated fatty acids. The presence of water as a dispersion medium separates the molecular species, further weakening the van der Waals attraction and resulting in lower melting temperatures. The polymorphic (solid-solid) transitions between crystalline states and the solid-liquid melt of the crystal are first-order phase transitions. Liquid crystalline to gel phase transitions may occur in cells when they are dried or cooled and may be deleterious in themselves or may result in phase separations that could be deleterious. Gel can also be used to describe a soap with a lamellar hexagonal liquid crystalline structure.

**glass**—A metastable state that exists below the equilibrium crystallization temperature. Glasses, which are often called amorphous solids, are distinguished from stable solid phases in that they lack crystalline structure. Glasses are distinguished from supercooled liquids in that they are highly viscous (viscosities of $10^{15}$ poise or more, compared with $10^{-2}$ poise for liquid water at 0°C). Clearly, these distinctions are somewhat arbitrary, since the number of defects (breaks in uniformity present in most crystals) and viscosity are relative terms. The highly viscous nature of glasses inhibits the molecular mobility required to form the molecular bonds of the equilibrium crystalline phase. Thus, glasses are kinetically, but not thermodynamically, stable. It is a matter of some debate whether water that is associated with macromolecular surfaces (bound water or unfrozen water) is metastable (and therefore, by our definition, a glass). While the motional characteristics of bound water suggest extreme viscosity, thermodynamic considerations (i.e., the strength of water and molecular surface interactions) suggest that a system in which water is associated with macromolecular surfaces may be at a lower energy state than water associated with other water molecules (i.e., the former situation may be the equilibrium phase).

**glass formers**—Substances that have the ability to avoid the crystallization process when cooled below the freezing point and eventually vitrify at the glass transition temperature. Glass formers can be defined as pure substances that form glasses (i.e., silica), or solutes that enhance the glass-forming characteristics of the solution. Crystallization is avoided by minimizing the difference between the freezing temperature and the glass transition temperature, and by reducing the rate at which crystals grow. Solutes can modify the phase transition behavior of solvents by lowering the temperature of freezing and by increasing the strength and number of solvent-solute intermolecular interactions (thereby increasing the viscosity of the solution). Glass formers in aqueous systems are hydrophilic. Effective aqueous glass formers are important to cryopreservation efforts because they obviate the need for cooling at extremely high rates and storage at extremely low temperatures. Glass formers also may be important for deep supercooling, since they may inhibit the lethal crystallization of water during extreme cold. Dimethyl sulfoxide, ethylene or propylene glycol, acetamide, glycerol, and sucrose-sugar mixtures are commonly used in cryopreservation protocols because of their glass-forming tendencies.

**glass stability**—The rate at which a metastable glass converts to the equilibrium crystalline condition (the process of devitrification). Since this transition requires molecular redistribution, it can be likened to a transport process; thus, the tendency to move is described by the difference in chemical potential between the glass and the equilibrium phase and the ability to move is described by the conductivity (a function of viscosity). Since viscosity is a function of the intermolecular interactions and the temperature, one can conclude that glass stability is a function of the chemical potential difference, the nature of intermolecular bonding and the temperature. The first two factors are functions of the concentration and properties of the glass former. Glasses become increasingly stable as the temperature is lowered below the glass transition temperature. A stable glass can remain for millennia and an unstable glass may last for a few hours. Presently, there is little documentation of the stability of aqueous glasses and this is a problem when trying to predict the longevity of cryopreserved tissues.

**glass transition**—A transition involving an abrupt change in the viscosity of a substance near the glass transition temperature ($T_g$), even though there is no abrupt change in the chemical potential or enthalpy of the substance at this temperature. There is some debate as to whether formation of an aqueous glass represents a discontinuity in the properties of water at all, and whether glasses formed by different methods can be considered equivalent. The abrupt increase in viscosity at the transition temperature has not been explained but is a function of the chemical composition of the solution and probably a function of the intermolecular bonding strength ($E_a$ in the Boltzmann distribution).

**hardiness promoter(s)**—Naturally occurring substance(s) synthesized in plants that induce(s) freezing resistance. (*See also* translocatable hardiness promoter).

**heterogeneous ice nucleation**—The formation of an ice embryo of critical size on a solid or liquid substrate, followed by free growth of the embryo from the supercooled liquid or supersaturated vapor phase. The substrate serves to increase the probability of critical embryo formation by reducing the number of molecules needed for critical size, and by the interfacial energy between the substrate and ice. The temperature or supersaturation needed for heterogeneous ice nucleation is largely determined by the properties of the substrate.

**homeohydric plant**—(also spelled homoiohydric). A plant that maintains a constant water status independent of its environment.

**homogenous ice nucleation**—The formation of a critical-size ice embryo by the random clustering of molecules in supercooled water or in supersaturated vapor. After achieving the critical size, the embryo becomes thermodynamically stable and grows freely. Temperature at which this occurs is –38°C.

**hysteresis protein**—*See* antifreeze protein.

**ice encasement**—The partial or complete covering of an organism by ice, resulting in anaerobic stress.

**ice nucleation**—An event triggering the conversion of supercooled water (a metastable phase) into ice (a stable phase). (*See* homogenous ice nucleation, heterogeneous ice nucleation, ice nucleus, ice-nucleating site.)

**ice nucleation-active (Ina$^+$, Ice$^+$, or INA)**—Organisms, e.g., bacteria, fungi, and lichens, that have ice nucleation activity at temperatures slightly below 0°C (i.e., –2 to –5°C).

**ice nucleation activity (INA)**—A measure of the ability of a substance—either biogenic (e.g., bacteria or lichens) or nonbiogenic (e.g., silver iodide)—to catalyze ice formation, thereby limiting the supercooling of water.

**ice nucleation threshold**—*See* threshold temperature.

**ice nucleus (ice embryo)**—Technically, the icelike cluster of water molecules that serves as the origin of crystallization. In practice, biologists also use this term synonymously with ice nucleation site.

**intercellular freezing**—*See* extracellular freezing.

**intrinsic ice nucleator**—Substances within plant tissues and organs capable of initiating heterogeneous ice nucleation.

**killing temperature**—The temperature at which an organism cannot recover from freezing injury as measured by a viability test.

**$LT_{50}$**—The minimum temperature at which 50% of the population survives.

**lamellar phase**—Orientation of phospholipids of membranes with the polar head groups exposed to an aqueous environment and the acyl side chain oriented toward the center of the phospholipid bilayer.

**liquid crystalline phase (polar lipid)**—A molecular crystal that has some long-range order (van der Waals), but these interactions are so weak that the crystal possesses virtually no rigidity and thus has properties of both liquids (ability to flow) and crystals (aligned in rodlike or lamellar configuration). In the smectic phase, the molecules align themselves in layers (the familiar bilayer construction of membranes, also known as $L_\alpha$ or lamellar phase). In the nematic phase, liquid crystals are not layered, but retain a parallel alignment. The type of liquid crystal is dependent on the molecule and the temperature. Liquid crystal formation can be induced by heating or adding water. In biology, the liquid crystalline phase usually refers to the $L_\alpha$-lamellar phase (the native bilayer configuration of membranes); however, hexagonal phases are also considered liquid crystalline.

**low-temperature exotherms (LTE)**—Exotherms that appear at a lower temperature than the large exotherm that represents freezing of intracellular water. The low-temperature exotherms represent a small percentage of the total water.

**low-temperature injury**—Ambiguous term used for injury occurring from temperatures above freezing (chilling injury) to temperatures as low as −196°C.

**melting**—A phase change in which a solid turns to a liquid. When the solid is a crystal, the reaction is endothermic and is a first-order transition. When the solid is a glass, the reaction is a second-order phase transition. Because there are no kinetic limitations with the formation of nuclei and growth of crystals, the temperature of fusion represents the equilibrium condition. That is, it is independent of the rate of cooling or warming or the presence of nucleators.

**melting point depression**—The decrease in melting point temperature of a solid due to the addition of solutes.

**membrane fluidity**—The physical state of membranes determined by fatty acid chain length, lipid saturation level, sterol components, and possibly proteins. Physiologically active membranes are in a relatively fluid, flexible liquid state that changes to a solid gel-like structure at lower temperatures.

**membrane permeability**—The degree to which a membrane will allow a solvent or solute to penetrate.

**metastable phase**—Liquid at a temperature below the melting point (supercooling), or vapor at higher vapor pressure than the saturation vapor pressure (supersaturated vapor).

**minimum survival temperature**—Lowest temperature above which cells either tolerate or avoid freezing and below which they are killed.

**nucleating site**—Specific locations on the surfaces of substrates where heterogeneous nucleation is promoted.

**nucleation rate**—The rate at which subcritical embryos of the condensed phase reach critical size in homogeneous nucleation. The same definition can be applied to heterogeneous nucleation if it is restricted to identical nuclei.

**nucleation temperature**—The temperature at which homogeneous or heterogeneous nucleation takes place in a supercooled liquid.

**orthodox seed**—A seed that survives almost complete dehydration and can therefore be stored under low relative humidity and low temperatures for extended times. Most agronomically important grain crops produce orthodox seeds (e.g., wheat, corn, soybean, bean). It is believed that the extreme viscosity of water in desiccated seeds prevents deteriorative reactions and allows seeds to be easily cryopreserved.

**phase**—The physical nature of matter, usually referring to substances in a solid, liquid, or gaseous form. Each phase is marked by distinct thermodynamic and motional properties that are a function of the intermolecular interactions of the substance.

**phase diagram**—A way to depict the combinations of temperature-pressure or temperature-concentration at which a phase is most stable. Generally, phase diagrams depict the equilibrium (i.e., most stable) condition, and the lines of the phase diagram represent the pressure-temperature-concentration conditions in which the chemical potential of the two phases is equal. However, more recently, phase diagrams have been used to give temperature and concentration relationships in which metastable states can be found.

**phase separation**—The process of dispersion or demixing a solution. Phase separations occur when immiscible substances are dispersed without an emulsifying agent (e.g., the difference between oil and vinegar salad dressing and mayonnaise is the proteinaceous emulsifier in the latter). Phase separations also occur when an aqueous solution of low molecular weight solutes freezes at temperatures above the eutectic temperature (pure water freezes leaving behind a more concentrated liquid solution). Phase separations are believed to occur in biological membranes, where lipids of the same species may interact and crystallize at the freezing temperature. This may result in the efflux of proteins, the crystallization of other lipid moieties (a cooperative process), and the general disruption of the semipermeability of the membrane. Phase separations eventually occur in glasses (this is observed in a stale lollipop where the sugar forms crystals.). The latter two examples of phase separations are believed to be important mechanisms in desiccation and freezing injury. Addition of emulsifying agents and vitrification are good ways to avoid phase separations.

**phase transition**—An abrupt change in the intermolecular association of a substance that gives rise to a change in structure or mobility. (*See also* freezing, melting, glass, and first- and second-order transitions.) For membranes, the temperature at which lipids adopt a hexagonally packed structure and enter a gel phase, thus losing their diffusional freedom.

**pit membrane**—Area of a cell wall that traverses a pit, which in turn is an area where there is an absence of secondary cell wall. In ray cells of xylem, the pit membrane is composed of three layers: 1) an outer layer of electron-dense material; 2) a middle layer of primary cell wall derived from both the ray cell and adjoining vessel element; and 3) an inner, amorphous layer.

**plasmolysis**—The withdrawal of water from a cell due to a lower water potential external to the cell, resulting in contraction of the protoplasm and separation from the cell wall.

**primary protein structure**—The number of polypeptide chains in a protein, their sequences of amino acids, and the location of inter- and intrachain disulfide bonds.

**protectants**—Molecules added to a biological material that help to prevent the deleterious effects of a stress. For freezing or desiccation stresses, protectants may prevent phase transitions or phase separations of the aqueous or lipid domains. Thus, protectants may be important in lowering the freezing temperature, forming glasses, acting as emulsifiers, and/or acting as plasticizers. Studies have also suggested that desiccation and freezing result in peroxidation. Protectants, in this case, would scavenge free radicals or protect molecules from free radical attack. The mode of action and nature of protectants may vary with different stresses; however, it is generally accepted that these molecules prevent injury rather than repair damages after the stress has been removed.

**quaternary protein structure**—The manner in which two or more folded polypeptide chains of proteins are held together and oriented in space with respect to one another.

**quench cooling**—Rapid cooling of a sample by immersion into a medium (e.g., Freon 22 or isopropane) cooled by liquid nitrogen (–196°C) or liquid helium (–268°C).

**second-order transition**—A phase transition in which there is no abrupt change in the first derivative of the chemical potential (*See* first-order transitions) before and after the transition temperature. Consequently, there are no abrupt changes in entropy or enthalpy and therefore no measurable transition enthalpy. However, the second derivative of the chemical potential [related to the heat capacity $(\Delta H/\Delta T)_p$] is discontinuous. This is observed as a shift in the baseline in differential scanning calorimetry thermograms. The change in heat capacity at the transition suggests an abrupt change in the internal energy and/or a change in the ability to expand and contract. Glass transition and transitions from conductivity to superconductivity are considered second order.

**secondary protein structure**—The location and extent of the two regular protein conformations (*See* alpha-helix and beta-sheets), as well as irregular conformations along the polypeptide chain.

**structural water**—A type of bound water that is believed to be important to the structural stability of the macromolecule. This water may be an integral part of the macromolecule and is therefore not considered in the argument of whether bound water is a glass.

**supercooling**—Maintenance of the liquid state at temperatures below the freezing point of the liquid. For example, water freezes at just below 0°C if nucleated with ice; however in the absence of active nucleators, water can remain in a metastable liquid state to –38°C. Many herbaceous plants supercool in the range of –2 to –10°C, flower buds at –10 to –40°C, and xylem ray parenchyma at –20 to –50°C.

**supercooling point (SCP)**—The temperature at which an organism, specifically an insect or other ectotherm, begins to freeze after various amounts of supercooling.

**symplast**—The living part of the plant that is contained within the plasma membranes of the cells.

**temperature of crystallization**—*See* supercooling point.

**tertiary protein structure**—The three-dimensional arrangement of all protein atoms in space, without regard to the relationship with neighboring molecules or subunits. As a rule, a protein has the unique three-dimensional arrangement determined by its amino acid sequence.

**thermal hysteresis**—The difference between freezing and melting temperatures.

**thermal hysteresis protein**—An alternative term for antifreeze protein.

**thermally stimulated depolarization current (TSDC)**—A technique that measures the electrical current generated during the thermal relaxation of a polarized substance. The technique involves polarizing a substance with an electric field at a given temperature (usually at room temperature) and then quench cooling. The field is then turned off. During controlled heating, a current is generated when ions migrate, dipoles disorient, or trapped molecules are released. Because the temperature of relaxation is specific for various water-solute interactions, the technique is useful in determining the mobility of different pools of water in a complex system.

**thermogram**—The data recorded by differential thermal analysis (DTA) and differential scanning calorimetry (DSC). In DTA, thermograms give temperature/change in temperature data, and in DSC, thermograms give temperature (or time) vs. power data. When substances undergo first-order transitions, DTA and DSC thermograms exhibit peaks in the exothermic or endothermic directions. When substances undergo a second-order transition, DSC thermograms sometimes show a change in the baseline (referring to a shift in power) but more often show endothermic (perhaps antemelting or cooperative transitions) and/or exothermic (devitrification) events before the first-order melting transition.

**threshold temperature**—The temperature at which the rate of heterogeneous ice nucleation at a particular ice nucleation site becomes active and results in the crystallization of water.

**transition enthalpy**—The energy required to drive a transition (endothermic) or the energy released during a transition (exothermic). Commonly denoted as $\Delta H_t$, the transition enthalpy is measured by integrating the area of a peak in a differential scanning calorimetry thermogram. In first-order transitions, $\Delta H_t \neq 0$, and in second-order transitions, $\Delta H_t = 0$.

**transition temperature**—The temperature at which a substance undergoes a phase change. Pure substances are identifiable by their equilibrium transition temperatures. Transition temperature is dependent on the composition of the system and the pressure. Generally, high concentrations of solutes decrease or increase the fusion or glass transition temperature. Transition temperatures for freezing and vaporization are dependent on the presence of nuclei or surfaces to form vapor; thus, they may not represent the equilibrium temperature. Rate of cooling is important to the freezing temperature, but it is not a determinant (except for instrument thermal conductivity limitations) in the melting transition.

**undercooling**—*See* supercooling.

**unfrozen water**—The portion of water that remains unfrozen at subzero temperatures. Unfrozen water can be detected in aqueous mixtures of most polymers, as well as some low molecular weight solutes, although the properties of the unfrozen phase may differ in the two situations. The presence of unfrozen water in polymeric mixtures suggests that the bound (or vicinal) water-surface interactions perturb the interactions of vicinal water with the water that is removed from the polymer surface (free water). Unfrozen water is often referred to as bound water; it may also be classified as a glass.

**vitreous**—In glassy state; a substance that is extremely viscous but has amorphous characteristics. (It is not crystalline.)

**vitrification**—The process by which a glass is formed. In biology, this term is usually used in the context of cryopreserving tissue when an aqueous glass is desired. To vitrify a substance, it must be cooled rapidly enough to temperatures well below the glass transition so that crystallization is inhibited. The required rate of cooling depends on the chemical composition of the substance (concentration of glass formers, water content for biological materials) and the size of the substance. For a hydrated unprotected substance, required cooling rates may be in excess of 20,000°C/second. With some dehydration and with high concentration of protectants, the cooling rate required to vitrify tissue may be dramatically reduced.

# Index